1a.

1b.

1c.

1d.

Plate 1a. Cassava common mosaic disease: light green patches on the leaf restricted by the veins.
Plate 1b. Cassava vein mosaic disease: mosaic and yellow vein banding.
Plate 1c. Cassava frogskin disease: hyperplasia of the root resulting in longitudinal fissures on the root.
Plate 1d. Cassava mosaic disease: mosaic and distortion of the leaf lamina.

2a.

2b.

2c.

2d.

Plate 2a. Cassava mosaic disease: stunting effect on a susceptible cultivar compared to a resistant one.

Plate 2b. Cassava brown streak disease leaf symptoms: 'feathery' chlorosis alongside the veins is common.

Plate 2c. Cassava brown streak disease leaf symptoms: a less common symptom in which the 'feathering' appears water-soaked.

Plate 2d. Cassava brown streak disease root symptoms: corky, yellow/brown necrotic patches in the starch-bearing tissue.

Cassava

Biology, Production and Utilization

Cassava

Biology, Production and Utilization

Edited by

R.J. Hillocks and J.M. Thresh

Natural Resources Institute, University of Greenwich, Kent, UK

and

A.C. Bellotti

Pest and Disease Management Project, Centro Internacional de Agricultura Tropical (CIAT), Cali, Colombia

CABI _Publishing_

CABI is a trading name of CAB International

CABI Head Office
Nosworthy Way
Wallingford
Oxfordshire OX10 8DE
UK

Tel: +44 (0)1491 832111
Fax: +44 (0)1491 833508
Email: cabi@cabi.org
Web site: www.cabi.org

CABI North American Office
38 Chauncey St
Suite 1002
Boston, MA 02111
USA

Tel: +1 617 395 4056
Fax: +1 617 354 6875
Email: cabi-nao@cabi.org

A catalogue record for this book is available from the British Library, London, UK.

Library of Congress Cataloging-in-Publication Data
Cassava: biology, production and utilization / edited by R.J. Hillocks and J.M. Thresh and A. Bellotti
     p.   cm.
    Includes bibliographical references (p.).
    ISBN 0-85199-524-1 (alk. paper)
      1. Cassava. I. Hillocks, R. J. II. Thresh, J. M. III. Bellotti, Anthony.
SB211 .C3 C37 2001
633.6'82—dc21

                          2001025502

ISBN-13: 978-0-85199-524-3

First printed 2002
Transferred to print on demand 2013

Printed and bound in the UK by CPI Group (UK) Ltd, Croydon, CR0 4YY

Contents

Contributors

A.C. Allem, EMBRAPA, Recursos Genéticos e Biotecnologia, C.P. 02372, 70849-970 Brasília, DF, Brazil

A.A.C. Alves, Embrapa Cassava and Fruits, Caixa Postal 007, 44.380-000, Cruz das Almas, Bahia, Brazil

C. Balagopalan, Crop Utilization and Biotechnology, Central Tuber Crops Research Institute, Sreekariyam, Trivandrum 695 017, Kerala, India

A.C. Bellotti, Pest and Disease Management Project, Centro Internacional de Agricultura Tropical (CIAT), A.A. 6713, Cali, Colombia

L.A. Calvert, Centro Internacional de Agricultura Tropical (CIAT), A.A. 6713, Cali, Colombia

M. Fregene, Centro Internacional de Agricultura Tropical (CIAT), A.A. 6713, Cali, Colombia

G. Henry, CIRAD-Amis, Rua Paulo Castro P. Nogueira 600, Campinas-SP 13092-400, Brazil

C. Hershey, 2019 Locust Grove Road, Manheim, PA 17545, USA

R.J. Hillocks, Natural Resources Institute, University of Greenwich, Chatham Maritime, Kent ME4 4TB, UK

R.H. Howeler, CIAT Regional Office in Asia, Department of Agriculture, Chatuchak, Bangkok 10900, Thailand

C. Iglesias, Weaver Popcorn Co., PO Box 20, New Richmond, IN 47967, USA

D.L. Jennings, 'Clifton', Honey Lane, Otham, Maidstone, Kent ME15 8JR, UK

D. Leihner, Research, Extension and Training Division, Sustainable Development Department, FAO, Via delle Terme di Caracalla, 00100 Rome, Italy

N.Q. Ng, International Institute of Tropical Agriculture (IITA), Oyo Road, Ibadan, Nigeria; International mailing address: IITA, c/o L.W. Lambourn & Co., 26 Dingwall Road, Croydon CR9 3EE, UK

S.Y.C. Ng, International Institute of Tropical Agriculture (IITA), Oyo Road, Ibadan, Nigeria; International mailing address: IITA, c/o L.W. Lambourn & Co., 26 Dingwall Road, Croydon CR9 3EE, UK

I.C. Onwueme, Fulton Center for Sustainable Living, Wilson College, Chambersberg, PA 17201, USA

J. Puonti-Kaerlas, Institute for Plant Sciences, ETH-Zentrum/LFW E 17, CH-8092 Zürich, Switzerland

J.M. Thresh, Natural Resources Institute, University of Greenwich, Chatham Maritime, Kent ME4 4TB, UK

A. Westby, Natural Resources Institute, University of Greenwich, Chatham Maritime, Kent ME4 4TB, UK

K. Wydra, Institut für Pflanzenpathologie und Pflanzenschutz, Georg August Universität, Grisebachstr. 6, D-37077 Göttingen, Germany

Preface

In the relatively short time since cassava arrived from South America, it has been introduced into all the tropical countries of Africa and Asia. In at least some parts of many of these countries, it is the main source of nutritional calories for the rural population. The epidemic of cassava mosaic disease in Uganda during the 1990s has focused attention on the reliance on cassava, of much of the rural population in northern Uganda and the neighbouring countries, now threatened by the epidemic.

In addition to its use as a food crop for local consumption, in some parts of the world cassava is widely used in processed foods. If new markets can be created for these products, as is happening in South America and to some extent in parts of Asia, cassava may increasingly be perceived as a cash crop with the potential to improve living standards for rural communities in the tropics which depend on agriculture for their livelihood.

Cassava can be very easily propagated and its capacity to produce some yield in a wide range of environments and under conditions of low rainfall and poor soil makes it an excellent subsistence crop. The increases in cassava cultivation that have been seen in Africa during the last decade of the 20th century may have been a response to declining soil fertility. It is almost certain that this, combined with the increased cost of fertilizer brought about by withdrawal of subsidies on agricultural inputs under structural adjustment policies, created the demand for the fourfold increase in cassava production reported in Malawi, for instance, during the 1990s. Although cassava can produce some yield under adverse conditions and with low levels of management, the increased yields that will be required to meet the expected population growth in the coming decades, particularly in Africa, will require a higher standard of agricultural management.

Alone among the world's major crops there, was until now, no monograph on cassava covering all the areas where it is grown. This is all the more surprising in view of its worldwide distribution and increasing importance in Africa as a food security crop. This then was the motivation in 1996 that brought the editors together to begin work on the book.

The global mandate for cassava research and breeding is divided between CIAT, Colombia and IITA, Nigeria. Several of the book's contributors are members, or, former members of one or, other of those international centres. The other contributors represent advanced research institutes or universities, with long histories of involvement in cassava research. The 15 chapters together provide a comprehensive review of cassava that will be a valuable resource for researchers, extensionists and students of tropical agriculture.

Rory Hillocks, Mike Thresh and Tony Bellotti
December 2000

Acknowledgements

We are grateful to the following: Peter Neuenschwander of IITA for his contribution to chapter 11 of the information on mealybug and whitefly in Africa. To Trudy Brekelbaum, Paul André Calatayud and Josephina Martinez at CIAT, Colombia, for their support and help in preparation of the manuscript. Chapter 2 drew significantly on the Review of Cassava in Latin America, carried out for IFAD by C. Hershey and co-workers. Chapter 3 used as its main source, Cassava in Africa, a report prepared for IITA by Dunstan Spencer and Associates. Preparation of the manuscript was partly supported by the UK Department for International Development (DFID) [*Crop Protection Programme* (CPP) – Project R7563 and *Crop Post-Harvest Research Programme* – Project R7497] and the colour plates in Chapter 12 were sponsored by DFID-CPP.

Chapter 1
The Origins and Taxonomy of Cassava

Antonio C. Allem

EMBRAPA, Recursos Genéticos e Biotecnologia, C.P. 02372, 70849-970 Brasília, DF, Brazil

Introduction

The origins of cassava (*Manihot esculenta* Crantz subspecies *esculenta*) have long been obscure. The three important questions to answer concern the botanical origin, (i.e. the wild species from which cassava descends), the geographical origin, (i.e. the area where the progenitor evolved in the geological past) and the agricultural origin (i.e. the area of initial cultivation of the wild ancestor by Amerindians). The core of the above argument rests on the assumption that if there is a living wild ancestor, its discovery would be likely to indicate the taxonomy of cassava. In turn, this would indicate the ancestry of the crop (the evolution of the ancestor and its phylogenetic relationships with related species) and the cradle of domestication. Current knowledge on the three topics shows that studies on the botanical origin of cassava have progressed far and stand on firm ground, those on the geographical origin have progressed and conjecture on the area where cultivation began has recently experienced a renewed surge following the appearance of novel ideas.

Studies on the taxonomy of the genus *Manihot* in Brazil led unexpectedly to progress in matters with a bearing on the origins and phylogeny of cassava. Accumulated empirical knowledge derived from field experience culminated with the formulation of a classification in which the Brazilian species were arranged in groups. The model of classification proposed was thought to mirror some degree of phyletic kinship between the species (Fig. 1.1). A number of the prospective clades proposed in Fig. 1.1 (e.g. the *Manihot nana* group; the *Manihot salicifolia* group, etc.) match remarkably closely with former taxonomic classifications advanced for the same groups in earlier classic treatments including that of Rogers and Appan (1973). Group VI in particular, the *M. esculenta* group, has been tested extensively by the scientific community through genetic studies and phylogenetic investigations, which in turn influenced cladistic and taxonomic classifications. This group is particularly highlighted in the present review. No attempt has been made in Fig. 1.1 to suggest that the sequence of groups presented represents evolution towards greater complexity, although there is a tendency in the model to progress from herbaceous to tree species.

The Botanical Origin of Cassava

Until 1982 there was a tendency in the literature to assume that cassava has no known ancestry (a comprehensive historical background on how this arose appears in

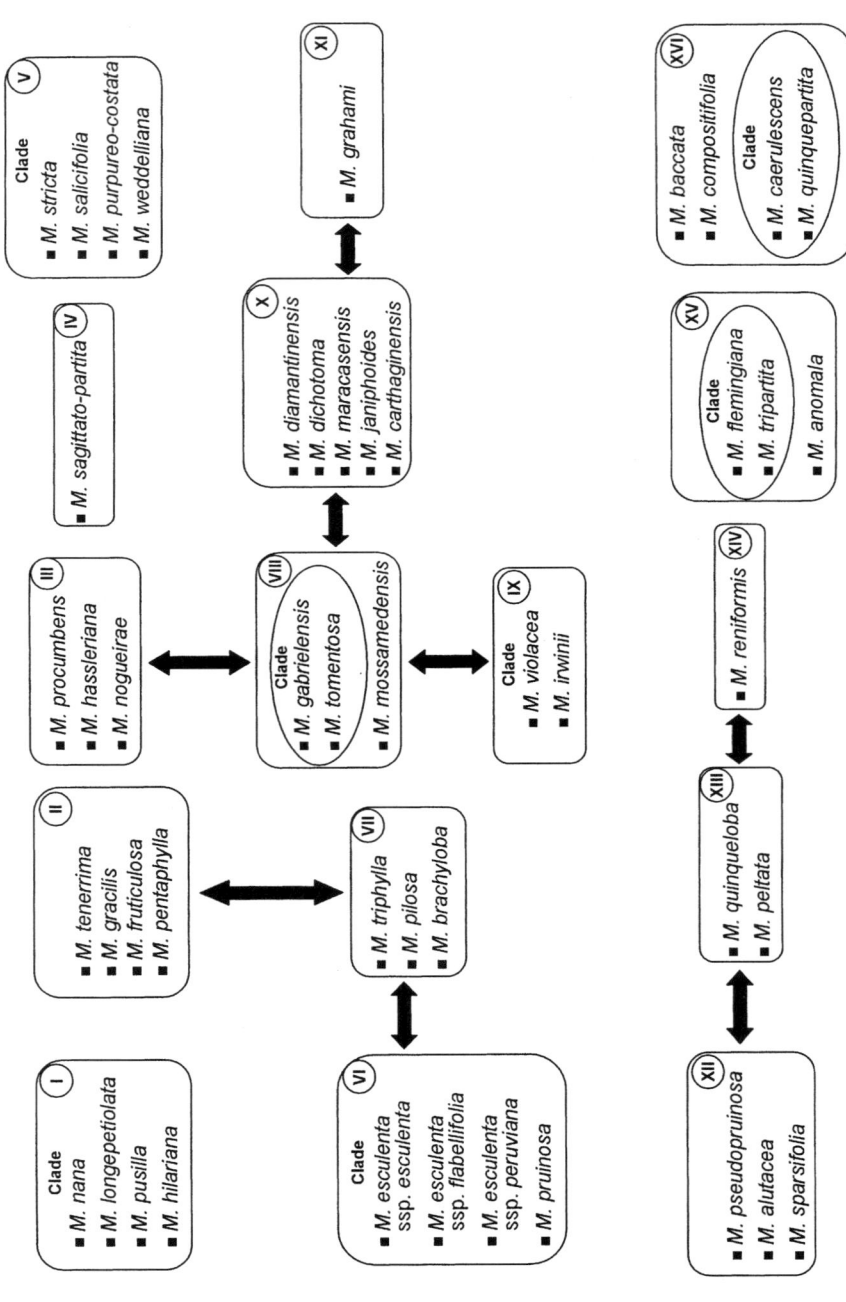

Allem, 1994a). In March 1982, however, a wild population indistinguishable on morphological grounds from the domesticate was found in the central Brazilian state of Goiás and led to an announcement that cassava grows in the wild (Allem, 1987). The communication proved a watershed in the historiography of cassava, since it established a link between that particular wild population and the botanical origin of the crop. Several publications doubted the find and raised the possibility that feral cassava had been regarded as wild material (Bretting, 1990; Heiser, 1990; Bertram, 1993). The conviction was that wild populations of *Manihot flabellifolia* are likely to have led to the genesis of modern commercial cultivars and landraces of cassava (Allem, 1994b). The author has recently reaffirmed this position (Allem, 1999) and other studies point to a similar conclusion (Roa *et al.*, 1997; Olsen and Schaal, 1999). In so doing, the author followed Alphonse de Candolle's 1882 view that the occurrence of a crop in wild habitats amounts to the find of the wild ancestor; a view not held by Vavilov and others (Harris, 1990).

A further elaboration of the 1987 announcement resulted in the formal recognition of three subspecies within the crop (Table 1.1). This view contrasts with that of others who considered that the crop had no traceable ancestor and was instead the by-product of indiscriminate introgression involving a number of wild relatives (Rogers, 1963, 1965; Harlan, 1965; Pickersgill, 1977; Bretting, 1990; Heiser, 1990; Sauer, 1994).

The original study of Rogers (1963) on the closest wild relatives of cassava did not greatly emphasize a hybrid origin for the domesticate (Table 1.2). However, this was stressed later and the postulation of introgression of wild genes into cassava became the norm in all subsequent publications on the subject (Rogers, 1965, 1972; Rogers and Appan, 1973; Rogers and Fleming, 1973). The paper of Rogers (1965) is particularly enlightening and offers insights into how interspecific hybridization came to be considered responsible for most of the difficulty found in establishing the proper separation of species in the genus. Rogers and Fleming (1973) stated that the amount of variation found in many native species of *Manihot* was due to introgression with cassava since the gene flow was certainly bi-directional. Rogers and Appen (1973) provided a number of amendments for species on the grounds that 'over a period of years phenotypic variation takes place [inside the introgressed hybrid] which expands the circumscription of such a species'. Rogers had been influenced by previous theories on the origins of crops, i.e. that crops could arise and evolve through periodic infusions of germplasm from adjacent weedy relatives (Harlan, 1961,

Fig. 1.1. (*opposite*) Phenetic relationships of Brazilian *Manihot* species. Empirically compelling phylogenetic kinships, deduced on the basis of phenetic and ecogeographic similarities, are presented as clades, which are thought to reflect common ancestry. From Allem (1995), modified. Arrows indicate strong phenetic similarity existing between groups of species. Within groups, a clade means strict morphological, geographical and ecological relationships recorded between the species concerned. The true number of Brazilian species is estimated around 47–50. A unique species of liana with simple and entire leaves from the Atlantic coast forest of the state of Bahia besides another kin species from the state of Mato Grosso (both belonging to Group V) await description. A number of new varieties and subspecies also await description. Some discrepancies are expected in phenetic and phylogenetic systems. For example, morphology determined the placement of *Manihot nogueirae* in Group III; however, the geography and ecology of the species recommend its inclusion in Group I and, in addition, *M. nogueirae* and *M. nana* seem to be hybridizing in the Federal District as discovered through isozyme studies (G. Second, Brasília, 1995, personal communication). A considerable number of traditional species (e.g. *Manihot orbicularis*, *Manihot glaziovii*, *Manihot stipularis*, etc.) do not appear in Fig. 1.1 as they will become synonymous of other species in the near future. Davis and Heywood (1963) remark that certain genera have undergone a modest level of differentiation where only groups of phenetically related species can be recognized. This applies to *Manihot*, a genus where infrageneric categories such as section are difficult to draw for many species. Pax (1910) recognized 11 sections for the genus while Rogers and Appan (1973) acknowledged 19 sections. The view is that a number of the sections devised by Rogers and Appan are artificial, thus echoing Croizat (1943) who denounced Pax's system as unworkable from the taxonomic standpoint.

Table 1.1.   The subspecies of cassava.[a]

Basionym	Novel status	Category
M. esculenta Crantz	*M. esculenta* Crantz ssp. *esculenta*	Cultivated stock
M. flabellifolia Pohl	*M. esculenta* Crantz ssp. *flabellifolia* (Pohl) Cifferi	Wild strain
M. peruviana Mueller	*M. esculenta* Crantz ssp. *peruviana* (Mueller) Allem	Wild strain

From Allem (1994a).
[a]Because cassava is known to include cultivated and wild forms, the species is no longer a cultigen but ranks as an indigen instead. The definitive infraspecific classification of the species is in preparation. There is a real possibility that two other species (e.g. *Manihot leptophylla*) join the extensive synonymy of the complex. In addition, a further wild variety with smooth fruits found in May 1994 and so far only known from the northern Brazilian Amazonian state of Tocantins awaits description.

Table 1.2.   Species of *Manihot* regarded close to cassava on the basis of morphology, ecology and geography.

Species	Range
M. carthaginensis	All countries bordering the Caribbean
M. aesculifolia	Mexico; Central America
M. grahami	Brazil, Paraguay, Uruguay, Argentina
M. flabellifolia	Brazil, Paraguay, Uruguay, Argentina
M. saxicola[a]	Guiana, Surinam, Venezuela

From Rogers (1963).
[a]A synonym of *M. esculenta* ssp. *flabellifolia* (see Allem, 1994a).

1970; Harlan and de Wet, 1963, 1965). Rogers assumed this had happened in *Manihot*. However, Harlan's concept of a weed meant either an intra-specific category, or a sibling species (Harlan, 1965) and not necessarily a distinct species. Harlan's standpoint focused mostly on intra-specific hybridization and drew on the fact that distinct populations of a species normally carry different peaks of adaptive norms. The record suggests that *persistent* artificial hybrids of *Manihot* are extremely difficult to obtain (Nichols, 1947; Jennings, 1963); Magoon *et al.* (1970) provide the sole available report on the unsuccessful fate of hybrid seeds. Natural hybrids are indeed a fact in the genus as assessed through the occurrence of individuals showing morphological intermediaries. This has been recorded in five instances involving populations in central Brazil (A.C. Allem, unpublished). However, these natural hybrids occur together with their parents which makes interpretation easier and there is no evidence from herbarium specimens that hybrid speciation (the spread of an intermediate form over a significant geographical area) occurs with *Manihot*.

Rogers and Appan (1973) regarded a species native to Mexico and Mesoamerica as the closest morphologically to cassava: 'as evident from the computer analyses the closest wild relative of *M. esculenta* is *M. aesculifolia* (H.B.K.) Pohl'. Bertram (1993) later commented that 'the section *Parvibracteatae* also contains the wild species *Manihot aesculifolia* which is closest to cassava in morphological terms' [and] 'the morphological similarity of *Manihot aesculifolia* to cassava is striking'. Bertram adds '*M. aesculifolia* and *M. carthaginensis* are genetically closest to cassava', and also 'several factors point clearly to these two species – *M. carthaginensis* and *M. aesculifolia* – as putative ancestors of the crop'. It has also been suggested that the cultivation of cassava in the Caribbean area resulted from the domestication of the wild species *Manihot carthaginensis* (Reichel-Dolmatoff, 1986).

Sophisticated research on the ancestry of cassava drew initially on North American species. Investigation on the phylogeny, through the use of restriction fragment length polymorphism (RFLP) analyses of chloroplast DNA, made use of 12 species from Central America and North America and *M. carthaginensis* from South America. The outcome was inconclusive (Bertram and Schaal, 1993). Subsequently, an increasing number of authors have become involved in the controversy involving the cassava species complex and novel data have appeared. Molecular studies carried out to unravel the ancestry of cassava discussed the hypothesis on the origin of the crop on a regular basis. The results were diverse but, overall, supported the view of *M. flabellifolia* as the progenitor of the crop (Table 1.3).

Species thought to be involved in the ancestry of cassava were reviewed more recently (Allem, 1999). One of them (*M. esculenta* ssp. *flabellifolia*) is regarded as the wild progenitor of modern cultivars and thus becomes part of the primary gene pool of the crop (Fig. 1.2). Another Brazilian species (*Manihot pruinosa*) is regarded as the nearest species to the GP1 of cassava and is difficult to separate from the wild strain *M. esculenta* ssp. *flabellifolia* on morphological grounds (Fig. 1.3). The study also included the Brazilian *Manihot pilosa* and *Manihot triphylla* as species close to cassava and pointed out the close vegetative and floral similarities between them. Earlier, *M. pilosa* had been found to be the closest wild relative of cassava (Grattapaglia *et al.*, 1987) but the checklist of species did not include *M. flabellifolia*. From parallel molecular studies it was concluded that *M. triphylla* stands closer to

cassava than the wild subspecies *M. esculenta* ssp. *peruviana* and *M. pilosa* is the most distant of all five species tested (Cabral *et al.*, 2000). If the biosystematic crosses now being done back up the above molecular results, *M. triphylla* may eventually join the wild GP1 of the crop.

Data on the taxonomic species concept of cassava came to entwine with earlier views which, through biosystematic crosses between cassava and its wild progenitor, had recorded the unusual high degree of fertility and genetic relationships holding between the two species (Bolhuis, 1953, 1969; Jennings, 1959; Roa *et al.*, 1997; Tables 1.4, 1.5 and 1.6). Such investigations provided a preliminary delimitation of the biological species of cassava. A synthesis became possible when results of molecular biology supplemented existing systematic and biosystematic data. Tests carried out with genetic and biochemical markers strongly suggested that a preliminary delimitation of the phylogenetic (cladistic) species of cassava was in prospect (cf. Table 1.3 below).

Table 1.3. Molecular and biochemical tests carried out to test Allem's (1987/1994a) hypothesis on the origin of cassava.[a]

Authority	Marker	Degree of support
Carvalho *et al.* (1993)	RFLP/RAPD	Moderate (+)
Fregene *et al.* (1994)	CpDNA/rDNA	Moderate (−)
Carvalho *et al.* (1995)	RAPD	Moderate (+)
Schaal *et al.* (1995)	rDNA/RAPDs	Moderate (−)
Brondani (1996)	Isozymes	Strong (+)
Schaal *et al.* (1997)	RAPDs	Strong (−)
Second *et al.* (1997)	AFLP	Moderate (−)
Bonierbale *et al.* (1997)	AFLP	Strong (+)
Roa *et al.* (1997)	AFLP	Strong (+)
Olsen and Schaal (1998)	Nuclear DNA	Strong (+)
Second (1998)	SSRs	Strong (+)
Roa *et al.* (1998a)	RFLP	Strong (−)
Roa *et al.* (1998b)	AFLP	Strong (+)
Olsen and Schaal (1999)	SSRs	Strong (+)
Cabral *et al.* (2000)	G3pdh	Strong (+)
Schaal and Olsen (2000)	G3pdh	Strong (+)
Roa *et al.* (2000)	SSRs	Strong (+)
Colombo *et al.* (2000)	RAPD/AFLP	Strong (+)
Olsen and Schaal (2001)	SSRs	Strong (+)
Olsen (2002)	G3pdh	Strong (+)

[a]Support for *M. flabellifolia* as the progenitor of cultivated cassava (Haysom *et al.* (1994) concluded that *M. flabellifolia* is not native to South America. The conclusion was in complete disagreement with the taxonomic history and geographical distribution of the species. The possibility cannot be ruled out that misidentified germ plasm used in the experiment, accounted for the flawed conclusion.

THE GP1 OF CASSAVA
(BIOLOGICAL SPECIES)

1. Cultivated Materials: *M. esculenta* ssp. *esculenta*
2. Wild Progenitors: *M. esculenta* ssp. *flabellifolia*
              M. esculenta ssp. *peruviana*
3. The Closest Wild Relative: *M. pruinosa*

THE GP2 OF CASSAVA
M. triphylla,[1] *M.pilosa*,[1] *M. brachyloba*,[1] *M. anomala*,[2]
M. epruinosa,[2] *M. gracilis*,[2] *M. tripartita*,[2] *M. leptophylla*,[2]
M. pohlii,[2] *M. glaziovii*,[3] *M. dichotoma*,[3]
M. aesculifolia,[4] *M. chlorosticta*[4]

Fig. 1.2. The primary (GP1) and secondary (GP2) gene pools of cassava. From Allem *et al*. (2000).
[1]These three species suggested by Allem (1999).
[2]These six species deduced from the text of Bai *et al*. (1993).
[3]These two species deduced from the text of Nichols *et al*. (1947).
[4]These two species deduced from the text of Roa *et al*. (1997).

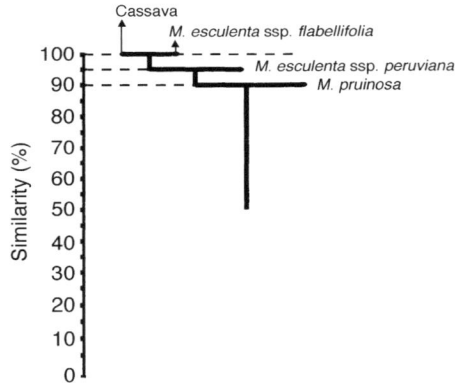

Fig. 1.3. Phenogram of cassava and its closest allies and empirically deduced phylogenetic relationships between species. Figures are merely illustrative and serve as a frame of reference. The assumption is that *M. pruinosa* and the two wild strains of cassava descend from a common stock. From Allem *et al*. (2000).

The Geographical Origin of Cassava

In this section geography is used in the sense of phylogeny, i.e. the area where evolution and divergence of the progenitor occurred and the pertinent phylogenetic relationships with other species. This standpoint differs from that of most authors who traditionally equate the geography of a crop with the area of domestication.

As argued in the preceding section, the view is that *M. pruinosa* and *M. esculenta* ssp. *flabellifolia* descend from a common primeval stock. Some pertinent questions apply. Did the ancestor of both species evolve in the Brazilian Cerrado savannah? Or was it the reverse, i.e. the ancestor evolved in the Amazon, spread to the Cerrado vegetation and then diverged to give *M. pruinosa* and *M. esculenta* ssp. *flabellifolia* and only then did the latter return to the Amazon? Both Cerrado and Amazonia are quite old types

Table 1.4. Results of crosses between cassava and *Manihot saxicola* in Indonesia between 1942 and 1949.

Female parent	Male parent	Flowers pollinated	Fruit-set	Success (%)	Seed-set	Success (%)
M. saxicola[a]	*M. esculenta*	335	125	37.3	76	20.2
M. esculenta	*M. saxicola*	96	47	48.9	80	56.7

Adapted from Bolhuis (1969).
[a]Seeds of *M. saxicola* came from Surinam. The species is a synonym of *M. esculenta* spp. *flabellifolia*.

Table 1.5. Results of crosses between cassava and *Manihot melanobasis* in Tanzania around 1955.

Female parent	Male parent	Flowers pollinated	Seed-set	Success (%)
M. melanobasis[a]	*M. esculenta*	125	225	60
M. esculenta	*M. melanobasis*	253	592	78

Adapted from Jennings (1959).
[a]The Tanzanian material originated in Surinam. *M. melanobasis* is a synonym of *M. esculenta* subspecies *flabellifolia*.

Table 1.6. Results of interspecific crosses between cassava and three *Manihot* species.

Female parent	Male parent	Crosses	Seed-set
M. esculenta ssp. *flabellifolia*	*M. esculenta* ssp. *esculenta*	16	Thousands[a]
M. aesculifolia[b]	*M. esculenta* ssp. *esculenta*	2	5
M. chlorosticta[c]	*M. esculenta* ssp. *esculenta*	14	1 to 148

Adapted from Roa *et al.* (1997).
[a]Crosses included backcrosses.
[b]North American and Meso-American species regarded by some writers as the closest wild relative of cassava.
[c]Mexican species; hybrid seeds showed strong dormancy.

of vegetation and the latter has experienced notable natural catastrophies over the last 100,000 years, during the Pleistocene period, which led to repeated extinctions (Prance, 1978; Simpson and Haffer, 1978). The answer to the question, 'is the Cerrado older than the Amazon?' may provide unique insights into the early evolution of the progenitor. If the Cerrado vegetation arose earlier than the Amazonian forest, this biome may hold the answer regarding the ancestry of cassava. The Brazilian state of Goiás, mainly composed of the Cerrado vegetation, is the primary centre of diversity of species in Brazil. *M. pruinosa* occurs exclusively in Goiás and Mato Grosso states and thrives in open forests ('Cerradão'), eventually overlapping sympatrically with *M. esculenta* ssp. *flabellifolia* in these sites. The working hypothesis is that, because most of the biological diversity of the genus *Manihot* in Brazil is concentrated in the Federal District and in the neighbouring state of Goías, the original stock that gave birth to *M. pruinosa* and *M. esculenta* ssp. *flabellifolia* may have arisen in the lax forests of the Central Brazilian savannah, afterwards differentiated into the two species and only then did the latter colonize the Amazon.

Discoveries concerning *M. pruinosa* and *M. esculenta* ssp. *flabellifolia* have a bearing on the phylogeny of cassava. The species share a common geography, ecology, morphology and life form, but no one disputed their taxonomic ranking or challenged them as distinct species for more than 150 years, since the original descriptions made by the Austrian botanist Johann Baptist Emanuel Pohl in 1827. The strict morphological kinship between *M. esculenta* ssp. *flabellifolia* and *M. pruinosa* is undisputable. Biosystematic trials will determine the degree of fertility between them and with cassava. Planned trials will also reveal whether *M. pruinosa* is part of the GP1 of cassava (and so joining *M. flabellifolia* and *M. peruviana*) or whether it belongs to the secondary (GP2) gene pool. If, as seems likely, *M. flabellifolia* is part of the gene pool of cassava, then *M. pruinosa* is certainly the wild species most closely related to the crop. Pioneering molecular data support the latter view (Second *et al.*, 1997; Olsen and Schaal, 1998) and later more elaborate studies further strengthened it (Olsen and Schaal, 1999). The fact that *M. pruinosa* and *M. esculenta* ssp. *flabellifolia* seem to descend from a common stock has implications both for the early evolution of cassava and for its agricultural origin. Among others, this may expand or narrow the areas eligible for consideration as the cradle of domestication.

The Agricultural Origin of Cassava

Cassava is an ancient crop species. Lathrap (1970) estimates that domestication began 5000–7000 years BC and the estimate receives support from archaeological finds in the Amazon (Gibbons, 1990). By the time the first Europeans reached the New World, the crop was already cultivated in all of neotropical America (Patiño, 1964). The antiquity of the domesticate receives further support from two related facts. First, cultivation from vegetative propagules (vegeculture) is assumed to be an older practice than seed-culture (Harris, 1967, 1971a; Lathrap, 1973, 1977). Second, the great importance of manioc in American aboriginal swidden type of cultivation is the result of a combination of ecological factors, which favoured the growth of the plant, and cultural factors – the latter principally in the form of tradition drift which played a determinant role in the retention and further spread of the crop among ancient Amerindian societies (Albuquerque, 1969; Harris, 1971b; Lathrap, 1973; Moran, 1975).

Since Pohl (1827) suggested Brazil as the place of origin of cassava, i.e. the place of initial domestication, investigators have spent over a hundred years considering the issue, without reaching a definitive and consensual conclusion. Attempts to review the evidence on the origin of the crop produced no hard facts (Renvoize, 1972). Historically, 'origin' meant the quest for the agricultural origin of the crop as the sole concern had to do with the original place of domestication. In other words, writers were neither concerned with the tracking down of a presumed ancestry of the crop nor with the raising of such a possibility. However, anything short of speculation on the cradle of domestication had to consider first the discovery of the wild progenitor. The consequence of this would be to narrow the candidacy of eligible areas for the initial cultivation of manioc (Table 1.7).

The view has been expressed that 'the domestication of manioc was perhaps even easier [relative to yams]. It can be reproduced by stem cuttings. All that is necessary is to cut off a branch and place it in the ground during the rainy season and tubers will be produced' (Harlan, 1975). This view is premature and somewhat misleading since it considers the propagation of the crop rather than its domestication

which, as a process, was certainly more complicated. Initially it is likely to have been difficult to select for easy rooting, since the percentage of survival of wild stem-cuttings of the ancestor averages only 10% (Allem and Goedert, 1991). Two examples can be cited on how cassava was probably domesticated in pre-Columbian America. On the 14 May 1986, Embrapa's collecting team stopped at a petrol station in the municipality of Cacoal (11°28′S, 61°21′W), Brazilian state of Rondônia. A pump attendant commented on seeing the wild stakes of the progenitor being carried and told the team that he knew the material by the names 'mandioca-brava' (wild cassava) and 'mandioca-do-mato' (bush cassava). It was a common weed by the roadsides and inside food crop plantations. The roots were reportedly highly toxic and fresh leaves could kill pigs. Shortly after the family arrived from the Brazilian southern state of Paraná in 1982, his father cleared a plot of virgin forest and this turned the wild *Manihot* into an aggressive weed. The father planted some stakes of the species and within a year the family was harvesting roots of considerable size. Mr Adelicio, the clerk, said that only flour could be made from the roots because of their toxicity, while stem-cuttings were said to root with relative facility. This may have been an exaggeration as wild conspecific forms of the domesticate have been reported to root with great difficulty (Lanjouw, 1939, with the Surinamese *Manihot saxicola* and Jennings, 1963, with *Manihot melanobasis* from Guyana). A second example on how manioc was possibly first domesticated came on the 26 May 1986 through a report obtained in the vicinity of the municipality of Vila Rica (09°58′S, 51°00′W), Brazilian state of Mato Grosso. Two riders came by and looked closely at the attempts of the team to dig up some roots of a wild population of *M. esculenta* ssp. *flabellifolia*. They were asked about the plant. Both replied it was a native of the area and thrived at the edge of woods and that it was common practice for local dwellers to prune it back as cattle appreciated it and consumed the fresh leaves. The plant produced swollen roots but people did not eat them. An interesting story followed. They had heard that a few local inhabitants had once planted cuttings of the wild cassava and that the F1 generation usually gave very unsatisfactory results. Starting with the F2 generation, however, the plant was already

Table 1.7. Areas suggested by authors as the likely cradle where the domestication of manioc first took place.[a]

Cradle of domestication	Source
Brazil	Pohl (1827)
Brazil	Mueller (1874)
Eastern tropical Brazil	de Candolle (1884)
Brazil	Pax (1910)
Peru	Cook (1925)
Brazil	Lanjouw (1932)
Northern Amazonia	Schmidt (1951)
Brazil – central Paraguay	Vavilov (1951)
Venezuelan savannahs	Sauer (1952)
South America	Anderson (1954)
Peru or Mexico	Rogers (1963)[a]
Brazil	Jennings (1963)
Southern Mexico, Guatemala, Honduras	Rogers (1965), Rogers and Appan (1970)
Eastern Venezuela	Reichel-Dolmatoff (1965)
Peru	Lanning (1967)
Mexico and Central America	Schwerin (1970)
Northern Amazonia	Lathrap (1970)
North America	Rogers (1972)
Amazonia	Spath (1973)
Central America and north-eastern Brazil	Purseglove (1976)
Amazonia	Schultes (1979)
Mesoamerica	Jennings (1979)
Brazilian states of Goiás, Mato Grosso and Rondônia	Allem (1997)
Brazilian states of Mato Grosso and Rondônia	Olsen and Schaal (1998, 1999)

[a]Bertram (1993, p. 98) reports 'He [Rogers] is reported to have backed away from this position later (Stone, 1984), but maintained a sense that Central American species may have had role in the evolution of cassava (D. Rogers, personal communication).'

'amansada' (tamed) and started producing tuberous roots of fair size. The riders remarked that people could only make flour from the roots because they were poisonous if not processed. An additional related example from Africa seems to give support to the above version. A rare ethnobotanical report on the agronomic behaviour of the conspecific *M. melanobasis* shows that domestication from the wild is a real possibility for woody forms of the genus: 'its roots are usually fibrous, but large tuberous roots have been obtained from plants that have been left in the ground for long periods' (Jennings, 1957). Jennings (1959) later specifies this time as 'many years'. A second edible taxon of *Manihot* (still undescribed nomenclaturally to science and apparently a hybrid between cassava and the Brazilian species *Manihot glaziovii*) is harvested after staying at least 3 years in the ground while optimal harvest is reached in 7 years, hence the

common name 'mandioca-de-sete-anos' (7-year cassava; Allem and Hahn, 1991).

Manipeba: the Transitional Link Between the Wild Ancestor and Cultivated Cassava?

What can properly be regarded as the transitional link between the two wild strains of cassava and the domesticate was found in northeastern Brazil in July–August 1985. The find offered for the first time insights on how domestication and evolution of the crop might have proceeded in pre-Columbian times. First, there was the record that a climbing cassava landrace called 'manipeba' existed in northeastern Brazil and yielded every 6 years (Albuquerque, 1969). It was also hinted that such a material would reveal an as yet undescribed second edible species in the genus. During a collection in the area (A.C. Allem, unpublished) it was discovered that the names 'manipeba-branca' and 'manipeba-preta' were used for modern cassava varieties in southern Ceará and northeastern Bahia. In the municipality of Mari, state of Paraíba, the name 'manipeba-graúda' was being applied to another distinct edible 'species' of *Manihot* (see the above remarks on 'mandioca-de-sete-anos'). However, in a few rural households of Paraíba the name manipeba was found associated with a distinctly primitive architectural indigenous landrace of cassava. Such a plant is virtually unknown to breeders of cassava. The following was found relative to this unique plant. In places of northeastern Brazil, manipeba nearly always refers to a primitive and rustic folk variety of cassava. Many agriculturalists know the name but few grow the plant in their backyards because it is uneconomic to do so. Those who do cultivate it do so on account of social tradition. In the few households where manipeba was found growing, the number of individual plants ranged from one to six (eventually eight). It seemed as if people were relying on the plant as reserve against times of famine in a region chronically plagued by drought. Soon it was realized that social tradition was the chief factor behind its cultivation. People raising manipeba showed a sort of special consideration for the plant. Invariably, with pride, they described the landrace as highly productive which, in turn, suggested optimal mass selection performance as decisive to its retention. What may be called a plantation of manipeba was found at only two sites. The first was in the municipality of Bento Fernandes (05°42′S, 35°48′W) state of Rio Grande do Norte, where about 100 individuals were being raised on a small farm. The second plantation was found in the municipality of Sapé (07°12′S, 35°14′W) state of Paraíba, where about 50 individuals were being grown. On botanical grounds, landraces of manipeba are exceedingly close to wild plants of *M. esculenta* ssp. *flabellifolia*. They share basically the same type of inflorescence and leaf morphology. Significantly, as observed for specimens of *M. esculenta* ssp. *flabellifolia* in the wild, the internodes of manipeba are widely spaced 6–8 cm apart along the stem. The discovery was significant in the path leading to domestication

since it was the first time that undomesticated characters were found present in a cultivated folk variety of the crop. The particularly strong evidence suggested that, under domestication, agronomic characteristics may differentiate prior to the differentiation of botanical characters (advanced cultivars of cassava normally show the leaf scars clustered closely together and very prominent, whereas those of manipeba and the wild strain are virtually absent). The habit of manipeba is invariably that of a low to medium tall shrub, highly branched and always semi-decumbent to decumbent, often crawling on the ground. Agronomically, landraces of manipeba correspond to bitter cassava, i.e. show high levels of cyanogenic glucosides and the roots are highly poisonous if not processed. The roots are used solely to make flour and a type of bread known as 'beiju'. It takes, on average, between 4 and 5 years to produce high-quality roots, another indication of its primitive state. This explains why the landrace is so little exploited regionally for few people are willing, or can afford, to wait so long. If harvested less than 3 years from planting, the roots of manipeba are said to be watery. The quality of its flour, however, according to local reports, is unsurpassed. Moreover, local people say that manipeba is 'boa de goma' (good starch yielder). Of relevance was the information that genotypes of manipeba, unlike those of modern cultivars of cassava, successfully withstand fire and competition from the surrounding native vegetation. The primitive variety persists in abandoned areas even though other varieties of cassava do not.

The existence of manipeba bridges the architectural and agronomic gap between wild forms of cassava and the crop. Its existence also suggests how domestication may have proceeded in primeval times and illustrates that a substantial part of the variation inherent within cassava may be the direct result of human selection. Manipeba stands as the perfect archetype of very primitive varieties of manioc. Such an ancient landrace, given its ethnobotanic importance in shedding light on likely routes leading to the domestication and cultivation of the domesticate, should necessarily be included in worldwide representative collections. Regrettably, there is a trend in northeastern Brazil for such interesting germplasm to be displaced by the distribution and spread of improved materials.

Summing up, if the hypothesis of Allem (1994a) is valid, i.e. the progenitor of cassava is restricted to parts of the South American neotropical mainland and a number of places thought of in the recent past as likely cradles and prospective centres of initial domestication are obligatorily ruled out as sites where cultivation first began (pertinent examples in Table 1.7 include, among others, Honduras, Guatemala and Mexico). The assumption is that cassava was domesticated straight from the wild and therefore, because the crop is vegetatively propagated, present cultivated forms should preserve a maximum purity from the early days, i.e. the cultivated forms are expected to have very similar botanical morphologies as the living ancestor. That is the main reason why *M. aesculifolia*, and also *M. carthaginensis*, are considered unlikely to have contributed genes to cassava, i.e. the wingless fruit sculpturing of these species and the morphology of their seeds (resembling a tick) are very different in shape from those of cassava. Cassava shares these very characters with the wild ancestor.

Other progenitors might have been involved in the evolution of cassava but evidence is lacking. For example, it is known that many pubescent cassava cultivars occur in Peru (S.K. Hahn, personal communication) and some of them may have descended from domesticated stocks of *M. peruviana*, thus conferring legitimacy to Cook's (1925) hypothesis that cassava may have been domesticated in eastern Peru, an area indeed inhabited by the subspecies *M. esculenta* ssp. *peruviana*. Dr Hahn reported to the author in 1993 that perhaps the true figure is 200–300 pubescent Peruvian cultivars but he did not remember whether any particular cultivar bore pubescent ovaries or fruits, traits associated with the wild strain *M. esculenta* ssp. *peruviana*.

As pertinently speculated by Spath (1973), cassava may have been domesticated asynchronously at several sites in the Amazon. Consistent with the view that cassava may have originated in the Brazilian Amazon, there have been two recent theories whose greater interest lies in the fact that reasons were for the first time advanced on why cassava is expected to have been domesticated in a particular area.

First, there was the report by Allem (1997), who argued in favour of the central Brazilian state of Goiás and the westernmost Brazilian states of Mato Grosso and Rondônia as likely cradles of initial domestication of the crop. This was on the grounds that populations of both wild strains are particularly dense in these areas; the idea was that stocks drew the attention of local Amerindians who proceeded to tame the wild plant, the stocks probably spotted inside the Indians' household plantations rather than in the bushwood. Second, Olsen and Schaal (1999) elected the Brazilian states of Mato Grosso and Rondônia as a putative cradle of domestication on the basis of the fact that strong genome congruity has been documented between local wild populations of *M. esculenta* ssp. *flabellifolia* and modern cultivars of *M. esculenta* ssp. *esculenta*. However, equally legitimate claims hold that the domestication of cassava may have taken place in northern Brazilian Amazonia through the activities of Arawak tribes (Schmidt, 1951; Stone, 1984; Fig. 1.4). Allem (1997) advanced the view that perhaps the ultimate key as to the cradle of domestication of cassava may lie with the study of migration patterns of South American Amerindians in pre-Columbian times and specific scholarly studies on this theme (e.g. Schmidt, 1951; Migliazza, 1982) may hold the answer to this intriguing question.

Whatever the future of this active area of research, the increasing prospect is that manioc was domesticated in part of the Amazon, possibly in the Brazilian forest areas, rather than in the savannah. One possibility is that the ancestor of cassava evolved in the Brazilian Cerrado before reaching the Amazon. Domestication, however, seems to have happened in the Amazon and a further element may shed additional light on the subject. An Amazonian origin in the rain forest might explain an apparent paradox recorded by investigators. Why is it that cassava stops all photosynthetic functions and the stomata close whenever atmospheric humidity decreases, even though there is plenty of water in the soil (El-Sharkawy and Cock, 1984; El-Sharkawy *et al.*, 1984)? An origin in the humid Amazon, where the relative humidity of the air is rarely less than 70% (Bastos, 1972), may explain such behaviour.

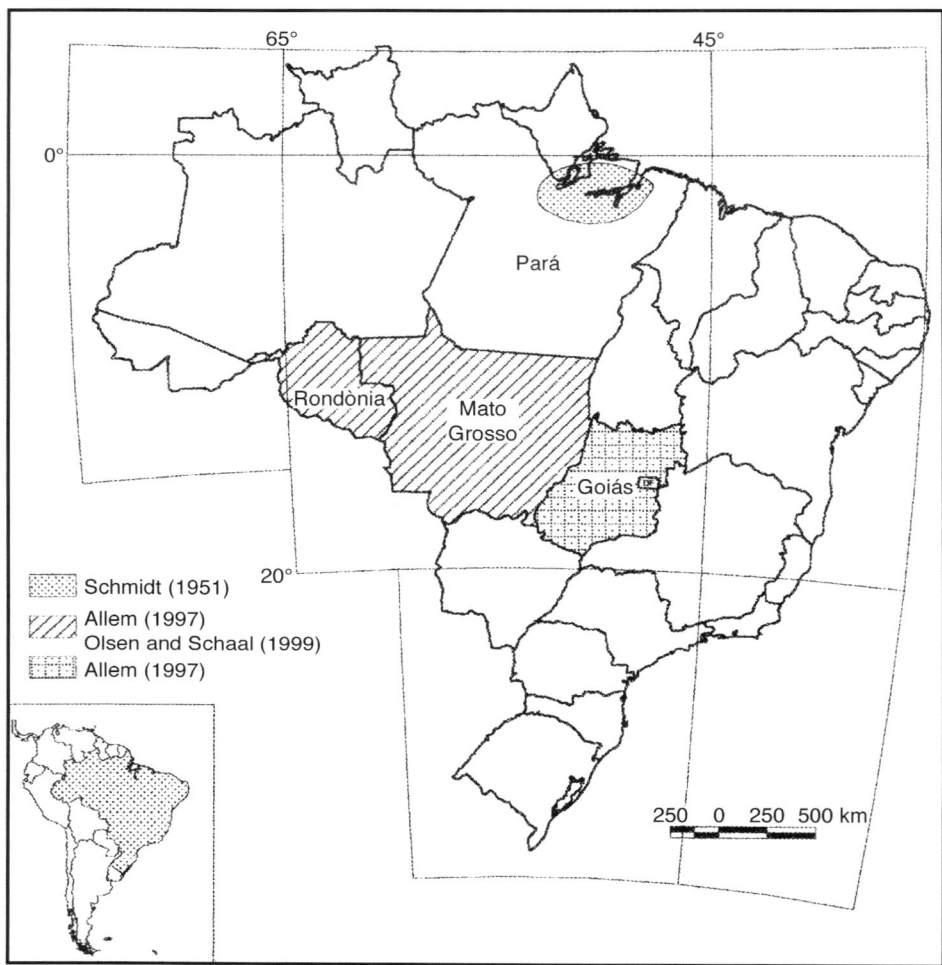

Fig. 1.4. Three putative areas of initial domestication in the Brazilian neotropics as suggested by Schmidt (1951), Allem (1997) and Olsen and Schaal (1999). Whether cassava was first domesticated in northern Brazilian Amazonia, or, somewhere in Brazil's westernmost states of Rondônia and Mato Grosso is controversial. However, if domestication did occur in Brazil it must have occurred along the periphery of the Amazon forest since the wild ancestor does not spread inside the tall forest of 'terra firme' (dry upland forest) which covers up to 90% of the Amazon. The referral of the collection Michael Goulding 80 (MG 86406) to *M. esculenta* ssp. *flabellifolia* in Allem's (1994a) study, supposedly the only specimen of the subspecies to have ever been collected in central Amazonia (Manaus area), was a mistake and is corrected here. The species in question is *Manihot brachyloba*, which ranges over much of Amazonia, in the interior and along its periphery. So far the wild progenitor of cassava has been collected in the Brazilian Amazon solely along the periphery of the forest (A.C. Allem, unpublished). Apparently, the ancestor did not penetrate the interior on account of soil type, i.e. it is absent in areas dominated by oxisols and ultisols, which occupy 90% of the Amazonian area. The large frequency of populations of *M. esculenta* ssp. *flabellifolia* and *M. esculenta* ssp. *peruviana* in northwestern Brazilian Amazonia (e.g. states of Mato Grosso and Rondônia) seems to associate with the presence of alfisols, in the Amazon they show a superficial sandy layer. The collection Allem and Silva 4477 (CEN), 31st Jan 1995, Brasil, state of Maranhão, 36 km (river Peritoró) SE of Peritoró along BR-316 highway heading for Caxias, is *M. esculenta* ssp. *flabellifolia*. The specimen was found close to the municipality of Caxias in Maranhão, an area inside northern Amazonia. However, Caxias is only 75 km away from Teresina,

Acknowledgements

I am indebted to Sérgio Eustáquio de Noronha and Abrahão Rezende Aidar (GIS laboratory) for the illustrations.

References

Albuquerque, M. (1969) *A Mandioca na Amazônia.* Superintendência do Desenvolvimento da Amazônia, Belém.

Allem, A.C. (1987) *Manihot esculenta* is a native of the neotropics. *Plant Genetic Resources Newsletter* 71, 22–24.

Allem, A.C. (1994a) The origin of *Manihot esculenta* Crantz (Euphorbiaceae). *Genetic Resources and Crop Evolution* 41, 133–150.

Allem, A.C. (1994b) *Manihot* germplasm collecting priorities. Report of the First Meeting of the International Network for Cassava Genetic Resources. International Plant Genetic Resources Institute, International Crop Network Series No. 10, Rome, pp. 87–110.

Allem, A.C. (1997) A reappraisal on the geographical origin of cassava (*Manihot esculenta*, Euphorbiaceae). In: Veiga, R.F. de A., Bovi, M.L.A., Betti, J.A. and Voltan, R.B.Q. (eds) *Anais do I Simpósio Latino-Americano de Recursos Genéticos Vegetais.* IAC/EMBRAPA-CENARGEN, Campinas, São Paulo, pp. 86–87.

Allem, A.C. (1999) The closest wild relatives of cassava (*Manihot esculenta* Crantz). *Euphytica* 107, 123–133.

Allem, A.C. and Goedert, C.O. (1991) Formação da base genética e manejo dos recursos genéticos de mandioca: o caso do Brasil. In: Hershey, C.H. (ed.) *Mejoramiento Genético de la Yuca en América Latina.* Centro Internacional de Agricultura Tropical, Cali, pp. 125–161.

Allem, A.C. and Hahn, S.K. (1991) Cassava germplasm strategies for Africa. In: Ng, N.Q., Perrino, P., Attere, F. and Zedan, H. (eds) *Crop Genetic Resources of Africa,* Vol. II. IITA, Ibadan, Nigeria, pp. 127–149.

Allem, A.C., Roa, A.C., Mendes, R.A., Salomão, A.N., Burle, M.L., Second, G., de Carvalho, P.C.L. and Cavalcanti, J. (2000) The primary gene pool of cassava (*Manihot esculenta* Crantz). In: Carvalho, L.J.C.B., Thro, A.M. and Vilarinhos, A.D. (eds) *Cassava Biotechnology: IV International Scientific*

Meeting. CBN, Embrapa Recursos Genéticos e Biotecnologia, Brazil, pp. 3–14.

Anderson, E. (1954) *Plants, Man and Life.* Andrew Melrose, London.

Bai, K.V., Asiedu, R. and Dixon, A.G.O. (1993) Cytogenetics of *Manihot* species and interspecific hybrids. In: Roca, W.M. and Thro, A.M. (eds) *Proceedings of the First International Scientific Meeting of the Cassava Biotechnology Network.* Centro Internacional de Agricultura Tropical, Cali, pp. 51–55. [CIAT Working Document no.123.]

Bastos, T.X. (1972) O estado atual dos conhecimentos das condições climáticas da Amazônia brasileira. *Boletim Técnico IPEAN,* Belém, 54, 68–122.

Bertram, R.B. (1993) Application of molecular techniques to genetic resources of cassava (*Manihot esculenta* Crantz, Euphorbiaceae): interspecific evolutionary relationships and intraspecific characterization. PhD thesis, University of Maryland.

Bertram, R.B. and Schaal, B.A. (1993) Phylogeny of *Manihot* and the evolution of cassava. *Cassava Biotechnology Network Newsletter* 1, 4–6.

Bolhuis, G.G. (1953) A survey of some attempts to breed cassava varieties with a high content of protein in the roots. *Euphytica* 2, 107–112.

Bolhuis, G.G. (1969) Intra and interspecific crosses in the genus *Manihot. Proceedings of the First International Symposium on Tropical Root Crops.* University of the West Indies, St Augustine, Trinidad, pp. 81–88.

Bonierbale, M., Roa, A.C., Maya, M.M., Duque, M.C. and Thome, J. (1997) Assessment of genetic diversity in *Manihot* species with AFLPs. *African Journal of Root and Tuber Crops* 2, 139.

Bretting, P.K. (1990) New perspectives on the origin and evolution of New World domesticated plants: introduction. *Economic Botany* 44 Supplement, 1–5.

Brondani, C. (1996) Variação isoenzimática de três espécies do gênero *Manihot* (Euphorbiaceae) relacionadas morfologicamente à mandioca (*Manihot esculenta* Crantz). *Pesquisa Agropecuária Brasileira* 31, 287–289.

Cabral, G.B., Carvalho, L.J.C.B. and Schaal, B.A. (2000) Relationship analysis of closely related species to cassava (*Manihot esculenta* Crantz) based on microsatellite-primed PCR. In: Carvalho, L.J.C.B., Thro, A.M. and Vilarinhos, A.D. (eds) *Proceedings of the Fourth International Scientific Meeting of the Cassava Biotechnology Network.* Embrapa Recursos Genéticos e Biotecnologia/CBN, Brazil, pp. 36–50.

Fig. 1.4. (*continued*) the capital of the neighbouring northeastern state of Piauí, a state largely covered by the xerophilous vegetation known as Caatinga. This suggests that collection 4477 comes from the eastern most part of the Brazilian Amazon, thus corresponding to its outskirts.

Carvalho, L.J.C.B., Cascardo, J.M.C., Limeira P.S., Ribeiro, M.C.M. and Fialho, J.F. (1993) Study of DNA polymorphism in *Manihot esculenta* Crantz and related species. In: Roca, W.M. and Thro, A.M. (eds) *Proceedings of the First International Scientific Meeting of the Cassava Biotechnology Network*. Centro Internacional de Agricultura Tropical, Cali, pp. 56–61. [CIAT Working Document no.123.]

Carvalho, L.J.C.B., Buso, G.M.C., Brondani, C., Allem, A.C., Fukuda, W.M.G. and Sampaio, M.J.A.M. (1995) Study on interspecific evolutionary relationships and intraspecific characterization of cassava germplasm at Cenargen/Embrapa. In: Thro, A.M. (ed.) *Proceedings of the Second International Scientific Meeting of the Cassava Biotechnology Network*. Centro Internacional de Agricultura Tropical, Cali, pp. 163–174. [CIAT Working Document no. 150.]

Colombo, C., Second, G. and Charrier, A. (2000) Genetic relatedness between cassava (*Manihot esculenta* Crantz) and *M. flabellifolia* and *M. peruviana* based on both RAPD and AFLP markers. *Genetics and Molecular Biology* 23, 417–423.

Cook, O.F. (1925) Peru as a center of domestication: tracing the origin of civilization through the domesticated plants. *Journal of Heredity* 16, 33–46.

Croizat, L. (1943) Preliminari per uno studio del genere *Manihot* nell' America meridionale. *Revista Argentina de Agronomia* 10, 213–226.

Davis, P.H. and Heywood, V.H. (1963) *Principles of Angiosperm Taxonomy*. Oliver and Boyd, London.

de Candolle, A. (1884) *Origin of Cultivated Plants*. Kegan Paul, Trench, London. [Translated from the French 'Origine des Plantes Cultivées', 1882, Germer Bailliere, Paris.]

El-Sharkawy, M.A. and Cock, J.H. (1984) Water use efficiency of cassava. I. Effects of air humidity and water stress on stomatal conductance and gas exchange. *Crop Science* 24, 497–502.

El-Sharkawy, M.A., Cock, J.H. and Held, K.A.A. (1984) Water use efficiency of cassava. II. Differing sensitivity of stomata to air humidity in cassava and other warm-climate species. *Crop Science* 24, 503–507.

Fregene, M.A., Vargas, J., Ikea, J., Angel, F., Tohme, J., Asiedu, R.A., Akoroda, M.O. and Roca, W.M. (1994) Variability of chloroplast DNA and nuclear ribosomal DNA in cassava (*Manihot esculenta* Crantz) and its wild relatives. *Theoretical and Applied Genetics* 89, 719–727.

Gibbons, A. (1990) New view of early Amazonia. *Science* 248, 1488–1490.

Grattapaglia, D., Nassar, N.M.A. and Dianese, J.C. (1987) Biossistemática de espécies brasileiras do gênero *Manihot* baseada em padrões de proteína da semente. *Ciência e Cultura* 39, 294–300.

Harlan, J.R. (1961) Geographic origin of plants useful to agriculture. *Publications of the American Association for the Advancement of Science* 66, 3–19.

Harlan, J.R. (1965) The possible role of weed races in the evolution of cultivated plants. *Euphytica* 14, 173–176.

Harlan, J.R. (1970) Evolution of cultivated plants. In: Frankel, O.H. and Bennett, E. (eds) *Genetic Resources in Plants: their Exploration and Conservation*. Blackwell, Oxford, pp. 19–32.

Harlan, J.R. (1975) *Crops and Man*. American Society of Agronomy/Crop Science Society of America, Madison, Wisconsin.

Harlan, J.R. and de Wet, J.M.J. (1963) The compilospecies concept. *Evolution* 17, 497–501.

Harlan, J.R. and de Wet, J.M.J. (1965) Some thoughts about weeds. *Economic Botany* 19, 16–24.

Harris, D.R. (1967) New light on plant domestication and the origins of agriculture: a review. *The Geographical Review* 57, 90–107.

Harris, D.R. (1971a) Agricultural systems, ecosystems and the origins of agriculture. In: Ucko, P.J. and Dimbleby, G.W. (eds) *The Domestication and Exploitation of Plants and Animals*. Duckworth, London, pp. 3–15.

Harris, D.R. (1971b) The ecology of agriculture in the tropics. *The American Scientist* 60, 180–193.

Harris, D.R. (1990) Vavilov's concept of centres of origin of cultivated plants: its genesis and its influence on the study of agricultural origins. *Biological Journal of the Linnean Society* 39, 7–16.

Haysom, H.R., Chan, T.L.C. and Hughes, M.A. (1994) Phylogenetic relationships of *Manihot* species revealed by restriction fragment length polymorphism. *Euphytica* 76, 227–234.

Heiser, C.B. (1990) New perspectives on the origin and evolution of New World domesticated plants: summary. *Economic Botany* 44 Supplement, 111–116.

Jennings, D.L. (1957) Further studies in breeding cassava for virus resistance. *Eastern African Agricultural Journal* 22, 213–219.

Jennings, D.L. (1959) *Manihot melanobasis* Muell. Arg – a useful parent for cassava breeding. *Euphytica* 8, 157–162.

Jennings, D.L. (1963) Variation in pollen and ovule fertility in varieties of cassava, and the effect of interspecific crossing on fertility. *Euphytica* 12, 69–76.

Jennings, D.L. (1979) Cassava. In: Simmonds, N.W. (ed.) *Evolution of Crop Plants*. Longman, London, pp. 81–84.

Lanjouw, J. (1939) Euphorbiaceae. In: Pulle, A. (ed.) *Flora of Surinam*, Vol. II, Part 1. Kon. ver.

Koloniaal Instituut te Amsterdam, Amsterdam, pp. 1–101.

Lanning, E.P. (1967) *Peru before the Incas*. Prentice-Hall, Englewood Cliffs, New Jersey.

Lathrap, D.W. (1970) *The Upper Amazon*. Thames and Hudson, London.

Lathrap, D.W. (1973) The antiquity and importance of long-distance trade relationships in the moist tropics of pre-Columbian South America. *World Archaeology* 5, 170–186.

Lathrap, D.W. (1977) Our father the cayman, our mother the gourd: Spinder revisited, or a unitary model for the emergence of agriculture in the New World. In: Reed, C.A. (ed.) *Origins of Agriculture*. Mouton, The Hague, pp. 713–751.

Magoon, M.L., Krishnan, R. and Bai, K.V. (1970) Cytogenetics of the F1 hybrid between cassava and ceara rubber, and its backcross. *Genetica* 41, 425–436.

Migliazza, E.C. (1982) Linguistic prehistory and the refuge model in Amazonia. In: Prance, G.T. (ed.) *Biological Diversification in the Tropics*. Columbia University Press, New York, pp. 497–519.

Moran, E.F. (1975) Food, development, and man in the tropics. In: Arnott, M.L. (ed.) *Gastronomy: the Anthropology of Food and Food Habits*. Mouton, The Hague, pp. 169–186.

Mueller, J. (1874) Euphorbiaceae. In: Martius, C.F.P. von (ed.) *Flora Brasiliensis* 11, 293–750.

Nichols, R.F.W. (1947) Breeding cassava for virus resistance. *Eastern African Agricultural Journal* 13, 184–194.

Olsen, K. (2002) Phylogeography of *Manihot esculenta* (Euphorbiaceae): II. Population History. *Evolution* (in press).

Olsen, K.M. and Schaal, B.A. (1998) Evolution in the cassava species complex: phylogeography and the origins of cultivated cassava. *Revista Brasileira de Mandioca* 17 Suplemento, 17.

Olsen, K.M. and Schaal, B.A. (1999) Evidence on the origin of cassava: phylogeography of *Manihot esculenta. Proceedings of the National Academy of Sciences of the USA* 96, 5586–5591.

Olsen, K.M. and Schaal, B.A. (2001) Microsatellite variation in cassava (*Manihot esculenta*, Euphorbiaceae) and its wild relatives: further evidence for a southern Amazonian origin of domestication. *American Journal of Botany* 88, 131–142.

Patiño, V.M. (1964) *Plantas Cultivadas y Animales Domesticos en America Equinoccial*, Vol. II. *Plantas Alimenticias*. Imprensa Departamental, Cali.

Pax, F. (1910) Euphorbiaceae-Adrianeae. In: Engler, A. (ed.) *Das Pflanzenreich*, IV. 147.II. 44, 1–111.

Pickersgill, B. (1977) Taxonomy and the origin and evolution of cultivated plants in the New World. *Nature* 268, 591–595.

Pohl, J.E. (1827) *Plantarum Brasiliae Icones et Descriptiones* 1, 1–136.

Prance, G.T. (1978) The origin and evolution of the Amazonian flora. *Interciencia* 3, 207–222.

Purseglove, J.W. (1976) The origins and migrations of crops in tropical Africa. In: Harlan, J.R., de Wet, J.M.J. and Stemler, A.B.L. (eds) *Origins of African Plant Domestication*. Mouton, The Hague, pp. 291–309.

Reichel-Dolmatoff, G. (1965) *Colombia*. Thames and Hudson, London.

Reichel-Dolmatoff, G. (1986) *Arqueologia de Colombia: un Texto Introductorio*. Fundación Segunda Expedición Botánica, Bogotá.

Renvoize, B.S. (1972) The area of origin of *Manihot esculenta* Crantz as a crop plant – a review of the evidence. *Economic Botany* 26, 352–360.

Roa, A.C., Maya, M.M., Duque, M.C., Tohme, J., Allem, A.C. and Bonierbale, M.W. (1997) AFLP analysis of relationships among cassava and other *Manihot* species. *Theoretical and Applied Genetics* 95, 741–750.

Roa, A.C., Maya, M.M., Chavarriaga, P., Duque, M.C., Mesa. E., Bonierbale, M.W., Tohme, J., Kochert, G. and Iglesias, C. (1998a) In search of the closest relatives of cassava: a morphological and molecular approach. *Revista Brasileira de Mandioca* 17 Suplemento, 18.

Roa, A.C., Chavarriaga, P., Duque, M.C., Bonierbale, M.W., Thome, J., Kochert, G. and Iglesias, C. (1998b) Microsatellites as a tool for assessing genetic diversity in *Manihot* species. *Revista Brasileira de Mandioca* 17 Suplemento, 18.

Roa, A.C., Chavarriaga-Aguirre, P., Duque, M.C., Maya, M.M., Bonierbale, M.W., Iglesias, C. and Tohme, J. (2000) Cross-species amplification of cassava (*Manihot esculenta*) (Euphorbiaceae) microsatellites: allelic polymorphism and degree of relationship. *American Journal of Botany* 87, 1647–1655.

Rogers, D.J. (1963) Studies of *Manihot esculenta* Crantz and related species. *Bulletin of the Torrey Botanical Club* 90, 43–54.

Rogers, D.J. (1965) Some botanical and ethnological considerations of *Manihot esculenta. Economic Botany* 19, 369–377.

Rogers, D.J. (1972) Some further considerations on the origin of *M. esculenta. Tropical Root and Tuber Crops Newsletter* 6, 4–10.

Rogers, D.J. and Appan, S.G. (1970) What's so great about cassava? *World Farming*, pp. 14, 16, 22.

Rogers, D.J. and Appan, S.G. (1973) *Manihot* and *Manihotoides* (Euphorbiaceae), a computer-assisted study. *Flora Neotropica*, Monograph No.13. Hafner Press, New York.

Rogers, D.J. and Fleming, H.S. (1973) Monograph of *Manihot esculenta* Crantz. *Economic Botany* 27, 1–114.

Sauer, C.O. (1952) *Agricultural Origins and Dispersals*, Series two. The American Geographical Society, New York.

Sauer, J.D. (1994) *Historical Geography of Crop Plants: a Select Roster*. CRC Press, Boca Raton, Florida.

Schaal, B.A. and Olsen, K.M. (2000) Gene genealogies and population variation in plants. *Proceedings of the National Academy of Sciences of the USA* 97, 7024–7029.

Schaal, B., Olson, P., Prinzie, T., Carvalho, L.J.C.B., Tonukari, N.J. and Heyworth, D. (1995) Phylogenetic analysis of the genus *Manihot* based on molecular markers. In: Thro, A.M. (ed.) *Proceedings of the Second International Scientific Meeting of the Cassava Biotechnology Network*. Centro Internacional de Agricultura Tropical, Cali, pp. 62–70. [CIAT Working Document no. 150.]

Schaal, B., Carvalho, L.J.C.B., Prinzie, T., Olsen, K., Hernandez, M., Cabral, G. and Moeller, D. (1997) Phylogenetic relationships and genetic diversity in *Manihot* species. *African Journal of Root and Tuber Crops* 2, 147–149.

Schmidt, C.B. (1951) A mandioca: contribuição para o conhecimento de sua origem. *Boletim de Agricultura, São Paulo* 52, 73–128.

Schultes, R.E. (1979) The Amazonia as a source of new economic plants. *Economic Botany* 33, 259–266.

Schwerin, K.H. (1970) Apuntos sobre la yuca y sus origenes. *Tropical Root and Tuber Crops Newsletter* 3, 4–12.

Second, G. (1998) *Manihot glaziovii*, an example for a proposed dynamic conservation of cassava genetic resources. *Revista Brasileira de Mandioca* 17 Suplemento, 18.

Second, G., Allem, A.C., Mendes, R.A., Carvalho, L.J.C.B., Emperaire, L., Ingram, C. and Colombo, C. (1997) Molecular markers (AFLP)-based *Manihot* and cassava numerical taxonomy and genetic structure analysis in progress: implications for their dynamic conservation and genetic mapping. *African Journal of Root and Tuber Crops* 2, 140–147.

Simpson, B.B. and Haffer, J. (1978) Speciation patterns in the Amazonian forest biota. *Annual Review of Ecology and Systematics* 9, 497–518.

Spath, C.D. (1973) Plant domestication: the case of *Manihot esculenta*. *Journal of the Steward Anthropological Society* 5, 45–67.

Stone, D. (1984) Pre-Columbian migration of *Theobroma cacao* Linnaeus, and *Manihot esculenta* Crantz, from northern South America into Mesoamerica: a partially hypothetical view. In: Stone, D. (ed.) *Pre-Columbian Plant Migration*. Papers of the Peabody Museum of Archaeology and Ethnology, Harvard University, Vol. 76, pp. 67–83.

Vavilov, N.I. (1951) The origin, variation, immunity and breeding of cultivated plants. *Chronica Botanica* 13, 1–364.

Chapter 2
Cassava in South America and the Caribbean*

Guy Henry[1] and Clair Hershey[2]

[1]CIRAD-Amis, Rua Paulo Castro P. Nogueira 600, Campinas-SP 13092–400, Brazil; [2]2019 Locust Grove Road, Manheim, PA 17545, USA

Origin and Distribution of Cassava in Latin America

Cassava and all its wild relatives have their genetic origins in Latin America (the term *Latin America* is used herein for the entire cassava-growing region of the New World). The crop was vital to the development of lowland tropical cultures throughout the New World. The Carib and Arawak Indians of the Caribbean and northern South America were probably some of the earliest cultivators of cassava, and many of their customs of cultivation and processing remain virtually intact today, in that region, and throughout the Amazon basin. Every tropical country of the region produces cassava, but its cultivation is most highly concentrated in four areas: northern and eastern coastal Brazil; southern Brazil and eastern Paraguay; northwestern South America (especially the Caribbean coast of Colombia); and the Greater Antilles (Cuba, Haiti, Dominican Republic). The Americas gave cassava to the rest of the world after the arrival of early European explorers. Along with the species itself, these explorers introduced cultivation and processing techniques from cassava's

homelands. This history has not only had a profound influence on the current status of the crop, but also on its potential for further development.

Cassava has numerous traits that confer comparative advantages in marginal environments, where farmers often lack the resources to improve the income-generating capacity of their land through purchased inputs. The species tolerates acid soils, periodic and extended drought, and defoliation by pests. It is highly compatible with many types of intercrops and flexible as to time of harvest. Furthermore, the crop serves a wide variety of food, feed and industrial purposes. These traits have combined to make cassava a significant sustaining force, benefiting the poor in the tropics.

Latin America currently represents less than one-fifth of the global cassava output of 166 million t. Of the continent's 28 million t, Brazil alone accounts for about 70%. Despite the historical importance of cassava, in recent years it has lagged behind other crops in growth rates for production and utilization. The reasons are many, with vital implications for projections of future crop development. Among the main factors, government policies and trends in food demand

---

* This chapter draws significantly on the work by Hershey *et al.* (1997) as the latter can be considered as the most complete and detailed assessment on this subject to date. Furthermore, the chapter incorporates cassava market information from consultancy reports by Henry *et al.* (1998) and Henry (1999).

resulting from urbanization have tipped the balance in favour of alternative food energy sources since the 1970s. Investment in cassava has not been adequate to keep it competitive in the agricultural and commercial worlds. As a crop predominantly grown and utilized by the poor, it has generally been relegated to a lower status by both public and private research institutions. The future of cassava in Latin America and the Caribbean (LAC) is defined most by its potential as a vehicle for linking the rural poor to growth markets. This potential follows from the complex, interacting effects related to urbanization, rising incomes, evolving trade policy and trends in other food and feed crops.

This chapter gives an overview of cassava production, utilization and market aspects in the principal cassava-growing regions of Latin America. This will be accomplished by presenting briefs summaries of production systems and production trends for cassava in several of the major producer countries. Finally, some implications for the future of cassava development in Latin America are presented.

Cassava in Selected Countries

Cassava systems in South America and the Caribbean are highly varied in all their aspects; hence it is useful to summarize cassava production in selected countries, before discussing continent-wide systems and trends. For this purpose we highlight the seven countries with the largest areas currently planted to cassava among producer countries in the region: Brazil, Colombia, Cuba, Haiti, Paraguay, Peru and Venezuela. These countries produce 97% of the region's cassava (Table 2.1).

Brazil

The region's largest country has been near the top (currently third place) in total cassava production globally, probably since the crop was first cultivated. Cassava is a major crop in three of the country's ecoregions: lowland humid north (19.5% of production), the dry northeast (46.3%) and the subtropical south (21.1%).

Table 2.1. Latin American cassava production trends, by country, 1983–1999.

Cassava production (t)	Year				
	1983	1987	1991	1995	1999
LAC	28,229,148	30,695,572	31,275,691	32,530,441	29,749,602
Argentina	139,000	148,300	150,000	160,000	165,000
Bolivia	180,385	424,248	414,598	295,700	400,006
Brazil	21,847,888	23,499,960	24,530,780	25,315,620	20,171,600
Colombia	1,554,700	1,260,390	1,645,213	1,751,899	1,956,051
Costa Rica	21,100	40,000	83,610	125,000	119,470
Cuba	325,000	305,000	300,000	250,000	250,000
Dominican Republic	92,514	97,836	137,422	136,821	155,755
Ecuador	194,794	131,190	90,279	75,683	138,172
El Salvador	23,322	27,887	32,080	32,495	30,000
Guatemala	9,100	9,832	14,000	15,952	16,000
Haiti	265,000	290,000	335,000	300,000	320,000
Honduras	6,554	7,400	8,215	8,730	10,081
Jamaica	17,188	17,021	12,111	17,447	14,972
Mexico	2,115	907	386	1,688	1,100
Nicaragua	72,680	56,000	52,000	51,500	51,000
Paraguay	2,610,000	3,467,700	2,584,900	3,054,394	3,500,000
Peru	485,443	537,033	410,693	547,439	885,100
Suriname	2,659	3,855	3,058	7,000	4,000
Trinidad and Tobago	2,000	717	1,107	696	1,400
Venezuela	324,733	317,776	381,069	299,233	487,685

Source: FAOSTAT, FAO (1999).

There is minor production in the acid soil, wet/dry savannahs of the central-west (*campo cerrado*; 4.4%) and in the subhumid southeast (8.7%). Over the last 20 years, national production has varied little, at about 2 million t year^{-1}. In the north (Amazon basin), however, production has more than tripled in the past 25 years, reflecting the role of cassava in 'frontier' agriculture (IBGE, 1992).

In Brazil *bitter* and *sweet* types of cassava are considered as different crops: *aipim* (sweet) and *mandioca* (bitter). Most of the cassava is of the latter type (high cyanogenic potential), which must be processed prior to consumption. The main product is a coarse, toasted flour (*farinha de mandioca*), the principal carbohydrate source of the poor and a complement to many other dishes. The south leads in starch production for food and industrial use (> 300,000 t year^{-1}), as well as for on-farm feeding (roots and leaves). There is a nascent animal feed market for dried chips in the northeast, where the cultivation of cereal crops is risky and shipping grain from the south (or Argentina) is relatively costly. Furthermore, during the last few years, some small- to medium-sized factories in the south have started to produce a line of frozen cassava-based snacks and convenience foods for national urban consumption and also for export.

Brazil has a strong national research programme and a network of state programmes working to improve cassava systems. The emphasis is on the production side, although in the past decade it has shifted towards greater emphasis on processing and marketing. The National Cassava and Fruit Research Centre (CNPMF) in Bahia State holds the world's largest national-programme collection of cassava germplasm. The National Centre for Genetic Resources and Biotechnology (CENARGEN) includes cassava and wild *Manihot* species within their mandates. Brazil has recently taken a strong leadership role in the adaptation of farmer participatory techniques for technology development and diffusion (Pires de Matos *et al.*, 1997).

Some of the world's more advanced cassava agriculture is found in Brazil's southern, subtropical region. Local research and extension programmes have been working to improve cassava since the early 1940s. The results are evident in their highly productive systems, with yields averaging 17–20 t ha^{-1} and up to 30–35 t ha^{-1} in intensive systems in Paraná and Mato Grosso do Sul states. This has been due mainly to better soils, larger farms and better managers, but also, to a strong demand by cassava processors for cheap raw material, expanding production technology demand and adoption.

Colombia

Perhaps the Latin American country with the highest agroecological diversity, Colombia hosts a wide range of systems for cassava cultivation and utilization. The highest proportion of production (45%) comes from the seasonally dry, semiarid Atlantic Coast region. Another 25% is produced in inter-Andean valleys of the eastern mountain range and 17% in the central part of the country. The eastern, acid-soil savannahs (*llanos orientales*) and the high-rainfall Pacific Coast are minor producers at 9 and 4%, respectively (Balcazar, 1997).

Along with this diversity of environments comes a wide range of biological problems. All but a few of the pests and diseases that affect cassava worldwide are endemic in Colombia. This not only represents a challenge for growers but also an opportunity for researchers to capitalize on 'hot-spot' environments when selecting for host-plant resistance. Moreover, many of the natural enemies of pests and pathogens thrive there and can be exploited in research and production.

Except for the Amazon and *llanos* regions, most cultivars have low cyanogenic potential and are consumed fresh. Many traditional Colombian food dishes include cooked cassava. In addition, sour cassava starch is an essential ingredient for several popular Colombian bakery products such as *pan de bono*. More recently, several cassava-based snack and convenience foods for urban consumers have appeared in supermarkets.

While cassava has traditionally been planted by small farmers (mostly intercropped with maize, yams, etc.), more recently, larger plantation-style plantings have been started in response to a boost in demand from cassava processors.

In the mid-1980s Colombia recognized the potential of cassava as a substitute for imported maize and sorghum in balanced animal feed rations and began a programme involving a range of R&D institutions and farmers' groups. This pilot project was built on the concept of the 'integrated cassava R&D project', a development model based on simultaneous work to improve production efficiency, develop new products and processing methods, and expand markets. These projects first concentrated on the animal feed market but later included fresh cassava, starch and flour. The combination of these initiatives contributed to an upturn in production from 1.3 million t in 1987 to 1.8 million t in 1996. While this first model concentrated on small farmer cooperatives on the Atlantic Coast, a more recent (1999) development, initiated through a consortium of private and public sector actors (CLAYUCA; *Consorcio Latin-Americano y del Caribe de Apoyo a la Investigación y Desarollo de la Yuca*), focuses on a larger scale agribusiness model in the Cauca Valley region.

Colombia is host country for the Centro Internacional de Agricultura Tropical (CIAT) and has not only contributed to the global cassava initiatives of this centre, but also benefited from its presence. As participants in testing new technology in the field, Colombian farmers have had the opportunity to be early beneficiaries. Some of the original work on basic agronomic practices (stake selection and treatment, planting position, plant density, herbicides) led to recommendations that were quickly and broadly adopted. Because of an extensive collaborative varietal testing network, Colombian institutions had an advance look at some of the new materials.

Cuba

Cassava production in Cuba follows two very distinct general forms: large state farms, where a relatively high level of technology is applied, and small private plots, which are becoming more common. The state-controlled system allows technology developed on experimental stations to be transferred almost immediately to production fields. These farms often have high-input systems for cassava, including mechanized land preparation, planting and harvesting, herbicide and fertilizer applications, even irrigation. Sometimes the use of fertilizer and pesticides has

not been economic, so the State subsidized them to help farms meet their production goals. More attention is now being given to the economics of production – the use of inputs to produce a profitable output. Despite this emphasis on technology, yield levels have been disappointing – some of the lowest in the region. This is partly because cassava is being grown on the poorer soils, and there is a shortage of inputs. Fertilizers and herbicides are increasingly diverted to higher value crops.

Cuba's research and extension system has been among the most consistently productive in the region, with a long-term, well-balanced interdisciplinary effort. The programme has developed packages of agronomic practices, new cultivars and pest control systems. Most Cuban production is used directly for fresh consumption. One of the early research successes was to develop a system to extend the period when fresh roots are available on the market, by combining a specific set of cultivars with differing maturities and staggered planting dates. Cuba has been promoting research on use of cassava in animal feed rations, but this is not yet a major market.

Haiti

Cassava is becoming more important in the Haitian's diet. Cassava is processed and baked to make the traditional Caribbean form of large flat bread *casabe*. Most cultivars are of the bitter type. From 1970 to 1995, annual per capita consumption increased from 32 to 35 kg, while all other countries of the region saw a decline. This increasing dietary role is, unfortunately, being driven largely by the effects of bringing more marginal soils under cultivation, degradation of existing cultivated land and very adverse economic conditions in the poorest and most populous country of the region.

Despite these pressing needs in a crop of increasing importance, Haiti has almost no research capacity. The language barrier, together with very volatile political interests (for R&D), has made it difficult for them to participate fully in the regional and international networks involved in cassava. The country desperately needs a substantial R&D effort in cassava as a means of raising living standards of the rural poor. What little research has been done has been mainly sponsored by non-governmental organizations (NGOs), with a short- to medium-term perspective and insufficient local support.

Paraguay

Paraguay maintains a very strong cultural attachment to cassava as part of a history that goes back to the Guarani Indians' reliance on this crop. Per capita consumption is the second highest in the world (after Democratic Republic of Congo). Most Paraguayans eat cooked cassava two or three times a day as part of the main dish or as a first course. Production nearly tripled in the period from 1961 to 1996. In the peak years of the late 1980s, production reached almost 4 million t. This relatively high level of production, and the strong agricultural sector in general, would seem to favour the move towards use of cassava as an industrial raw material for the production of animal feed or starch, for instance. Cassava must compete in several of these markets with maize, cotton and soybean, which are also major crops there.

Paraguay is the most rural of South American countries. Over 60% of the land area is agricultural, and nearly as many people live in rural areas as in the cities, in contrast to most of the continent where between three and seven times more people live in cities. It also has the highest share of agriculture in the GDP–26% – compared to its neighbours (Brazil, 9.6%; Argentina, 6.0%). Currently, however, it has one of the continent's highest urban growth rates (4.5% year^{-1}), so the dynamics of urbanization are likely to drive some of the same trends as elsewhere on the continent – a move towards more industrial uses of cassava, with a consequent decline in food uses.

The cultural importance of cassava has not been accompanied by a concerted institutional interest. Resources for research have been most irregular, depending very much on the personal interest of individual ministers of agriculture. The extension service has been relatively more active both in research and extension, attempting to fill a void where no separate research effort was organized for cassava. During the 1980s, one of the most active periods for research on cassava in Paraguay, substantial work was done on germplasm collection, agronomic practices,

pest control and utilization. Towards the end of the 1990s, there has been a renewed interest in cassava, especially regarding the potential of adding value to increase demand and farmer income, as well as to popularize cassava as a commercial crop in areas where decreasing interest in growing cotton has left a major gap.

Peru

Most of Peru's production is in the eastern part of the country, in the rainforest and on the lower slopes of the Andes. The highly populated coastal area relies almost totally on irrigation and therefore grows higher-value crops. As elsewhere in the Amazon basin, cassava is a staple. Production has been relatively stable for the past 35 years, with a rising trend in the 1990s. As pressure on land increases, slash-and-burn systems are more difficult to sustain; thus those populations near urban markets seek to intensify and commercialize their agriculture. Some cassava is shipped across the Andes to Lima, mainly for recent immigrants to the city, who retain a preference for cassava over more accessible potatoes. Because most production is isolated from major markets, future development will need to focus on internal markets close to areas of cultivation. As most cassava is grown in humid rainforest environments, opportunities for drying chips for animal feed are limited if natural drying is to be used. Starch and flour are possible options for value-added products.

Peru has a very limited research capacity in cassava and does not have any nationally coordinated effort. Projects in processing for animal feed, flour for partial substitution in bakery products and marketing of treated fresh cassava have been some of the principal thrusts in the last part of the 1990s.

Venezuela

With the global oil boom in the 1970s, Venezuela was not motivated to pursue agricultural development, and rural areas not linked to spill-over from oil income suffered the consequences. The share of agriculture in the country's GDP is quite low – only 5.0%. Land use is also low with only about 25% of the total area in agriculture.

Nevertheless, the area planted to cassava and production have been relatively stable in the past 35 years. This reflects the fact that most farmers who rely on cassava do not have many other options for income. Most of the growing areas are drought-prone (coastal) or have acid, low-fertility soils (savannah and rainforest). Various private companies have tried to establish a starch industry based on cassava, but with limited success. Currently, other agroindustries are applying the lessons of past failures and are working in a more integrated manner to coordinate production with processing capacity and market demand. These industries manage cassava plantations, with technology and performance levels similar to those in the state of Paraná in Brazil.

During the 1960s and 1970s, Venezuela had a strong cassava research and training programme based at the Central University at Maracay. This group initiated countrywide work in germplasm collection and evaluation, production practices, developing expertise in the areas of utilization in animal feeding and pest management. Currently, the private sector is sponsoring a modest but effective research programme on cassava, aimed mainly at production for starch.

Production Systems in Latin America

Cassava is nearly always part of a farming system that includes other crops or animal components. System characteristics are associated with environmental influences, economic constraints and opportunities, and cultural traditions. The cassava plant can tolerate long periods of drought after it is established, but it must be planted during a period of adequate soil water. In most systems, growers plant cassava near the beginning of the rainy season. Cassava is slow to develop a canopy, so early weed control is crucial. This is accomplished mainly by hand weeding, but use of herbicides (pre- and post-emergence) is increasing. Farmers rarely apply inputs to control pests or diseases.

As there is no sharply defined maturity period, harvest may extend over several weeks or even months, depending upon the end use. As the level of drought, soil infertility and/or acidity stress increases, cassava tends to become a more dominant component of the cropping system. In fertile inter-mountain valleys of the Andes, for example, cassava is one of many crops. In the semiarid interior of northeast Brazil, or, in the acid-soil rainforests, cassava can play a dominant role.

In some regions, production and harvesting are seasonal, determined by low temperature, drought or excessive rain. In the highlands where plant growth is slow, the production cycle is typically 18–24 months. Similarly, in the subtropics, farmers often leave the crop in the ground over winter and harvest after the second growing season. Where rainfall is very low, growth may be so slow that reasonable production is obtained only after the second or third rainy season. In areas of seasonal flooding (*varzeas* of the Amazon region), harvest may be as early as 5–6 months because cassava does not tolerate water-saturated soils.

The more traditional systems tend to be more complex and rely on labour rather than purchased inputs. System complexity has evolved out of the complementary interaction effects of individual components to provide a balance between stability and productivity. In more modern systems, farmers incorporate purchased inputs to achieve greater productivity and reasonable stability.

The labour-intensive nature of cassava husbandry is an area of concern as the labour force in agriculture declines. This is most notable in South America, less so in the Caribbean. From 1970 to 1990, the average number of labourers per hectare of agricultural land in Brazil, Colombia and Paraguay declined from 0.29 to 0.21. In Haiti and Cuba the decline in absolute numbers was the same, but went from 0.71 to 0.63 labourers. By comparison, in Thailand, Indonesia and India, the number of agricultural labourers remained stable in the same period, at about 1.25 labourers ha^{-1}. As the cost of rural labour increases, mechanization becomes a more urgent issue, especially for planting and harvesting, a top priority for more high-input high-output farmers. While mechanized planters are becoming more popular in southern Brazil, appropriate harvesters are still in a developmental phase. Different prototypes exist, but relatively high purchase costs and too high harvest losses (breakage of roots) still need additional research.

Production trends

In comparison with Asia or Africa, production trends for cassava in Latin America have been quite stable over the past 25 years. Brazil accounts for most of the aggregate variations. This country strongly dominates the Latin American cassava production. Most countries have had gradual tendencies to increase or decrease production, but few have realized dramatic shifts due to major production or market forces (Table 2.2). This is to be expected in the traditional production systems and constrained markets that characterize most of the region.

Aggregate production since the 1960s can be broadly characterized into three phases. From 1961 to 1972, there was a marked increase in area and production, mainly in response to continued population growth. Area harvested peaked at 2.85 million ha in 1977. Between 1977 and 1984, area planted steadily declined, as the full impact of wheat import subsidies (in Brazil) and other policy disincentives were translated into reduced consumer demand for cassava. Since the mid-1980s, the area planted has been relatively stable with some tendency to decline. Production climbed at a higher rate than increases in planted area during the 1960s but then continued at constant levels for the next 25 years. In the past few years, there has been a trend of increasing yields, as the adoption of productivity-enhancing technology accelerates in the region. However, the true impact of these technologies has been masked to some extent due to the negative climatic impacts on yields caused by El Niño and La Niña. In the past 15 years, there have been wider cyclical variations in area and production as compared to the previous two decades. Specific causes of this fluctuation are difficult to pinpoint but may be

Table 2.2. Cassava production, area and yield in LAC, 1990 and 1999.

Cassava-producing countries	Production (t)		Area harvested (ha)		Yield (kg ha^{-1})	
	1990	1999	1990	1999	1990	1999
LAC	32,154,182	28,578,126	2,738,936	2,353,252	117,397	121,441
Antigua and Barbuda	45	40	9	9	50,000	44,444
Argentina	140,000	165,000	14,000	16,000	100,000	100,000
Bolivia	393,590	400,006	36,358	40,000	108,254	100,002
Brazil	24,284,700	20,171,600	1,933,620	1,539,180	125,592	131,054
Colombia	1,939,020	1,956,051	207,310	184,718	93,532	99,856
Costa Rica	65,000	119,470	4,700	6,000	138,298	199,117
Cuba	300,000	250,000	72,000	65,000	41,667	38,462
Dominican Republic	132,027	155,755	20,476	24,000	64,479	64,898
Ecuador	134,245	138,172	24,590	19,760	54,593	69,925
El Salvador	28,600	30,000	1,800	1,900	158,889	155,263
French Guyana	18,967	10,375	1,983	1,690	95,648	61,391
Guadeloupe	1,654	1,460	120	130	137,833	112,308
Guatemala	15,700	16,000	5,000	5,000	31,400	32,000
Guyana	21,800	25,957	2,000	2,200	109,000	117,986
Haiti	330,000	320,000	82,000	74,418	40,244	43,000
Honduras	7,968	10,081	1,000	1,100	79,680	91,645
Jamaica	11,803	14,972	991	780	119,102	191,949
Mexico	3,073	1,100	407	145	75,504	75,862
Nicaragua	53,000	51,000	4,800	4,700	110,417	108,511
Panama	29,965	30,309	6,040	5,400	49,611	56,128
Paraguay	3,549,947	3,500,000	239,900	240,000	147,976	145,833
Peru	381,069	885,100	40,794	80,000	93,413	110,638
Puerto Rico	2,377	281	300	39	79,233	72,051
Venezuela	301,647	487,685	37,795	40,000	79,811	108,060

Source: FAOSTAT, FAO (1999).

related to uncertainties in the cassava market-place as agricultural and trade policy in the region undergo reform and adjustment. For a more detailed treatise on trends by individual country, refer to Henry and Gottret (1996).

Several less-aggregated trends exist although they are less obvious. For example, while cassava yields in north and northeast Brazil are struggling to overcome natural calamities, in the south and southwest, both area and yields continue to rise. Furthermore, during the last decade (Table 2.2), cassava area and yields are rising markedly in many of the Central American and Caribbean countries. This phenomenon can be explained partly by the bullish export demand for fresh and frozen cassava from the European Union (EU) and USA, mainly supplied by Costa Rica.

Utilization, Market Systems and Trends

Fresh cassava roots and flour for human consumption

More than half the cassava produced in the region is used directly for human food and the remainder for animal feed or industrial uses. This aggregate picture, however, masks regional variations. In Brazil and Paraguay – the two largest producers – 50 and 65%, respectively, of production is destined for animal feed. (These official figures may not reflect reality; the authors believe that 30% for Brazil and 40% for Paraguay may be more realistic.) Much of this is for on-farm use in non-intensive systems for pigs and chickens. In nearly all the other producing countries, the food market predominates, and only 10–20% of production goes for animal feed.

Previous fresh cassava and flour production trends and the current situation in LAC, have been analysed extensively by Henry and Gottret (1996) and Hershey et al. (1997). Consumption of fresh cassava in Colombia and Paraguay, and farinha in northeast Brazil will increase with decreasing cassava prices (relative to its major substitutes) for the lowest income groups in both rural and urban areas. Furthermore, studies (Henry, 1996) have shown evidence that the average urban consumer in Brazil is willing to pay more for better quality farinha; thus, higher quality cassava products may expand traditional

demand in these areas. The traditional farinha de mandioca industry in southern Brazil has been under increasingly heavy competition for raw materials by the growing starch industry. Drought conditions in northeast Brazil have boosted the demand for farinha (from the south) for the past several years, but this is not sustainable. At present, it is not clear what the future prospects are for these industries (CERAT, 1997).

In Colombia, Peru, Brazil (Ceará) and Ecuador, experiences from integrated cassava projects show that there is some potential for cassava to substitute partially for wheat flour in bakery, pastry and snack food industries (Henry, 1996; Ospina et al., 1996; Eguez, 1996). To benefit from some of these opportunities, appropriate socio-economic and political conditions are necessary and detailed ex-ante feasibility studies are required. Currently, there is renewed interest in Brazil for developing cassava flour-based products for urban and export markets.

Chips and leaves for animal feed

On-farm feeding of fresh or dried cassava has a long tradition, but mainly in very non-intensive systems. With rapidly increasing demand for animal products – meat, milk and eggs – cassava is finding markets in balanced rations for animal feeds. The technical details for managing dried cassava in these rations are well established in terms of both the milling and blending process, as well as the animal nutrition side (Buitrago, 1994). The main constraints for continued expansion of this market are constancy of raw material supply throughout the year, stability of product quality and price competitiveness.

Ospina et al. (1996), Henry and Best (1994) and Hershey et al. (1997) have reported extensively on the cassava chip experiences and its future potential for animal feed in Brazil and Colombia. Gottret et al. (1997) report a calculated demand potential by the feed industry in Colombia of > 500,000 t year^{-1} at certain relative prices and quality levels. Actual cassava chip utilization averages 30,000–50,000 t. However, as mentioned earlier, a recent plan for an industrial-size integrated production/processing plant in the Cauca Valley is taking shape. Similar and higher figures have been reported for Ceará State, Brazil (Henry, 1996), depending on the level at which cassava is included. In Ceará the potential demand for chicken and pig feed rations is augmented by the demand from dairy farmers for supplementing with cassava chips during the dry season.

Besides utilizing cassava roots and leaves for animal feed, cassava starch and flour processing by-products have traditionally been valued for feed use. While this practice at the farm level seems to offer good returns, the larger-scale industries face the constraint of cost-effective drying options. In Thailand the existence of large open-air drying floors has reduced this problem.

Starch-based applications

Starch is not a major product from cassava in the region overall, but it is important in local economies, especially in Colombia (northern Cauca Province), Brazil (south) and Paraguay, and its production is increasing. The two basic forms are native and modified. One of the popular forms of starch modification is fermentation for a variety of bakery goods. Fermentation and sun drying combine to give cassava starch the capacity to trap air and expand. Baked products have a consistency similar to the gluten-containing wheat flour. The cassava/cheese breads are the commonest products from sour starch. Native and modified non-sour starch are used in an array of food and industrial products: food processing, adhesives, paper and textile manufacturing, and others.

In Brazil, cassava starch production increased from 200,000 t in 1990 to approximately 300,000 t in 1997 (Vilpoux, 1998). Roughly 70% of Brazil's starch utilization is based on domestic maize starch, currently bringing the total industry an estimated 1 million t year^{-1} (Vilpoux, 1998). Hence Brazil's starch expansion has been typically maize-based. Maize starch manufacturing is concentrated in two large international (US origin) companies: CPC International/Refinação de Milho Brasil and Cargill, both based in southern Brazil. The cassava starch industry represents small- to medium-sized companies, distributed in the states of Sao Paulo, Minas Gerais, Santa Catarina, Paraná and lately, Mato Grosso do Sul.

Current utilization of starch is detailed in Table 2.3. This shows 69% of total starch for the food sector, 17% for the paper industry and 5% for the textile industry. It also shows that 43% is native, 46% is hydrolysed (sweeteners) and 11% is other modified starch. Vilpoux (1998)

Table 2.3. Utilization (t) of Brazilian starch and starch derivatives by industrial sector, 1997.

Starch type	Food sector				Paper sector		Textile sector	Other sectors	Total
	Sweeteners	Bakery/ pastry	Powder products	Others	Paper	Cardboard			
Native starch	2,100	26,500	93,000	109,100	66,300	43,500	20,000	77,000	437,500
Modified									113,250
Acid modified	2,600			1,500	29,900	4,300	30,000		68,300
Cationic					1,800	200			2,000
Anfoteric					24,300				24,300
Dextrins/pregel.			100	300	100	50		100 18,000	18,650
Hydrolysed									472,200
Glucose syrups	141,200	800	3,100	30,400			200	1,000	176,700
Glucose powder	200	100	300	5,100			100		5,800
Maltose syrups				271,500					271,500
Malto dextrins	400	300	2,800	14,400			300		18,200
Total	146,500	27,700	99,300	432,300	122,400	48,050	50,700 96,000		1,022,950

Source: Henry *et al.* (1998).

notes that, in 1997, the food industries that increased their starch utilization the most were the frozen and dehydrated foods sectors (with 18.2%). Furthermore, the same source notes that future growth in demand for starch (modified and native) in the food sector will be especially strong for the ready and semi-ready product lines. Other US private sector information (PROAMYL, 1996) notes the potential increasing demand for cationic starches for the high-quality paper industry.

There are several constraints for cassava to compete against maize as a starch source crop. One of these is market concentration. Two companies in Brazil account for the 700,000 t of maize starch production, whereas the cassava industry is divided among more than 60 firms. Big maize starch companies can invest in product research, reach bigger customers and reduce production costs, which is more difficult for cassava starch firms. The other major constraint regards the relatively higher production costs of cassava (as raw material for starch), as shown in Table 2.4.

Few hard data exist regarding the cassava starch situation in Venezuela. Scattered first-hand information reports that there are currently two large-scale integrated (with root production) starch factories. One of these operates a 7000-ha cassava farm, partly irrigated, with an average productivity of 25–30 t ha^{-1} year^{-1}. The roots are processed into native starch and glucose syrup. While the latter still represents a small share, the immediate objective is to increase this product output. The primary market is Venezuela, but native starch exports for the Colombian paper industry have also been reported at a very competitive price compared to Colombian starches. The main starch source in Venezuela remains maize starch, mostly imported from the USA.

The main cassava starch products in Colombia are sour and native starches. There are reports of a recent investment in the province of Cauca for a cassava-based glucose syrup factory (Gottret et al., 1997). However, no data are available on production or capacity figures. Cassava sour starch production is mainly concentrated in the Cauca Valley, with a total average production of 23,000 t from approximately 200 small-scale processing units. Several larger units producing native cassava starch operate in the Atlantic Coast region. Colombian starch utilization is principally satisfied by starch imports from the USA (maize), Venezuela (cassava), Brazil (cassava/maize) and sometimes Ecuador (cassava). Several maize-based starch factories (Maizena) exist, but at least one seems to be in the process of closing down. Gottret et al. (1997) report the relatively high prices of Colombian cassava-based starch. In 1997, Colombian native starch was priced at US$500–550 t^{-1} versus US$450–480 t^{-1} for imported maize starch. At these prices, Thai and even Brazilian starch possibly could be imported at a significant profit. It should be noted that the Colombian starch market is in the hands of only a very few operators, dictating imports and market prices.

Few hard data on cassava starch are available for Paraguay. Henry and Chuzel (1997) have noted that small volumes of cassava starch have traditionally been manufactured in small-scale household processing units, for manufacturing of *chipas*, a typical snack. More

Table 2.4.   Main cassava and maize production costs.

	Maize		Cassava[a]	
	US$[b]	%	US$[b]	%
Mechanized activities	80.08	35.8	85.36	18.4
Input	223.86	61.2	34.61	7.4
Labour force	62.18	17.0	345.02	74.2
Total	366.12		465.00	

Source: Vilpoux (1998).
[a]One year-old cassava with 20–25 t ha^{-1} productivity.
[b]1997 US$.

recently, however, there is growing interest among Brazilian starch manufacturers across the border (Paraná and Mato Grosso do Sul) for joint-venture investments in large-scale cassava starch manufacturing (> 200 t day^{-1}), taking advantage of relatively lower land and labour prices. Most starch utilized in Paraguay currently originates from Brazil and, to a lesser extent, from the USA (maize starch).

Cassava-based snacks and convenience foods

Fast foods made from cassava in the form of chips are commercialized in Europe and in some Latin American countries. In Europe these chips are sold in supermarkets as a snack food, very similar to extruded maize products. There is a prawn-flavoured product made in France with Thai cassava starch produced by the Tai-Yang company. Similar products (Fritopan and Mandiopan) exist in Colombia and Brazil, respectively. These products are not ready to eat and have to be fried, to allow expansion of the product. The necessity of frying makes consumption of this product difficult, which affects its marketability.

The fast growth of the urban areas, the distance between work and home, and accelerated life styles are determinants of the constant expansion of frozen food markets. In Brazil the 3.6- million-t frozen food market is still relatively small, compared to the US market of 14.5 million t. Data from the Brazilian Food Industry Association (ABIA) show that frozen and dehydrated foods were the segments that grew most in 1997 (Gazeta Mercantil, 1998). Five years ago, the Agricultural Cooperative of Cotia (CAC) was the only big enterprise selling a frozen cassava product similar to potato chips. Today, there are several frozen cassava and cassava-based products in the market, produced and distributed by different-sized enterprises.

Latin American cassava-product exports

While limited volumes of cassava starch are exported from Brazil, Latin America's main cassava export product remains fresh/frozen cassava roots for human consumption, targeted to ethnic population groups in the European Community (EC) and USA. Table 2.5 summarizes EC fresh cassava imports between 1993 and 1997. Note that the figures for 1993 and 1994 relate to the EC with 12 members, while 1995/96/97 figures relate to the EC with 15 members. No data are currently available to assess how much more cassava was imported to the EC as a result of Austria, Sweden and Finland's entrance to the community. However, none of these countries has large ethnic populations from developing countries (those most likely to consume fresh cassava) and consequently we can safely assume that the enlargement of the EC had little effect on fresh cassava imports. The same table indicates that imports have increased both in value and quantity over recent years. Costa Rica is the primary supplier, with Ecuador, Surinam and Ghana supplying much smaller, but still significant quantities.

In 1997 the UK imported approximately 940 t of fresh cassava (estimated from data supplied by the Home Grown Cereals Authority, UK). At 23% of the estimated 1997 EC imports, this figure indicates that the UK is one of the major buyers within the EC. As consumers in the UK tend to come from ethnic minorities, the market size is limited. Cassava enters the country either as fresh whole roots that have been preserved in clear wax and fungicide, or, as frozen pieces that arrive in refrigerated containers. The UK market is currently oversupplied. Traders either predict a decline in the market or, at most, a continuation of the current level of sales (personal communications, various traders, New Spitalfield Market, London, 1997). Prospective entrants to the EC market would have to be competitive with exporters from Costa Rica, who operate highly efficient market channels.

US Department of Commerce trade figures (summarized in Table 2.6) reveal significant imports of cassava to the USA. The figures relate to frozen, fresh or dried cassava, although they import very little or no dried cassava (personal communication, Linda Wheeler, USDA Foreign Agricultural Service, 1997). The figures in the table therefore, can be assumed to relate almost entirely to fresh or frozen cassava, again coming mostly from Costa Rica.

Table 2.5. EC imports of fresh cassava[a] by country of origin.

	1993[b]		1994[b]		1995[c]		1996[c]		1997[d]	
	Quantity	Value	Quantity	Value	Quantity	Value	Quantity	Value	Quantity	Value
	t	'000 US$	t	'000 US$	t	'000 US$	t	'000 US$	t	'000 US$
EC Total	3409	1914	3480	2509	4022	3015	5001	3571	4147	3187
Costa Rica	2502	1532	2747	2015	3485	2590	4089	2807	3658	2699
Ecuador	0	0	5	3	76	50	219	161	230	219
Surinam	133	68	411	213	188	133	272	205	26	18
Ghana	91	45	124	63	89	75	220	210	152	134
Malaysia	8	7	7	6	17	16	34	27	36	31
Barbados	0	0	0	0	17	13	22	15	1	1
Brazil	20	12	0	0	0	0	34	41	5	5
St Vincent	4	3	49	62	29	30	4	5	6	6
Dominican R.	0	0	8	2	28	10	10	8	0	0
Vietnam	2	3	10	10	7	7	22	16	7	17
Philippines	0	0	0	1	10	12	8	10	11	14
Honduras	131	86	63	45	20	18	0	0	0	0
Singapore	11	9	6	5	14	13	2	7	0	0
Ivory Coast	7	7	0	0	14	9	0	0	2	29
India	0	0	2	4	0	0	15	7	0	0
Guatemala	0	0	0	0	0	0	10	10	3	2
Indonesia	15	32	35	67	9	21	2	5	0	0
Trinidad and Tobago	0	0	0	0	0	0	11	13	0	0
El Salvador	0	0	0	0	0	0	0	0	9	7
Guyana	0	0	0	0	0	0	8	5	0	0
Grenada	0	0	4	4	7	6	0	0	0	0
Thailand	424	63	6	6	0	0	3	4	1	2
Jamaica	0	0	0	0	3	2	0	0	0	0
Venezuela	32	23	0	0	0	0	0	0	0	0
USA	18	9	0	0	0	0	0	0	0	0
Dominica	9	10	0	0	0	0	0	0	0	0

Source: Henry and Westby (2000).
[a]Definition: fresh and whole or without skin and frozen manioc, whether or not sliced, for human consumption.
[b]EC12.
[c]EC15.
[d]EC15 preliminary figures.

The Institutional Resource Base

During the 1970s Latin America in general was committed to improving agriculture as a strategy for broad-based development. Many countries sent key scientists, or whole teams, for advanced training and strengthened their research system in expanded and improved facilities. The cassava sector benefited from this broad investment in agricultural research. Several countries that previously had no cassava programme at all, or very minor efforts, developed national plans for cassava and established research teams to carry them out.

These national programmes were complemented by the establishment of CIAT in Colombia. The CIAT Cassava Programme became a major institutional force for cassava research and training, as well as for acting as a convenor to bring together national scientists in forums for international exchange and collaboration. The strong interdisciplinary orientation of this programme became an operational model for many national programmes in the following years.

This surge in interest and investment was followed by an economic downturn for much of the region by the mid-1980s. This was especially

Table 2.6. USA imports of fresh cassava by country of origin.

	1996		1997[a]	
	Quantity	Value	Quantity	Value
	t	'000 US$	t	'000 US$
USA total	32,343	16,070	34,285	21,044
Colombia	39	18	0	0
Costa Rica	31,744	15,691	32,953	20,317
Dominican Republic	78	26	170	142
Ecuador	31	11	221	118
Egypt	4	10	4	12
Fiji	0	0	2	12
Ghana	64	24	52	16
Honduras	21	7	26	14
Hong Kong	0	1	8	4
India	0	0	2	1
Indonesia	20	44	0	0
Ivory Coast	0	0	0	2
Jamaica	0	3	19	25
Malaysia	5	4	0	0
Mexico	66	0	154	31
Nicaragua	0	0	4	4
Nigeria	18	19	0	0
Panama	0	0	102	35
Peru	9	8	0	0
Philippines	198	188	201	199
Thailand	3	4	0	0
Tonga	40	11	12	13
Venezuela	0	0	344	94
Vietnam	3	1	12	4

Source: Henry and Westby (2000).
[a]Estimated values.

acute for countries that had borrowed heavily, were experiencing runaway inflation and had difficulty making loan payments. Paring back on government expenditures often hit agriculture hardest, with its declining political power. Within agriculture, the cassava sector was among the least important. The once-strong or moderate programmes of Mexico, Panama, Dominican Republic, Ecuador and Venezuela were phased out, or reduced to very low levels of operation.

Currently, the core of countries with strong institutions in cassava R&D is very limited – only Brazil and Cuba retain an interdisciplinary team in the context of a programme with national responsibility for cassava research. Cassava programmes are plagued by a high turnover of scientists although some programmes have

very experienced staff. In a reorganization in late 1996, CIAT replaced its commodity-oriented programmes with a project structure that gives less emphasis to commodity development and higher priority to integrating commodities with resource management. Hence this institution's ability to support national cassava programmes has been somewhat diluted. On the other hand, CIAT becomes more of a resource for integrating key components into broader agricultural development.

During the mid-1980s to early 1990s, CIAT gave high priority to promoting network development. Several semi-formal and informal networks were formed with missions and activities relevant to Latin America. In reality, most of these networks are a latent resource rather that actual functioning entities. Many depended heavily on CIAT for operational support and have not been able to find other resources to continue their activities.

- The *Cassava R&D Network*, while never given a formal network structure, is a broad association of cassava scientists working across all disciplines and areas, linked by a regional newsletter published at CIAT, by attendance at various cassava-related meetings, by communication, by visits and interchange of technology components.
- The *Cassava Breeding Network* held its first meeting in Cali, Colombia in 1987 and reconvened for triennial meetings thereafter. Cassava does not lend itself well to the types of international cultivar-testing programmes that are often the main thrust of breeders' networks. None the less, the interchange of information and germplasm fostered by the network, has contributed significantly to upgrading the quality and uniformity of genetic improvement activities in the region.
- The *Manihot Genetic Resources Network* (MGRN) was established in 1992 under the auspices of the International Plant Genetic Resources Institute (IPGRI). Latin America, with its position as a centre of origin for cassava, clearly should be taking a lead role in assuring the viability and productivity of MGRN. As for the other networks, poor funding and a diminishing core of cassava scientists are making this nearly impossible. The breeders' network and the MGRN have now informally merged in view of their overlapping functions and interests.
- The *Cassava Biotechnology Network* (CBN) functions globally and includes active participation from several Latin American countries, especially Colombia, Brazil, Cuba and Venezuela. This is the only network with strong involvement of advanced research institutions in developed countries. As CBN evolves towards a regionalized structure, the Latin American participants will intensify their contacts and interchange, possibly to collaborate on more region-specific issues.
- A *Southern Cone Network* was established in the late 1980s to address some of the specific problems of subtropical environments,

with participation by Paraguay, northern Argentina and southern Brazil. The activities of this network have since been absorbed by the more discipline-oriented networks.

- Plant protection practitioners have functioned in a sort of consortium of regional efforts to address the highly eco-regional nature of pests and diseases. This network has not held regional meetings but has been involved in cross-institutional training and implementation of pest management strategies.
- A global *Postharvest Network* brings together a large group of scientists, mainly from universities and private industry, who previously had little contact with each other. The interchange in meetings and informal communication have been a major contribution to setting the stage for the innovations and initiatives needed to bring expanded market-led benefits to the cassava sector.
- In 1999, as a result of additional cassava R&D resource reductions at CIAT, coupled with increased demands for R&D support from cassava-sector representatives, the regional private/public sector consortium CLAYUCA was formed. Institutional partners CIAT and CIRAD joined public agencies and private groups (feed, food and industrial sectors) from five (still increasing) countries in the region to co-finance this novel network to offer concrete solutions to common high-priority sector constraints.

Projections and Future Perspectives

Several projections exist regarding future cassava production and utilization levels (Henry and Gottret, 1996; FAO, 1997; Rosegrant and Gerpacio, 1997). However, the different time periods and data sets used, applied to very different models, generated results that are very hard to compare (or validate). It is sufficient for our purpose to discuss some summarized results from FAO (1997) regarding projected production/utilization growth rates to the year 2005. Table 2.7 shows that total Latin American cassava utilization (or production) is projected

Table 2.7. Global cassava utilization growth rates (past and projected) and shares by continent, 1983/93–1993/05.

Region	World (%)	Africa (%)	Asia (%)	LAC (%)	Share of total use (%)
Total use					
1983–1993	2.4	4.3	1.6	0.7	100
1993–2005	1.8	2.4	2.5	1.5	100
Food					
1983–1993	2.4	3.9	0.1	0.7	59
1993–2005	2.2	2.5	2.0	0.8	58
Feed					
1983–1993	1.1	7.6	4.7	0.6	24
1993–2005	−0.2	1.8	2.5	1.3	22
Other use					
1983–1993	4.7	5.3	6.8	1.1	17
1993–2005	3.1	2.3	4.2	3.4	20

Source: FAO (1997), as cited in Henry and Westby (2000).

to increase significantly from an earlier annual growth rate of 0.7% to 1.5% by the year 2005. Furthermore, the feed utilization annual growth rate is projected to double, while starch utilization (other uses) is projected to triple its annual growth to the year 2005. Rosegrant and Gerpacio (1997) and Henry and Gottret (1996), on the other hand, project annual growth rates to be in the order of 0.8 and 0.6–0.8%, respectively. In addition, these authors assign future production growth largely to yield increases, while FAO assigns similar shares to area and yield, contributing to future growth.

Supply-side interventions

A constrained market for cassava in much of Latin America does not mean that work on the production side is unwarranted. Market viability and farmers' ability to earn a fair profit follow closely from production efficiency. This is true for all markets but is increasingly decisive in industries where cassava competes in global markets with other carbohydrate sources. There is a long lead time for many technology components, especially varietal improvement. The simplest new production practices normally entail at least a 5-year development, testing and diffusion period until impact at the farm level

can be expected. Economic benefits from new cultivars can easily take 15–20 years from the time of making a cross in the breeder's nursery. The design of production research has to anticipate and be coordinated with planning for market expansion or new market development.

The fact that experimental yields easily reach levels three to five times the national average suggests that some quite effective yield-increasing technologies already exist. Furthermore, farmers who adopt these technologies are able to realize significant yield gains. Most farmers, however, are constrained from realizing the full potential of new technologies by their economic and environmental conditions. In theory, purchased inputs can alleviate most stresses including drought, low soil fertility, pests and diseases. However, the application of these inputs may not be economical, may simply not be available, or, the credit systems to allow farmers to invest in these inputs are unavailable or unsatisfactory. This review therefore concentrates on those technologies with applicability for resource-poor farmers, following on the previous discussion of constraints and opportunities.

Environmental resources

Broad priorities for environmental protection in cassava-production areas are similar across continents: soil erosion control and fertility maintenance, protection of fragile or ecologically significant natural habitats and minimizing environmental contamination from farm chemicals or pollutants from processing. The relative importance of each varies from one region to another. The Americas have the additional responsibility of protecting the habitats for diversity of wild *Manihot* species.

Approaches to controlling soil erosion are very much linked to cropping systems and it is appropriate that research be directed specifically at the unique features of cassava-based systems, while drawing on more general knowledge about erosion. Farmers already apply several traditional practices to control erosion, and new methods are available at the experimental level. The first challenge is to demonstrate to farmers the extent and the consequences of erosion under current practices. There are simple, inexpensive ways of capturing soil runoff and measuring losses. These have been used mainly in research but can also be an effective tool in demonstration plots for farmers and in participatory research. Given that adoption of suggested practices has usually been disappointing, farmer participation in research design is an important step forward. Several Colombian and international institutions are collaborating in pioneering work in the Andean hillside systems of Colombia, and this effort needs to be expanded to a range of agroecosystems.

Genetic resources

Cassava genetic resources available in the Americas are of critical global importance. This evolutionary homeland of cassava and its wild relatives includes the major part of the crop's genetic diversity, as well as the inter- and intra-species diversity of the natural enemies of many cassava pests and diseases. The region holds two of the principal cassava germplasm collections in the world: at CNPMF/CENARGEN, Brazil, with about 2000 accessions and at CIAT in Colombia with over 6000 accessions.

Managing these resources adequately for long-term future use must be a research priority. An important step toward this end was formation of the MGRN in 1992. Several working groups identified research priorities in germplasm collection (wild and cultivated);

conservation and regeneration techniques, especially for the wild species; safe exchange of germplasm; documentation and evaluation; and utilization (IPGRI, 1994). Since its establishment, the network has had limited activity, despite the pressing needs it faces.

Most of the currently held collections in the Americas were made in the 1960s and 1970s, with periodic small additions in later years. There is no comprehensive catalogue of the existing collections in the Americas. The two principal collections (CIAT and CNPMF) are well characterized for basic morphological and agronomic traits, but there is no reliable way to relate this to the total genetic diversity. Some experts consider the existing *ex situ* collections to represent a large proportion of the total diversity, while others believe much more needs to be collected. The first priority should be to pursue a path towards consensus. The MGRN is the obvious forum for this discussion. Agreement is needed on methodology for measuring genetic diversity reliably, a comprehensive inventory of existing information on *in situ* and *ex situ* diversity, and identification of methodology and resources for filling information gaps.

Conservation of *Manihot esculenta* is refined to a point of quite high security with a combination of *in vitro* and field techniques. CIAT (Latin America and Asia), IITA (Africa) and a few national programmes maintain their local germplasm *in vitro*. The global needs for germplasm security certainly do not require that every country have *in vitro* laboratory facilities. A more efficient, cost-effective approach would be an internationally coordinated, secure system that holds a base collection and one or two duplicates at key sites. This should not be a disincentive for any country to manage its germplasm properly, but is an acknowledgement of the practical reality of many countries' financial and technical difficulties in developing secure systems.

Varietal development

New cultivars have long benefited both large and small growers. Specifically targeting benefits to small and medium resource-poor farmers, however, is a possible option for cassava programmes. Cultivars that rely on unavailable or expensive inputs to express their potential, are not suitable for most cassava growers. Breeders in the past few decades have generally sought adaptation to stressful environments as a means to benefit resource-poor farmers. Pest and disease resistance, drought tolerance, adaptation to acid soils and nutrient-use efficiency, are some of the key traits that will increase yields and farmer income with moderate input use. At the same time, reasonable responsiveness to improved soil fertility allows farmers to take advantage of inputs when conditions permit. Exploration of novel traits for new production systems can have substantial long-term payoff. Changes in plant and root architecture to meet the demands of mechanization, to improve nutrient-use efficiency, or to increase plant density need to be introduced into plant breeding schemes 15–20 years before on-farm demand is anticipated.

The basis of new cultivars is the broad array of farmer-selected landraces. Most cassava-growing countries of the region have identified superior local germplasm. Recommendations of these to local growers and transfer to other regions are some of the quickest and most effective means of deploying superior genetic materials. With the application of scientific principles, the evaluation process is now more systematic and the interchange broader in scope.

CIAT has played a prominent role in supplying improved germplasm for evaluation by national programmes. The international centres in general are reducing their investment in varietal development on the basis of national programmes' acquiring capacity in genetic improvement over the past few decades. National programmes did indeed develop capacity in cassava improvement, but much of that has been lost in budget-cutting for both personnel and operations. Today there are few programmes in Latin America with the institutional capacity to implement a full breeding programme; most have only the most rudimentary capacity of evaluating finished cultivars. R&D planners must combat the reality that there is a serious erosion of capacity in germplasm management and varietal development in the Latin American public sector, with no current prospects for investment by private companies. Strengthening existing programmes and extending their benefits through networking are clear needs for the region.

Crop management

Because New World farmers have cultivated cassava for thousands of years, they have been able to optimize resources to a remarkable degree within traditional cultivation systems.

Cassava is often known as a crop that will yield reasonably even when given suboptimum care. Other more sensitive crops may fail completely unless more attention is given to management. In this context it makes sense for farmers to give a lower priority to cassava in multiple-crop systems. It also means that new management practices will have to be relatively simple and inexpensive to be successful. Science has had limited success in improving these traditional practices unless some change is introduced from outside the system. Recommendations to change planting position, plant density or plant arrangements, by themselves, rarely provide more than minor yield advantages. On the other hand, when any new technology component is introduced, such as a new cultivar, chemical weed control or chemical fertilizer, concomitant changes in other components will probably be required to re-optimize the system. This has long been known by crop scientists – hence the typical recommendation that farmers should adopt technology *packages* rather than individual components. This continues to be a major challenge for research and extension workers.

The principal crop management opportunities for sustainable increases in production profitability lie in increasing labour productivity, improved quality of planting material, improved soil fertility and better weed control.

● *Labour productivity.* Rising wages, driven by advancing economies, and tighter profit margins from competition with other carbohydrate sources, will drive farmers to strive continually for higher labour productivity. Land preparation, weeding and harvesting occupy the largest share of production labour inputs. Farmers at any scale of operation are usually economically rational when choosing production methods that are labour-intensive versus labour-saving. In most areas where terrain is amenable, mechanization is making inroads. Most of this is non-crop-specific, such as land preparation or mechanical weeding. The private sector will manage quite well in offering non-crop-specific mechanization to cassava growers, who in turn will make economically rational decisions about adoption.

- Cassava-specific mechanization is very little used. This tends to be quite expensive because the market will not as yet support mass production. Certainly there are some inherent complexities in mechanization. With much of cassava produced on moderate or steep slopes, conventional machinery may be inappropriate. There is a special need to design small-scale machinery adapted to irregular terrain. Mechanization would probably force a move towards monoculture, given the complications of mechanized intercropping. Currently, there are a few planters and harvesters on the market, but these are used almost exclusively in large plantation-type operations. There should be potential for custom planting and harvesting businesses, or, for farmer cooperatives to pool resources to purchase machinery.
- Typically, mechanization and breeding objectives evolve in parallel – breeders adapt crop characteristics to limitations or possibilities of machinery, and engineers design machinery to fit changing varietal traits. One might envisage this phenomenon in cassava, especially for harvest machinery. Breeders may need to produce more erect plant types to accommodate row-crop harvesters and select for root forms compatible with mechanical lifting mechanisms.
- Another practical need for mechanization is for sowing cover crops of small-seeded species within cassava plantations. Farmers may be enthusiastic about the benefits of cover crops but are reluctant to adopt the practice if seeding management is too difficult.
- *Quality of planting material and novel propagation systems.* Planting material, in the form of stem pieces, can be improved through either management or genetics. On the management side, the critical research entry points should be in establishing criteria for culture of mother plants (e.g. seed banks), storage conditions and treatments to enhance viability vigour.

There is already a large body of knowledge about planting material management, which needs to be adapted and complemented by national programmes for local conditions. As this has always been a key link in the production process, there is relevant indigenous farmer knowledge that has not been documented or tapped.

- In the longer term, non-conventional types and systems of planting material will be able to contribute substantially to the economics of cassava production. Alleviating the constraints imposed by bulkiness and perishability of planting material will become increasingly important for adding even greater flexibility to production systems. This can be done either with variations on vegetative propagation systems or with true seed. The possibility of true-seed propagation of cassava was proposed seriously more than 10 years ago. A broad, integrated initiative to look at both agronomic and genetic aspects should be undertaken. Given the long lead time required – certainly more than the typical 10–15 years for cultivar development – this type of research already needs to be anticipating the needs of a very different cassava sector a few decades into the 21st century. The main advantages could be a lower level of disease transmission (especially viruses) from one generation to the next, ease of handling, storability and added flexibility in production system design. Problems to overcome include seed harvest, seedling germination and vigour, and genetic variability of seed-derived populations.
- *Soil fertility.* Technically, the solution to low-soil fertility is straightforward – nutrients added at recommended levels. The first step to efficient fertility management is farm-level soil testing to define nutrient needs. Few cassava farmers have ready access to this service and can understandably be reluctant to add fertilizer when the soil nutrient status is unknown. Access to soil analyses on a regular basis must be the foundation of economic decisions on fertilizer use. In some countries this service is offered by fertilizer-supply companies, but recommendations may be considered suspect because of obvious interests in

promoting sales. Partnerships between private companies and extension services could go a long way towards providing timely and credible soil analyses for cassava growers.

- Fertilizer is often the most cost-effective way to add required nutrients, but it is not the only way. Farmers in traditional systems have generally succeeded in achieving stable, albeit low, yield levels by various management systems. Fallow periods, crop rotation, intercropping, green manures and nutrient-efficient cultivars contribute to soil fertility. Some of these methods may not meet the needs of high productivity agriculture to support society's growing demands adequately, but understanding the principles behind the traditional systems is a prerequisite to rational change.

- Mycorrhizae, soil-borne fungi associated with some plant roots, play a major role in P uptake in cassava. These fungi are present in virtually all cassava plantations. In the absence of these associations, cassava will, in fact, produce reasonably only if fertilized at very high rates of P. There are known variations in the efficiency of different strains, but preliminary work in this area has been constrained by difficulties of controlled multiplication and inoculation of these organisms. While a considerable amount of basic research has been done, as well as some attempts to move technology to the practical field level applications, the work has not received the long-term support needed to realize farm-level socio-economic impact.

- *Pest management.* As cassava production practices gradually move ever farther from the equilibrium between an ancient crop and its pest environment, some of the control agents that were once broadly effective in traditional systems now need to be managed carefully. It is critical that they not be destroyed by unwise use of pesticides that affect non-target species. Beyond this, the population levels and their biotype makeup often need to be managed artificially for full effectiveness. Continuing the pursuit of basic and applied knowledge of these systems will be critical to timely deployment of environmentally sound pest control methods.

- There are already some good examples of managed, enhanced biocontrol systems in the Americas, and others in Africa. Benefits to Africa from controlling mealybugs and cassava green mites with predators and parasites introduced from the Americas have already come to billions of dollars. There are still untapped biocontrol resources that will be exploited in the future for the benefit of all cassava-growing regions.

- The cassava hornworm is a migratory pest with highly unpredictable movements from one season to the next. The larvae are voracious feeders and can completely defoliate a plantation in a matter of days. The young larvae are susceptible to a potent, naturally occurring baculovirus, easily prepared from infected late-instar larvae and stored dry or frozen. By artificial application of this virus, hornworm populations can be controlled effectively with no risk to humans or the environment. The techniques are commonly used in southern Brazil. Early work on whiteflies and burrowing bugs is promising. We may expect that continued intensification of cassava systems will place further pressures on the balance between cassava pests and their natural enemies.

- CIAT, IITA and national programmes in Latin America and Africa are now involved in developing model systems for integrated pest management that span the range from farmer input into research design to advanced technology for biotype identification of natural enemies by genetic fingerprinting. These programmes will make extensive use of the biological resources of the Latin American cassava systems.

- *Weed control.* Latin America does not have the same tradition as much of Asia for intensive input to cropping systems that keep weeds under very close control. Weeding consumes a major part of labour inputs in cassava production and is often inadequate. Some of the options are improved mechanical control, herbicides, cultivars with rapid canopy development or intercropping systems to achieve rapid shading and competition. In general farmers have already made optimum use of intra- or interspecific canopy characteristics for weed control. Breeders could easily produce very vigorous cultivars that would make an even greater contribution to controlling weeds; however, these gains would probably not come without an offsetting sacrifice to production potential. The better option is to focus cropping system and varietal traits on more productivity-oriented alternatives and control weeds by other means.

- In many cropping systems herbicides are becoming the most economical means of controlling weeds, health and ecological concerns notwithstanding. Some broad-spectrum, pre-emergent herbicides [e.g. metolachlor (Dual) and diuron (Karmex)] can be used effectively on cassava. Herbicide development has been largely in the private sector and very much concentrated on crops with potential for high sales volumes. Cassava has not been a focus of chemical company research for the simple reason of low market share. This will change only gradually, but eventually more cassava-oriented herbicides will reach the market.
- A medium-term possibility is to incorporate herbicide-resistance genes into the cassava genome as is already being done commercially with several species, most notably maize and soybean. For example, glyphosate-resistant cassava could be sprayed post-emergence with no damage to the crop, greatly reducing labour inputs. This technology will best be developed in partnerships between the public and private sector. The legal issues of patent rights and farmer-produced seed will need to be debated jointly by scientists, producers and policymakers. The risk and complexity may make chemical/biotechnology company investment unattractive unless a form of public institution support can be integrated into the commercialization process.

Institutional support

The declining cassava R&D capacity within national and international programmes in Latin America is alarming. While Brazil and Cuba continue to support comprehensive research programmes, no other country has a multi-disciplinary research team with national responsibility. One of the highest priorities for a global cassava development strategy needs to be to reverse this decline. This does not mean investment to re-create capacity in the same model of previous decades.

Support for cassava R&D has historically been almost exclusively in the public domain. Some new models for private investment are beginning to emerge, and other alternative possibilities for strengthening the cassava sector need to be considered. Neither the public nor the private sector alone will be able to come up with the resources for sustaining an adequate long-term R&D effort. Creative and practical public/private partnerships will be the key operational and funding mode for the coming years. Cassava farmers are generally all too aware of the limitations of past public-supported research. Budgets are stretched thin, and it is nearly impossible for many institutions to address more than a few high-priority areas.

One of the principal emerging forms of research support is from the processing/marketing sector. In the past cassava reached a level of commercialization to attract private research investment in a few areas, such as southern Brazil (alcohol, starch) and Venezuela (starch). In Colombia there are now several models based on 'processing poles', where entrepreneurs and public research come together to design and implement integrated production and processing systems for animal feed and starch.

Commercialization will attract private investments only when there are reasonable expectations of short- to medium-term profit. Cultivars, often the first production component for private research, are too easily multiplied on farm for a seed company to profit from sales. Agrochemicals are a lucrative business in many crops. Cassava could attract chemical company interest as a research area, but the merits of this interest from a producer viewpoint could be questioned. Public institutions would be challenged to provide unbiased information about ecologically and economically sound pest management alternatives to balance the potential promotion of chemical use by private companies.

The private sector will slowly but increasingly invest in cassava research, but it will not be motivated to cover all the research areas of cassava relevant to meeting development goals. Universities and research centres must be supported in their responsibilities for training and technology development that contribute to each country's broad goals for its citizens.

Demand-side interventions

Processing is at the interface between supply- and demand-side interventions. It is foremost a means of converting a highly perishable and bulky product into ones that are easily stored and transported. Beyond these basic functions, processing adds value, from which the processor earns income and consumers obtain a more desirable product. Processes that generate income directly or indirectly for the producer can make a significant contribution to development objectives.

The Americas are home to many of the innovations that transform cassava from a fresh root to a multi-use processed product. While there is considerable diversity across these processes, tradition probably has had a significant role in limiting the exploration of new uses in any given locality. Most of the current processed forms of cassava are practically unchanged from those used hundreds or thousands of years ago. In both Asia and Africa many of these forms were adopted, but they also added many new processes. The global experience clearly shows the high potential for expanding the product range for cassava. Success in doing so entails parallel development of processing and markets. Interventions in process development are needed both to improve efficiency and quality of current processes and to develop new products with high market potential. Many technologies are specific to the process leading to a given end product; others have broader application.

The fresh market

The patterns of consumption of fresh roots are changing, and this warrants a new look at how this product is managed. The main challenge is to conserve roots economically while conforming to the needs of marketing in urban environments. CIAT developed inexpensive techniques for prolonging the shelf-life of fresh roots by means of a preservative treatment and storage in plastic bags. The techniques have been subjected to several semi-commercial pilot studies and launched in a few commercial markets by private entrepreneurs. In higher income neighbourhoods, frozen cassava is popular, but costs are still prohibitive for poorer consumers.

The Caribbean and parts of Latin America are near-neighbours to one of the fastest growing Latin populations in the world – in the USA. Many of these residents have retained some of their tropical dietary customs including a taste for cassava. This is a specialized and lucrative market. Fresh roots for export are commonly coated with a thin film of paraffin to prevent deterioration for up to a few weeks. Costa Rica has established a near monopoly on this market, but its potential growth should allow a broadened participation in the benefits. This commerce is driven almost wholly by private enterprise and would be a good opportunity to promote private/ public complementary R&D.

Flour

Brazil, with its large market share of processed cassava, has been the Latin American leader in research on processing. The largest volume is converted to *farinha*, consumed especially in the northeast. There is a wide range of levels of sophistication for *farinha* processing – from the primitive family units to large mechanized factories – but by far the most is processed in small units. Except for progressive small improvements in processing, this traditional product in its current form, with its low-income elasticity, does not have a high potential to impact demand for cassava. The private sector will continue to develop and apply innovations to this industry. The public sector may play a role in adapting and transferring technologies from larger industries to small rural industries in order to encourage their competitive status. Adding further value by modifications to processing to create a greater diversity of flour-based products is also possible.

A potentially more dynamic market is for refined flour for partial substitution of wheat in bakery products. This is not a new product but has been mainly an artisanal enterprise. To develop this market at significant volumes, cassava flour must be of consistently high quality and at a lower price than the product it replaces.

Consistency of quality is a challenge, given the inherent nature of cassava cultivation. Wheat is cultivated in highly managed systems and is harvested at low moisture content. Cassava roots are exposed to highly variable environments, are in contact with high microbiological populations in the soil and have a high water content until processed. Cassava flour contains residual cyanogenic compounds, whose level varies depending on inherent levels and processing technologies. Currently few official standards exist for levels acceptable in flour for human consumption. These will need to evolve with the product (Jones *et al.*, 1996). Early indications from Peru, Ecuador and Colombia are positive in terms of appropriate technology development, market demand and product quality.

Starch

Starch is a growing commodity in Latin America but still absorbs only a very small part of total production. In 1992 the region produced only 4% of the world's starch: 330,000 t from cassava and 1 million t from maize. Brazil produces about two-thirds of the region's cassava starch, of which about 68% is used as native starch, 28% as modified starch (10% as sour starch) and 3% as tapioca (Cereda *et al.*, 1996).

Most starch is processed in small- and medium-sized, community-level factories in labour-intensive techniques. Large, modern factories are found mainly in southern Brazil, with a few in Colombia, Paraguay and Venezuela. There is a wide range of opportunities that should be pursued in starch processing. The main considerations are water quality and use, efficiency of extraction, consistency of quality and waste management.

Cereda *et al.* (1996) cite the difficulties of competing with maize starch, whose prices are stable and quality is high and consistent. Native starch from maize and cassava are commercialized in virtually the same markets: foodstuffs (cheese breads, cookies, ice cream, chocolate, processed meats), paper and cardboard, textiles, pharmaceutical products, glues and adhesives, and modified starches. Major constraints of the industry are: (i) consistent supplies of raw materials (Brazilian cassava starch factories

shut down for 4.5 months of the year when roots are unavailable); (ii) operational capital; (iii) markets; and (iv) technology and quality. Some of the large industries that use starch are investing in the starch production sector to solve these problems.

Fermented starch is a more complex process, and the end users normally require some quite specific traits. Most is used in baking, where consistent flavour and texture are fundamental to meeting consumer demand. Efficiency of starch extraction may be important but is secondary to producing a consistent, quality product. Three critical components impinge on this quality: (i) fresh root characteristics; (ii) quality of the water used in starch extraction; and (iii) microbial environment. Any one of these can be difficult to control in the artisanal factories where most sour starch is produced. It is probably the unique combination of all these variables that give the specific traits to the starch from any given area. This location-specificity of starch characteristics is in a sense a value-added trait that can command a market premium. Consumers can readily identify quality differences in the starch from different regions. More research needs to be directed at identifying the factors that impinge on product quality, finding means to stabilize these variables and to capitalize further on region-specific quality traits with a market premium. These highly location- and process-specific traits may allow small-scale producers and processors to compete with larger factories.

Cassava residue and waste water from starch extraction are becoming increasing environmental concerns. Small factories typically have small enough quantities of waste that it can be used as backyard animal feed and the waste water discharged without major environmental impact. This is not to say waste management is optimal or that the effects are not damaging; but there is usually little incentive for the private sector to invest in pollution-reducing strategies, except where some payoff from recycling or from by-product utilization is feasible. The public sector institutions need to take the lead role in educating processors about environmental degradation, working with governments to define reasonable regulations and finding economically viable alternatives.

Animal feed

Use of cassava in balanced rations is a well-developed science as a result of an extensive research background and long-term use in some countries; however, it is still a nascent industry in the Americas. There is localized experience in chipping and drying for this industry in Colombia and Brazil, but not elsewhere. The tools and techniques are extremely simple in environments that allow sun drying – basically a chipper and a cement patio for sun drying. As this market develops and expands throughout the Americas, local adaptation of this process will need to be developed. In some environments this will involve artificial drying or combined artificial and sun drying. There are a wide range of chipping machines on the market, driven by pedal power, electric motor or gasoline/diesel engine. The fine-tuning process for each region can best be a private–public joint venture. While the technology exists for drying under nearly any conditions, the focus needs to be on economic viability to produce a commodity that will compete in very tight markets with the coarse grains.

This market depends on up-to-the-minute price and supply information to optimize purchasing for lowest cost rations. The information deficiencies in the cassava sector are a serious detriment to competitiveness. Upgrading this capacity needs to be part of development planning.

References

Balcazar, V.A. (1997) Desarrollos del cultivo de la yuca en Colombia. Paper presented at the Global Cassava Development Strategy Progress Review Workshop, 10–11 June, Working Doc. 5, IFAD, Rome.

Buitrago, J. (1994) *La Yuca en la Alimentacion Animal.* CIAT, Cali, Colombia.

CERAT (1997) Unpublished information from meetings with industry representatives. UNESP CERAT, Botucatu-SP, Brazil.

Cereda, M.P., Takitane, I.C., Chuzel, G. and Vilpoux, O. (1996) Starch potential in Brazil. In: Dufour, D., O'Brien, G.M. and Best, R. (eds) *Cassava Flour and Starch: Progress in Research and Development.* CIRAD/CIAT, Cali, Colombia, pp. 19–24.

Eguez, C. (1996) Cassava flour and starch in Ecuador: its commercialisation and use. In: Dufour, D., O'Brien, G.M. and Best, R. (eds) *Cassava Flour and Starch: Progress in Research and Development.* CIAT/CIRAD, Cali, Colombia.

FAO (1997) Workshop on Global Cassava Strategy, IFAD, Rome 10 -11 June. Draft working notes on selected chapters of the world cassava economy: recent trends and medium term outlook. Basic Foodstuffs Service (ESCB) of the Commodities and Trade Division, Rome.

FAO. FAOSTAT database, various years, www.fao.org/waicent/faoinfo/economic/giews/

Gazeta Mercantil (1998) Novidades na indústria de alimentos. Ano LXXVII, no. 21.250, 12 January.

Gottret, V., Ostertag, C., Alonso, L. and Laing, D. (1997) *Estudio de Mercado de los Diferentos Usos de la Yuca en Colombia.* CIAT in collaboration with CCI, Cali, Colombia.

Henry, G. (1996) Etudes de marchés et débouchés pour les nouveaux produits dérivés du manioc. In: Griffon, D. and Zakhia, N. (eds) *Valorisation des Produits, Sous-produits et Déchets de la Petite et Moyenne Industrie de Transformation du Manioc en Amerique Latine* (Rapport Scientifique Final, CEE STD3). CIRAD, Montpellier, pp. 284–332.

Henry, G. (1999) A integração do Brasil nos mercados internacionais. Invited paper presented at the X Congreso Brasileiro de Mandioca, 11–16 October, Manaus-AM, Brazil.

Henry, G. and Best, R. (1994) Impact of integrated cassava projects among small-scale farmers in selected Latin American countries. In: Ofori, F. and Hahn, S.K. (eds) *Proceedings of the 9th Meeting of the International Society for Tropical Root Crops,* 20–26 October. Accra, Ghana, pp. 304–310.

Henry, G. and Chuzel, G. (1997) Internal CIRAD trip report of Paraguay, March 1997, CIRAD, Montpellier, France.

Henry, G. and Gottret, V. (1996) *Global Cassava Trends. Reassessing the Crop's Future.* CIAT Working Doc. No. 157. CIAT, Cali, Colombia (also published in Thai).

Henry, G. and Westby, A. (2000) Global cassava starch markets: Current situation and outlook. In: Howeler, R.H., Oates, C.G. and O'Brien, G.M. (eds) *Proceedings of the International Symposium on Cassava, Starch and Starch Derivatives,* 7–12 November, Nanning, Guangxi, China, pp. 79–100.

Henry, G., Westby, A. and Collinson, C. (1998) Study of Global Cassava products and markets, phase 1. Report of a consultancy, FAO-ERS, Rome.

Hershey, C., Henry, G., Best, R. and Iglesias, C. (1997) Cassava in Latin America and the Caribbean:

resources for global development. Regional Review report, IFAD, Rome.

IBGE (1992) Brazilian statistical data sets, by state. Sao Paulo, Brazil.

IPGRI (1994) International Network for Cassava Genetic Resources. Report of the First Meeting of the International Network of Cassava Genetic Resources, CIAT, Cali, Colombia, 18–23 August, 1992. International Crop Network Series No. 10. International Plant Genetic Resources Institute (IPGRI), Rome, Italy.

Jones, D.M., Trim, D.S. and Wheatley, C.C. (1996) Improving processing technologies for high-quality cassava flour. In: Dufour, D., O'Brien, G.M. and Best, R. (eds) *Cassava Flour and Starch: Progress in Research and Development.* CIRAD/CIAT, Cali, Colombia, pp. 276–288.

Ospina, B., Poats, S. and Henry, G. (1996) Integrated cassava research and development projects in Colombia, Ecuador and Brazil. In: Dufour, D.,

O'Brien, G. and Best, R. (eds) *Cassava Flour and Starch: Progress in Research and Development.* CIAT-CIRAD, Cali, Colombia, pp. 324–332.

Pires de Matos, A., Eloy Canto, A.M.M., Ospina P., da Silva Souza, J. and Fukuda, W.M.G. (1997) Farmer participatory research: the turning point for the cassava development in Northeast Brazil. Paper presented at Global Cassava Development Strategy. Progress Review Workshop. Working Doc. 5, IFAD, Rome, 10–11 June.

PROAMYL (1996) Unpublished data. CIRAD-AMIS, Montpellier, France.

Rosegrant, M.W. and Gerpacio, R.V. (1997) Roots and tubers in the 21st century: their role and importance in the global food market. IFPRI Discussion Document, IFPRI, Washington, DC.

Vilpoux, O. (1998) O mercado de amido, *Fax Jornal No. 71,* 20 January, UNESP-CERAT, Botucatu, Brazil, pp. 1–2.

Chapter 3
Cassava in Africa

Rory J. Hillocks

Natural Resources Institute, University of Greenwich, Chatham Maritime, Kent ME4 4TB, UK

Origins of Cassava in Africa

South America, probably the Amazon region, may have been the centre of origin for species that gave rise to *Manihot esculenta*. While there is some controversy regarding the exact botanical origins of the progenitors of modern cultivated cassava, the archaeological evidence points to the Amazon region as the centre of domestication (see Chapter 1).

In the 16th century, Portuguese navigators took cassava from Brazil to the west coast of Africa (Jones, 1959) and later to East Africa through Madagascar and Zanzibar (Jennings, 1976). Although cassava appears to have been grown in Fernando Po in the Gulf of Benin and around the mouth of the Congo River by the end of the 16th century, it did not spread much in West Africa until the 20th century. Cassava was unknown north of the river Niger before 1914 (Purseglove, 1968). Some local spread seems to have taken place along rivers by Africa traders and travellers in the 17th century.

Cassava was taken from Brazil to Reunion off the East African coast in 1736 and was recorded in Zanzibar in 1799 (Purseglove, 1968). With the exception of the coastal region, cassava was not widely grown in East Africa until the late part of the 18th or early 19th century. The explorer Speke found no cassava on the western shore of Lake Victoria when he went there in 1862, but Stanley recorded it in Uganda in 1878. Cassava may have reached Lake Victoria along trade routes from the east (Jameson and Thomas, 1970) or from the west (Purseglove, 1968). The crop became established in Uganda during the 19th century and its value as a food security crop was soon realized. Records show that in 1963/64, around 175,000 ha were grown in Uganda, where the drier regions to the east and north were the largest producers (Jameson and Thomas, 1970).

Most of the spread of cassava in Africa away from the coast and riverside trading posts took place during the 20th century due to the colonial powers encouraging its cultivation as a reserve against famine and the ability of the crop to survive locust attack. Cassava is now grown in all African countries south of the Sahara and north of the Limpopo River. In 1972, the International Institute of Tropical Agriculture (IITA) was inaugurated with its headquarters in Ibadan, Nigeria, under the auspices of the Consultative Group on International Agricultural Research (CGIAR). IITA shares the global mandate for cassava with the Centro Internacional de Agricultura Tropical (CIAT) in Colombia and is responsible for developing the crop in Africa.

Production Trends

Africa now produces more cassava than the rest of the world combined (see Table 3.1) and the

Table 3.1.   Mean yield and production statistics for cassava-producing countries in Africa 1996–1998.

Country	Yield (kg ha^{-1}) 1998	Production (1000 million t) 1996	1997	1998
Angola	5,573	2,500	2,326	3,211
Benin	8,747	1,452	1,625	1,625
Burkina Faso	2,000	2	2	2
Burundi	8,882	549	603	622
Cameroon	16,667	1,700	1,700	1,500
Cape Verde	11,538	3	3	3
Central African Republic	3,046	526	579	579
Chad	6,111	268	250	275
Comoros	5,556	50	50	50
Congo Democratic Republic	7,500	16,800	16,800	16,800
Congo Republic	7,191	791	780	791
Côte d'Ivoire	5,075	1,653	1,699	1,700
Eq. Guinea	2,579	49	49	49
Gabon	5,000	210	215	215
Gambia	3,000	6	6	6
Ghana	11,389	7,111	7,000	7,172
Guinea	5,800	667	732	812
Guineabissau	16,200	17	16	16
Kenya	9,286	880	900	910
Liberia	6,527	213	283	313
Madagascar	6,678	2,353	2,418	2,404
Malawi	2,778	190	200	200
Mali	7,273	1	1	1
Mozambique	5,556	4,734	5,337	5,639
Niger	7,667	230	225	230
Nigeria	11,274	31,418	30,409	30,409
Reunion	7,200	2	2	2
Rwanda	6,250	250	250	250
Sao Tome	10,714	3	3	3
Senegal	2,336	36	37	47
Sierra Leone	4,992	281	310	310
Somalia	10,000	50	52	52
Sudan	1,800	10	10	10
Tanzania	8,933	5,992	5,704	6,193
Togo	5,184	548	596	579
Uganda	6,681	2,245	2,291	2,285
Zambia	4,951	620	702	817
Zimbabwe	4,231	150	160	165
Africa	8,223	84,559	84,326	85,945
World	9,798	164,711	164,045	158,620

Source: FAO Production Yearbook, Vol. 52 (1998).

largest producing nations are Nigeria (35% of total African production and 19% of world production), Democratic Republic of Congo (DRC; 19% of African production), Ghana (8%), Tanzania (7%) and Mozambique (6%). The four largest producers have increased their share from about 70% to 80% of total African production over the last two decades. The biggest increase has been in Nigeria which increased its share from 22% to 35%, and Ghana which increased its share from 4% to 8% (IITA, 1997). The share of other producers has declined, and DRC has moved from being the largest to the second largest producer in Africa, after Nigeria. However, it is in the DRC, Tanzania and Zambia (possibly also Mozambique [northern], but no figures available) that cassava is the most important crop to the largest proportion of farming households (Table 3.2).

Total production of cassava in Africa increased from *c*. 35 million t in 1965 to over 80 million t in 1995 (Fig. 3.1), an annual growth rate of 2.9%. This is roughly the same as the population growth rate, so that average per capita production did not increase during the period. However, during the last decade, per capita production has increased as total production has grown faster (3.8%) than in the preceding decade.

Increases in the cultivation of cassava during the 1990s occurred, at least partly, in response to declining soil fertility and increased cost of inorganic fertilizers. Although Malawi for example, is not one of the major producers of cassava, national production increased from 20,000 million t to over 80,000 million t in the decade between 1989 and 1999 (Teri *et al.*, 1999). This increase in cassava production was probably achieved, at least to some extent, by replacing maize with cassava.

Most of the increases in cassava production in Africa have been due to increases in area under cultivation, rather than increases in yield per hectare. Average yields have only increased by 33% over the last two decades, but area under production has increased by about 70%. While the annual rate of growth in area under production has increased to 3.2% from 1.3% during the previous decade, that of yield has declined from 1.2% to 0.6%. Only in Ghana has yield increased significantly between 1990 and 1995 (Fig. 3.2). As cassava production is expanding in Africa, the crop is to a large extent replacing fallow, confirming that most of the production increase has been due to increase in crop area. Cassava is often planted just before land is allowed to go into fallow, indicating that the crop is also being used to increase land use intensity. Cassava is replacing other root crops, especially yam in the humid zone, maize in the non-humid zone, and other food crops in the sub-humid zone (IITA, 1997).

Results from surveys conducted with funding from the Rockefeller Foundation; the 'Collaborative Study of Cassava in Africa' (COSCA), show that for farmers across all agroecological zones, the main reason why cassava production

is increasing is in response to famine, hunger and drought. The second most important reason is the resistance of the crop to pests and diseases. These findings confirm that cassava is planted as a food security crop. Increases in production as a consequence of population growth, higher prices and increased market access, as well as increasing yield of cassava, are more important in the humid than in other zones (except for high yield in the highland-humid zone). This points to the importance of these market-related factors in driving farmers to increase cassava production. This is significant, as such factors are expected to increase in intensity over the next two decades, as a result of further urbanization.

Production Constraints

Shortened fallow periods and declining soil fertility

In Africa the predominance of various fallow systems differs between villages, depending on soil fertility status and on pest/disease, market and demographic pressures. It is often reasoned that as fallow periods decline, cassava will increasingly replace crops which require higher soil fertility and production labour. However, although cassava is well adapted to growing under continuous cultivation, it is not as frequently grown under that system as other major staples.

The farmers' ability to respond to declining fallow periods, due to demographic, market, pest/disease and other pressures by replacing more demanding crops with cassava, is constrained by its long cropping cycle. Cassava can be harvested from 6 months after planting, but most of the available local varieties do not attain maximum yield before 18 months. Currently, improved varieties attain their maximum yield at 12–15 months. Under intensive cultivation, where the fallow period is often less than 1 year, long-duration varieties are not ideally suited because they are usually harvested before they attain maximum yield. However, early-bulking varieties are not likely to reduce this pressure unless they are combined with agronomic practices for greater water and nutrient-use efficiency. Shortening fallow periods require varieties selected for efficient nutrient

Table 3.2. Percentage distribution of villages in which farmers reported that selected crops are the most important crop in cassava growing areas of sub-Saharan Africa, by country, 1991.

Important crop	Côte d'Ivoire	Ghana	Nigeria	Tanzania	Uganda	DRC	Zambia	Malawi	Burundi	Kenya	Weighted mean
Cassava	8.4	35.0	20.5	51.6	42.5	80.0	60.6	12.4	9.6	7.6	32.8
Yams	17.8	21.3	35.8	–	–	–	–	–	–	–	7.5
Cocoyam	–	–	–	–	–	–	–	–	2.9	–	0.3
Potato	–	–	–	–	–	–	–	–	5.8	–	0.6
Plantain	–	16.3	–	–	–	–	–	–	–	–	1.6
Bananas	–	–	–	5.7	1.9	–	–	–	–	–	0.8
Maize	–	12.5	15.3	21.7	13.2	12.1	12.1	82.8	31.7	68.2	27.0
Rice	16.8	–	3.2	2.8	–	3.6	–	–	–	4.6	3.1
Other food crops	–	3.8	25.3	12.3	34.9	4.3	27.3	4.8	50.0	4.6	16.7
Cash crops	57.0	11.3	–	5.7	7.6	–	–	–	–	15.2	9.7

Source: IITA (1997).

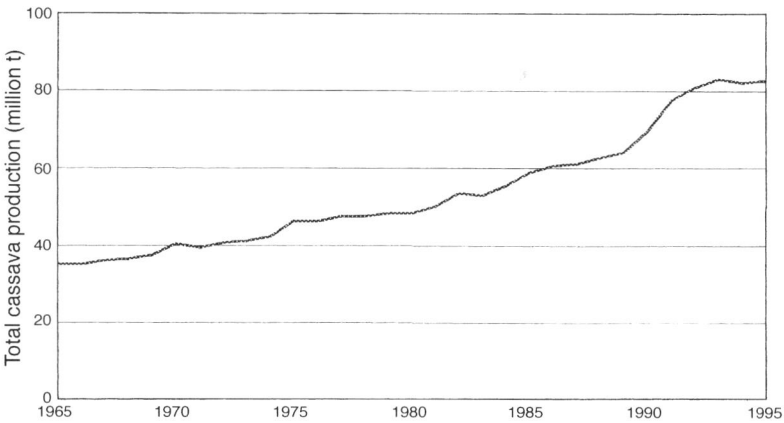

Fig. 3.1. Trend in total cassava production in Africa 1965–1995. *Source*: IITA (1997).

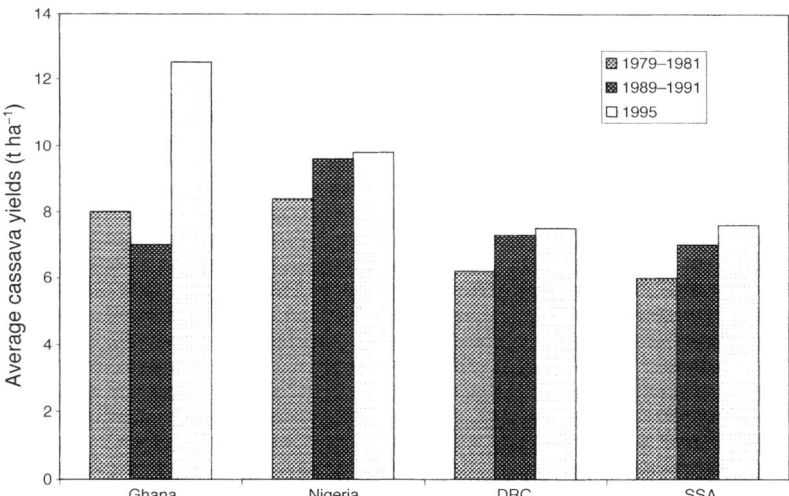

Fig. 3.2. Average cassava yields in three of the top producing countries in Africa compared to the average for sub-Saharan Africa (SSA). DRC, Democratic Republic of Congo. *Source*: IITA (1997).

assimilation, and for better ability to be inter-cropped with legumes.

Access to good quality planting material

Cassava production is dependent on an adequate supply of vegetative propagules (i.e. stem cuttings). The multiplication rate of these materials is very low in comparison with crops grown from true seed. In addition, cassava planting materials are bulky and highly perishable as they soon dry after harvest, unless carefully

stored. Multiplication and distribution of cassava planting material are expensive therefore, relative to conventional seed services. The yield stability and environmental development of cassava is highly dependent on the quality of planting materials, and there is evidence that the initial use of healthy cuttings is an important factor in the subsequent attainment of good yields. Conversely, cuttings with low vigour, and which are infested/infected by pests and pathogens, often limit cassava production. However, there is insufficient knowledge concerning criteria appropriate for selection of vigorous and clean cuttings, and on the optimal conditions for their propagation and maintenance. Pests and diseases, together with poor cultural practices, combine to contribute to yield losses that may be as high as 50%.

In dry agroecosystems, where biomass production is usually low in comparison with more humid areas, and in areas where new materials such as improved varieties are being introduced for the first time, the production of planting material in sufficient quantities is a major restriction to the widespread and rapid adoption of the crop or a new variety.

Lack of well-adapted varieties

In the countries included in the COSCA survey, representing around 80% of cassava production in Africa, farmers are continually abandoning old cultivars and introducing new ones. For instance, among the 20 most popular local cultivars grown in southern Tanzania during the 1970s, only eight of these could be identified 20 years later (R.J. Hillocks, unpublished). This indicates farmers' need for better varieties, but also highlights the danger of loss of genetic diversity. While it is increasingly evident that cassava is expanding into the semi-arid and mid-altitude zones, the available improved germplasm is mostly adapted to the lowland humid tropics. Therefore, germplasm adapted to other agroecological zones is needed. Moreover, expansion of the utilization of cassava for new industrial uses requires germplasm with high yield as well as quality that is suited to specific end-uses.

Plant pests and diseases

As cassava cultivation in Africa intensified, indigenous pests attacked the crop and exotic pests were introduced. Although it is now widely accepted that cassava in Africa is attacked by a number of serious pests, few in-depth studies of the ecological and economic importance of any of these species have been carried out. The major cassava pests in Africa include relatively few phytophagous arthropods, pathogens and weeds, compared to the pest complex found in the neotropics. The most severe pests are the exotic species accidentally introduced into areas where the local germplasm is susceptible to attack, where effective natural enemies/ antagonists are absent and where a tradition of practices to cope with the introduced pests had not had sufficient time to evolve. In addition, pest problems are being created where intensification of cassava production erodes the environmental stability inherent in balanced agroecosystems.

The major pests are cassava green mites (CGM; *Mononychellus* spp.), elegant grasshopper (*Zonocerus elegans* L. and *Zonocerus variegatus* Thunb.), cassava mealybug (CM; *Phenacoccus manihoti* Matile-Ferrero), root mealybug (*Planococcus citri* Risso) in the rainforest ecozones, cassava mosaic viruses (CMVs), cassava bacterial blight (CBB; *Xanthomonas axonopodis* pv. *manihotis* Berthet and Bondar), cassava anthracnose disease [CAD; *Colletotrichum gloeospoioides* f. sp. *manihotis* Henn. (Penz.) Sacc.], and root rots in the humid lowlands. The role of termites, nematodes and certain weed species particular to specific ecozones, has been reported as constraints but have not received adequate attention. See chapter 10 for more detailed information on insect pests and chapters 11 and 12 for diseases.

The appearance of CM and CGM as introduced pests in the 1970s in Africa had a devastating effect in farmers' fields. In particular, CM attack was so severe that it threatened the future of cassava in Africa. Massive efforts spanning several continents and involving numerous international and national research institutions under the leadership of IITA, led to the development of a successful continent-wide biological control programme. Natural enemies of CM were identified in South America, and the parasite *Apoanagyrus lopezi*, has been released in many countries in Africa. Biological control, along with improved varieties and cultural practices, provides a cost-effective, sustainable and environmentally friendly technology for the control of CM without using insecticides. The widespread establishment and documented impact of exotic predatory mite species offers good prospects for biological control of CGM as well.

The trend towards increasing the shelf-life of fresh cassava and an extended storage of dried cassava (chips/flour), will aggravate the problems caused by postharvest pests. These include the devastating larger grain borer (LGB;

Prostephanus truncatus) and a number of root rots. LGB can consume as much as 74% of cassava chips within only 4 months of harvest. Fungi are also known to infest cassava chips during processing and handling, in the field or during storage; they may lead to the formation of mycotoxins, making the chips unable to meet trade and health standards.

Variety Improvement and Adoption

Since the initial cassava germplasm introductions by the Portuguese, subsequent introductions and breeding programmes have generated high-yielding, disease-resistant genotypes. The early challenge for improvement was to produce varieties resistant to cassava mosaic disease (CMD) and later to CBB. During the 1970s, building on work done earlier in East Africa, improved cultivars incorporating resistance to these diseases were developed. Genotypes resistant to CMD and CBB are available at IITA as virus-indexed plantlets, ready to be shipped to any institution.

Several clones that combine good levels of resistance to CGM in addition to the other diseases and pests, and have low cyanogenic potential, have been developed at IITA. Many cassava improvement programmes in Africa have received these materials in tissue culture and true seed forms. These clones are tested under local environmental conditions and those that outperform local varieties are released to farmers. For example, in West Africa, over 25 improved cassava varieties have been released, or, are recommended by the National Programme in Nigeria. In Sierra Leone up to eight varieties have been released by the Institute of Agricultural Research and four in Ghana. In Central and East Africa, Cameroon has released seven varieties, Democratic Republic of Congo five, Uganda nine, Tanzania ten, and Zimbabwe four. At least three varieties have been released by each of the other cassava-producing countries in Africa (IITA, 1997).

Procedures for meristem culture and virus indexing for CMD in cassava have been established. Since the early 1980s, an agreement has been reached with African phytosanitary regulatory agencies to permit the movement of *in vitro* virus-tested cassava germplasm in Africa. The

meristem culture technique has been used successfully to transfer cassava germplasm from the field gene banks at IITA, and in Ghana and Republic of Benin to the *in vitro* gene bank at IITA, and from IITA to a number of countries in Africa and to CIAT, in Colombia.

Despite these achievements, adoption of improved varieties of cassava was not widespread by 1991, except in Nigeria (Nweke, 1994a,b). However, more recent evidence points to increased adoption rates in Ghana, Uganda and Sierra Leone. During COSCA surveys, the most frequent reason given by farmers for discarding varieties was 'late bulking' (Table 3.3). The implication is that in villages where the cultivation of those genotypes was abandoned, the farmers were selecting for early bulking. Where farmers were selecting for high root yield, weed suppression, good in-ground storability, disease and pest tolerance, good processing qualities, desirable branching habit, low cyanogenic potential, good cooking qualities, good planting material yield, etc., the varieties that did not have those desired traits were abandoned (Table 3.3).

Crop Production Systems

Cassava in Africa is usually grown in mixed stands with other crops. The most frequent companion crops are maize, sorghum and pigeonpea. Results from COSCA show that individual farmers grow an average of six to seven different companion crops, with a range of one to 15 crops. Only about 25% of the fields were planted to a single crop. Rice, yam and cassava were the crops grown most often as sole crops (IITA, 1997).

The amount of labour used in field production of cassava differs between African countries, with the highest in Nigeria and lowest in Côte d'Ivoire. Farmers allocate more labour to all farm operations, except land clearing, in high- than in low- population density areas (Tshiunza, 1996). The total amount of labour allocated to the production of cassava is highest under recurrent cultivation and statistically the same between shifting and continuous cultivation systems. The amount of labour allocated to each farm operation under recurrent cultivation is greater than that under shifting cultivation. The difference in total field production labour between recurrent

Table 3.3.   Reasons given by farmers in Africa for abandoning cassava varieties.

Reason	% of farmers
Late bulking of roots	20
Low yield	16
Weed competition	11
Poor in-ground storability	10
Susceptible to pests and diseases	8
Poor processing quality	7
Undesirable branching habit	5
High cyanide content	5
Poor cooking quality	2
Poor yield of planting material	1
Introduction of better varieties	1
Susceptible to drought	1
Low leaf yield	1
Others	12

Source: COSCA and Nweke (1994a).

and continuous cultivation systems, and between shifting and continuous cultivation is due to land clearing and weeding operations. More labour is allocated to land clearing operations in shifting and recurrent than continuous cultivation. The opposite holds for weeding operations.

Hired labour is the external input used most frequently in cassava production systems in Africa. The COSCA showed that labour was employed for use in one or more farm tasks in 41% of the fields of the major food crops including cassava (Nweke, 1994b). The proportion was highest for yam (67%), lowest for sweet potato (13%) and close to the overall average for cassava (40%). Hired labour is used in various combinations with family labour for land clearing, seedbed preparation, planting, weeding, harvesting and field-to-home transportation operations. It is used most often in land clearing and seedbed preparation.

In some countries, farmers may use machinery in cassava production. In Africa, only three operations are likely to be mechanized: land clearing, land preparation and field-to-home transportation. Of the three farm operations, transportation is the most frequently mechanized, with transport of cassava from 30% of the fields being by motorized means (Tshiunza, 1996). Pingali *et al.* (1987) also report that transport is usually the first farm operation to be transferred from human to animal power

and the second is ploughing. Primary tillage and transport are extremely energy demanding and are usually transferred to a new source of power even when wages are low.

Generally, the cultivation of cassava is thought to require less labour per unit of output than most other major staples (Goering, cited in Berry, 1993). In Sierra Leone for instance, cassava requires less labour per unit of output than upland rice and maize. Expansion of cassava production in Africa seems to be leading to greater labour productivity in the region.

Cassava can grow and give reasonable yields in soils of low fertility but fertilizer is often required for the crop to reach its maximum production potential. Cassava requires relatively little nitrogen to achieve high yields, so that it responds to no more than 100 kg ha^{-1} of nitrogen, after which there are diminishing returns (Tshiunza, 1996). Phosphorus is the most important nutrient for obtaining yield increases in cassava, with yield response to applications of as high as 400 kg ha^{-1}, although levels of 100 to 150 kg ha^{-1} are frequently recommended. Cassava extracts more potassium from the soil than any other element. A high-yielding crop extracts 100 kg ha^{-1}, or more of K. To maintain high yields when cassava is grown in an area continuously, potassium fertilization is essential. Potassium availability also affects tuberous root quality, since its deficiency leads to lower dry matter and starch content, and a higher cyano-genic potential.

Despite the potential benefits, chemical fer-tilizers are applied to only about 3% of cassava fields, and manure to about 7% of the fields. This compares to 2% of banana/plantain, 11% of rice, 15% of maize, 20% of yam and 5% of crops overall (Nweke, 1994b). Lageman (cited in Tshiunza, 1996), observes that, in the densely populated village of Umuokele in Nigeria, farm-ers applied mulch and manure to their outer fields. While fallow periods decline as population density increases, the use of organic manure, livestock grazing and other agricultural land-use intensification cultural practices become more frequent in the cassava-producing zones of sub-Saharan Africa.

Crop Utilization

Cassava plays a food security role in areas prone to drought, famine and in periods of civil dis-turbances. The crop's ability to provide a stable food base is a function of its flexibility in terms of planting and harvesting strategies and because of its relative tolerance of poor soil and pest/disease problems. It is also widely appreciated as a low-cost carbohydrate source for urban con-sumers, especially where it is available in conve-nient forms for working urban housewives.

Cassava is a major source of dietary energy for low income consumers in many parts of tropical Africa, including major urban areas (Dahniya et al., 1994; Berry, 1993; Nweke, 1994a,b). Table 3.2 shows that farmers in a third of villages in the cassava growing areas of Africa rated it as the most important crop. In half of the countries covered it was rated as the most important food crop. Maize was chosen as the most important crop in three countries (Malawi, Burundi and Kenya), while yam was rated as most important only in Nigeria. Cash crops as a group were chosen as most important only in Côte d'Ivoire (IITA, 1997).

Cassava makes a greater contribution to total calorie intake in Africa than maize or sorghum. FAO statistics indicate that cassava makes a much smaller contribution than cereals to protein supplies, partly because they do not consider cassava leaves a food item, and conse-quently a source of protein. Nevertheless, the growth in supplies of protein between 1990 and 1995 has been the same for cassava and rice (20%). Only maize had a higher growth rate (28%), while sorghum and millet increased their contribution to protein supplies by only 8% (IITA, 1997).

One of the main obstacles to the expansion of cassava has been the limited understanding of cyanogenesis in cassava. Recent studies have clarified much of the confusion around the toxic potential of cassava and have clarified the mech-anisms of the removal of cyanogenic compounds from cassava during processing (see Chapter 14).

Most cassava in Africa is used domestically, so cassava has played little role as a foreign exchange earner or in import substitution. How-ever, there now appears to be an opportunity to export cassava products, as the traditional Asian exporters appear to be having difficulties in satisfying demand, particularly in the European Union market, due to changes in the relative costs of production. Some African countries are already taking advantage of this trend. For example, exports of cassava chips from Ghana which commenced with 500 t in 1993, reached 29,000 t in 1996 (IITA, 1997).

Thirty per cent of the cassava root produced in Africa is for fresh consumption. In addition, cassava leaves are a preferred vegetable in many countries. 'Shelf-life' of fresh cassava roots rarely exceeds 2 days. Storage and packaging tech-nologies to extend shelf-life will contribute to increasing cassava root availability and reliab-ility, stabilizing prices and facilitating export. However, there is little reported research on ways of extending shelf-life and reducing postharvest losses.

The highly perishable nature of harvested cassava and the presence of cyanogenic gluco-sides call for immediate processing of the storage roots into more stable and safer products. The extent to which the potential market for cassava may be expanded depends largely on the degree to which the quality of various processed prod-ucts can be improved to make them attractive to various markets, local and foreign, without significant increases in processing costs.

Traditionally, cassava roots are processed by a variety of methods into many different food products, depending on locally available process-ing resources, local customs, and preferences (see Chapters 14 and 15). Processors are mainly

located in rural areas and obtain their supplies
of cassava roots in the same way as itinerant
traders. Less than 2% of cassava is processed
through factories that are sometimes owned by
cooperatives.

Ugwu and Ay (1992) classified cassava
products in Africa into nine groups as follows:

- cooked fresh roots;
- cassava flours: fermented and unfermented;
- granulated roasted cassava (*gari*);
- granulated cooked cassava (*attieke*, *kwosai*);
- fermented pastes;
- sedimented starches;
- drinks (with cassava components);
- leaves (cooked as vegetable); and
- medicines.

Flours are the most widely used cassava product
in Africa and are processed in a variety of ways.
Drying and milling are the most essential steps.
Flours from unfermented cassava roots are more
common in areas where sweet cassava varieties
dominate.

Over 50% of all villages in the COSCA survey
had some form of cassava processing centre, and
the use of processing machinery, particularly
mills, was found to be widespread, except in
Central African countries. Mechanized processing has been found to be positively associated
with population density. Also, as access to
market improves, cassava processing tends to
become more mechanized (Ugwu and Ay, 1992).

Improved cassava processing equipment
has been designed and tested by research institutions for use at farm and village levels with the
objective of reducing postharvest losses, increasing labour productivity and improving product
quality. The equipment includes a peeling tool,
manually operated chipping, grating and grinding machines, an efficient multiple fuel-type
frying stove, a de-watering device and a tray-
type dryer. Simple processes, allowing farmers to
convert the highly perishable cassava roots into
dry, easily stored and freely traded commodities
such as chips and flour, are available. They make
it possible for high-volume users, such as agroindustries, to develop cassava-based operations.

High labour requirements appear to be the
only resource constraint in cassava processing,
which none of the traditional techniques can
circumvent. Evidence from eastern Nigeria

shows that high root yields attained through
the adoption of improved cassava varieties would
not have substantial economic advantage using
manual processing technology (Nweke *et al.*,
1991). The economic advantage of yield-
increasing technology may not result in
expanded production, and hence into expanded
cash income opportunities, if there is no match-
ing investment in cost-saving technology at the
processing stage. Some Nigerian farmers using
IITA's high-yielding TMS 30572 variety to produce *gari*, have been observed during certain
seasons to cut back drastically on planting
because they were unable to process the previous
season's plantings (Nweke *et al.*, 1991).

There are some possibilities for import substitution of cassava flour and industrial starch.
For example, Ouraga-Djoussou and Bokanga
(1998) report that around the city of Ibadan in
Nigeria, cassava flour produced by a new method
is in high demand by four large biscuit manufacturers. A high premium price has been put on
the flour compared to the traditional '*lafun*' type
of flour. Women's groups that were processing
cassava into *gari* have organized themselves to
take advantage of this new market, particularly
since the price offered by the biscuits factories
is high and the returns much better than
those obtained from *gari*. Results of an economic
analysis by Ouraga-Djoussou and Bokanga
(1998) show that with a 15% substitution rate of
wheat flour with cassava, Nigeria could save up
to US$14.8 million in foreign exchange annu-
ally, with US$12.7 million going to cassava pro-
cessors and US$4.2 million to cassava farmers.

The potential of cassava as a foreign
exchange earner in Africa needs to be assessed
carefully. Preliminary indications are that the
average cost of production of pellets and chips
may be too high to allow West Africa to compete
with Asian countries in the international
market. For example, a survey of 11 cassava
pellet producers in southwest Nigeria in 1996
showed that their mean production cost for dry
pellets was 22,500 Naira t^{-1} (80 Naira was at the
time approximately US$1). Only about 10% of
establishments produced at less than or equal
to the monthly mean world market price (FOB)
of 13,000 Naira t^{-1} for January and February
1996 (F.I. Nweke, J.K. Lynam and S.A. Folayan,
unpublished).

Research and Extension

The International Agricultural Research Centres (IARCs), National Agricultural Research Systems (NARS) and regional networks conduct research on cassava in Africa. The most active IARC is the IITA which is funded through the CGIAR. The CIAT, the other CGIAR centre active in Africa, operates through the IITA. Smaller international programmes are executed by the French Centre de Coopération Internationale en Recherche Agronomique pour le Développement (CIRAD) and the Natural Resources Institute (NRI) of the UK.

IITA conducts a full research programme on cassava, from breeding and pest management to farming systems and postharvest technology. CIAT mainly provides germplasm to IITA's breeding programme and collaborates on biocontrol programmes. CIRAD and Institut de Recherches Scientifiques pour le Développement en Coopération (ORSTOM) have programmes for genetic improvement and physiology.

National programmes for cassava research exist in most African countries in which cassava is grown. There are particularly strong programmes in Cameroon, Congo, Côte d'Ivoire, Malawi, Nigeria, Sierra Leone, Tanzania, Uganda and DRC. All these national programmes are linked together through networks or regional organizations, the most important of which are the Eastern Africa Root Crops Research Network (EARRNET), the Southern Africa Root Crops Research Network (SARRNET), the conference of the African leaders of agricultural research (Conférence des Responsables de Recherche Agronomique Africains; CORAF), the Association for Strengthening Agricultural Research in Eastern and Central Africa (ASARECA), the Southern African Centre for Cooperation in Agricultural Research and Training (SACCAR), and the Institut du Sahel (INSAH).

Many of the NARS face constraints due to human resource problems, shortage of funding and problems of management and coordination of research programmes. The Special Programme for African Agricultural Research (SPAAR) has developed four Frameworks For Action (FFA) in order to provide guidelines and outline strategies that will enable NARS to fulfil their role. These frameworks are based on the building of coalitions between national and regional institutions in technology generation and transfer. The key features are strengthening of the NARS in order to make them more responsible to their clients and to develop their capacity for policy and economic analysis. Implementation of the FFAs generally calls for a number of principles including:

1. Preparation of a national agricultural research strategy or master plan.
2. The establishment of adequate, sustainable, stable and timely funding through transparent and accountable funding mechanisms that pool the collective efforts of donors to address priority national and agricultural research.
3. Enhancement of institutional and management capacity of NARS.
4. The strengthening of research – extension – farmer linkages.
5. The formation of research advisory groups consisting of coalition of all the major research partners.
6. Effective regional collaboration among national, regional and international research institutions.

The FFAs are to be implemented through regional coordinating mechanisms, which include CORAF for West and Central Africa, INSAH for the Sahel, ASARECA for East and Central Africa and SACCAR in South Africa.

Future Needs

Variety improvement

Cassava genetic resources in Africa are under the threat of erosion. The COSCA survey indicates that farmers in Africa are abandoning old cultivars in favour of new ones. While this is a necessary part of agricultural development, traditional cassava cultivars must be collected to prevent further loss of desirable genetic diversity. Additional variability should be sought, particularly if adapted to harsh environments. Moreover, the potential contribution of wild relatives of cassava should not be ignored. The early cassava introductions were limited and not sufficient to transfer the wide genetic base existing at the centre of origin of the species.

Further introductions from Latin America hold promise to broaden the germplasm base of cassava, by providing unique sources of variability not currently available in Africa.

The search for, and utilization of, new sources of resistance to each of the major pests and diseases need to be intensified, with the aim of diversifying resistance that would prove difficult for pathogens and pests to circumvent. Problems, which are recalcitrant to conventional methodologies, need to be addressed using the tools of biotechnology.

Protocols for identifying and eliminating pests, diseases and poor plant vigour in cassava cuttings are needed by NARS involved in plant quarantine and plant protection activities, and for selection, propagation, and management of pathogen-free cuttings by extension agents and farmers in a sustainable manner. The objective would be to develop and implement a strategy to produce and maintain clean and vigorous cassava planting material, given the specific requirements of the major cassava-growing ecozones in Africa.

Information from COSCA has provided much valuable information on the characteristics of improved varieties most sought after by farmers in SSA. Prominent among them are high yield, early bulking and high dry matter content. Breeding programmes need to develop varieties, which reach maximum bulk around 9 months, rather than the current average of 15 months.

Integrated pest management

Pests and diseases still take their toll on cassava production. Cassava brown streak disease (CBSD; caused by *Cassava brown streak virus*) and insect pests are increasingly becoming problems of economic importance. Plant protection research needed to address these issues can broadly be grouped into characterization and adaptive/strategic activities. The major characterization themes include yield loss due to grasshoppers, green mite, mealybug and plant diseases such as CMD, CAD and root rots in specific ecozones, resistance screening, and soil nutrient trials. These investigations should provide a quantitative basis for deciding whether or not to develop specific pest intervention technologies.

Another area of plant protection in cassava that is often ignored concerns the ecosystem-specific pests. During extensive plant protection diagnostic surveys carried out by Ecologically Sustainable Cassava Plant Protection, a number of important pest constraints were identified as being associated with specific ecozones, but not exclusively on cassava. These include the variegated grasshopper in the transition forest and moist savannah, termites in the moist and dry savannah, nematodes in all ecozones, and vertebrate pests and root rots in the rain and transition forests.

Adaptive/strategic research themes for cassava plant protection and production should include the classical biological control of LGB and CGM, and the identification and integration of sustainable control methods for CMD, CBB, CAD, weeds, termites, nematodes and vertebrates. The development of packages of integrated control methods for root rots (including CBSD) will have a direct impact on the quantity and quality of marketable tuberous roots. Methods to protect cuttings from infection by root and stem rot pathogens would be part of the development of protocols for producing and managing clean and vigorous cuttings. Integrated participatory on-farm trials, where appropriate, and the elucidation of indigenous knowledge systems should also be pursued.

Improved postharvest management

To be competitive and to increase their income, cassava farmers need to sell high-quality processed products with a long shelf-life. One of the major advantages cassava has over other starchy crops is the variety of uses to which the roots can be put. In addition to being a major staple food for humans, it also has an excellent potential as livestock feed, and in textile, plywood, paper, brewing, chemical and pharmaceutical industries. The major constraint, however, is that cassava deteriorates rapidly. Fresh roots must be transformed into more stable products within 2 or 3 days from harvest. This transformation requires technology for peeling, grating, boiling, fermenting, drying, frying and milling.

Timely harvests and efficient postharvest operations play a crucial role in the lives of farmers. Appropriate equipment to carry out these operations reduces crop waste and enables more complete utilization of the food crops grown. This is especially true of farmers who are moving from subsistence agriculture to large-scale commercial production. IITA has devoted attention to designing and fabricating improved processing equipment. While the primary goals of these technologies have been to minimize crop losses and improve labour productivity and product quality, activities that have evolved recently include on-farm testing and demonstration of the equipment, training of manufacturers and networking to promote the use of improved equipment. It should be borne in mind that by improving the quality of farm products, incomes and standards of living are also raised for farming families who use these technologies.

High labour requirement appears to be the main resource constraint in cassava processing. Cassava processing is most often the responsibility of women. High root yields attained through the adoption of improved cassava varieties would not have substantial cost saving advantage under manual processing technology. Consequently, there is a need to give greater attention to cassava processing technologies in Africa. A better understanding of the distribution of marketing margins in the cassava chain from producers to middlemen to processors to end-users will point the way to better targeting of technology dissemination.

Industrialists and entrepreneurs often shy away from using cassava in their applications because of the absence of a local example to follow and the uncertainty of success. Therefore, product development research needs to be strongly promoted and the private sector should be encouraged to participate. Issues that need to be addressed include raw material import substitution, promotion of a positive image for cassava, development of products for existing and new markets, identification of the functional characteristics of cassava genotypes in relation to various end uses, utilization of cassava plant parts (e.g. leaves, peel, etc.) for livestock feeding and the leaves for human consumption.

References

Berry, S.A. (1993) Socio-economic aspects of cassava cultivation and use in Africa: implications for development of appropriate technology. COSCA Working Paper No. 8. Collaborative Study of Cassava in Africa, International Institute of Tropical Agriculture, Ibadan, Nigeria.

Dahniya, M.T., Akoroda, M.O., Alvarez, M.N., Kaindaneh, P.M., Ambe-Tumanteh, J., Okeke, J.E. and Jalloh, A. (1994) Development and dissemination of appropriate root crops packages to farmers in Africa. In: Ofori, F. and Hahn, S.K. (eds) *Proceedings of Ninth Symposium of the International Society of Tropical Root Crops*, 20–26 October 1991, Accra, Ghana. International Society for Tropical Root Crops, Wageningen, The Netherlands, pp. 2–9.

FAO Production Yearbook (1998) FAOSTAT database. http://apps.fao.org/cgi-bin/nph_db.pl

IITA (1997) Cassava in Africa: past, present and future. Report prepared for IITA by Dunstan Spencer and Associates. International Institute of Tropical Agriculture, Ibadan, Nigeria.

Jameson, J.D. and Thomas, D.G. (1970) Cassava. In: Jameson, J.D. (ed.) *Agriculture in Uganda*. Oxford University Press, Oxford, pp. 247–251.

Jennings, D.L. (1976) Cassava, *Manihot esculenta* (Euphorbiaceae). In: Simmonds, N. (ed.) *Evolution of Crop Plants*. Longman, London, pp. 81–84.

Jones, W.O. (1959) *Manioc in Africa*. Stanford University Press, Stanford, Connecticut.

Nweke, F.I. (1994a) Cassava distribution in Africa. COSCA Working Paper No. 12. Collaborative Study of Cassava in Africa, International Institute of Tropical Agriculture, Ibadan, Nigeria.

Nweke, F.I. (1994b) Farm level practices relevant to cassava plant protection. *African Crop Science Journal* 2, 563–582.

Nweke, F.I., Ugnu, B.O., Asadu, C.L.A. and Ay, P. (1991) Production costs in the yam based cropping systems of southeastern Nigeria. *Research Monograph No. 6*. Resource and Crop Management Program, International Institute of Tropical Agriculture, Ibadan, Nigeria.

Ouraga-Djoussou, L.H. and Bokanga, M. (1998) Cassava and wheat consumption in Africa: new opportunities for cassava in the 21st century. In: Akoroda, M.O. and Ekanayake, I.J. (eds) *Proceedings of the Sixth Triennial Symposium of the International Society for Tropical Root Crops – Africa Branch*. Lilongwe, Malawi, 22–28 October 1995. International Society of Tropical Root Crops, Wageningen, The Netherlands, pp. 328–333.

Pingali, P., Bigot, Y. and Binswanger, H.P. (1987) *Agricultural Mechanisation and the Evolution of*

Farming Systems in sub-Saharan Africa. Johns Hopkins University Press, Baltimore, Maryland.

Purseglove, J.W. (1968) *Tropical Crops: Dicotyledons.* Longman, London.

Teri, J.M., Sandifolo, V.S., Kaeya, E.H., Moyo, C.C., Chipungu, F.P. and Benesi, I.R.M. (1999) Root crops revolution in Malawi: lessons learned. *Roots* 6(2), 24–39.

Tshiunza, M. (1996) Agricultural intensification and labour needs in the cassava-producing zones of sub-Saharan Africa. Doctoraatsproefschrift No. 326 aan de Faculteit Landbouwkunfige en Toegpaste Biologische Wetenschappen van de Universeteit Leuven.

Ugwu, B.O. and Ay, P. (1992) Seasonality of cassava processing in Africa. COSCA Working Paper No.9. Collaborative Study of Cassava in Africa, International Institute of Tropical Agriculture, Ibadan, Nigeria.

Chapter 4
Cassava in Asia and the Pacific

I.C. Onwueme

*Fulton Center for Sustainable Living, Wilson College, Chambersburg,
PA 17201, USA*

Origin and Distribution of Cassava in Asia

Cassava was introduced to most parts of Asia in the late 18th and early 19th centuries. Some of the locations for the early arrival of cassava were in India, Java and the Philippines. Most of these introductions were done by European explorers, who in turn had obtained cassava from South America. By the time cassava was introduced to various parts of Asia, it had already been a major commodity of trade between South America and Europe for over a century. After its introduction, cassava was initially used primarily as a food for local consumption. It was cultivated mainly on marginal lands, by poor farmers who often used it as an emergency crop. By the second half of the 19th century, domestic use of cassava was firmly established in various part of Asia. From domestic use, cassava progressed to become a cash crop for export, and by the first half of the 20th century, cassava products from Asia had already begun to compete on world markets with similar products from Latin America.

Today, cassava is grown to some extent in all the tropical and subtropical countries of Asia. However, the countries of major cassava production in Asia are Indonesia, Thailand, India, China, Philippines and Vietnam. In Indonesia, most of the production is in East Java and Lampung Province. In Thailand, the northeast

of the country and the central plain account for most of the cassava production. In China, production is mostly in the southeastern provinces of Guangdong and Guanxi, which account for over four-fifths of China's cassava production. In India, Kerala and Tamil Nadu states are the main cassava producers, with some additional production in the northeastern part of the country. In both the Philippines and Vietnam, production is concentrated in the central and southern parts of each country.

Generally speaking, the most fertile lands, which are usually the lowlands, are reserved for rice, so that cassava cultivation occurs on the less fertile hillsides. The most common soil types in the cassava areas are Ultisols (Howeler, 1988). This is true in Thailand, Indonesia, China and Philippines. Inceptisols, Entisols and Alfisols are also used in parts of Java and southeast Thailand. As implied above, the soils used for cassava are usually low in fertility, and there is a frequent need to apply fertilizers and organic manures. In most of the countries, the soils used for cassava have a high clay content, although in Thailand they are sandy loam in texture. Most of these soils are acidic, with pH ranging from 4.5 to 6.5, but their organic matter content is usually low.

With the intensification of production, low organic matter and undulating topography, it is not surprising that most of the Asian cassava soils have a high potential for erosion. This potential has indeed become a reality in

Thailand, especially because of the light sandy-loam texture of its cassava soils. Soil erosion is also a serious problem in cassava soils in Philippines, China, Java and Kerala.

The Asian cassava regions are nearly all tropical, with monsoon climates. Except for China and northeast India, temperatures are usually in the range of 26–27°C throughout the year. Rainfall is generally high (1500–3000 mm year^{-1}). However, the high-intensity cassava regions of Thailand, east Java and Tamil Nadu are much drier (900–2000 mm), a situation that gives cassava an ecological comparative advantage over other crops.

Agronomy and Utilization of Cassava in Asia

Cropping systems for cassava vary widely from one part of Asia to another. In commercialized production, sole cropping predominates, especially in Thailand, Malaysia and Sumatra. Where cassava is intercropped, the intercrop combinations usually include groundnut, rice, maize and vegetables. Cassava is also grown under plantations of coconut or rubber in parts of Kerala, Philippines, Thailand, Java and Malaysia. In such situations, yields are usually relatively low. Except for Tamil Nadu where cassava irrigation is practised, virtually all Asian cassava is grown under rain-fed conditions.

There is virtually no commercial trade in cassava planting materials in Asia. Cassava planting material is usually obtained from the farmers' own fields (or from neighbours' fields) at or before the previous harvest. Stems destined for planting are usually cut into long lengths, tied in bundles and stored in shady conditions. Just before planting, they are cut to the normal planting lengths of 15–20 cm.

Planting is done on the flat or on ridges, although mounds are sometimes used in India. Each stem cutting is usually placed vertically or slanted, with some of it sticking out of the soil. Horizontal placement, with the cutting entirely buried, is common in China, Malaysia and the Philippines. Plant spacing in the field is usually about 90×90 cm, but where intercropping is practised, the spacing between cassava plants is wider.

Because of its requirement for a long growing season, cassava in Asia is nearly always planted at the start of the major rainy season. The exact month that this corresponds to varies from place to place. In areas where rain occurs throughout the year as in Sri Lanka, Sumatra and parts of Malaysia, cassava planting occurs all year round.

Weed control is most commonly accomplished with hand-held hoes or bullock-drawn ploughs. Weeding twice or three times early in the season, is usually sufficient, since little or no weeding is needed once effective canopy closure has occurred. Weed control with herbicides is practised in Malaysia and parts of Thailand. Harvesting occurs 9–12 months after planting. The most common practice is hand-pulling, aided by hand tools.

In Indonesia, Philippines, Sri Lanka and Kerala (India), fresh cassava roots are utilized for direct human consumption. In Thailand and Indonesia, much of the production is processed into chips and pellets which are exported for animal feeding. Cassava processed into starch or starch-based products is important in most of the countries, especially Malaysia, Thailand, Indonesia, India (Tamil Nadu), China and Vietnam.

Area, Production and Yield

Table 4.1 shows that in 1999, about 3.3 million ha of cassava were harvested in Asia. Thus, Asia accounted for 20% of the area of cassava harvested in the world. This was more than the area harvested in South America where cassava originated. The area of cassava harvested in Asia declined by around 13%, over the 4 years from 1996 to 1999.

About 46 million t of cassava were produced in Asia in 1999. This was 27% of world production. This was less than the production from Africa (51%), but more than the production from South America (17%). Despite minor fluctuations, Asian cassava production has remained relatively steady from 1996 to 1999. Average cassava yields in Asia were approximately 14 t ha^{-1} in 1999. This is higher than the average yields in Africa, South America or Oceania.

Within Asia, the largest areas devoted to cassava cultivation are in Indonesia and

Table 4.1.   World production, area and yield of cassava from 1996–1999.

	1996	1997	1998	1999
Production				
World	165,324,042	164,303,125	164,004,441	168,054,531
Central America and Caribbean	1,040,749	999,869	1,017,149	931,876
Africa	84,568,300	84,405,250	85,959,190	92,119,233
Asia	48,837,430	47,547,650	46,445,280	46,057,280
Oceania	198,683	168,092	168,592	183,292
South America	30,678,879	31,182,262	27,414,220	28,763,123
Area (ha)				
World	16,753,288	16,283,307	16,244,075	16,579,480
Central America and Caribbean	202,109	194,521	197,600	196,729
Africa	10,290,280	10,097,360	10,331,450	10,823,616
Asia	3,757,735	3,477,549	3,482,549	3,276,649
Oceania	16,920	15,920	15,920	15,848
South America	2,486,247	2,497,953	2,216,553	2,266,638
Yield (t ha^{-1})				
World	9.87	10.09	9.91	10.14
Central America and Caribbean	5.15	5.14	5.15	4.74
Africa	8.22	8.36	8.32	8.51
Asia	13.00	13.67	13.34	14.06
Oceania	11.74	10.56	10.59	11.57
South America	12.34	12.48	12.37	12.69

Source: FAO database (2000).

Thailand, with each country having about 1.2 million ha of cassava (Table 4.2). India, Philippines, Vietnam and China follow in that order, each harvesting about a quarter of a million ha of cassava in 1999.

Thailand produced 16.5 million t and Indonesia produced 15.5 million t of cassava in 1999. India, China, Philippines and Vietnam are the next highest producers in that order, although as a group, their production is far below those of Indonesia and Thailand.

Yield levels vary considerably within Asia, depending on the cropping systems used and the general production conditions. Among the major producers, India stands out as having the highest yield levels of about 24 t ha^{-1}. China, Thailand and Indonesia also have relatively high average yields of 15.9, 15.5 and 12.8 t ha^{-1}, respectively. Yields in Vietnam, Philippines and Sri Lanka are relatively low at 8–8.5 t ha^{-1}.

Cassava Trade in Asia

Table 4.3 shows that Asia is by far the world leader in the export of cassava products. The main commodities exported are dried cassava (mainly in the form of chips and pellets for animal feeding), cassava starch and cassava tapioca. In 1998, the various countries of Asia exported about 4 million t of dried cassava, 0.5 million t of cassava starch and 58,000 t of cassava tapioca.

For dried cassava, Thailand is by far the greatest exporter, followed distantly by Indonesia and Vietnam (Table 4.4). Thailand again is the outstanding leader in cassava starch exportation, followed by China, Malaysia and Indonesia. The largest exporters of tapioca are Indonesia, Thailand, China and India in that order.

Most of Asia's cassava exports go to countries of the European Union. The tariff situation changed dramatically in the late 1990s, and has made cassava less competitive. This casts doubt on the long-term sustainability of this cassava trade. The various exporting countries are now exploring diverse strategies for tackling the problem.

While most cassava exports from Asian countries go to other continents (especially Europe), there are countries within Asia that

Table 4.2. Asian production, area and yield of cassava from 1996–1999.

	1996	1997	1998	1999
Production				
Asia	48,837,430	47,547,650	46,445,280	46,057,280
Cambodia	69,656	69,656	69,656	67,500
China	3,600,744	3,600,744	3,600,744	3,650,658
India	5,979,000	5,979,005	5,979,000	6,000,000
Indonesia	17,002,460	15,072,050	16,052,830	15,421,885
Laos	70,000	70,000	70,000	70,000
Malaysia	400,000	400,000	400,000	400,000
Myanmar	77,148	80,000	80,000	88,144
Philippines	1,910,780	1,958,050	2,000,000	1,786,710
Sri Lanka	271,000	250,000	250,000	257,153
Thailand	17,387,800	18,083,600	15,958,500	16,506,625
Vietnam	2,067,300	1,983,000	1,983,000	1,983,000
Area (ha)				
Asia	3,757,735	3,477,549	3,482,549	3,276,649
Cambodia	13,000	13,000	13,000	7,000
China	230,060	230,060	230,060	230,045
India	244,000	244,000	244,000	250,000
Indonesia	1,415,100	1,233,550	1,233,550	1,205,330
Laos	5,100	5,100	5,100	5,100
Malaysia	39,000	39,000	39,000	39,000
Myanmar	6,785	7,000	7,000	7,736
Philippines	231,855	235,000	240,000	210,000
Sri Lanka	32,000	32,000	32,000	30,064
Thailand	1,265,096	1,200,000	1,200,000	1,065,435
Vietnam	275,600	238,700	238,700	226,800
Yield (t ha^{-1})				
Asia	13.00	13.67	13.34	14.06
Cambodia	5.36	5.36	5.36	9.64
China	15.65	15.65	15.65	15.87
India	24.50	24.50	24.50	24.00
Indonesia	12.01	12.22	13.01	12.79
Laos	13.72	13.72	13.72	13.76
Malaysia	10.26	10.26	10.26	10.26
Myanmar	11.37	11.43	11.43	11.39
Philippines	8.24	8.33	8.33	8.59
Sri Lanka	8.47	7.81	7.81	8.55
Thailand	13.74	15.07	13.30	15.49
Vietnam	7.50	8.31	8.31	7.97

Source: FAO database (2000).

import significant quantities of cassava products, mainly from other Asian countries. Korea (South), Turkey and China import substantial quantities of dried cassava, while China, Indonesia and Japan are notable importers of cassava starch. Importers of cassava tapioca within Asia include Philippines, Malaysia, Bangladesh and Singapore in that order.

Cassava Industry in the Asia-Pacific Region

Cassava in Thailand

Cassava was introduced into Thailand about 1850, and during its first century, became established as a food crop in the eastern

Table 4.3. World cassava exports by continents, from 1995–1998.

	1995	1996	1997	1998
Dried cassava (t)				
World	4,090,063	4,487,984	3,327,385	4,716,407
Central America and Caribbean	47,813	55,129	42,158	42,136
Africa	27,925	31,321	43,041	43,034
Asia	3,737,849	4,149,940	3,001,917	4,282,297
Oceania	945	945	965	511
South America	3,490	1,211	4,841	3,514
Cassava starch (t)				
World	525,195	539,298	454,355	510,450
Central America and Caribbean	16	4	30	67
Africa	4	47	41	80
Asia	501,736	512,952	439,802	495,193
Oceania	1	0	0	0
South America	19,552	15,738	13,259	13,391
Cassava tapioca (t)				
World	63,085	40,320	46,005	59,720
Central America and Caribbean	0	0	0	4
Africa	431	422	427	409
Asia	61,909	38,956	44,562	58,412
Oceania	0	0	0	0
South America	592	649	765	730

Source: FAO database (2000).

seaboard province (Titapiwatanakun, 1990). Since about 1956, the crop has spread to other parts of the country, especially the northeast which is now the major producing area. The dramatic increase in cassava hectarage in the northeast was achieved through partial replacement of *kenaf* fields, and by opening up new previously forested land. Today, most of Thailand's cassava production occurs in the northeast and in the central plains. Very little production takes place in the north and the south regions.

The topography in the cassava producing areas is generally undulating and the soils are mostly Ultisols of loamy sand or sandy loam texture, and pH 5–6.5. Some of these soils are erosion-prone, and most of them have been degraded due to erosion and long-term intensive cropping. Mean temperature in the cassava regions is about 27°C and the annual rainfall is 1100–1500 mm in the central plain and 900–1400 mm in the northeast. The rainy season commences in April.

In Thailand, most of the cassava is grown as a sole crop (Howeler, 1988). Occasionally, it is intercropped with maize, groundnut, rubber or coconuts. Normal spacing in sole cropped cassava is about 1 × 1 m. Planting occurs from May to November, with most of the planting being done in May to June. Land preparation is with tractors or by bullocks. Planting is done on the flat or on ridges, with the cutting placed in a vertical position. Harvesting occurs January–March or in October and most is harvested before it is 1 year old. Very little harvesting is done in the rainy months from May to September. The problems of harvesting in the rainy months include the low starch content of the tubers at that time, low prices, reduced demand from buyers and increased difficulty of sun drying, which is the main method for producing the chips (Tongglum *et al.*, 1988). Chips for export must be below 14% moisture.

The first major industrial use of cassava was for processing into starch which started in the mid-1940s. The starch was used domestically as well as for export. However, in the 1960s, the processing of cassava into chips and pellets for animal feeding gained momentum, stimulated by the emerging market for these products in Europe. For the next two decades, the cassava industry witnessed a very rapid and successful expansion. By the 1980s, there were over 2500

Table 4.4.   Quantities of cassava exported from various Asian countries from 1995–1998.

	1995	1996	1997	1998
Dried cassava (t)				
Asia	3,737,849	4,149,940	3,001,917	4,282,297
China	163	87	78	18
Indonesia	481,483	388,591	247,001	–
Malaysia	33	21	21	52
Philippines	267	587	328	221,404
Singapore	0	222	673	–
Sri Lanka	1,180	1,010	1,166	1,166
Thailand	3,224,191	3,728,747	2,722,114	3,981,000
Vietnam	30,500	30,500	30,500	78,000
Cassava starch (t)				
Asia	501,736	512,952	439,802	495,193
China	2,650	1,469	2,772	831
China, Hong Kong	15,197	26,943	32,116	27,482
India	1,954	65	65	65
Indonesia	30,870	17,924	7,338	82,803
Malaysia	202	14,292	14,292	18
Pakistan	0	0	102	22
Singapore	3,231	1,040	856	1,718
Thailand	447,625	451,126	328,246	382,246
Cassava tapioca (t)				
Asia	61,909	38,956	44,562	58,412
China	4,953	6,278	6,682	7,719
China, Hong Kong	4,901	1,185	1,717	1,645
India	28,238	5,311	5,311	5,311
Indonesia	1,510	9,810	18,863	31,617
Malaysia	1,448	1,930	1,100	1,215
Philippines	7,721	30	18	43
Singapore	823	927	689	670
Thailand	12,313	13,448	10,181	10,181

Source: FAO database (2000).

chip factories, 600 pellet factories and 125 cassava starch factories in Thailand.

About 50% of the country's cassava now goes to starch production. Of the total starch production, nearly half is exported to Taiwan and Japan. Domestic cassava consumption in Thailand is negligible, and over 90% of cassava production is destined for export.

For over two decades, the cassava industry was sustained by favourable price and quota policies of the European Community (EC). Starting in the 1980s, EC policy changes started to work towards reducing the imports of cassava pellets from Thailand. In 1993, EC grain prices fell, thereby putting further competitive pressure on cassava. In response to the increasingly unfavourable European market for cassava, the Thai authorities have developed a policy aimed

at: (i) reducing the total land area devoted to cassava; (ii) diversifying cassava products and markets; and (iii) reducing the cost of cassava production through improved varieties and agronomic packages (Henry *et al.*, 1994).

Active cassava research in Thailand is conducted at the Rayong Field Crops Research Center of the Department of Agriculture, and at the Sriracha Research Station of Kasetsart University. The Centro Internacional de Agricultura Tropical (CIAT, with headquarters in Cali, Colombia) is collaborating in much of the research, and has been instrumental in the dissemination of improved cassava cultivars and technological packages, not only in Thailand, but in neighbouring countries also. The cassava industry in Thailand, and indeed in all of South-East Asia, has benefited tremendously from the research and extension activities of CIAT. The Rayong series of |cultivars which they have released have contributed to the boosting of yields among commercial producers. Most of these cultivars are high glucoside types.

Cassava in Indonesia

Cassava is the fourth most important food crop in Indonesia after rice, maize and soybeans. In 1999, Indonesia produced approximately 15.4 million t of cassava, harvested from 1.2 million ha. These figures are similar to those for Thailand. However, a larger percentage of Indonesia's cassava production is used domestically for human consumption. Indeed, about 70% of Indonesia cassava production is used for human food, in both the fresh form and in a dried chip form called *gaplek* (Wargiono *et al.*, 1995). Consequently, a smaller percentage is exported.

Within Indonesia, the most important cassava-producing areas are Java, south Sumatra and Kalimantan. Unlike in Thailand, the soils are mostly clay textured and the rainfall is relatively heavy (1500–3000 mm year^{-1}). The rainy season commences in September/October. Planting occurs all year round, but with October–November being favoured. Similarly, harvesting occurs all year round, but over 60% of the crop is harvested during June–October. Land preparation is done with hand tools, bullocks or occasionally with tractors. Planting is done on the flat or on levies. The cuttings are placed vertically. Intercropping with maize, groundnut, rice or soybean is the most common practice, but in the larger plantations, sole cropping is practised.

Indonesia exported about 247,000 t of dried cassava and 7300 t of cassava starch in 1997. Although Indonesia still ranks second in the world in terms of cassava exports, these quantities are much lower than those of Thailand, and the general trend from 1993 to 1997 has been a decline in the cassava export trade.

The Indonesian government has instituted a policy of food diversification and increasing the productivity of the uplands (Poespodarsono and Widodo, 1995). This policy has favoured the intensification of cassava cultivation. Much cassava and extension work in Indonesia is carried out by the Bogor and Malang Research Institutes

for Food Crops and by Brawijaya University in Malang. CIAT has also collaborated in some of the research, and has been instrumental in the release of some promising cultivars.

Cassava in India

India is the third largest producer of cassava in Asia. In 1999, about 6 million t were produced from 250,000 ha. India has also consistently recorded the highest yields per hectare (24.5 t ha^{-1}) for cassava among the Asian countries.

Cassava production in India is concentrated in the states of Kerala and Tamil Nadu, with some production in the northeast of the country. In Kerala, rainfall is sufficient (1800–2000 mm year^{-1}), but in Tamil Nadu it is 900 mm year^{-1}, and supplemental irrigation is practised. In Tamil Nadu, the rain-fed crop is planted in October/November, while the irrigated crop is planted in January–April. In Kerala, planting is done in April/May or in September/October. In Kerala, planting is done on raised mounds or levies, but in Tamil Nadu, flats or ridges are used. Vertical placement of the cutting is the most common practice. Sole cropping of cassava is common, as well as intercropping with groundnut, vegetables or coconuts. Harvesting occurs 10–11 months after planting.

Most of the cassava produced in India is used for human consumption. In Kerala, for example, about 70% of the cassava goes for human consumption (Padmaja *et al.*, 1992).

Cassava research in India is done mostly by the Central Tuber Crops Research Institute in Trivandrum, as well as in various State Agricultural Universities. There is a Cassava Technology Transfer Program (CTTP) with emphasis on both production and processing technologies.

Cassava in China

In Asia, China ranks fourth after Thailand, Indonesia and India in cassava production. In 1999, about 230,000 ha were harvested, producing 3.6 million t (Table 4.2). Cassava production in China is concentrated in the tropical/subtropical extreme southeast corner of the country, south of the Yangze river. This is in the provinces of Guangxi, Guangdong and Hainan Island. The

Chinese cassava zone is cooler (mean temperature 20–24°C) than any of the other cassava-producing regions in Asia, and occasional frosts occur. In Guangxi, the slopes are steep and terraces are used. Soils in China, used for cassava production are clay Oxisols, or Alfisols with low fertility, and pH 4.5–6.5. Annual rainfall is 1800–2500 mm in Hainan Island, but slightly less (1200–1800 mm) in Guangxi. The rainy season starts in March, and planting occurs in March/April. The crop is grown on the flat or on ridges. It is usually intercropped with groundnut or rubber. The stem cuttings are usually planted horizontally. Harvesting is carried out in December/January.

In China, cassava is used mostly for animal feeding (as chips) and for industrial purposes such as the manufacture of starch, fructose and monosodium glutamate. Cassava research in China is conducted by the Guangxi Subtropical Crops Research Institute, the South China Academy of Tropical Crops, and the Upland Crops Research Institute, in collaboration with CIAT (Yinong *et al.*, 1995). Research emphasis is on soil fertility maintenance, erosion control, planting methods, planting time and harvesting time. A Cassava Cooperation Network has recently been established.

Cassava in the Philippines

In the Philippines, cassava ranks third among the food crops after rice and maize. The country produces 1.8–2.0 million t year^{-1} of cassava on a harvested area of 210,000–240,000 ha. Production and area under cassava declined between 1998 and 1999 but average yield increased (Table 4.2). Average yields are only about 8.4 t ha^{-1}, one of the lowest in Asia. Most of the production is in the southern (Mindanao) and central (Visayas) parts of the country. The soils are clay Ultisols with pH 5–7. Rainfall is heavy (2000–3000 mm year^{-1}).

Land is often prepared using bullocks, and planting is done on the flat or on ridges. Sole cropping is most common, although cropping in coconut plantations is also done. Planting takes place in June or November, and harvesting occurs in January/February or in July/August. The stem cuttings are usually planted horizontally.

Of the total cassava production in the Philippines, about a third is used for human food, another third for animal feed and a third for industrial purposes (Den *et al.*, 1992). The animal feed is mainly in the form of dried chips, some of which is exported. Industrial products made from cassava include starch and glucose.

Active cassava research and promotion work is carried out by the Philippines Root Crops Research and Training Center in Baybay, and at the University of the Philippines, with collaboration from CIAT. Extension work on cassava is part of the Integrated Root Crop Extension Program.

Cassava in Vietnam

Vietnam produces about 2 million t year^{-1} of cassava each year, harvested from an area of about 230,000 ha. Yields are low, averaging about 8 t ha^{-1}, but with the recent introduction of improved cultivars, these yields are expected to improve. Cassava production occurs in all parts of the country, but with a greater concentration in the northern, mountainous region and the central parts of the country.

Vietnam's cassava/soils are mostly Ultisols of pH 5–6 (Bien and Kim, 1992). The soils are low in fertility and the terrain is undulating. The richer soils are reserved for rice cultivation. Rainfall is 1500–2500 mm year^{-1}. Planting occurs in February–May, and harvesting takes place 10–12 months later. In most areas, the stems are planted vertically, but in the sandier soils, horizontal planting is practised. Field spacing is about 1 × 1 m. Because of the high price of fertilizer compared with the low market return for cassava, most producers find it uneconomical to apply fertilizers. Cassava is usually grown as a sole crop, but occasionally it is intercropped with groundnut, mungbean, maize or winged bean.

For decades, cassava was a crop of last resort in Vietnam. Recently it has become increasingly used as a raw material for industry. Starch processing occurs in rural households as well as in large factories. While most of the starch is used for domestic food processing, some of it is used in the textiles and paper industries. In general, 30–40% of Vietnam's cassava is used for industrial purposes, 30% for animal feed and 10–20% for human consumption (Ngoan *et al.*, 1995).

Cassava research in Vietnam has been supported by the Vietnam Root and Tuber Crops Research Program which was initiated in 1989. This program has been working in collaboration with CIAT. Some cassava research is done at the Hung Loc Agricultural Research Center.

Cassava in Malaysia

Malaysia harvests about 400,000 t year^{-1}of cassava from an area of about 39,000 ha. In Peninsula Malaysia, production is concentrated in the state of Perak, with Ultisol soils. There is also potential for cassava production in the peat soils (Histosols) of Malaysia.

Rainfall is heavy, averaging 2000–3000 mm year^{-1}. The rainy season commences in October, but planting and harvesting occur all year round. Planting is done on tractor-made ridges. The stem cuttings are usually laid horizontally and entirely buried, at a spacing of 1×1 m, but vertical planting is done on the peat soils. Sole cropping is the common practice, but cassava is sometimes intercropped with groundnut or maize.

Most of the cassava is for starch production, while a small quantity is processed into chips for animal feed. The use of cassava for direct human consumption is negligible. Malaysia also imports some cassava starch and dried cassava from Thailand. Cassava research is done at the Malaysian Agricultural Research and Development Institute (MARDI), but a relatively low priority is given to the crop.

Cassava in Sri Lanka

Cassava was first introduced to Sri Lanka from Mauritius in 1796. Peak production of 0.85 million t was achieved in 1974. Since then, production has declined steadily (Bandara and Sikurajapathy, 1992). In 1999, Sri Lanka harvested 257,000 t of cassava on an area of 30,000 ha. Average yields are 7.8–8.6 t ha^{-1}, one of the lowest in Asia.

Production occurs in all parts of the country, but it is more concentrated in the southwestern part. The land is generally sloping, and the soils are Ultisols with low fertility and pH 5–6. Rainfall is heavy and monsoonal, averaging 2000–3000 mm year^{-1}. Planting and harvesting occur year round. Planting is done on mounds prepared with hand tools, and the cuttings are planted vertically or slanted.

Virtually all the Sri Lanka cassava production is used for human consumption, with a small quantity being used as feed. Some cassava research is done at the Central Agricultural Research Institute (CARI) in Peradeniya.

Cassava in Oceania

Cassava is not a particularly important crop in Oceania. Even though the Pacific Islands constitute one of the most intensive users of root crops in the world, other root crops such as sweet potato, taro, tannia and yam usually come ahead of cassava in terms of culinary preference or cultural attachment. However, cassava remains an important dietary staple and is often produced in greater quantities than the preferred root crops.

In 1999, all of Oceania harvested only 16,000 ha of cassava, producing 183,000 t (see Table 4.5). The average yield was 11.6 t ha^{-1}. The total cassava production from Oceania is only about 1% of the production of Indonesia alone or Thailand alone.

Within Oceania, the largest areas devoted to cassava are in Papua New Guinea (10,000 ha), Fiji (2610 ha) and Tonga (2100 ha). The largest annual production of cassava is from Papua New Guinea (75,000 t) and Fiji (50,336 t).

Papua New Guinea

Cassava production in Papua New Guinea is an extremely casual affair. Even though cassava is usually the cheapest root crop in markets (on a fresh or dry weight basis), its popularity lags far behind that of sweet potato, yam, taro or even sago. There are no large scale plantations or processing plants, and what is produced is offered in local markets as fresh tuberous roots for human consumption. With virtually all the grain for animal feeding being imported, there is potential for expansion of the cassava industry in the country. Cassava research and extension is carried out by the Department of Agriculture and Livestock, and at the University of Technology in Lae.

Table 4.5.   Oceania production, area and yield of cassava from 1996–1999.

	1996	1997	1998	1999
Production (t)				
Oceania	198,683	168,092	168,592	183,292
Cook Islands	3,000	2,500	3,000	3,000
Fiji	40,427	50,336	50,336	27,136
French Polynesia	5,500	5,500	5,500	5,500
Micronesia	11,800	11,800	11,800	11,800
New Caledonia	2,800	2,800	2,800	2,800
Papua New Guinea	115,000	75,000	75,000	112,000
Samoa	400	400	400	400
Solomon Islands	1,100	1,100	1,100	2,000
Tonga	28,000	28,000	28,000	28,000
Wallis and Futuna Islands	2,400	2,400	2,400	2,400
Area (ha)				
Oceania	16,920	15,920	15,920	15,848
Cook Islands	170	170	170	170
Fiji	1,856	2,610	2,610	1,938
French Polynesia	300	300	300	300
Micronesia	1,100	1,100	1,100	1,100
New Caledonia	400	400	400	400
Papua New Guinea	11,000	10,000	10,000	10,500
Samoa	30	30	30	30
Solomon Islands	70	70	70	70
Tonga	2,100	2,100	2,100	2,100
Wallis and Futuna Islands	230	230	230	230
Yield (t ha^{-1})				
Oceania	11.74	10.56	10.59	11.57
Cook Islands	17.65	14.71	17.65	17.65
Fiji	15.49	19.29	19.29	13.68
French Polynesia	18.33	18.33	18.33	18.33
Micronesia	10.73	10.73	10.73	10.73
New Caledonia	7.00	7.00	7.00	7.00
Papua New Guinea	10.45	7.50	7.50	10.67
Samoa	13.33	13.33	13.33	13.33
Solomon Islands	15.71	15.71	15.71	16.00
Tonga	13.33	13.33	13.33	13.33
Wallis and Futuna Islands	10.43	10.43	10.43	10.43

Source: FAO database (2000).

Fiji

In Fiji, cassava is the predominant root crop in terms of quantity produced and consumed, although not in terms of culinary preference. It is grown mostly as the last crop in the ginger–taro–cassava–fallow rotation, usually on sloping hillsides.

In 1994, cassava accounted for about half of the total root crop production in Fiji, with taro being a distant second with about a quarter of the total (Onwueme, 1996). In the same year, 166 t of cassava was exported to Australia and New Zealand, mostly as frozen tubers. This represented less than 1% of the total production. The rest (99%) was used domestically for human consumption, mostly marketed in the form of fresh tubers. There is virtually no cassava processing into dried forms for human or animal use. However, given the vigorous domestic livestock industry which imports large quantities of feed, there is potential for cassava to be used for animal feed. Cassava research and extension work are carried out at the Koronivia Research Station, and at various stations of the Ministry of Agriculture, Fisheries and Forests.

Tonga

In Tonga, cassava is produced in greater quantity than all the other root crops combined. It is usually grown as the last crop in the rotation before the land is allowed to revert to fallow. Many farmers grow cassava between coconut stands, but sole cropping is practised by the more commercially oriented producers. There is a small export trade in frozen cassava (1508 t in 1993), but the overwhelming proportion of the cassava produced is traded locally as fresh tubers for human consumption. There is virtually no processing of cassava into dried products. Its greater quantity notwithstanding, culinary and social preference for cassava in Tonga is very low. In this respect, it falls far behind yam, taro and, of course, kava. Cassava research and extension work is carried out at the Vaini Research Station.

Australia

There is some latent interest in cassava in the northern tropical parts of Australia, but the economics of production have not permitted this interest to be realized. However, cassava continues to attract research interest at the University of Queensland and a few other institutions in the country.

References

Bandara, W.M.S.M. and Sikurajapathy, M. (1992) Recent progress in cassava varietal and agronomic research in Sri Lanka. In: Howeler, R.H. (ed.) *Cassava Breeding, Agronomy and Utilization Research in Asia. Proceedings of the 3rd Regional Workshop*, Malang, Indonesia, 22–27 October 1990, pp. 96–105.

Bien, P.V. and Kim, H. (1992) Cassava production and research in Vietnam: historical review and future direction. In: Howeler, R.H. (ed.) *Cassava Breeding, Agronomy and Utilization Research in Asia. Proceedings of the 3rd Regional Workshop*, Malang, Indonesia, 22–27 October 1990, pp. 106–123.

Den, T.V., Palomar, L.S. and Amestoso, F.J. (1992) Processing and utilization of cassava in the Philippines. In: Howeler, R.H. (ed.) *Cassava Breeding, Agronomy and Utilization Research in Asia. Proceedings of the 3rd Regional Workshop,*
Malang, Indonesia, 22–27 October 1990, pp. 339–354.

FAO database (2000) FAOSTAT database. http://apps.fao.org/cgi-bin/nph_db.pl

Henry, G., Klakhaeng, K. and Gottret, M.V. (1994) *Maintaining the Edge.* Centro Internacional de Agricultura Tropical (CIAT), Cali, Colombia.

Howeler, R. (1988) Agronomic practices for cassava production in Asia. In: Howeler, R.H. and Kawano, K. (eds) *Cassava Breeding and Agronomy Research in Asia.* Proceedings of workshop held in Thailand, 26–28 October 1987, pp. 313–340.

Ngoan, T.N., Quyen, T.N., Kim, H. and Kazuo, K. (1995) Recent progress in cassava varietal improvement in Vietnam. In: *Cassava Breeding, Agronomy Research and Technology Transfer in Asia.* CIAT, Cali, Colombia, pp.253–261.

Onwueme, I.C. (1996) Root and Tuber crops in Fiji, W. Samoa, Tonga and Vanuatu. Mission Report, Food and Agriculture Organization of the United Nations, Bangkok, 42pp.

Padmaja, G., Balagopalan, C., Kurup, G.T., Moorthy, S.N. and Nanda, S.K. (1992) Cassava processing, marketing and utilization in India. In: Howeler, R.H. (ed.) *Cassava Breeding, Agronomy and Utilisation Research in Asia. Proceedings of the 3rd Regional Workshop*, Malang, Indonesia, October 22–27 1990, pp. 327–338.

Poespodarsono, S. and Widodo, Y. (1995) Recent progress of cassava varietal improvement in Indonesia. In: *Cassava Breeding, Agronomy Research and Technology Transfer in Asia.* CIAT, Cali, Colombia, pp. 175–182.

Titapiwatanakun, B. (1990) Significant factors contributing to the success of the Thai cassava industry. In: Howeler, R.H. (ed.) *Proceedings of the 8th Symposium of the International Society for Tropical Root Crops*, 30 October–5 November, Bangkok, Thailand, p. 724.

Tongglum, A., Tiraporn, C. and Sinthuprama, S. (1988) Cassava cultural practices research in Thailand. In: Howeler, R.H. and Kawano, K. (eds) *Cassava Breeding and Agronomy Research in Asia.* Proceedings of workshop held in Thailand, 26–28 October 1987, pp. 131–144.

Wargiono, J., Guritno, B., Sugito, Y. and Widodo, Y. (1995) Recent progress in cassava agronomy research in Indonesia. In: *Cassava Breeding, Agronomy Research and Technology Transfer in Asia.* CIAT, Cali, Colombia, pp. 147–174.

Yinong, T., Jun, L., Weite, Z. and Baiping, F. (1995) Recent progress in cassava agronomy research in China. In: *Cassava Breeding, Agronomy Research and Technology Transfer in Asia.* CIAT, Cali, Colombia, pp. 195–216.

Chapter 5
Cassava Botany and Physiology

Alfredo Augusto Cunha Alves

Embrapa Cassava and Fruits, Caixa Postal 007, 44.380–000, Cruz das Almas, Bahia, Brazil

Introduction

Cassava is a perennial shrub of the family Euphorbiaceae, cultivated mainly for its starchy roots. It is one of the most important food staples in the tropics, where it is the fourth most important source of energy. On a worldwide basis it is ranked as the sixth most important source of calories in the human diet (FAO, 1999). Given the crop's tolerance to poor soil and harsh climatic conditions, it is generally cultivated by small farmers as a subsistence crop in a diverse range of agricultural and food systems. Although cassava is a perennial crop, the storage roots can be harvested from 6 to 24 months after planting (MAP), depending on cultivar and the growing conditions (El-Sharkawy, 1993). In the humid lowland tropics the roots can be harvested after 6–7 months. In regions with prolonged periods of drought or cold, the farmers usually harvest after 18–24 months (Cock, 1984). Moreover, the roots can be left in the ground without harvesting for a long period of time, making it a very useful crop as a security against famine (Cardoso and Souza, 1999).

Cassava can be propagated from either stem cuttings or sexual seed, but the former is the commonest practice. Propagation from true seed occurs under natural conditions and is widely used in breeding programmes. Plants from true seed take longer to become established, and they are smaller and less vigorous than plants from

cuttings. The seedlings are genetically segregated into different types due to their reproduction by cross-pollination. If propagated by cuttings under favourable conditions, sprouting and adventitious rooting occur after 1 week.

Morphological and Agronomic Characteristics

Cassava, which is a shrub reaching 1–4 m height, is commonly known as tapioca, manioc, mandioca and yuca in different parts of the world. Belonging to the dicotyledon family Euphorbiaceae, the *Manihot* genus is reported to have about 100 species, among which the only commercially cultivated one is *Manihot esculenta* Crantz. There are two distinct plant types: erect, with or without branching at the top, or spreading types.

The morphological characteristics of cassava are highly variable, which indicate a high degree of interspecific hybridization. There are many cassava cultivars in several germplasm banks held at both international and national research institutions. The largest germplasm bank is located at Centro Internacional de Agricultura Tropical (CIAT), Colombia, with approximately 4700 accessions (Bonierbale *et al.*, 1997), followed by EMBRAPA's collection in Cruz das Almas, Bahia, with around 1700 accessions (Fukuda *et al.*, 1997), representing the germplasm of the following Brazilian eco-systems: lowland and highland semiarid tropics, lowland humid subtropics, lowland subhumid tropics, lowland humid tropics, and lowland hot savannah. The cassava genotypes are usually characterized on the basis of morphological and agronomic descriptors. Recently, the International Plant Genetic Resources Institute (IPGRI) descriptors (Gulick *et al.*, 1983) were revised, and a new version was elaborated, in which 75 descriptors were defined, 54 being morphological and 21 agronomic (Fukuda *et al.*, 1997). Morphological descriptors (for example, lobe shape, root pulp colour, stem external colour) have a higher heritability than agronomic (such as root length, number of roots per plant and root yield). Among morphological descriptors, the following were defined as the minimum or basic descriptors that should be considered for identifying a cultivar: (i) apical leaf colour; (ii) apical leaf pubescence; (iii) central lobe shape; (iv) petiole colour; (v) stem cortex colour; (vi) stem external colour; (vii) phyllotaxis length; (viii) root peduncle presence; (ix) root external colour; (x) root cortex colour; (xi) root pulp colour; (xii) root epidermis texture; and (xiii) flowering.

Given the large number of cassava genotypes cultivated commercially and the large diversity of ecosystems in which cassava is grown, it is difficult to make a precise description of the morphological descriptors as there is a genotype-by-environmental conditions interaction. Thus, in addition to morphological characterization, molecular characterization, based mainly on DNA molecular markers, has been very useful in order to evaluate the germplasm genetic diversity (Beeching *et al.*, 1993; Fregene *et al.*, 1994).

Roots

Roots are the main storage organ in cassava. In plants propagated from true seeds a typical primary tap root system is developed, similar to dicot species. The radicle of the germinating seed grows vertically downward and develops into a taproot, from which adventitious roots originate. Later, the taproot and some adventitious roots become storage roots.

In plants grown from stem cuttings the roots are adventitious and they arise from the basal-cut surface of the stake and occasionally from the buds under the soil. These roots develop to make a fibrous root system. Only a few fibrous roots (between three and ten) start to bulk and become storage roots. Most of the other fibrous roots remain thin and continue to function in water and nutrient absorption. Once a fibrous root becomes a storage root, its ability to absorb water and nutrients decrease considerably. The storage roots result from secondary growth of the fibrous roots; thus the soil is penetrated by thin roots, and their enlargement begins only after that penetration has occurred.

Anatomically, the cassava root is not a tuberous root, but a true root, which cannot be used for vegetative propagation. The mature cassava storage root has three distinct tissues: bark (periderm), peel (or cortex) and parenchyma. The parenchyma, which is the edible portion of the fresh root, comprises approximately 85% of total weight, consisting of xylem vessels radially distributed in a matrix of starch-containing cells (Wheatley and Chuzel, 1993).

The peel layer, which is comprised of sclerenchyma, cortical parenchyma and phloem, constitutes 11–20% of root weight (Barrios and Bressani, 1967). The periderm (3% of total weight) is a thin layer made of a few cells thick and, as growth progresses, the outermost portions usually slough off. Root size and shape depend on cultivar and environmental conditions; variability in size within a cultivar is greater than that found in other root crops (Wheatley and Chuzel, 1993). Table 5.1 lists morphological and agronomic characteristics of the root and its variability in cassava.

Stems

The mature stem is woody, cylindrical and formed by alternating nodes and internodes. On the nodes of the oldest parts of the stem, there are protuberances, which are the scars left by the plant's first leaves. A plant grown from stem cuttings can produce as many primary stems as there are viable buds on the cutting. In some cultivars with strong apical dominance, only one stem develops.

The cassava plant has sympodial branching. The main stem(s) divide di-, tri- or tetrachotomously, producing secondary branches that produce other successive branchings. These

Table 5.1.   Some morphological and agronomic characteristics of roots and their variability in cassava.

Root characteristic	Variability	Reference
Morphological		
External colour	White or cream; yellow; light brown; dark brown	Fukuda and Guevara (1998)
Cortex colour	White or cream; yellow; pink; purple	Fukuda and Guevara (1998)
Pulp (parenchyma) colour	White; cream; yellow; pink	Fukuda and Guevara (1998)
Epidermis texture	Smooth; rugose	Fukuda and Guevara (1998)
Peduncule	Sessile; pedunculate; both	Fukuda and Guevara (1998)
Constriction	None or little; medium; many	Fukuda and Guevara (1998)
Shape	Conical; conical–cylindrical; cylindrical; irregular	Fukuda and Guevara (1998)
Agronomic		
No. storage roots/plant	3–14	Dimyati (1994); Wheatley and Chuzel (1993); Ramanujam and Indira (1983); Pinho et al. (1995)
Weight of storage roots/plant	0.5–3.4 kg FW	Dimyati (1994); Wheatley and Chuzel (1993); Ramanujam and Indira (1983)
Weight of one storage root	0.17–2.35 kg	Barrios and Bressani (1967); Ramanujam and Indira (1983)
Length of storage root	15–100 cm	Wheatley and Chuzel (1993); Barrios and Bressani (1967); Pinho et al. (1995)
Diameter of storage root	3–15 cm	Wheatley and Chuzel (1993); Barrios and Bressani (1967); Pinho et al. (1995)
Diameter of fibrous root	0.36–0.67 mm	Connor et al. (1981)
Depth of fibrous root	Up to 260 cm	Connor et al. (1981)
Amylose in root starch	13–21% FW	O'Hair (1990)
Protein in whole root	1.76–2.68% FW	Barrios and Bressani (1967)
Protein in pulp (parenchyma)	1.51–2.67% FW 1.0–6.0% DW	Barrios and Bressani (1967) Wheatley and Chuzel (1993)
Protein in peel	2.79–6.61% FW 7.0–14.0% DW	Barrios and Bressani (1967) Wheatley and Chuzel (1993)
DM in whole fresh root	23–43%	Barrios and Bressani (1967); Ghosh et al. (1988); Ramanujam and Indira (1983); O'Hair (1989)
DM in peel	15–34%	Wheatley and Chuzel (1993); Barrios and Bressani (1967); O'Hair (1989)
DM in pulp	23–44%	Wheatley and Chuzel (1993); Barrios and Bressani (1967); O'Hair (1989)
Carbohydrates in whole root	85–91% DW	Barrios and Bressani (1967)
Carbohydrates in peel	60–83% DW	Barrios and Bressani (1967)
Carbohydrates in pulp (parenchyma)	88–93% DW	Barrios and Bressani (1967)
Starch in whole root	20–36% FW	Wholey and Booth (1979); O'Hair (1989); Ternes et al. (1978)
	77% DW	Ghosh et al. (1988)
Starch in peel	14–25% FW	O'Hair (1989)
	44–59% DW	Wheatley and Chuzel (1993)

Continued

Table 5.1. *Continued.*

Root characteristic	Variability	Reference
Starch in pulp (parenchyma)	26–40% FW	O'Hair (1989)
	70–91% DW	Wheatley and Chuzel (1993)
Peel in whole root	11–20% FW	Barrios and Bressani (1967)
Crude fibre in whole root	3.8–7.3% DW	Barrios and Bressani (1967)
Crude fibre in peel	9.2–21.2% DW	Barrios and Bressani (1967)
	5.0–15.0% DW	Wheatley and Chuzel (1993)
Crude fibre in pulp	2.9–5.2% DW	Barrios and Bressani (1967)
(parenchyma)	3.0–5.0% DW	Wheatley and Chuzel (1993)
Total sugars	1.3–5.3% DW	Wheatley and Chuzel (1993)

DM, dry matter; FW, fresh weight; DW, dry weight.

branchings, which are induced by flowering, have been called 'reproductive branchings'.

Stem morphological and agronomic characteristics are very important to characterizing a cultivar (Table 5.2). The variation of these characteristics depends on cultivar, cultural practice and climatic conditions.

Leaves

Cassava leaves are simple, formed by the lamina and petiole. The leaf is lobed with palmated veins. There is generally an uneven number of lobes, ranging from three to nine (occasionally 11). Only a few cultivars are characterized by having three-lobed mature vegetative leaves, which may represent the primitive ancestral form (Rogers and Fleming, 1973). Leaves near the inflorescence are generally reduced in size and lobe number (most frequently three-lobed), but the one closest to the base of the inflorescence is frequently simple and unlobed.

Leaves are alternate and have a phyllotaxy of 2/5, indicating that from any leaf (leaf 1) there are two revolutions around the stem to reach the sixth (leaf 6) in the same orthostichy as leaf 1. In these two revolutions there are five successive intermediate leaves (not counting leaf 1).

The main leaf morphological and agronomic characteristics and their variation are given in Table 5.3. Many of them (mainly the morphological ones) are used to characterize cultivars and may vary with environmental conditions and plant age.

Mature leaves are glabrous and each leaf is surrounded by two stipules (approximately 0.5–1.0 cm long), which remain attached to the stem when the leaf is completely developed (CIAT, 1984). The petiole length of a fully opened leaf normally varies from 5 to 30 cm, but may reach up to 40 cm.

The upper leaf surface is covered with a shiny, waxy epidermis. Most stomata are located on the lower (abaxial) surface of the leaves; only a few can be found along the main vein on the upper (adaxial) surface (Cerqueira, 1989). Of 1500 cultivars studied, only 2% had stomata on the adaxial surface (El-Sharkawy and Cock, 1987a). The stomata on the upper surface are also functional and bigger than those on the undersurface. Both are morphologically paracytic, with two small guard cells surrounded by two subsidiary cells (Cerqueira, 1989). The number of stomata per leaf area range from 278 to 700 mm^{-2}, and all stomatal pores can occupy from 1.4 to 3.1% of the total leaf area (Table 5.3).

Flowers

Cassava is a monoecious species producing both male (pistillate) and female (staminate) flowers on the same plant. The inflorescence is generally formed at the insertion point of the reproductive branchings; occasionally inflorescences can be found in the leaf axils on the upper part of the plant. The female flowers, located on the lower part of the inflorescence, are fewer in number than male flowers, which are numerous on the upper part of the inflorescence. On the same inflorescence, the female flowers open 1–2 weeks before the male flowers (protogyny).

Table 5.2.   Some morphological and agronomic characteristics of stems and their variability in cassava.

Stem characteristic	Variability	Reference
Morphological		
Cortex (collenchyma) colour	Yellow; light green; dark green	Fukuda and Guevara (1998)
External colour	Orange; green–yellow; gold; dark brown; silver; grey; dark brown	Fukuda and Guevara (1998)
Phyllotaxis length	Short (< 8 cm); medium (8–15 cm); large (> 15 cm)	Fukuda and Guevara (1998)
Epidermis colour	Cream; light brown; dark brown; orange	Fukuda and Guevara (1998)
Growth habit	Straight; zigzag	Fukuda and Guevara (1998)
Apical stem colour	Green; green-purple; purple	Fukuda and Guevara (1998)
Branching habit	Erect; dichotomous; trichotomous; tetrachotomous	Fukuda and Guevara (1998)
Agronomic		
Diameter of mature stem	2–8 cm	CIAT (1984); Ramanujam and Indira (1983)
Plant height	1.20–3.70 m	Ramanujam and Indira (1983); Ramanujam (1985); Veltkamp (1985a); Pinho *et al.* (1995)
No. of nodes from planting– 1st branch level	22–96	Veltkamp (1985a)
No. of days from planting– 1st branch level	49–134	Veltkamp (1985a)
No. of apices/plant	2.8–27.5	Pinho *et al.* (1995)

Male and female flowers on different branches of the same plant can open at the same time. Normally, cassava is cross-pollinated by insects; thus it is a highly heterozygous plant.

The flowers do not have a calyx or corolla, but an indefinite structure called perianth or perigonium, made up of five yellow, reddish or purple tepals. The male flower is half the size of the female flower. The pedicel of the male flower is thin, straight and very short, while that of the female flower is thick, curved and long. Inside the male flower, there is a basal disk divided into ten lobes. Ten stamens originate from between them. They are arranged in two circles and support the anthers. The five external stamens are separated and longer than the inner ones, which join together on the top to form a set of anthers. The pollen is generally yellow or orange, varying from 122 to 148 μm in size, which is very large compared to other flowering plants (Ghosh *et al.*, 1988). The female flower also has a ten-lobed basal disk, which is less lobulated than the male flower. The ovary is tricarpellary with six ridges and is mounted on the basal disk. The three locules contain one ovule each. A very small style is located on top of the ovary, and a stigma with three undulated, fleshy lobes originates from the style.

Fruit and seeds

The fruit is a trilocular capsule, ovoid or globular, 1–1.5 cm in diameter and with six straight, prominent longitudinal ridges or aristae. Each locule contains a single carunculate seed. The fruit has a bicidal dehiscence, which is a combination of septicidal and loculicidal dehiscences, with openings along the parallel plane of the dissepiments and along the midveins of the carpels, respectively. With this combination of dehiscences, the fruits open into six valves causing an explosive dehiscence, ejecting the seeds some distance (Rogers, 1965). Fruit maturation generally occurs 75–90 days after pollination (Ghosh *et al.*, 1988). The seed is ovoid–ellipsoidal, approximately 100 mm long, 6 mm wide and 4 mm thick. The weight varies from 95 to 136 mg per seed (Ghosh *et al.*, 1988). The smooth seed coat is dark brown, mottled with grey. The seeds usually germinate soon after collection, taking about 16 days for germination.

Table 5.3. Some morphological and agronomic characteristics of leaves and their variability in cassava.

Leaf characteristic	Variability	Reference
Morphological		
Apical leaf colour	Light green; dark green; green–purple; purple	Fukuda and Guevara (1998)
Apical pubescence	Absent; present	Fukuda and Guevara (1998)
Shape of central lobe	Ovoid; elliptic–lanceolate; obovate–lanceolate; oblanceolate; lanceolate; linear; pandurate; linear–pyramidal; linear–pandurate; linear–hostatilobada	Fukuda and Guevara (1998)
Petiole colour	Green–yellow; green; green–red; red-green; red; purple	Fukuda and Guevara (1998)
Mature leaf colour	Light green; dark green; green-purple; purple	Fukuda and Guevara (1998)
Protuberance of leaf scars	No protuberance; protuberant	Fukuda and Guevara (1998)
No. of lobes	3; 5; 7; 9; 11	Fukuda and Guevara (1998)
Agronomic		
Petiole length	5–30 cm	Ghosh et al. (1988)
	9–20 cm	CIAT (1984)
Total chlorophyll	2.18–2.86 mg g^{-1} leaf FW	Ramanujam and Jos (1984)
Central lobe length	4–20 cm	CIAT (1984)
Central lobe width	1–6 cm	CIAT (1984)
No. of stomata/leaf area in adaxial epidermis	278–700 mm^{-2}	Ghosh et al. (1988); Cerqueira (1989); Splittstoesser and Tunya (1992); Connor and Palta (1981)
Relative area of stomata pore (% from leaf area)	1.4–3.1%	Cerqueira (1989); Pereira and Splittstoesser (1990)
Stipule length	0.5–1.0 cm	CIAT (1984)
Leaf thickness	100–120 μm	Pereira and Splittstoesser (1990)
DM in mature leaf	25%	Barrios and Bressani (1967)
Fibre in mature leaf	4.58% DW	Barrios and Bressani (1967)
Ash in mature leaf	8.28% DW	Barrios and Bressani (1967)
Protein in mature leaf	7.1–8.9% FW	Barrios and Bressani (1967)
	28.8% DW	Barrios and Bressani (1967)
Soluble carbohydrates	11.36% FW	Barrios and Bressani (1967)
	44.84% DW	Barrios and Bressani (1967)

DM, dry matter; FW, fresh weight; DW, dry weight.

Growth and Development

Plant developmental stages

As cassava is a perennial shrub it can grow indefinitely, alternating periods of vegetative growth, storage of carbohydrates in the roots, and even periods of almost dormancy, brought on by severe climatic conditions such as low temperature and prolonged water deficit. There is a positive correlation between the total biomass and storage root biomass (Fig. 5.1; Ramanujam, 1990). During its growth, there are distinct developmental phases. The occurrence, duration and existence of each phase depend on several factors related to varietal differences, environmental conditions and cultural practices. The initial growth (at 15-day intervals) from emergence to 150 days is presented in Fig. 5.2. Growth at 60-day intervals during the first cycle (0–360 days after planting; DAP) is shown in Fig. 5.3. The results in these two figures are consistent with other authors (Howeler and Cadavid, 1983; Ramanujam and Biradar, 1987; Távora et al., 1995; Peressin et al., 1998). The periods and main physiological events during the growth of a cassava plant under favourable conditions in the field can be visualized in these figures and are summarized below:

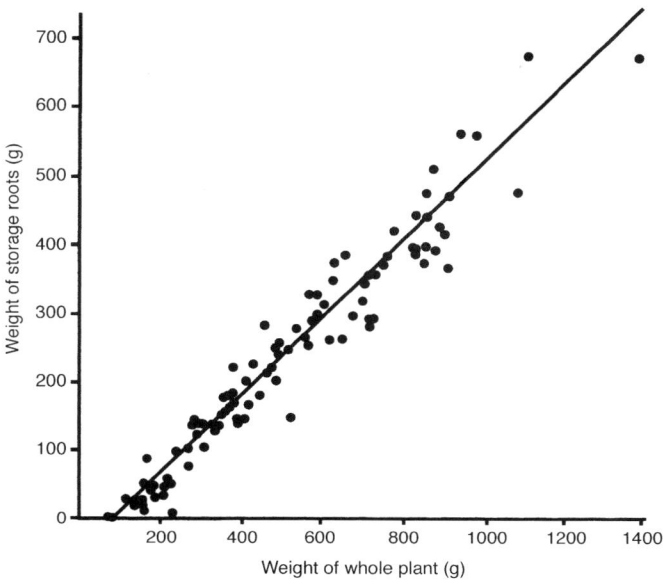

Fig. 5.1. Relationship between dry weight (DW) of whole plant (x) and DW of storage roots (y) for individual plants of a field trial at the University of the West Indies: $y = 0.56x–34$; $r^2 = 0.96$; $n = 112$. (Source: Boerboom, 1978.)

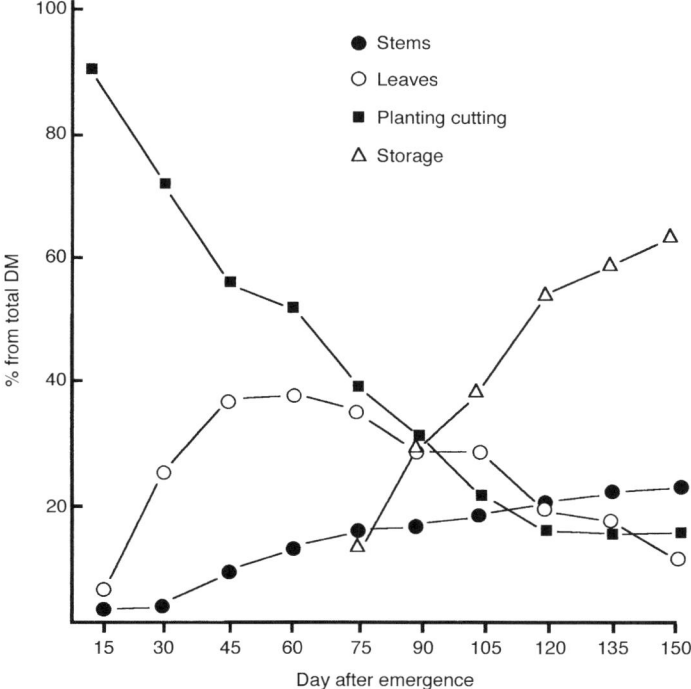

Fig. 5.2. Partitioning of dry matter during the initial development of cassava cv. Cigana, Cruz das Almas, Bahia, Brazil. (Source: Porto, 1986.)

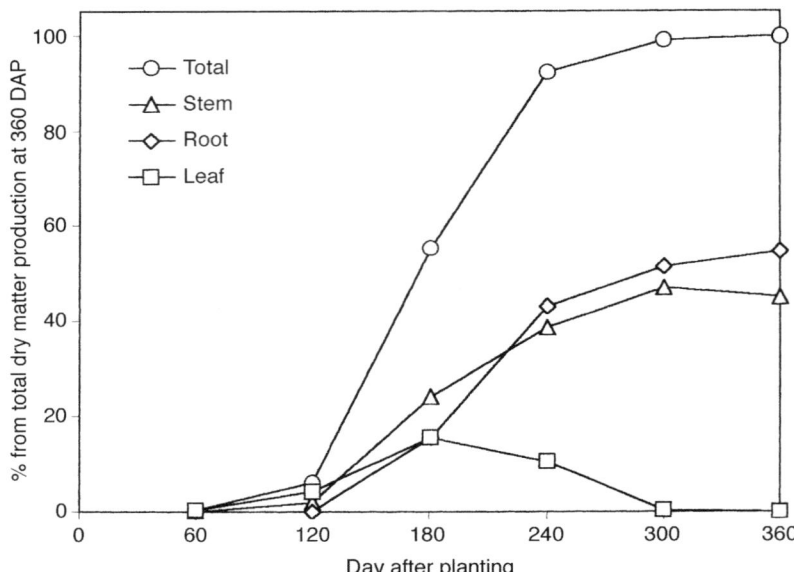

Fig. 5.3. Growth of cassava plant during the first cycle (12 months). Average of two varieties. DAP, days after planting. Graph made from data of Lorenzi (1978).

Emergence of sprouting – 5–15 DAP

- From 5–7 DAP the first adventitious roots arise from the basal cut surface of the stake and occasionally from the buds under the soil.
- 10–12 DAP the first sprouting occurs, followed by small leaves which start to emerge (Conceição, 1979).
- Emergence is achieved at 15 DAP.

Beginning of leaf development and formation of root system – 15–90 DAP

- The true leaves start to expand around 30 DAP (Fig. 5.2) when the photosynthetic process starts to contribute positively to plant growth.
- Until 30 DAP, shoot and root growth depends on the reserves of the stem cutting.
- The fibrous roots start to grow, replacing the first adventitious roots. These new roots start to penetrate in the soil, reaching 40–50 cm deep, and function in water and nutrient absorption (Conceição, 1979).
- Few fibrous roots (between three and 14) will become storage roots, which can be distinguished from fibrous roots from 60 to 90 DAP (Cock *et al.*, 1979). At 75 DAP the

storage roots represent 10–15% of total dry matter (DM; Fig. 5.2).

Development of stems and leaves (canopy establishment) – 90–180 DAP

- Maximum growth rates of leaves and stems are achieved in this period, and the branching habit and plant architecture is defined (Fig. 5.3).
- From 120 to 150 DAP the leaves are able to intercept the most of the incident light on canopy (Veltkamp, 1985c).
- Maximum canopy size and maximum DM partition to leaves and stems are accomplished (Howeler and Cadavid, 1983; Ramanujam, 1985; Távora *et al.*, 1995).
- The storage root continues to bulk.
- The most active vegetative growth for cassava occurs in this period (Ramanujam, 1985).

High carbohydrate translocation to roots – 180–300 DAP

- Photoassimilate partition from leaves to roots is accelerated, making the bulking of storage roots faster (Fig. 5.3).

- The highest rates of DM accumulation in storage roots occur within this period (Fig. 5.3; Boerboom, 1978; Távora *et al.*, 1995; Peressin *et al.*, 1998).
- Leaf senescence increases, hastening rate of leaf fall (Fig. 5.3).
- Stem becomes lignified (Conceição, 1979).

Dormancy – 300–360 DAP

- Rate of leaf production is decreased.
- Almost all leaves fall and shoot vegetative growth is finished.
- Only translocation of starch to root is kept, and maximum DM partition to the roots is attained.
- This phase occurs mainly in regions with significant variation in temperature and rainfall.
- The plant completes its 12-month cycle, which can be followed by a new period of vegetative growth, DM accumulation in the roots and dormancy again.

Leaf area development

The analyses of crop growth and yield are usually evaluated on the basis of two parameters: leaf area index (LAI), i.e. leaf area per unit ground area, and net assimilatory rate (NAR), i.e. the rate of DM production per unit leaf area.

In cassava a positive correlation between the leaf area or leaf area duration and yield of storage roots has been reported, indicating that leaf area is crucial in determining crop growth rate and the storage bulking rate of cassava (Sinha and Nair, 1971; Cock, 1976; Cock *et al.*, 1979).

For cassava the leaf area per plant depends on the number of active apices (branching pattern), the number of leaves formed/apex, leaf size and leaf life. Given that there are significant varietal variations and influence of environmental conditions (Veltkamp, 1985a), it is important to characterize the development of cassava leaf area and its components. Table 5.4 lists values of some parameters related to leaf growth that have been found in cassava. These parameters are discussed below.

After leaf emergence (folded, 1 cm long) and under normal conditions, the cassava leaf reaches its full size on days 10–12. Leaf life (from emergence to abscission) depends on cultivar, shade level, water deficit and temperature (Cock *et al.*, 1979; Irikura *et al.*, 1979). It ranges from 40 to 210 days (Table 5.4), but is commonly 60–120 days (Cock, 1984).

There are marked differences in leaf size among the different cultivars, and the size varies with the age of the plant. The leaves produced from 3 to 4 MAP are those that become the largest; maximum total leaf area is reached from 4 to 5 MAP (Cock *et al.*, 1979; Irikura *et al.*, 1979). Leaf size is influenced by changing the branching

Table 5.4. Parameters related to leaf growth during the first cycle (12 months) and some values found in cassava.

Leaf growth parameter	Value	Reference
Individual leaf area	50–600 cm^2	Ramanujam (1982); Splittstoesser and Tunya (1992); Veltkamp (1985a)
Expansion period (from emergence to full size)	12 days	Conceição (1979)
Leaf longevity (from fully expanded to abscission)	36–100 days	Ramanujam and Indira (1983); Conceição (1979)
Leaf life (from emergence to abscission)	40–210 days	Splittstoesser and Tunya (1992); Irikura *et al.* (1979); Ramanujam (1985); Veltkamp (1985a)
No. leaves retained/plant	44–146	Ramanujam and Indira (1983)
Leaf production rate/shoot	4–22 week^{-1}	Ramanujam and Indira (1983)
Leaf shedding rate/plant	10–24 leaves week^{-1}	Ramanujam and Indira (1983)
Total leaf area	1.24–3.38 m^2	Ramanujam and Indira (1983)
Cumulative no. of leaves/apex	117–162	Veltkamp (1985a); Cock *et al.* (1979)

pattern. Larger leaves are produced when the number of active apices is reduced (Tan and Cock, 1979). The rate of leaf formation decreases with plant age and is lower at low temperatures (Irikura *et al.*, 1979).

Differences in mean LAI are closely related to the rate of root bulking. The optimal LAI for storage root bulking rate is 3–3.5 (Cock *et al.*, 1979) and exists over a wide range of temperature (Irikura *et al.*, 1979). Initial leaf area development is slow, taking 60–80 DAP before an LAI of 1.0 is reached. From 120 to 150 DAP the light interception by the canopy is around 90% with an LAI of 3 (Veltkamp, 1985a). In order to obtain high storage root yields, the crop should reach an LAI of 3–3.5 as quickly as possible and maintain that LAI for as long as possible (Cock *et al.*, 1979; Veltkamp, 1985a). Substantial leaf abscission began at LAI values of 5.0–6.0 (Keating *et al.*, 1982a).

Temporal development of cell division and cell expansion

Leaf area development is largely dependent on cell division and cell expansion processes, which determine the number of cells per mature leaf and cell size, respectively. Thus the final leaf area is directly affected by the rate and duration of cell division and expansion (Takami *et al.*, 1981; Lecoeur *et al.*, 1995). A temporal development of cell division and expansion for adaxial epidermal cells in cassava has been proposed by Alves (1998; Fig. 5.4), in which the transition from leaf cell division to cell expansion processes is discrete and occurs when leaf area reached 5% of its final size, corresponding to the first folded leaf toward the top. Thus when the leaf starts to unfold, almost all cell division stops and rapid cell expansion starts.

Dry matter partitioning and source–sink relationship

During cassava growth the carbohydrates from photosynthesis have to be distributed to assure good development of the source (active leaves) and provide DM to the sink (storage roots, stem and growing leaves). Cassava DM is translocated mainly to stems and storage roots, and DM

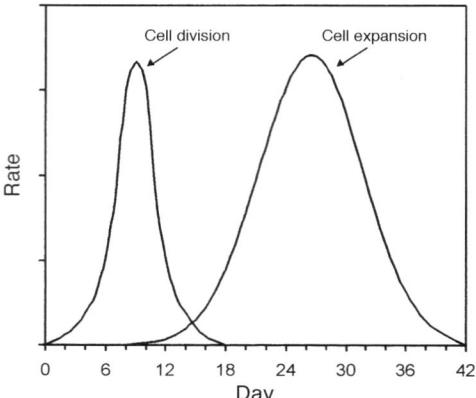

Fig. 5.4. Temporal development of cell division and cell expansion processes during cassava leaf development in adaxial epidermal cells. (Source: Alves, 1998.)

accumulation in the leaves decreases during the crop cycle. Until 60–75 DAP, cassava accumulates DM more in leaves than in stems and storage roots, not including the stem cutting (Fig. 5.2). Then the storage roots increase rapidly, reaching 50–60% of the total DM around 120 DAP (Fig. 5.2; Howeler and Cadavid, 1983; Távora *et al.*, 1995). After the fourth month, more DM is accumulated in the storage roots than the rest of the plant. At harvest (12 months) DM is present mainly in roots, followed by stems and leaves (Fig. 5.3; Howeler and Cadavid, 1983). Thus, during the growth cycle, the DM distribution to the different parts is constant with a high positive linear correlation of the total DM with shoot and root DM (Fig. 5.1; Veltkamp, 1985d).

The period of maximum rates of DM accumulation depends on genotypes and growing conditions. Lorenzi (1978) at high latitude and Oelsligle (1975) at high altitude reported maximum rates of DM accumulation at 4–6 and 7 months, respectively. Under more tropical conditions, where growth is faster, Howeler and Cadavid (1983) found an earlier period of maximum rates, at 3–5 MAP. The distribution of DM to the economically useful plant parts is measured by harvest index (HI). In cassava HI represents the efficiency of storage root production and is usually determined by the ratio of storage root weight to the total plant weight. Significant differences in HI have been reported among cultivars, indicating that it can be used as a selection criterion for higher yield potential in cassava. HI values of 0.49–0.77 have been reported after 10–12 MAP (Lorenzi, 1978; Cavalcanti, 1985; Pinho *et al.*, 1995; Távora *et al.*, 1995; Peressin *et al.*, 1998). Although DM distribution is constant, its accumulation depends upon photoassimilate availability (source activity) and sink capacity of the storage parts. The number of storage roots and their mean weight are yield components that determine sink capacity. The significant positive correlation of photosynthetic rate with root yield and total biomass, as well as the correlations between LAI, interception of radiation and biomass production (Williams, 1972; Mahon *et al.*, 1976; El-Sharkawy and Cock, 1990; Ramanujam, 1990), indicates that demand for photoassimilates by roots increases the photosynthetic activity.

The balance between 'source' and 'sink' activity is essential for the plant to reach its maximum productivity. Studies have shown that up to 25% reduction in the number of storage roots did not affect total or root DM and the IAF (Cock *et al.*, 1979). On the other hand, Ramanujam and Biradar (1987) observed that reduction of 50–75% in storage roots did affect root growth rate without changing shoot growth rate, indicating that shoot growth is independent of storage root growth. Influence of source size on DM production shows that the NAR and storage root growth rate is reduced when the source size is increased from LAI 3.0 to 6.0 (Ghosh *et al.*, 1988).

Flowering

Little is known about flowering in cassava, and some clones have never been known to flower. Flowering can start 6 weeks after planting although the precise flowering time depends on cultivar and environment. It appears that cassava flowers best at moderate temperatures (approximately 24°C). It has been suggested that forking is related to the onset of flowering, which is promoted by long days in some cultivars. Usually, the apical meristem becomes reproductive when branching occurs, but the abortion of flowers is very common.

Keating *et al.* (1982a) evaluated cassava at 12 different planting dates at a high latitude (27° 37′ S), where photoperiods range from 14.8 to 11.2 h. Concentration of first flowering and forking occurred in photoperiods > 13.5 h. This result is consistent with Bruijn (1977) and Cunha and Conceição (1975), who suggested flowering in cassava may be promoted by increasing day length.

Photosynthesis

Cassava photosynthesis follows a C_3 pathway (Veltkamp, 1985e; Edwards *et al.*, 1990; Angelov *et al.*, 1993; Ueno and Agarie, 1997) with maximum photosynthetic rates varying from 13 to 24 μmol CO_2 m^{-2} s^{-1} under greenhouse or growth chamber conditions (Mahon *et al.*, 1977b; Edwards *et al.*, 1990) and from 20 to 35 μmol CO_2 m^{-2} s^{-1} in the field (El-Sharkawy and Cock, 1990). It exhibits a high CO_2 compensation point, from 49 to 68 μl l^{-1}, typical of C_3 plants (Mahon *et al.*, 1977a; Edwards *et al.*, 1990; Angelov *et al.*, 1993). In field-grown cassava, photosynthesis has high optimum temperature (35°C) and wide plateau (25–35°C; El-Sharkawy and Cock, 1990) and is not light saturated up to 1800 μmol PAR m^{-2} s^{-1} (El-Sharkawy *et al.*, 1992b) or 2000 μmol PAR m^{-2} s^{-1} (Angelov *et al.*, 1993). Thus cassava is adapted to a tropical environment, requiring high temperature and high solar radiation for optimal leaf development and for expression of its photosynthetic potential. Both storage root yield and total biomass show positive correlation with photosynthesis rate (El-Sharkawy and Cock, 1990; Ramanujam, 1990).

Morphologically, cassava leaves combine some novel characteristics related to high productivity and drought tolerance and, consequently, to photosynthesis. The lower mesophyll surface is populated with papillose-type epidermal cells, while the upper surface is fairly smooth, with scattered stomata and trichomes. The papillae appear to add about 15% to leaf thickness and to lengthen the diffusion path from the stomatal opening to the bulk air perhaps two- to threefold (Angelov *et al.*, 1993). Cassava leaves have distinct green bundle-sheath cells, with small, thin-walled cells, spatially separated below the palisade cells (different from Kranz-type leaf anatomy). In addition to performing C_3 photosynthesis, these cells may function in transport of photosynthates in the leaf. In Kranz anatomy, typical of C_4 plants, the bundle sheaths are surrounded by and all in direct contact with many mesophyll cells (Edwards et al., 1990).

Cyanide Content

All cassava organs, except seeds, contain cyanogenic glucoside (CG). Cultivars with < 100 mg kg^{-1} fresh weight (FW) are called 'sweet' while cultivars with 100–500 mg kg^{-1} are 'bitter' cassava (Wheatley et al., 1993). The most abundant CG is linamarin (85%), with lesser amounts of lotaustralin. Total CG concentration depends on cultivar, environmental condition, cultural practices and plant age (McMahon et al., 1995). The variation found in some parts of cassava is shown in Table 5.5.

Linamarin, which is synthesized in the leaf and transported to the roots, is broken down by the enzyme linamarase, also found in cassava tissues (Wheatley and Chuzel, 1993). When linamarin is hydrolysed, it releases HCN, a volatile poison (LD_{50-60} mg for humans; Cooke and Coursey, 1981); but some cyanide can be detoxified by the human body (Oke, 1983). In intact roots the compartmentalization of linamarase in the cell wall and linamarin in cell vacuoles prevents the formation of free cyanide. Upon processing, the disruption of tissues ensures that the enzyme comes into contact with its substrate, resulting in rapid production of free cyanide via an unstable cyanhydrin intermediary (Wheatley and Chuzel, 1993). Juice extraction, heating,

fermentation, drying or a combination of these processing treatments aid in reducing the HCN concentration to safe levels (O'Hair, 1990).

Physical Deterioration of Storage Roots

Cassava roots have the shortest postharvest life of any of the major root crops (Ghosh et al., 1988). Roots are highly perishable and usually become inedible within 24–72 h after harvest due to a rapid physiological deterioration process, in which synthesis of simple phenolic compounds that polymerize occurs, forming blue, brown and black pigments (condensed tannins; Wheatley and Chuzel, 1993). It is suggested that polyphenolic compounds in the roots oxidize to quinone-type substances, which is complexed with small molecules like amino acids to form coloured pigments that are deposited in the vascular bundles (Ghosh et al., 1988). The accumulation of the coumarin, scopoletin, is especially rapid, reaching 80 mg kg^{-1} dry weight (DW) in 24 h (Wheatley and Chuzel, 1993). Tissue dehydration, especially at sites of mechanical damage to the roots encourages the rapid onset of deterioration. The phenolic compounds may be released on injury. The changes during storage of roots depend upon condition and duration of storage, physiological state of the stored material and varietal characteristics (Ghosh et al., 1988).

Polyphenol oxidase (PPO) is an enzyme that oxidizes phenols to quinone. Any process that inhibits PPO, such as heat treatment, cold storage, anaerobic atmosphere and dipping

Table 5.5. Cyanide concentration in different parts of the cassava plant.

Part of plant	Total cyanide concentration	Source
Root pulp (parenchyma)	3–121 mg 100 g^{-1} DW	Barrios and Bressani (1967)
	3–135 mg 100 g^{-1} DW	Wheatley and Chuzel (1993)
	1–40 mg 100 g^{-1} FW	Barrios and Bressani (1967)
Root peel	6–55 mg 100 g^{-1} DW	Wheatley and Chuzel (1993)
	5–77 mg 100 g^{-1} DW	Barrios and Bressani (1967)
	17–267 mg 100 g^{-1} FW	Barrios and Bressani (1967)
Leaf	1–94 mg 100 g^{-1} DW	Barrios and Bressani (1967)
	0.3–29 mg 100 g^{-1} FW	Barrios and Bressani (1967)

DW, dry weight; FW, fresh weight.

roots in solutions of inhibitors (e.g. ascorbic acid, glutathione and KCN) prevents vascular streaking (Ghosh *et al.*, 1988).

Secondary deterioration can follow physiological or primary deterioration 5–7 days after harvest. This is due to microbial infection of mechanically damaged tissues and results in the same tissue discoloration with vascular streaks spreading from the infected tissues (Wheatley and Chuzel, 1993).

Environmental Effects on Cassava Physiology

Cassava is found over a wide range of edaphic and climatic conditions between 30°N and 30°S latitude, growing in regions from sea level to 2300 m altitude, mostly in areas considered marginal for other crops: low-fertility soils, annual rainfall from < 600 mm in the semiarid tropics to > 1500 mm in the subhumid and humid tropics. Given the wide ecological diversity, cassava is subjected to highly varying temperatures, photoperiods, solar radiation and rainfall.

Temperature

Temperature affects sprouting, leaf size, leaf formation, storage root formation and, consequently, general plant growth. The behaviour of cassava under the temperature variations that usually occur where cassava is normally cultivated indicates that its growth is favourable under annual mean temperatures ranging from 25 to 29°C (Conceição, 1979), but it can tolerate from 16 to 38°C (Cock, 1984). Table 5.6 summarizes the ranges of temperature and their principal physiological effects on cassava development.

At low temperatures (16°C) sprouting of the stem cutting is delayed, and rate of leaf production, total and storage root DW are decreased (Cock and Rosas, 1975). Sprouting is hastened when the temperature increases up to 30°C but is inhibited with temperatures > 37°C (Keating and Evenson, 1979). As temperature decreases, leaf area development becomes slower because the maximum size of individual leaves is smaller, and fewer leaves are produced at each apex although leaf life is increased (Irikura *et al.*, 1979). At a temperature of 15–24°C, the leaves remain on the plant for up to 200 days (Irikura *et al.*, 1979), while at higher temperatures leaf life is 120 days (Splittstoesser and Tunya, 1992).

There is a genotype-by-temperature interaction for yield ability. Irikura *et al.* (1979) evaluated four cultivars under different temperatures and found that higher yields were obtained at different temperatures according to the cultivar, indicating that the effect of natural selection is highly significant on varietal adaptation (Table 5.7).

Table 5.6. Effect of temperature on cassava development.

Air temperature (°C)	Physiological effects
< 17 or > 37	Sprouting impaired
28.5–30	Sprouting faster (optimum)
< 15	Plant growth inhibited
16–38	Cassava plant can grow
25–29	Optimum for plant growth
< 17	Reduction of leaf production rate, total and root DW
20–24	Leaf size and leaf production rate increased; leaf life shortened
28	Faster shedding of leaves; reduction in no. branches
25–30	Highest rates of photosynthesis in greenhouse
30–40	Highest rates of photosynthesis in the field
16–30	Transpiration rate increases linearly and then declines

Sources: Wholey and Cock (1974); Cock and Rosas (1975); Mahon *et al.* (1977b); Conceição (1979); Irikura *et al.* (1979); Keating and Evenson (1979); El-Sharkawy *et al.* (1992b).
DW, dry weight.

Table 5.7.   Fresh root yield (t ha^{-1}) of four contrasting cassava types at 12 months after planting (MAP) under three different temperature regimes.

Variety	Temperature (°C)		
	20	24	28
M Col 22	9.3	27.7	39.4
M Mex 59	22.8	38.8	30.4
M Col 113	24.2	26.1	23.9
Popayán	28.9	15.7	9.4

Source: Irikura *et al.* (1979).

The main effect of temperature is on biological production, as DM partitioning does not change much when cassava is cultivated under different temperatures (Cock and Rosas, 1975). Higher temperatures are associated with a greater crop growth rate (CGR) and high photosynthetic rate. El-Sharkawy *et al.* (1992b) evaluated the potential photosynthesis of three cultivars from contrasting habits under different growing environments and verified that photosynthetic rate increased with increasing temperature, reaching its maximum at 30–40°C. In all cultivars photosynthesis was substantially lower in leaves that had developed in the cool climate than in those from the warm climate. The high sensitivity of photosynthesis to temperature points to the need for genotypes more tolerant to low temperature, which could be used in the highland tropics and subtropics.

Photoperiod

Day length affects several physiological processes in plants. The differences in day length in the tropical region are very small, varying from 10 to 12 h throughout the year. Thus photoperiod may not limit cassava root production in this region. On the other hand, the restrictions regarding cassava distribution outside the tropical zone can be due to effects of day length variation on its physiology. Although studies about day length effect in cassava are scarce, tuberization, photoassimilates partitioning and flowering are reportedly affected.

Experiments in which the day length was artificially changed have shown that the optimal light period for cassava is around 12 h, with probable varietal differences in the critical day length (Bolhuis, 1966). Long days promote growth of shoots and decrease storage root development, while short days increase storage root growth and reduce the shoots, without influencing total DW (Fig. 5.5). The increase in shoot DW under long days is a result of significant increases in plant height, leaf area per plant, number of apices per plant, and number of living leaves per apex (Veltkamp, 1985b). This suggests an antagonistic relationship between shoot growth and development of the storage roots in response to variation in day length.

There are varietal differences in sensitivity to long days (Carvalho and Ezeta, 1983). Under field conditions, Veltkamp (1985b) submitted three genotypes to two day lengths (12 and 16 h) during the whole growth period. He observed that the percentage decrease in storage root yield under the long day was greatest in M Col 1684 (47%) and least in M Col 22 (13%; Table 5.8). Yield differences resulted mainly from decreased efficiency of storage root production or HI under 16-h days because total DM was greater under 16-h days. No day-length effect was found for the crop weight at which the starch accumulation in the roots apparently started (AISS values; Table 5.8). Thus the storage root yield reduction seems to be more related to a change in the distribution patterns of DM rather than to a delay in storage root initiation. Considering that photoperiod primarily affects shoots and a secondary response occurs in the roots (Keating *et al.*, 1982b) and that shoot growth has preference over root growth (Cock *et al.*, 1979; Tan and Cock, 1979), long photoperiods may increase the growth requirements of the shoots, thereby reducing the excess carbohydrates available for root growth (Veltkamp, 1985b).

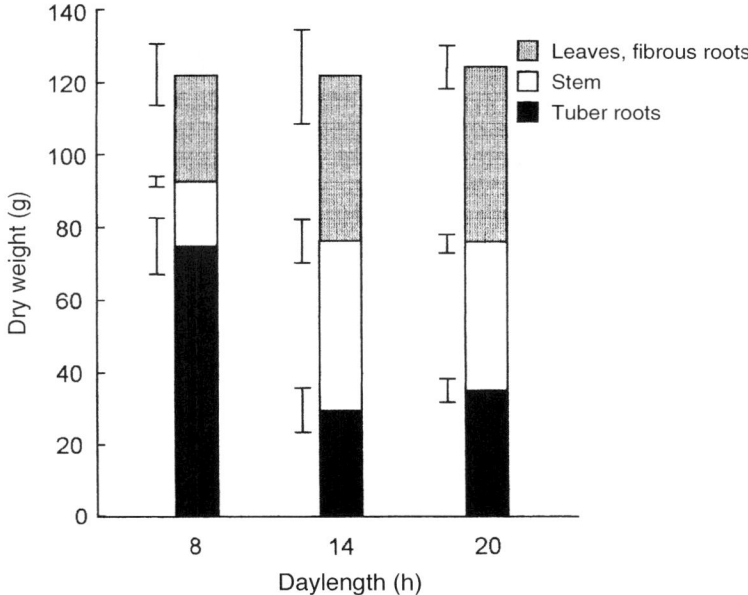

Fig. 5.5. Effect of day length on cassava dry matter distribution 16 weeks after planting. Vertical lines indicate twice the standard error of total dry weight (DW), stem DW and storage root DW.

Table 5.8. Cassava dry matter (DM) production and distribution 272 days after planting (DAP) under 16 h and natural day length (approximately 12 h).

Genotype	Day length	Total DM (t ha^{-1})[a]	DW storage root (t ha^{-1})	Harvest Index (HI)	AISS[b] (t ha^{-1})
M Col 1684	Natural[c]	16.7	9.1	0.54	0.55
	16 h[d]	17.3	4.6	0.27	0.25
M Ptr 26	Natural	14.5	8.1	0.61	0.60
	16 h	15.9	4.9	0.42	0.36
M Col 22	Natural	15.5	9.5	0.56	0.65
	16 h	19.5	8.3	0.31	0.45

Source: Veltkamp (1985b).
[a]Includes weight of fallen leaves.
[b]AISS, apparent initial start of starch accumulation; corresponds to crop weight at which starch accumulation apparently starts in roots.
[c]Natural = approximately 12 h.
[d]16 h = 12 h of natural day length + 4 h of artificial light.
DW, dry weight.

Solar radiation

The commonest cassava production system is intercropping with other staple crops. In Latin America and Africa, cassava is usually associated with an earlier maturing grain crop such as maize, rice or grain legumes (beans, cowpeas or groundnuts; Mutsaers *et al.*, 1993). Cassava is also intercropped with perennial vegetation (Ramanujam *et al.*, 1984). Cassava is usually planted after the intercropped species. Even when it is planted at the same time, the associated crop such as maize is established faster than cassava. Thus in an associated cropping system

cassava is always subjected to different degrees
of shading and low light intensity in the early
stages of development. Considering that cassava
is a crop that requires high solar radiation
to perform photosynthesis more efficiently (El-
Sharkawy *et al.*, 1992b), it is very important to
know the effect of shade on cassava development and production. Ramanujam *et al.* (1984)
evaluated 12 cassava cultivars under the shade
in a coconut garden (85–90% shading). Under
shading, the root bulking process started about
3 weeks after that in plants grown without
shading, and the number of storage roots per
plant and NAR was reduced under shading.
Okoli and Wilson (1986) submitted cassava to
six shade regimes and observed that all levels of
shade delayed storage root bulking and at 20,
40, 50, 60 and 70% shade reduced cassava yield
by 43, 56, 59, 69 and 80%, respectively.

In relation to shoots, under field conditions,
shading increases plant height and the leaves
tend to become adapted to low light conditions by
increasing leaf area per unit weight (Fukai *et al.*,
1984; Okoli and Wilson, 1986; Ramanujam
et al., 1984) and shortening leaf life only under
severe shading. Under ideal growing conditions,
cassava leaves have a life of up to 125 days
(Splittstoesser and Tunya, 1992). Levels of shade
up to about 75% have very little effect on leaf life,

but under 95–100% shade, leaves abscise within
10 days (Cock *et al.*, 1979).

Aresta and Fukai (1984) observed that only
22% shade decreased both fibrous root elongation rate (53%) and storage root growth rate
(36%) without altering shoot growth rate, which
was significantly decreased (32%) only under
68% shade. Thus under limited photosynthesis
caused by low solar radiation, most of the photosynthates are utilized for shoot growth, affecting
storage root development significantly, showing
that the shoots are a stronger sink than roots.

Water deficit

Cassava is commonly grown in areas receiving
< 800 mm rainfall year^{-1} with a dry season of
4–6 months, where tolerance to water deficit is
an important attribute. Although it is a drought-
tolerant crop, growth and yield are decreased by
prolonged dry periods. The reduction in storage
root yield depends on the duration of the water
deficit and is determined by the sensitivity of
a particular growth stage to water stress. The
critical period for water-deficit effect in cassava
is from 1 to 5 MAP – the stages of root initiation
and tuberization. Water deficit during at least 2
months of this period can reduce storage root

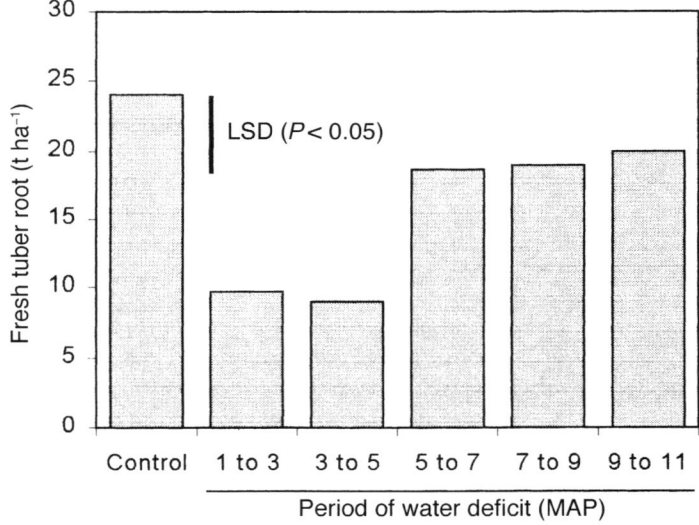

Fig. 5.6.   Effect of water deficit during different growth periods on cassava yield.

yield from 32 to 60% (Connor *et al.*, 1981; Porto *et al.*, 1988). Figure 5.6 shows the root yield reduction caused by water deficit imposed for 2 months during successive 2-month periods from 1 to 11 MAP. Clearly, they found that the severer effect corresponded to stress occurring from 1 to 5 MAP (i.e. period of rapid leaf growth and tuberization) compared with the later period of storage root bulking.

Drought tolerance

Plants respond to water deficit at many different levels: morphological, physiological, cellular and metabolic. The responses are dependent upon the duration and severity of stress, the genotype of the stressed plant, the stage of development, and the organ and cell type in question (Bray, 1994). Multiple responses allow the plant to tolerate water stress. Some of these responses and the current status of knowledge with regard to cassava are discussed here.

Control of stomatal closure and leaf growth

A primary response to water stress is stomatal closure, which decreases photosynthetic CO_2 assimilation and, in turn, growth. Stomata have a high capacity to respond to changes in the water status of the plant and atmosphere. They close as the leaf water potential decreases and when the vapour pressure deficit between the leaf and the air increases (generally due to a decrease in relative humidity). As stomata are the route by which CO_2 enters the leaf, stress-induced decreases in stomatal aperture can limit the rate of CO_2 diffusion into the leaf and, therefore, the rate of photosynthesis.

When water is available, cassava maintains a high stomatal conductance and can keep internal CO_2 concentration high; but when water becomes limiting, the plant closes stomata in response to even small decreases in soil water potential (El-Sharkawy and Cock, 1984). The rapid closure of the stomata and the resulting decline in transpiration lessens the decrease in leaf water potential and soil water depletion, thereby protecting leaf tissues from desiccation (Ike, 1982; Palta, 1984; El-Sharkawy and Cock,

1984; Cock *et al.*, 1985). This response to early stages of soil water depletion has been described as isohydric, a behaviour shared by cowpeas, maize and several other crops (Tardieu and Simonneau, 1998). Leaf area growth is also decreased in response to water stress but is rapidly reversed following the release from stress (Connor *et al.*, 1981; Palta, 1984; El-Sharkawy and Cock, 1987b; Baker *et al.*, 1989). This response limits the development of plant transpirational surface area during water deficit and keeps sink demand well balanced with plant assimilatory capacity.

Leaf conductance to water vapour has been evaluated as an indicator of the capacity of different cassava genotypes to prevent water loss under prolonged drought. Considerable variation has been observed in leaf conductance (Porto *et al.*, 1988), and this parameter seems to be useful for pre-selecting sources of germplasm conferring adaptation to prolonged dry periods.

Abscisic acid accumulation

Many authors have reported that a substantial number of drought responses in plants can be mimicked by external application of abscisic acid (ABA) to well-watered plants (Davies *et al.*, 1986; Trewavas and Jones, 1991). This treatment promotes characteristic developmental changes that can help the plant cope with water deficit, including decrease of stomatal conductance (MacRobbie, 1991; Trejo *et al.*, 1993), restriction of shoot growth (Creelman *et al.*, 1990) and leaf area expansion (Van Volkenburgh and Davies, 1983; Lecoeur *et al.*, 1995) and stimulation of root extension (Sharp *et al.*, 1993, 1994; Griffiths *et al.*, 1997). All these effects of ABA application, together with the observation that environmental stress stimulates ABA biosynthesis and ABA release from sites of synthesis to action sites, suggest a role for ABA as a stress hormone in plants. Moreover, studies have indicated that for certain plant responses, sensitivity to water deficit is correlated with changes in ABA concentrations (Trejo *et al.*, 1995; Borel *et al.*, 1997) and genotypic responsiveness to ABA (Blum and Sinmena, 1995; Cellier *et al.*, 1998). Information about

ABA accumulation in cassava has not been reported in the literature.

Alves and Setter (2000) published the first report concerning ABA in cassava. They cultivated five cassava genotypes in pots in a greenhouse and evaluated the temporal patterns of ABA accumulation in mature leaves and in immature expanding leaves, during water deficit and after release from stress to determine the extent to which the stress and recovery response of leaf area growth is associated with the temporal pattern of ABA accumulation. At 3 and 6 days after withholding irrigation, all genotypes accumulated large amounts of ABA in both expanding and mature leaves, but these high ABA levels were almost completely reversed in respect of control levels after 1 day of rewatering (Fig. 5.7). Correspondingly, young leaves halted leaf expansion growth and transpiration rate decreased. Young leaves accumulated more ABA than mature leaves both in control and stressed treatments (Fig. 5.7). This rapid return to control ABA levels corresponded with a rapid recovery of leaf area growth rates. This rapid reduction in leaf area growth and stomatal closure may be due to cassava's ability to synthesize rapidly and accumulate ABA at an early phase of a water-deficit episode.

Osmotic adjustment

One means of increasing drought tolerance is by accumulation of osmotically active solutes, so that turgor and turgor-dependent processes may be maintained during episodes of drydown. Osmotic adjustment (OA), defined as the difference in osmotic potential between control and stressed plants, allows cell enlargement and plant growth at high water stress and allows stomata to remain partially open and CO_2 assimilation to continue at low water potentials that are otherwise inhibitory (Pugnaire et al., 1994).

In cassava OA has not been thoroughly examined. Although leaf water potential remains relatively unchanged during water deficit episodes (Connor and Palta, 1981; Cock et al., 1985; El-Sharkawy et al., 1992a), a result suggestive of little or no OA, such observations do not rule it out. Furthermore, these studies involved mature leaves, not young organs, which might especially benefit from accumulation of osmolytes. Alves (1998) evaluated the role of OA in cassava by measuring the osmotic component of leaf water potential in mature, expanding and folded leaves. The largest increase in solutes caused by water deficit occurred in the youngest tissue (folded leaves)

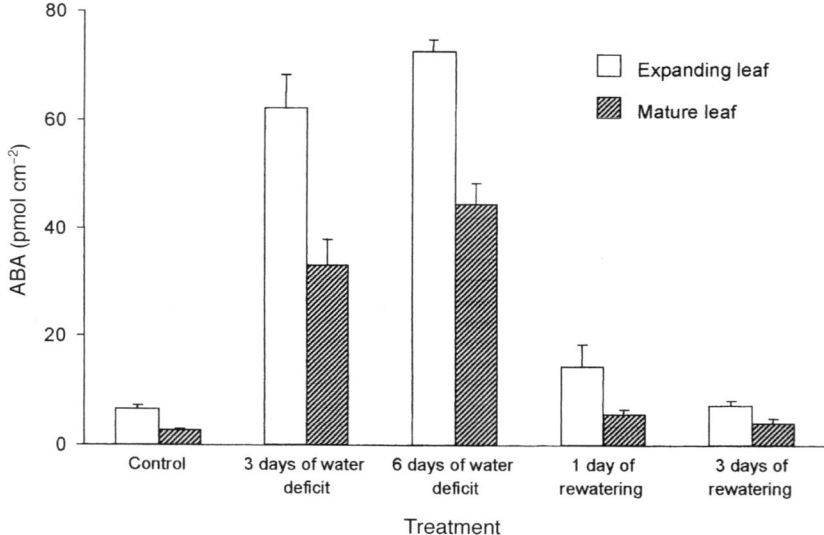

Fig. 5.7. Abscisic acid (ABA) concentration in expanding and mature cassava leaves in control after 3 and 6 days of water deficit, followed by 1 and 3 days of rewatering; average of five genotypes with three replicates; bars represent SEM ($n = 15$); data from Alves and Setter (2000).

and the extent of OA increased progressively from mature to folded leaves. As young tissues (meristems) are involved in regrowth and recovery after drought, further research is needed to give a fuller picture of cassava's responses to water deficit.

References

Alves, A.A.C. (1998) Physiological and developmental changes in cassava (*Manihot esculenta* Crantz) under water deficit. PhD thesis, Cornell University, Ithaca, New York, 160 pp.

Alves, A.A.C. and Setter, T.L. (2000) Response of cassava to water deficit: leaf area growth and abscisic acid. *Crop Science* 40, 131–137.

Angelov, M.N., Sun, J., Byrd, G.T., Brown, R.H. and Black, C.C. (1993) Novel characteristics of cassava, *Manihot esculenta* Crantz, a reputed C3–C4 intermediate photosynthesis species. *Photosynthesis Research* 38, 61–72.

Aresta, R.B. and Fukai, S. (1984) Effects of solar radiation on growth of cassava (*Manihot esculenta* Crantz). II. Fibrous root length. *Field Crops Research* 9, 361–371.

Baker, G.R., Fukai, S. and Wilson, G.L. (1989) The response of cassava to water deficits at various stages of growth in the subtropics. *Australian Journal of Agricultural Research* 40, 517–528.

Barrios, E.A. and Bressani, R. (1967) Composición química de la raíz y de la hoja de algunas variedades de yuca *Manihot*. *Turrialba* 17, 314–320.

Beeching, J.R., Marmey, P., Gavalda, M.C., Noirot, M., Haysom, H.R., Hughes, M.A. and Charrier, A. (1993) An assessment of genetic diversity within a collection of cassava (*Manihot esculenta* Crantz) germplasm using molecular markers. *Annals of Botany* 72, 515–520.

Blum, A. and Sinmena, B. (1995) Isolation and characterization of variant wheat cultivars for ABA sensitivity. *Plant, Cell and Environment* 18, 77–83.

Boerboom, B.W.J. (1978) A model of dry matter distribution in cassava (*Manihot esculenta* Crantz). *Netherlands Journal of Agricultural Science* 26, 267–277.

Bolhuis, G.G. (1966) Influence of length of the illumination period on root formation in cassava (*Manihot utilissima* Pohl). *Netherlands Journal of Agricultural Science* 14, 251–254.

Bonierbale, M., Guevara, C., Dixon, A.G.O., Ng, N.Q., Asiedu, R. and Ng, S.Y.C. (1997) Cassava. In: Fuccillo, D., Sears, L. and Stapleton, P. (eds) *Biodiversity in Trust*. Cambridge University Press, Cambridge, pp. 1–20.

Borel, C., Simonneau, T., This, D. and Tardieu, F. (1997) Stomatal conductance and ABA concentration in the xylem sap of barley lines of contrasting genetic origins. *Australian Journal of Plant Physiology* 24, 607–615.

Bray, E.A. (1994) Alterations in gene expression in response to water deficit. In: Basra, A.S. (ed.) *Stress-Induced Gene Expression in Plants*. Hardwood Academic Publishers, Chur, Switzerland, pp. 1–23.

Bruijn, G.H. (1977) Influence of daylength on the flowering of cassava. *Tropical Root Tuber Crops Newsletter* 10, 1–3.

Cardoso, C.E.L. and Souza, J. da S. (1999) Aspectos Agro-Econômicos da Cultura da Mandioca: Potencialidades e Limitações. Documentos CNPMF no. 86, EMBRAPA/CNPMF, Cruz das Almas, BA, Brazil, 66pp.

Carvalho, P.C.L. de and Ezeta, F.N. (1983) Efeito do fotoperíodo sobre a 'tuberização' da mandioca. *Revista Brasileira de Mandioca* 2, 51–54.

Cavalcanti, J. (1985) Desenvolvimento das raízes tuberosas em mandioca (*Manihot esculenta* Crantz). MSc thesis, Universidade Federal do Ceará, Fortaleza, Brazil, 66 pp.

Cellier, F., Conéjéro, G., Breitler, J.C. and Casse, F. (1998) Molecular and physiological responses to water deficit in drought-tolerant and drought-sensitive lines of sunflower. Accumulation of dehydrin transcripts correlates with tolerance. *Plant Physiology* 116, 319–328.

Cerqueira, Y.M. (1989) Efeito da deficiência de água na anatomia foliar de cultivares de mandioca (*Manihot esculenta* Crantz). MSc thesis, Universidade Federal da Bahia, Cruz das Almas, Brazil.

CIAT (1984) *Morphology of the Cassava Plant; Study Guide*. Centro Internacional de Agricultura Tropical, Cali, Colombia.

Cock, J.H. (1976) Characteristics of high yielding cassava varieties. *Experimental Agriculture* 12, 135–143.

Cock, J.H. (1984) Cassava. In: Goldsworthy, P.R. and Fisher, N.M. (eds) *The Physiology of Tropical Field Crops*. John Wiley & Sons, Chichester, pp. 529–549.

Cock, J.H. and Rosas, S. (1975) Ecophysiology of cassava. In: (eds) *Symposium on Ecophysiology of Tropical Crops*. Communications Division of CEPLAC, Ilhéus, Bahia, Brazil, pp. 1–14.

Cock, J.H., Franklin, D., Sandoval, G. and Juri, P. (1979) The ideal cassava plant for maximum yield. *Crop Science* 19, 271–279.

Cock, J.H., Porto, M.C.M. and El-Sharkawy, M.A. (1985) Water use efficiency of cassava. III. Influence of air humidity and water stress on gas exchange of field grown cassava. *Crop Science* 25, 265–272.

Conceição, A.J. da (1979) *A Mandioca.* UFBA/ EMBRAPA/BNB/BRASCAN NORDESTE, Cruz das Almas, BA.

Connor, D.J. and Palta, J. (1981) Response of cassava to water shortage. III. Stomatal control of plant water status. *Field Crops Research* 4, 297–311.

Connor, D.J., Cock, J.H. and Parra, G.E. (1981) Response of cassava to water shortage. I. Growth and yield. *Field Crops Research* 4, 181–200.

Cooke, R.D. and Coursey, D.G. (1981) Cassava: a major cyanide-containing food crop. In: Vennesland, B., Conn, E.E. and Knowles, C.J. (eds) *Cyanide in Biology.* Academic Press, New York.

Creelman, R.A., Mason, H.S., Bensen, R.J., Boyer, J.S. and Mullet, J.E. (1990) Water deficit and abscisic acid cause differential inhibition of shoot versus root growth in soybean seedlings. Analysis of growth, sugar accumulation, and gene expression. *Plant Physiology* 92, 205–214.

Cunha, H.M.P. and Conceição, A.J. da (1975) Indução ao florescimento da mandioca (*Manihot esculenta* Crantz) – Nota prévia. In: *Projeto Mandioca.* Universidade Federal da Bahia, Escola de Agronomia, Cruz das Almas, Convênio UFBA/BRASCAN NORDESTE, Brazil, pp. 11–14.

Davies, W.J., Metcalfe, J., Lodge, T.A. and Da Costa, A.R. (1986) Plant growth substances and the regulation of growth under drought. *Australian Journal of Plant Physiology* 13, 105–125.

Dimyati, A. (1994) Cassava genetic resources in Indonesia: status and future outlook. In: IPGR, International Network for Cassava Genetic Resources. Report of the first meeting of the International Network for Cassava Genetic Resources, CIAT, Cali, Colombia, 18–23 August 1992. International Crop Network Series No. 10. International Plant Genetic Resources Institute, Rome, pp. 66–70.

Edwards, G.E., Sheta, E., Moore, B.D., Dai, Z., Franceschi, V.R., Cheng, S.-H., Lin, C.-H. and Ku, M.S.B. (1990) Photosynthetic characteristics of cassava (*Manihot esculenta* Crantz), a C3 species with chlorenchymatous bundle sheath cells. *Plant and Cell Physiology* 31, 1199–1206.

El-Sharkawy, M.A. (1993) Drought-tolerant cassava for Africa, Asia and Latin America. *BioScience* 43, 441–451.

El-Sharkawy, M.A. and Cock, J.H. (1984) Water use efficiency of cassava. I. Effects of air humidity and water stress on stomatal conductance and gas exchange. *Crop Science* 24, 497–502.

El-Sharkawy, M.A. and Cock, J.H. (1987a) C3-C4 intermediate photosynthetic characteristics of cassava (*Manihot esculenta* Crantz). I. Gas exchange. *Photosynthesis Research* 12, 219–236.

El-Sharkawy, M.A. and Cock, J.H. (1987b) Response of cassava to water stress. *Plant and Soil* 100, 345–360.

El-Sharkawy, M.A. and Cock, J.H. (1990) Photosynthesis of cassava (*Manihot esculenta*). *Experimental Agriculture* 26, 325–340.

El-Sharkawy, M.A., Hernandez, A.D.P. and Hershey, C. (1992a) Yield stability of cassava during prolonged mid-season water stress. *Experimental Agriculture* 28, 165–174.

El-Sharkawy, M.A., Tafur, S.M.D. and Cadavid, L.F. (1992b) Potential photosynthesis of cassava as affected by growth conditions. *Crop Science* 32, 1336–1342.

FAO (1999) www.apps.fao.org/lim500/nph-wrap.pl?FS.CropsAndProducts&Domain=FS&servlet=1. Consulted on October 3, 1999.

Fregene, M.A., Vargas, J., Ikea, J., Angel, F., Tohme, J., Asiedu, R.A., Akoroda, M.O. and Roca, W.M. (1994) Variability of chloroplast DNA and nuclear ribosomal DNA in cassava (*Manihot esculenta* Crantz) and its wild relatives. *Theoretical and Applied Genetics* 89, 719–727.

Fukai, S., Alcoy, A.B., Llamelo, A.B. and Patterson, R.D. (1984) Effects of solar radiation on growth of cassava (*Manihot esculenta* Crantz). I. Canopy development and dry matter growth. *Field Crops Research* 9, 347–360.

Fukuda, W.M.G. and Guevara, C.L. (1998). *Descritores Morfológicos e Agronômicos para a Caracterização de Mandioca (Manihot esculenta Crantz).* Documentos CNPMF no.78. EMBRAPA/CNPMF, Cruz das Almas BA, Brazil.

Fukuda, W.M.G., Silva, S.O. and Porto, M.C.M. (1997) *Caracterização e Avaliação de Germoplasma de Mandioca (Manihot esculenta Crantz).* Catalog. EMBRAPA/CNPMF, Cruz das Almas BA, Brazil.

Ghosh, S.P., Ramanujam, T., Jos, J.S., Moorthy, S.N. and Nair, R.G. (1988) *Tuber Crops.* Oxford & IBH Publishing Co., New Delhi, pp. 3–146.

Griffiths, A., Jones, H.G. and Tomos, A.D. (1997) Applied abscisic acid, root growth and turgor pressure response of roots of wild-type and the ABA-deficient mutant, *Notabilis*, of tomato. *Journal of Plant Physiology* 151, 60–62.

Gulick, P., Hershey, C. and Esquinas Alcazar, J. (1983) *Genetic Resources of Cassava and Wild Relatives.* Rome.

Howeler, R.H. and Cadavid, L.F. (1983) Accumulation and distribution of dry matter and nutrients during a 12-month growth cycle of cassava. *Field Crops Research* 7, 123–139.

Ike, I.F. (1982) Effect of water deficits on transpiration, photosynthesis and leaf conductance in cassava. *Physiologia Plantarum* 55, 411–414.

Irikura, Y., Cock, J.H. and Kawano, K. (1979) The physiological basis of genotype-temperature interactions in cassava. *Field Crops Research* 2, 227–239.

Keating, B.A. and Evenson, J.P. (1979) Effect of soil temperature on sprouting and sprout elongation of stem cuttings of cassava. *Field Crops Research* 2, 241–252.

Keating, B.A., Evenson, J.P. and Fukai, S. (1982a) Environmental effects on growth and development of cassava (*Manihot esculenta* Crantz). I. Crop development. *Field Crops Research* 5, 271–281.

Keating, B.A., Evenson, J.P. and Fukai, S. (1982b) Environmental effects on growth and development of cassava (*Manihot esculenta* Crantz). III. Assimilate distribution and storage organ yield. *Field Crops Research* 5, 293–303.

Lecoeur, J., Wery, J., Turc, O. and Tardieu, F. (1995) Expansion of pea leaves subjected to short water deficit: cell number and cell size are sensitive to stress at different periods of leaf development. *Journal of Experimental Botany* 46, 1093–1101.

Lorenzi, J.O. (1978) Absorção de macronutrientes e acumulação de matéria seca para duas cultivares de mandioca. MSc thesis, Universidade de São Paulo, Escola Superior de Agricultura Luis de Queiroz, Piracicaba, Brazil, 92 pp.

MacRobbie, E.A.C. (1991) Effect of ABA on ion transport and stomatal regulation. In: Davies, W.J. and Jones, H.G. (eds) *Abscisic Acid: Physiology and Biochemistry*. BIOS, Oxford, pp. 153–168.

Mahon, J.D., Lowe, S.B. and Hunt, L.A. (1976) Photosynthesis and assimilate distribution in relation to yield of cassava grown in controlled environments. *Canadian Journal of Botany* 54, 1322–1331.

Mahon, J.D., Lowe, S.B. and Hunt, L.A. (1977a) Variation in the rate of photosynthesis CO_2 uptake in cassava cultivars and related species of *Manihot*. *Photosynthetica* 11, 131–138.

Mahon, J.D., Lowe, S.B., Hunt, L.A. and Thiagarajah, M. (1977b) Environmental effects on photosynthesis and transpiration in attached leaves of cassava (*Manihot esculenta* Crantz). *Photosynthetica* 11, 121–130.

McMahon, J.M., White, W.L.B. and Sayre, R.T. (1995) Cyanogenesis in cassava (*Manihot esculenta* Crantz). *Journal of Experimental Botany* 46, 731–741.

Mutsaers, H.J.W., Ezumah, H.C. and Osiru, D.S.O. (1993) Cassava-based intercropping: a review. *Field Crops Research* 34, 431–457.

Oelsligle, D.D. (1975) Accumulation of dry matter, nitrogen, phosphorus, and potassium in cassava (*Manihot esculenta* Crantz). *Turrialba* 25, 85–87.

O'Hair, S.K. (1989) Cassava root starch content and distribution varies with tissue age. *HortScience* 24, 505–506.

O'Hair, S.K. (1990) Tropical root and tuber crops. *Horticultural Reviews* 12, 157–166.

Oke, O.L. (1983) The mode of cyanide detoxification. In: Nestel, B. and MacIntyre, R. (eds) *Chronic Cassava Toxicity*. International Development Research Centre, Ottawa, pp. 97–104.

Okoli, P.S.O. and Wilson, G.F. (1986) Response of cassava (*Manihot esculenta* Crantz) to shade under field conditions. *Field Crops Research* 14, 349–360.

Palta, J.A. (1984) Influence of water deficits on gas-exchange and the leaf area development of cassava cultivars. *Journal of Experimental Botany* 35, 1441–1449.

Pereira, J.F. and Splittstoesser, W.E. (1990) Anatomy of the cassava leaf. *Proceedings International Society of Tropical Horticulture* 34, 73–78.

Peressin, V.A., Monteiro, D.A., Lorenzi, J.O., Durigan, J.C., Pitelli, R.A. and Perecin, D. (1998) Acúmulo de matéria seca na presença e na ausência de plantas infestantes no cultivar de mandioca SRT 59 – Branca de Santa Catarina. *Bragantia* 57, 135–148.

Pinho, J.L.N. de, Távora, F.J.A.F., Melo, F.I.O. and Queiroz, G.M. de (1995) Yield components and partitioning characteristics of cassava in the coastal area of Ceará. *Revista Brasileira de Fisiologia Vegetal* 7, 89–96.

Porto, M.C.M. (1986) Fisiologia da mandioca. Monograph presented at VI Curso Intensivo Nacional de Mandioca, 14–24 October 1986, EMBRAPA/ CNPMF, Cruz das Almas BA, Brazil, 21 pp.

Porto, M.C.M., Bessa, J.M.G. and Lira Filho, H.P. (1988) Diferenças varietais no uso da água em mandioca, sob condições de campo no Estado de Pernambuco. *Revista Brasileira de Mandioca* 7, 73–79.

Pugnaire, F.I., Endolz, L.S. and Pardos, J. (1994) Constraints by water stress on plant growth. In: Pessarakli, M. (ed.) *Handbook of Plant and Crop Stress*. Marcel Dekker, New York, pp. 247–259.

Ramanujam, T. (1982) Leaf area in relation to petiole length in cassava. *Turrialba* 32, 212–213.

Ramanujam, T. (1985) Leaf density profile and efficiency in partitioning dry matter among high and low yielding cultivars of cassava (*Manihot esculenta* Crantz). *Field Crops Research* 10, 291–303.

Ramanujam, T. (1990) Effect of moisture stress on photosynthesis and productivity of cassava. *Photosynthetica* 24, 217–224.

Ramanujam, T. and Biradar, R.S. (1987) Growth analysis in cassava (*Manihot esculenta* Crantz). *Indian Journal of Plant Physiology* 30, 144–153.

Ramanujam, T. and Indira, P. (1983) Canopy structure on growth and development of cassava (*Manihot esculenta* Crantz). *Turrialba* 33, 321–326.

Ramanujam, T. and Jos, J.S. (1984) Influence of light intensity on chlorophyll distribution and anatomical characters of cassava leaves. *Turrialba* 34, 467–471.

Ramanujam, T., Nair, G.M. and Indira, P. (1984) Growth and development of cassava (*Manihot esculenta* Crantz) genotypes under shade in a coconut garden. *Turrialba* 34, 267–274.

Rogers, D.J. (1965) Some botanical and ethnological considerations of *Manihot esculenta*. *Economic Botany* 19, 369–377.

Rogers, D.J. and Fleming, H.S. (1973) A monograph of *M. esculenta* with an explanation of the taximetrics methods. *Economic Botany* 27, 1–113.

Sharp, R.E., Voetberg, G.S., Saab, I.N. and Bernstein, N. (1993) Role of abscisic acid in the regulation of cell expansion in roots at low water potentials. In: Close, T.J. and Bray, E.A. (eds) *Plant Response to Cellular Dehydration During Environmental Stress.* American Society of Plant Physiologists, Rockville, Maryland, pp. 57–66.

Sharp, R.E., Wu, Y., Voetberg, G.S., Saab, I.N. and LeNoble, M.E. (1994) Confirmation that abscisic acid accumulation is required for maize primary root elongation at low water potentials. *Journal of Experimental Botany* 45, 1743–1751.

Sinha, S.K. and Nair, T.V. (1971) Leaf area during growth and yielding capacity of cassava. *The Indian Journal of Genetics and Plant Breeding* 31, 16–20.

Splittstoesser, W.E. and Tunya, G.O. (1992) Crop physiology of cassava. *Horticultural Reviews* 13, 105–129.

Takami, S., Turner, N.C. and Rawson, H.M. (1981) Leaf expansion of four sunflower (*Helianthus annuus* L.) cultivars in relation to water deficits. I. Patterns during plant development. *Plant, Cell and Environment* 4, 399–407.

Tan, S.L. and Cock, J.H. (1979) Branching habit as a yield determinant in cassava. *Field Crops Research* 2, 281–289.

Tardieu, F. and Simonneau, T. (1998) Variability among species of stomatal control under fluctuating soil water status and evaporative demand: modeling isohydric and anisohydric behaviours. *Journal of Experimental Botany* 49, 419–432.

Távora, F.J.A.F., Melo, F.I.O., Pinho, J.L.N. de and Queiroz, G.M. de (1995) Yield, crop growth rate and assimilatory characteristics of cassava at the coastal area of Ceará. *Revista Brasileira de Fisiologia Vegetal* 7, 81–88.

Ternes, M., Mondardo, E. and Vizzotto, V.J. (1978) *Variação do teor de amido na cultura da mandioca em Santa Catarina.* Indicação de Pesquisa no. 23, Empresa Catarinense de Pesquisa Agropecuária, Florianópolis, SC, Brazil.

Trejo, C.L., Davies, W.J. and Ruiz, L.D.M.P. (1993) Sensitivity of stomata to abscisic acid. An effect of the mesophyll. *Plant Physiology* 102, 497–502.

Trejo, C.L., Clephan, A.L. and Davies, W.J. (1995) How do stomata read abscisic acid signals? *Plant Physiology* 109, 803–811.

Trewavas, A.J. and Jones, H.G. (1991) An assessment of the role of ABA in plant development. In: Davies, W.J. and Jones, H.G. (eds) *Abscisic Acid: Physiology and Biochemistry.* BIOS, Oxford, pp. 169–188.

Ueno, O. and Agarie, S. (1997) The intercellular distribution of glycine decarboxylase in leaves of cassava in relation to the photosynthetic mode and leaf anatomy. *Japanese Journal of Crop Science* 66, 268–278.

Van Volkenburgh, E. and Davies, W.J. (1983) Inhibition of light-stimulated leaf expansion by abscisic acid. *Journal of Experimental Botany* 34, 835–845.

Veltkamp, H.J. (1985a) Canopy characteristics of different cassava cultivars. *Agricultural University Wageningen Papers* 85, 47–61.

Veltkamp, H.J. (1985b) Growth, total dry matter yield and its partitioning in cassava at different daylengths. *Agricultural University Wageningen Papers* 85, 73–86.

Veltkamp, H.J. (1985c) Interrelationships between LAI, light interception and total dry matter yield of cassava. *Agricultural University Wageningen Papers* 85, 36–46.

Veltkamp, H.J. (1985d) Partitioning of dry matter in cassava. *Agricultural University Wageningen Papers* 85, 62–72.

Veltkamp, H.J. (1985e) Photosynthesis, transpiration, water use efficiency and leaf and mesophyll resistance of cassava as influenced by light intensity. *Agricultural University Wageningen Papers* 85, 27–35.

Wheatley, C.C. and Chuzel, G. (1993) Cassava: the nature of the tuber and use as a raw material. In: Macrae, R., Robinson, R.K. and Sadler, M.J. (eds) *Encyclopaedia of Food Science, Food Technology, and Nutrition.* Academic Press, San Diego, California, pp. 734–743.

Wheatley, C.C., Orrego, J.I., Sanchez, T. and Granados, E. (1993) Quality evaluation of cassava core collection at CIAT. In: Roca, W.M. and Thro, A.M. (eds) *Proceedings of the First International Scientific Meeting of the Cassava Biotechnology Network.* CIAT, Cartagena, Colombia, pp. 255–264.

Wholey, D.W. and Booth, R.H. (1979) Influence of variety and planting density on starch accumulation in cassava roots. *Journal of the Science of Food and Agriculture* 30, 165–170.

Wholey, D.W. and Cock, J.H. (1974) Onset and rate of root bulking in cassava. *Experimental Agriculture* 10, 193–198.

Williams, C.N. (1972) Growth and productivity of tapioca (*Manihot utilissima*). III. Crop ratio, spacing and yield. *Experimental Agriculture* 8, 15–23.

Chapter 6
Agronomy and Cropping Systems

Dietrich Leihner

Research, Extension and Training Division, Sustainable Development Department, FAO, Via delle Terme di Caracalla, 00100 Rome, Italy

Introduction

Cassava is often grown under low-input/low-output production systems, particularly when it is grown as a food crop. Planting material is easily obtained from the plant stems available from the farmers' own or neighbouring fields. Although the crop is affected by a number of arthropod pests, diseases and by weed competition, it generally requires little attention once established. Nevertheless, attention to a few simple aspects of agronomic management can result in a doubling or tripling of output at low cost. In this chapter principles of good agronomic management are described, dealing with the preparation and handling of planting material, soil preparation, planting techniques, weed control, intercropping techniques and soil conservation systems.

Planting Material Production and Handling

Rapid multiplication and selection

Propagation of cassava through true seed is feasible, but no commercially viable seed propagation system is yet available. Cassava continues to be propagated vegetatively through stem cuttings or stakes (as they should be called). The number of commercial stakes obtained from a single mother plant in a year ranges from three to 30, depending upon growth habit, climate and soil conditions. This is considerably less than the propagation rate that can be achieved with other commercial crops that are propagated through true seed. Thus the development of improved cassava production technology should include more effective propagation schemes. A system using small two-node cuttings from which a number of successively growing shoots are obtained, rooted in boiled water and planted in the field was devised by Cock *et al.* (1976). This system produces 12,000–24,000 commercial stakes after 1 year. A much more productive method was devised later (Cock, 1985) using cassava leaves excised with their axillary buds, transferred to a mist chamber for sprouting and root formation of the propagules, which are transferred to a peat pot and 2–3 weeks later, to the field. Although this system is more labour intensive, 100,000–300,000 commercial stakes can be produced in about 18 months.

Stems must be transported with care to prevent bruising and peeling. Stakes should be cut at a right angle without placing stems on a base to prevent breaking or splitting that provides entry points for pathogens and insect pests. A stake should be at least 20 cm long and have a minimum of 4–5 nodes with viable buds to ensure crop establishment. Stems should be sufficiently lignified to ensure that stakes do not dry out too fast after planting, but overlignified tissue should be avoided. Stakes have the right degree of maturity when their pith diameter measures approximately half the total stake diameter.

Visual inspection of mother plants prior to cutting the planting material is an effective practice for reducing phytosanitary problems. Although infection with a viral, fungal or bacterial pathogen, producing no visible symptoms, can never be ruled out completely, many of these phytosanitary problems produce clearly visible signs of infection on leaf or stem tissue. Externally adhering insects can also be detected easily. If plants with visible symptoms are excluded from stake production, then a first important step towards a healthy new crop has been made (Lozano *et al.*, 1977).

Stake treatment

Even if utmost care is taken to select planting material from apparently healthy mother plants, the presence of adhering pathogens or insects can never be fully avoided, either from carryover, or, from new infestation with soil-borne pathogens and insects. The best way to reduce these problems and to protect stakes is to reduce soil infestation by means of crop rotations and cultural practices such as drainage or planting on ridges. Treatment of stakes with chemical disinfectants and protectants has a number of advantages. Mixtures of contact and systemic fungicides, with an occasional insecticide when necessary, can protect planting material, which in turn may enhance sprouting, root formation and growth. If stakes have to be stored, this treatment also provides a certain degree of protection and the period of viability under storage may be extended. Pesticide combinations for stake treatment have been suggested by Lozano *et al.* (1981).

Storage

In cassava-growing areas with dry, cool or flooded periods, during which planting is not recommended or feasible, planting material may have to be stored for several months. During storage the stems gradually deteriorate, leading eventually to a total loss of viability. The type of planting material to be stored, storage time and conditions can, however, retard this deterioration process.

Selection of well-developed and well-nourished mature and healthy stems from mother plants, and adequate storage conditions are the first steps towards minimizing detrimental storage effects. Mother plants whose stems are to be stored should have a well-balanced nutritional status to ensure good stand establishment after storage (Leihner, 1984b). Stems for storage should be as long as possible and not cut into stakes as this greatly accelerates dehydration.

Physiological deterioration of the planting material is principally linked to two processes: respiration and dehydration. Freshly cut cassava stems consist of living tissue that continues to metabolize during storage, losing mostly soluble carbohydrates for up to 60 days or more after cutting (Leihner, 1984b; Oka *et al.*, 1987). This means that valuable reserves are being lost, reducing resprouting vigour after planting. Moist, hot storage conditions will enhance this process more than cool or dry conditions. Dehydration of stored cassava stems reduces metabolic activity of the tissue and may reduce respiration, but it leads to a progressive loss of viability, rendering stems unsuitable for planting. A minimum level of 60% moisture in the stakes has been identified as the threshold for satisfactory preservation of viability (Wholey, 1977; Leihner, 1984b, 1986).

Storing cassava stakes under inadequate conditions may lead to a drastic loss in viability even after rather short periods. Leihner (1984b) reported a drop in percentage sprouting from 100 to 30% when short stakes were stored for just 15 days at 24°C average temperature under sun exposure and without the possibility of reabsorbing moisture from soil, rain or dew. In contrast, stakes stored as long stems under shady conditions with 72% average relative humidity (RH) and chemical protection reached over 95% sprouting even after 201 days of storage (Leihner, 1986). Improvement of sprouting was reached through rehydrating stakes for 4 h in water or a nutrient solution. If the stored material is of high quality and storage conditions are right, long-term storage of cassava planting material is possible without losing viability.

Extensive research has been carried out on storage conditions (e.g. Silva, 1970; Correa and Vieira, 1978; Sales Andrade and Leihner, 1980; Centro Internacional de Agricultura Tropical (CIAT), 1980, 1982), making it possible to identify practices that keep cassava planting material viable for several months. Long stems (50–100 cm) should be treated with fungicides and insecticides before storage and kept in a shady place with high RH (70–80%) and moderate ambient temperature (20–23°C). Excessive heat and direct sun accelerate metabolic activity and dehydration. If longer-term storage is envisaged, stored stems may be buried 5–10 cm in the ground with their basal end allowing root formation below and sprouting of the apical buds above. Stems stored under those conditions may need watering if conditions get overly dry. Although this system keeps stems viable over long periods, a large portion on either end of the stem has to be discarded to ensure stakes come from parts of the stem that have not previously rooted or sprouted. All stored stakes should be re-treated chemically before planting to provide extra protection and stimulate rooting and sprouting.

Land Preparation

Tillage versus no till

Cassava needs a sufficiently loose-textured soil, not only for initial fibrous root penetration, but also to allow for root thickening. This may not always require a thorough manual or mechanized soil preparation. When cassava was domesticated, it was probably cultivated principally by slash-and-burn practices that eliminated competing vegetation but did not alter soil structure. The friable, high organic matter soil conditions that can be found in nondegraded slash-and-burn systems give cassava roots good growing conditions. The only soil preparation probably used by early planters was loosening of the soil locally with a planting stick to bury the stake. These ideal conditions essentially allowed a no-till soil preparation for cassava planting. Under more degraded slash-and-burn conditions, or, with permanent agriculture, a thorough loosening of the soil is normally required to allow the introduction of

the stake and provide well-drained, aerated conditions for the root system. Cassava is a hardy crop withstanding many types of stress, but it easily succumbs to excessive soil moisture and root rot, resulting in extensive yield losses. To prevent these losses, soil preparation is necessary to allow good drainage and aeration.

Ridges, raised beds or mounds

Ezumah and Okigbo (1980) pointed out that in the humid and subhumid climates of West Africa, drainage conditions often determine the type of land preparation required, as well as the size of ridges or mounds and the location of crops on them. In the Democratic Republic of Congo, for example, there was no yield advantage from ridges as compared with flat or untilled plots whenever the field was mulched. Lowest root yields occurred in unmulched, untilled fields. In Cuba, a revolution in cassava production was achieved when traditional planting techniques similar to those used in sugarcane were abandoned for slanted planting on top of 40-cm high ridges (Rodriguez Nodals, 1980). In an erosion study, Reining (1992) compared mechanized soil preparation (flat, contour ridges) with a minimum tillage system, where cassava was planted in an existing grass sod by just loosening the soil with a shovel where stakes were to be inserted. Flat and ridged preparations gave no significant root yield differences over three growing seasons, while the minimum tillage system yielded less than 30% of that obtained in the other two systems. The higher bulk density of the soil under the grass sod and its quick hardening under dry conditions, together with competition from the grass, were thought to be responsible for the negative result of minimum tillage which, however, minimized soil erosion. A number of other researchers (reviewed by Toro and Atlee, 1980) agree that in most cases manual or mechanized soil preparation is preferred and that in areas of high rainfall or heavy soils, good drainage must be provided by preparing ridges, beds or mounds, although the exact configuration is not so important. There is evidence, however, that soil preparation intensity can be reduced when collateral practices improving soil structure and drainage, such as mulching, are implemented.

Planting Techniques

Stake position

Cassava cultivar, soil characteristics and climate together determine whether there is an advantage to vertical, inclined or horizontal planting, or, whether any position may be used. Since the first reports from Indonesia (Koch, 1916), extensive experimentation on positioning of the stake has been carried out in Latin America, Africa and Asia, a thorough review of which was presented by Toro and Atlee (1980). In tests conducted at CIAT by Castro *et al.* (1978) and Castro (1979), stake sprouting and emergence under field conditions were always more rapid with vertical planting than with any other method (Fig. 6.1). Vertical placement results in fast crop establishment and soil cover development, together with good anchorage provided by a deep root system and less risk of lodging. Under extremely adverse climatic conditions, placing a stake vertically 10–15 cm in the ground reduces heat damage and exposure of roots to erosion effects. Fast crop emergence also reduces weed competition. Horizontal planting has the advantage that there is no need to worry about planting stakes upside down. The planting operation itself does not require stooping or bending over, and the shallower root system resulting from horizontal planting allows for greater ease of harvest.

Based on experiences in many cassava-growing areas around the world, the following criteria for planting position should be considered: in regions with medium-to-heavy soils and adequate rainfall (1000–2000 mm year^{-1}), stake position does not matter because the moisture will be adequate for sprouting. In areas with sandy soils or erratic rainfall, however, vertical planting is safest. In this case 20-cm stakes should be planted at a depth of 10–15 cm in the soil to ensure better contact with available moisture. If stakes are planted horizontally, the buds will rot because of high soil temperature, while in vertical planting, the stake might serve as a heat defuser.

Planting depth

A literature review by Tan and Bertrand (1972) suggests that decisions on depth of planting, similar to position of planting, should be based upon the characteristics of locally planted cultivars as well as climatic and soil conditions. A too-shallow planting depth may expose stakes to less-than-optimum conditions of moisture and temperature, resulting in poor crop stands and low root yields. Celis and Toro (1974) define the conditions for deciding which planting depth should be adopted. On dry sandy soils, stakes should be

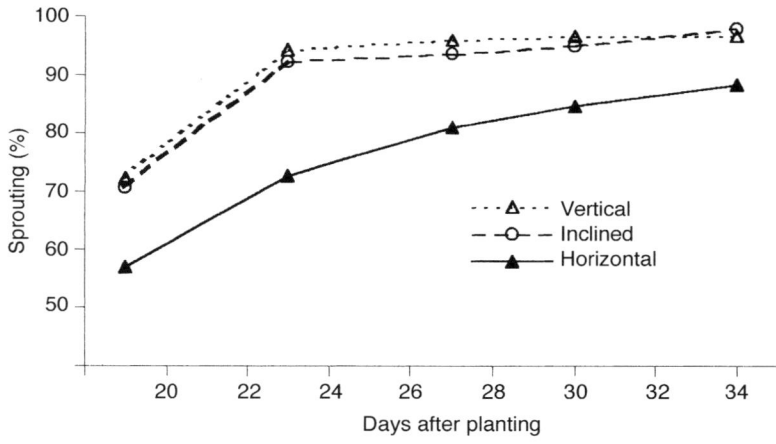

Fig. 6.1. Effect of planting position on rate of emergence and final percentage sprouting of cassava; mean of ten cultivars and four planting dates at CIAT, Cali, Colombia. (Source: CIAT, 1979.)

planted at greater depths than on wet and heavy soils where shallower planting is indicated. In the former case, however, deeper planting may make harvesting more difficult and raise costs, particularly when harvesting is done manually. These observations are supported by Normanha and Pereira (1950), who planted cassava at depths of 5, 10 and 15 cm in two seasons per year over a 3-year period. Under hot, dry conditions, stakes planted 15 cm deep sprouted more rapidly than those planted at shallower depths, probably due to more available moisture at the greater planting depth. Under better moisture and lower temperature conditions, however, the opposite occurred; moreover, harvesting was easier and yields greater when stakes had been planted at only 5 cm depth. For mechanized planting, which is common in Brazil, planting depths of 10–20 cm are common as this is the operating depth of the planters.

When moisture and temperature conditions at planting time are optimal and high-quality planting material is used, planting stakes vertically at a depth of 5–15 cm has little influence on emergence, crop growth and final root yield. Placing a 20-cm stake in vertical position to approximately half its length into the ground appears to be the most appropriate for both planting and harvesting operations.

Planting density

Information on optimum planting density for maximum root yield varies enormously from country to country and even from one ecological zone to another within the same country. Factors such as growth habit (early, late or non-branching), soil fertility, moisture regime and temperature all have an important influence on the size of the cassava canopy that has to be accommodated in the field as the crop matures.

Other aspects influencing plant density are cropping system and production objectives. In a survey dealing with cassava research by 37 institutions in 11 South and Central American countries, Leihner and Castro (1979) found that sole-cropped cassava is planted at an average density of 11,300 plants ha^{-1}; intercropped cassava at a lower density of 8900 plants ha^{-1}. When root production is the sole objective, densities around 10,000 plants ha^{-1} are normally adequate for producing a large number of commercial-size roots (Fig. 6.2), which are preferred for fresh consumption. In cases where root size is of no concern, higher planting densities can be used, resulting in a higher total production of small roots. For a combined objective of root and stake production, planting densities around 20,000 are adequate. If the sole objective is stake production, densities up to 40,000 plants ha^{-1} are optimal (Leihner, 1984a).

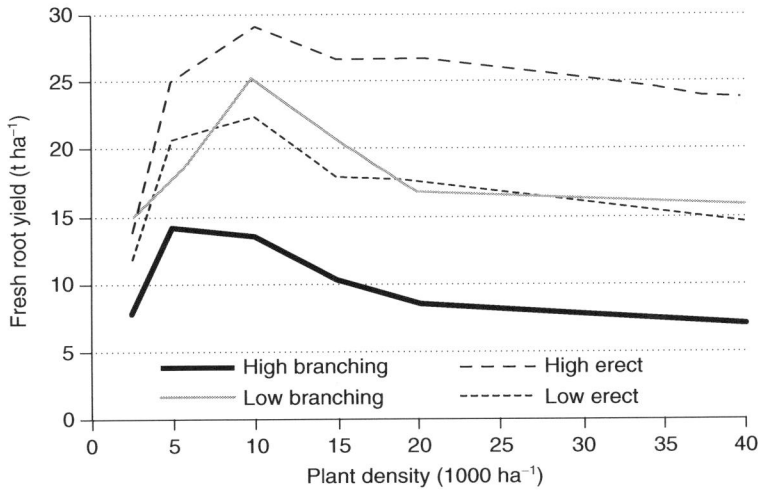

Fig. 6.2. Effect of planting density and growth habit on production of commercial cassava roots. (Source: CIAT, 1976.)

Many authors report that for most cassava genotypes no significant commercial root yield increases are to be obtained with planting densities much greater than 10,000 plants ha^{-1} (Tardieu and Fauche, 1961; CIAT, 1976, 1977; Castro *et al.*, 1978). The evaluation of new germplasm and other field experimentation as well as commercial production in many regions is thus carried out using a standard 10,000 plant ha^{-1} population. Significantly lower plant populations (5000 plants ha^{-1}) are justified when very tall vigorous and profusely branching genotypes are used (CIAT, 1976). Higher populations (up to 20,000 plants ha^{-1}) are recommended when less-vigorous genotypes are grown under low-fertility soil conditions (Santos *et al.*, 1972; Mattos *et al.*, 1973).

Planting pattern

In seeded, vegetatively planted or transplanted crops, the term 'spacing' includes both planting density and the spatial distribution of plants in the field. In order not to confound effects of either parameter, planting density and planting pattern are dealt with separately herein.

Whilst changes in planting density – in the range 2500–10,000 plants ha^{-1} – have usually produced a clear effect on cassava root yield, the crop appears to react much less to changes in planting pattern. A separate effect of spatial arrangement has seldom been reported but is relevant when cassava is intercropped, grown in agroforestry systems or when mechanization is introduced, requiring a spatial arrangement of plants other than the most frequently used square configuration.

There is little specific research on this topic, but available information suggests that cassava is a rather flexible crop, maintaining the same yield level, even when the strictly square arrangement is replaced by a variety of rectangular configurations. CIAT (1977) reports no significant yield differences either in total or commercial root production when three cultivars were grown at a standard 10,000 plants ha^{-1} density in spatial arrangements with 1–2 stakes per planting site ranging from the quadratic 1 × 1 m to a strongly square pattern of 2 × 0.5 m (Fig. 6.3).

Similarly, Leihner (1983) found no differences in root yield when comparing cultivars with different growth habit in three different ecological zones of Colombia using spatial arrangements from 1 × 1 to 2 × 0.5 m whilst maintaining planting density at around 10,000 plants ha^{-1}. Cock *et al.* (1978) tested mechanical harvesters and found that the standard spacing of 1 × 1 m was a problem for centrally mounted harvesters. At this spacing, two cassava rows had to be harvested simultaneously to prevent tractor wheels running over the unharvested crop. With one-row harvesters, however, this proved to be impossible. Changing to a 1 × 1.6 m row spacing, whilst maintaining the same plant

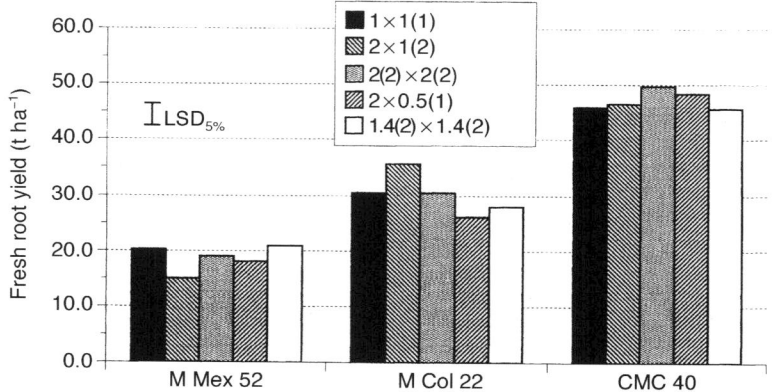

Fig. 6.3. Effect of planting patterns on commercial root production of three cassava cultivars grown at CIAT, Cali, Colombia, during 1976/77; standard planting density 10,000 plants ha^{-1}; no. of plants per site in parentheses. (Source: CIAT, 1977.)

population, allowed the operation of the one-row harvester with no sacrifice in yield.

Different spatial arrangements can thus be adopted to satisfy specific production system needs, without compromising yield potential. Considering the vast proportion of cassava grown in polycultural systems and its potential as an industrial crop requiring mechanization, this is an important point in the flexible management of planting systems.

Weed Control

Weed competition

Similarly to other crops, cassava suffers from competition with weeds for space, light, water and nutrients. In other annual crops there are critical periods during which weed competition causes significant yield decline. Until canopy closure, the earliest growth stages are normally the most susceptible, so that keeping crops weed-free during this period is a pre-condition for high productivity. Studies were carried out at CIAT by Doll et al. (1982) to determine the duration of this critical period for cassava on a fertile soil at a standard 10,000 plants ha^{-1} planting density, with high pressure from particularly aggressive weeds. From one to four hand weedings were carried out during the first 4 months, after which canopy closure was reached. Weed competition during the first 60 days after planting reduced yields to approximately 50% of the weed-free control. Weeding after 120 days did not increase root production. Thus, under conditions at CIAT, the critical period for weed competition in cassava lasts until 4 months after planting. Only if a good level of weed control is achieved during this period can acceptable root yields be obtained. In contrast, late weed infestations that occur when leaf area is gradually reduced before harvest appear to have little influence on root yield; but they may have a negative effect on the harvest operation itself and on stake quality.

Mechanical control

Worldwide, hand weeding is still the most frequent method of weed control. Hoes, machetes or sharpened shovels are used. Where these are lacking or where weeds are used for food or feed purposes, pulling them out by hand is the preferred method. In more technological production systems, mechanical weed control is also practised, using animal-drawn implements or tractors. In this case weeding should start as soon as competition begins and weeds are still easy to control. Montaldo (1966) and Delgado and Quevedo (1977) suggest that weeding should start 21 or at the latest 28–35 days after planting and should be repeated as necessary until canopy closure. On the other hand, too early mechanical weeding could damage young cassava plants and their superficially developing root system (CIAT, 1973). Whilst hand weeding is probably the most effective and least damaging weed-control method, it is also the most expensive, representing up to half the total production cost. Thus farmers decide on the number of hand weedings, not solely based on agronomic necessity, but also on the relationship between the number (and cost) of hand weedings and potential yield increase. Doll et al. (1982) found that with just two timely hand weedings carried out at 30 and 60 days after planting, 77% of maximum yield could be obtained at a relatively moderate cost. In this way, a small number of timely weedings, well spaced within the critical period, may give good yields at low cost and may thus be the most profitable option.

Chemical control

To date, no herbicide has been developed specifically for cassava. Most instructions for herbicide use do not even mention cassava and information on how to use them in cassava is not usually available. Comprehensive screening was therefore carried out with a large number of commercially available herbicides when chemical weed control in cassava was developed as a component of improved production technology. Doll and Piedrahita (1976) tested a number of herbicides in cassava for selectiveness and effectivity. They classified 18 products as highly selective and 12 as moderately so. As a group, the substituted ureas (diuron, linuron, fluometuron) were found suitable, being classified as moderately selective for cassava, particularly for controlling broadleaf weeds effectively. Mixtures with highly selective herbicides of the acetanilide group (alachlor, butachlor) are recommended for their extended effectiveness against grassy weeds. It is also possible to use wide-spectrum herbicides such as oxyfluorfen, which control both grassy and broadleaf weeds adequately.

Given the importance of early weed control in cassava, the use of pre-emergence herbicides is indicated. For cassava, as for other crops, this means that land preparation and planting should be done prior to herbicide application. Even if the vertical planting position is used, leaving cassava stakes partially exposed, this is not a problem if overhead herbicide application is done immediately after planting (up to 3 days later) because stakes suffer no damage from contact with herbicides if axillary buds have not started to sprout. If the application cannot be made at this early stage, then broadcast applications should be replaced by directed or banded applications, using protective shields to avoid herbicide contact with sprouting plants.

In the majority of cassava-growing areas, very little or no herbicides are used, either because of their unavailability or high cost. For this reason the further development of mechanical and cultural methods should have high priority. On the other hand, commercial plantations require simple-to-use, low-cost weed control methods. Chemical control has the greatest potential for fulfilling these requirements; thus further development of chemical methods should be continued for these conditions.

Cultural control

Cultural weed control makes use of non-mechanical and non-chemical practices that help suppress weeds by increasing the competing ability of the crop. Practices that contribute to good crop establishment and growth – such as selection of adapted cultivars, use of high-quality stakes, the correct planting density and plant protection – will in most cases significantly favour cultural control. With cassava, the exclusive use of cultural weed control methods is difficult during the first 3–4 months after planting because of its slow initial growth, even if agronomic practices are optimal. Supporting cultural measures such as the use of mulches, green covers or intercrops are, however, possible.

Both plant type and planting density determine the number of days needed by cassava to reach complete ground cover. The more vigorous, early-branching and leafy the plant type, the shorter will be the time to reach ground cover. Similarly, at higher planting densities, the cassava will reach ground cover earlier than at lower densities. To establish the cultural weed control potential of contrasting plant types and densities, Leihner (1980) carried out studies in the Colombian Atlantic Coast region, using both a vigorous and a non-vigorous cultivar planted at densities of 7500 and 15,000 plants ha^{-1} at three different manual weed-control levels (no control, intermediate or optimum control). Results (Fig. 6.4) showed that vigorous cultivars are less sensitive to deficiencies in weed control than non-vigorous cultivars. Unfortunately, however, the former are usually too leafy and therefore have a low root-yield potential. Although vigorous cultivars may achieve an acceptable yield with poor weed control, their yield will not reach that of less-vigorous cultivars under good weed-control conditions. By planting less-vigorous cultivars at high densities, rapid ground cover can be achieved, thereby improving the crop's ability to compete with weeds and attain high yield levels.

The possibility of preventing or reducing weed growth by using live or dead soil covers has been the subject of several studies in cassava-growing regions. In Bali, Nitis (1977) and Nitis and Suarna (1977) reported undersowing cassava with a *Stylosanthes guianensis* cover crop. The beneficial effect of *Stylosanthes* on root yields was, however, attributed more to its N-fixing ability than to cultural control of weeds. At CIAT, Leihner (1980) compared the use of a perennial legume (*Desmodium heterophyllum*) green cover, an annual legume (*Phaseolus vulgaris*) intercrop and sugarcane bagasse mulch as cultural weed control methods with manual weeding and chemical control. Manual weeding produced greatest cassava yields. Annual and perennial green covers, and also the mulch, produced somewhat lower cassava yields but at a much lower cost. The pre-emergent herbicide alone was the least effective method. More recent research on the effect of perennial legume cover crops on cassava sheds a more critical light on this practice as a whole (Leihner *et al.*, 1996a,b; Müller-Sämann and Leihner, 1999). Cassava undersown with *Pueraria phaseoloides*, *Centrosema macrocarpum*, *Centrosema acutifolium* or *Zornia glabra* suffered root yield reductions of up to 40% due to the competitive effect of these legumes. Moreover, the legumes were not able to control erosion effectively in their year of establishment because of slow initial ground cover, which made additional weeding operations necessary. Even with agronomic practices such as increased planting density of cassava, more vigorous cassava genotypes and less-competitive legumes

such as *Chamaecrista rotundifolia*, a species with outstanding soil-cover capacity, the legume covers decreased cassava yields considerably. They were thus considered attractive only to farmers who can make efficient use of the 3–4 t ha^{-1} forage dry matter produced in these systems.

Leihner (1980) examined the weed-control effectiveness of intercropping cassava with common beans. Under good weed-control conditions, intercropped cassava yielded 15% less than the corresponding sole crop; but when no weed control was practised, a 44% greater root yield was observed in intercropped compared to sole cropped cassava (Fig. 6.5). These

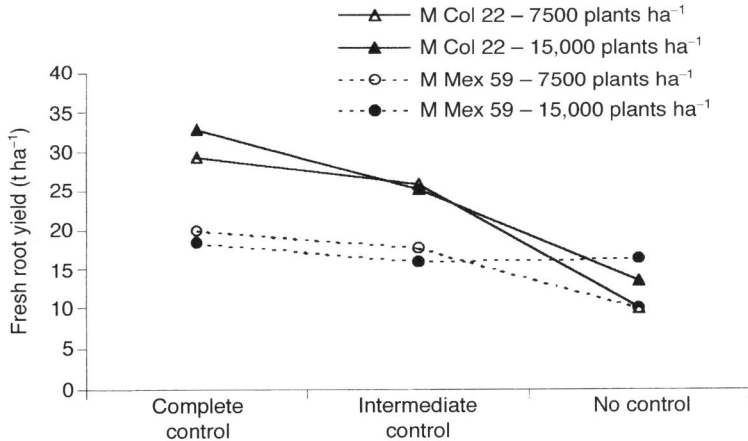

Fig. 6.4.   Effect of plant type and planting density on cassava yield at different weed control levels; M Col 22 (non-vigorous) and M Mex 59 (vigorous), planted at ICA-Caribia, Atlantic Coast, Colombia, 1978/79. (Source: Leihner, 1980.)

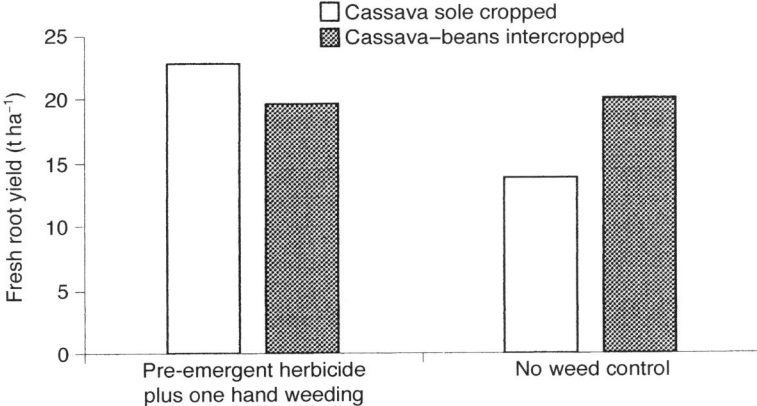

Fig. 6.5.   Effect of two weed control levels on root yield in sole cropped cassava and a cassava–common bean intercrop. (Source: CIAT, 1979.)

data confirm the excellent cultural weed control potential of intercropping, particularly under marginal, low-input conditions.

Although these data show the effectiveness of specific cultural weed control methods in cassava, their exclusive use may entail a number of problems. The establishment of green covers or intercropping is usually more labour intensive than planting cassava alone. Cultural weed control alone may not be as effective as chemical or mechanical methods. The requirement of timeliness – a particularly critical aspect of weed control in cassava – may not always be fulfilled. On the other hand, cultural control is always ecologically sound and, depending on the method adopted and local availability of materials, it can also be low cost in terms of purchased inputs. The possibilities of combining cultural control with other weed-control methods are numerous and provide farmers with a variety of choices of either labour- or capital-intensive practices. This adds great flexibility to weed management, enabling cassava producers to adopt the system that best fits their means and thus obtain optimum results in terms of both crop productivity and economics.

Intercropping

Ecological, socioeconomic and nutritional aspects

Cassava adapts to a wide range of ecological conditions and is known for its tolerance of low soil fertility, drought and pests. This is why the crop holds an important position in traditional tropical cropping systems, particularly those of the small-farm and subsistence sectors. In these cropping systems, cassava is often found in mixed stands, together with a variety of other food or cash crops. For generations the traditional farmer has adopted intercropping as a production system in order to reduce the risk of crop failure, obtain production at different times during the year, make the best use of available land and labour resources, and provide the family with a balanced diet. Estimates indicate that at least one-third of the cassava grown worldwide is intercropped (Cock, 1985). Continents

and regions reflect their own characteristic crop combinations and sequences, with cassava often being found at the end of the cycle. The greatest complexity of cassava intercropping systems is probably found in homestead gardens of rural farming families in Africa.

When farmers adopt cassava intercropping as a production system, a relatively small plot suffices to provide the family with the basic dietary elements. Sources of carbohydrates such as cassava, sweet potato (*Ipomea batatas*), yams (*Dioscorea* spp.), taro (*Colocasia* sp., *Xanthosomas* sp.) and plantains (*Musa* sp.) provide the primary caloric component. The intercrops such as common beans (*P. vulgaris*), cowpea (*Vigna unguiculata*), mung beans (*Vigna radiata*), groundnut (*Arachis hypogaea*) and pigeon pea (*Cajanus cajan*) contribute the necessary protein. Based on traditional farmers' yield levels, a very conservative estimate shows that 1 ha of cassava intercropped with black common bean can produce 10 t ha^{-1} of fresh cassava roots with 30% starch and 600 kg ha^{-1} of beans with 28% protein. At a caloric value of 4480 kcal kg^{-1} of starch, this would provide the following amounts of food energy and protein:

$$10,000 \text{ kg of cassava} = 13.44 \times 106 \text{ kcal}$$
$$= 56,270 \text{ MJ}$$
$$600 \text{ kg of beans} = 168 \text{ kg of protein.}$$

Assuming that the daily requirement of an adult person is 10.5 MJ (2500 kcal) and 100 g of protein, then 1 ha would supply 5376 caloric rations and 1680 protein rations, i.e. 1680 complete rations and a surplus of 3696 caloric rations or 38,686 MJ (9.24 × 106 kcal), without considering the protein content of cassava or the caloric value of beans. Thus 1 ha of a cassava–common bean intercrop supplies the annual food requirement for approximately five adults, leaving a surplus of about 6 t of cassava to be fed to animals or sold.

Although this is by no means a complete diet, it shows the enormous potential of cassava intercropping to provide a solid nutritional foundation on which to base a complete diet with minerals and vitamins added through vegetable and fruit consumption. Furthermore, there are still many poor around the world whose daily calorie and protein intake is far below the amounts quoted above.

Species and genotype selection

Cassava is intercropped with both long- and short-season crops. In plantation crops such as coconut palm, oil palm or rubber, the unproductive juvenile period of trees can last 5–7 years (Enjalric *et al.*, 1999). When intercropping newly established rubber with cassava and other food or cover crops in Gabon, early growth and ground cover development of rubber was so slow that four consecutive cycles of food crops were feasible before the trees started to compete seriously with the other crops for light (Leihner and Ziebell, 1998).

Cassava is also intercropped under mature coconut palms or rubber trees in regions of India or China, where arable land is extremely scarce (Cock, 1985). Under trees, the cassava tends to suffer from insufficient sunlight; hence productivity is very low. A selection of more shade-tolerant cultivars may, however, be feasible, yielding at least some extra carbohydrate without requiring additional land.

Intercropping cassava with perennial species is not widespread and the vast majority of systems involve cassava as a long-season crop, combined with short-season annual food or cash crops. Maize, cowpea, common bean and groundnut are the commonest intercropping partners. Associations with grain legumes are particularly promising, not only because of their aforementioned nutritional advantages but also for their soil-improving potential. Some agronomic implications of cassava–legume intercropping were discussed by Leihner (1979), and a comprehensive treatment of the issue was provided by CIAT as a cassava intercropping monograph (Leihner, 1983). Based on this

information, it has been established that in cassava, genotypic traits such as vigour and branching habit (sometimes termed 'leafiness') are important determinants of suitability for intercropping. Cultivars with an erect growth habit (late branching) and medium vigour possibly produce less shade over an intercrop than those with early branching and high initial vigour. Furthermore, cultivars with medium vigour and late branching more closely resemble the ideal plant type for maximum yield in single culture described by Cock *et al.* (1979). It thus appears that medium-vigour genotypes with an erect growth habit are the most suitable for association with low-growing intercrops as they impose little competition on the intercrop initially and also have high yield potential. Only when cassava is intercropped with tall-growing maize, more vigorous plant types may be required to compete favourably with the maize.

When selecting grain legumes for intercropping at the beginning of the cassava growth cycle, an important characteristic of the legume is early flowering and maturity. With early maturity, the period of competition with cassava is reduced and excessive shading of the legume during pod filling is avoided. When both crops grow together in the field for a longer period of time, the interaction between them becomes more accentuated, and yields are mutually affected (Table 6.1). In associations of cassava with early-maturing legumes (common bean, cowpea), yield formation of both crops occurred largely independent from each other. An increasingly negative mutual influence was noticed, however, when the legume growth cycle exceeded 100 days.

Table 6.1. Correlations between yields of cassava and associated legumes with varying number of days to physiological maturity.

	Days to physiological maturity	Correlation of cassava/legume yields (r)
Bean	80	0.01
Cowpea	90	0.05
Groundnut	106	−0.14
Soybean	125	−0.35[a]

[a]Significant at $P = 0.05$.
Source: Leihner (1983).

Relative planting time

Relative planting time – i.e. planting the intercrop before, at the same time or after cassava – has both biological and practical implications. The biological implications include the fact that cassava does not impose much competition at the beginning of its growth cycle, but it does not tolerate much competition either. As a result, cassava yield can be drastically reduced if the intercrop is planted earlier than cassava, creating strong competition for light, water and nutrients at a time when cassava is still a weak competitor. On the other hand, if cassava is planted earlier than the intercrop, shading and competition for other growth factors may affect growth and yield of the latter. Thung and Cock (1979) established that simultaneous planting of cassava and common bean produced greatest total yields. This practice has been verified by growing cassava with various other grain legumes and maize. A practical implication of simultaneous planting is that it requires only one operation instead of two separate procedures to establish the association. To a certain degree, this facilitates the use of mechanization in intercropping systems if already existing machinery is adapted for that purpose.

While relative planting time can help regulate light competition when the associated crops initiate their growth cycle together, the situation is different for an intercrop sown into a fully developed cassava stand. Here, light may be the most limiting factor for the intercrop; nevertheless, observations made at CIAT showed that cassava intercepted less light towards the end of its growth cycle. This allowed the production of common bean intercrops during the last months prior to the cassava harvest. Comparing results of interplanting at 7, 8 and 9 months after cassava, bean yield was reduced the least when interplanting was done at 9 months, beans reaching up to 50% of their yield as a sole crop. It was concluded that the later an intercrop is sown into an already established cassava crop, the better is its yield. Nevertheless, the productivity of an intercrop grown under these conditions is much below that of an association where both crops begin their growth cycle together.

Planting density

In traditional cassava intercropping, farmers tend to use lower planting densities than in sole crops. The reduced number of plants per unit area, together with the competition imposed by one or several intercrops, may partially explain the low productivity of cassava in traditional intercropping systems. There is however, ample scope for improvement. The flexibility of cassava with regard to spatial arrangement (see Fig. 6.3) allows use of a wider-than-usual spacing between rows and still maintain optimum planting density by using smaller plant-to-plant distances within the row. Such a rectangular planting pattern has no adverse effect on cassava yield, but it facilitates the accommodation of intercrops and reduces competition. Different optimum planting densities for genotypes with different growth habits that have been found for sole crops appear to be valid for intercropped cassava as well. With leafy and early-branching cultivars, maximum sole crop yields are obtained at relatively low densities of 5000–8000 plants ha^{-1}. These densities produced the best yields when these cultivars were intercropped with common beans (Thung and Cock, 1979). Cultivars with less foliage and late branching do not show the same degree of coincidence. Nevertheless, this type of cassava when sole cropped still produces up to 92% of maximum yield at intermediate planting densities of 7000–9000 plants ha^{-1}, and also gives acceptable yields (75–90% of maximum) in association with common beans. This suggests that near-optimum planting densities for sole-cropped cassava may also be used in intercropping to obtain best results.

The yield of grain legumes does not vary greatly in response to planting densities within a relatively wide range. Trials with common bean, cowpea and groundnut grown as sole crops and intercropped with cassava showed either constant yields or not very accentuated responses when planting density of the legumes ranged from 50 to 200% of optimum sole-crop density (Leihner, 1983). Using the optimum density for sole-cropped legumes or only slightly increased densities, for cassava–grain legume intercrops frequently results in maximum grain legume yield when legumes are planted simultaneously with cassava. Similar observations have been reported for maize when optimum sole-crop densities were used in intercropping systems with cassava, with rectangular spacings that allowed an easy accommodation of both crops in the field and reduced competition (CIAT, 1981; Meneses, 1980).

Nutrient management

Nutrient requirements of cassava and the crops most frequently intercropped with it are well studied for sole culture conditions (Jacob and von Uexküll, 1973; Andrew and Kamprath, 1978; Asher *et al.*, 1980; Howeler, 1981).

There is, however, little information on nutrient requirements and response to fertilization of cassava and intercrops when grown in association. Intercropping represents an intensification of the demand for nutrients, particularly when each associated crop is planted at its normal single culture density. In this situation the removal of elements from the soil is greater in the intercropping system than in single culture. If these nutrients are not replaced by an adequate nutrient supply, soil fertility deteriorates. Results from research in Colombia (Leihner, 1983) point to a contrasting response of cassava to NPK under sole cropping as opposed to

intercropping conditions. When a cassava–cowpea intercrop and its respective sole crops were amended with 0–300 kg ha^{-1} N, sole cropped cassava root yield only responded positively up to the first increment (50 kg ha^{-1}), then gradually declined below the control level. Sole cropped cassava normally requires only modest quantities of N to reach optimum leaf area index (LAI) for maximum root yield. Any amount of N in excess results in overly heavy top growth to the detriment of root filling. In contrast, intercropped cassava took advantage of an increase in N supply, producing the highest root yields only with the final increment of N (Fig. 6.6). Under competition from cowpeas, top growth and canopy development of cassava was below optimum at low N rates, reaching optimum LAI for maximum root yield only when large amounts of N were applied. Cowpeas, on the other hand, did not show a response to N fertilization under either sole- or intercropped conditions. The response pattern was the same for both cassava and cowpeas when the reaction to K fertilizer was tested in a similar sole crop–intercrop trial.

A totally different behaviour of cassava and cowpea was observed by the same author when the sole- and intercrop response to P was tested on a highly P-deficient soil. Cassava responded with root yield increases up to the last increment of P (132 kg ha^{-1} P) only when sole cropped.

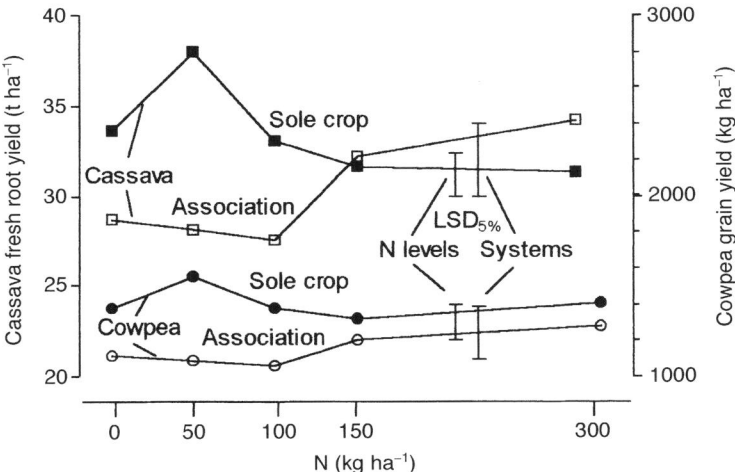

Fig. 6.6. Yield response from cassava and cowpea when sole cropped or intercropped and amended with different levels of nitrogen.

Under intercropped conditions, a negative response was noted, starting with the second increment (44 kg ha^{-1} P). On the other hand, cowpea responded positively from the first to last increment of P, irrespective of cultivation system (Fig. 6.7). At higher P rates, cowpea became so competitive that it reduced growth and yield of cassava whilst taking full advantage of the improved P nutrition for grain yield formation.

These examples demonstrate that the same nutrient management may lead to sharply contrasting responses in mixed crop systems and single cultures. In all cases changes in nutrient supply also changed the competitive ability of cropping-system components. Nutrient management in cassava intercropping systems must therefore be based on specific knowledge of inter-cropping-system responses to a given nutrient as the interaction between crops may lead to different growth and nutrient requirements compared to the sole crops.

Soil Conservation in Cassava-based Systems

Background of soil degradation and conservation in cassava

The problem of soil degradation currently is of worldwide importance, tropical regions being more seriously affected than temperate zones. The erosive nature of tropical rainstorms and the increasing cultivation of marginal and steep lands have reduced the depth of fertile topsoil in many regions of the tropics. In parts of the tropical lowlands, sandy soils have been exhausted and eroded through permanent cultivation. Under these conditions high-value cash crops such as vegetables or grain legumes, can no longer be grown and they are replaced by less demanding but low-value crops, such as cassava. With its slow early growth and poor initial soil cover, cassava creates conditions favourable to water erosion and soil degrada-tion, particularly when cultivated without fertil-izer. Thus agricultural lands degrade further, and environmental constraints build up to such a degree that it may be increasingly difficult for small-scale farmers to grow even cassava, their 'crop of last resort'. Hence there is a need to incorporate soil-conservation components into cassava-production systems.

There are differences among crops regard-ing their tendency to hinder or enhance soil erosion, related to root system development, growth habit and canopy dynamics. A number of studies appear to suggest that cassava is more erosion-enhancing than other crops due to its wide spacing and slow initial canopy development (Howeler, 1998; Putthacharoen *et al.*, 1998), but contradictory results have also

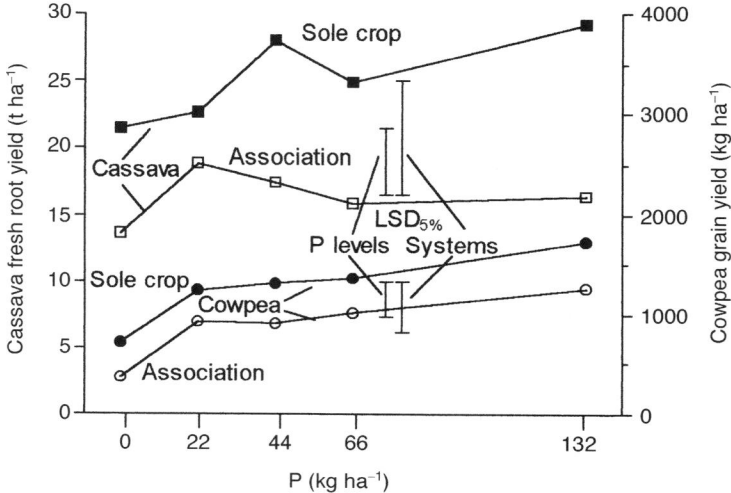

Fig. 6.7. Yield response from cassava and cowpea when sole cropped or intercropped and amended with different levels of phosphorus.

been reported (Howeler, 1991b). In most situations, however, appropriate crop management practices are more important to achieve soil conservation than crop selection. Soil conservation in cassava as in other crops should therefore follow a set of practices that have at least one of the two following objectives in common: (i) maintain soil infiltration rates at sufficiently high levels to reduce runoff to a negligible amount; and (ii) dispose of runoff water safely from the field should rainfall exceed the infiltration capacity of the soil (Lal, 1977). Cultural practices that help maintain a high infiltration capacity frequently involve conservation tillage practices, the use of mulch or live vegetation cover. The safe disposal of runoff is mostly achieved through physical manipulation of the surface by constructing contour bunds or terraces, surface drainage ways, and contour or tied ridges. Some of these practices have been tested successfully in cassava and will be discussed below.

Conservation tillage

Lal (1990) reviewed a wide range of conservation tillage options. He described the two key concepts to be included in conservation tillage for soil and water conservation as: (i) residue mulching; and (ii) an increase in random soil surface roughness. Whilst residue mulching can be implemented in no tillage, minimum tillage or mulch farming systems, an increase in random roughness can be achieved through various forms of soil tillage including chisel ploughing, strip tillage, ridge–furrow systems and tillage methods that cause soil inversion.

Lal (1977, 1990) has pointed out the many advantages of crop production involving a soil cover with mulch. This practice reduces water erosion by reducing raindrop impact, decreases crusting and surface sealing, increases surface storage of runoff water, decreases runoff velocity, improves soil structure and porosity, and improves the biological activity of the soil, favouring the formation of macropores, which maintain high infiltration rates and keep runoff low. Calculating the cropping factor (C-Factor) of the Universal Soil Loss Equation for different cropping systems in Indonesia, Abdurachman et al. (1984, cited in Lal, 1990) found a high C-Factor (little soil protection) of 0.588

for a rice + maize + cassava intercrop without additional soil protection. The same intercrop managed with residue mulch reached a C value of 0.357, whilst the use of rice straw mulch at a rate of 6 t ha^{-1} reduced the C-Factor to 0.079, demonstrating the good soil-protection potential of this management practice. On a sandy, low-fertility soil of the Colombian Atlantic Coast, Cadavid et al. (1998) grew cassava over an 8-year period with an annual application of 12 t ha^{-1} of dry Panicum maximum mulch. Mulch applications significantly increased root and top biomass, increased root dry matter (DM) content whilst reducing its yearly variation, and decreased root HCN, particularly in the absence of fertilizer. Mulch applications also significantly reduced soil temperatures within the top 20 cm and increased soil organic carbon, K, P, Ca and Mg. Without mulch, soil pH and root yield decreased over the years.

Although mulching is a useful concept and has its well-documented virtues in many crops including cassava, its practical use in farmers' fields is minimal. There are a variety of possible reasons: (i) the desired amount of mulch may not be available in all ecological regions and for all farming systems; (ii) labour for harvesting, transporting and applying the mulch may not be at hand or too costly; and (iii) on small farms there may be competing uses for mulching material, which might be needed as cattle feed, for roof thatching or other purposes. In these situations conservation tillage may include other tillage techniques such as contour ridges, tied ridges, raised bed systems and broad-bed furrow systems.

On either sloping or flat land, ridges increase surface roughness and help reduce runoff. On sloping land of two southwest Colombian test sites, Reining (1992) examined runoff control and soil-conservation effectiveness of contour ridges in cassava. Among six systems tested, ridges together with contour grass strips had the lowest average total runoff and the lowest average as well as maximum runoff rates at both test sites. Soil loss of 3 t ha^{-1} across test sites and years was amongst the lowest of all treatments, being similar to the minimum tillage treatment where cassava was planted in an existing grass or weed cover. With contour ridging, cassava fresh root yields of up to 31 t ha^{-1} were among the best of all soil-conservation treatments, whereas the minimum tillage system produced an average yield of just above 10 t ha^{-1}.

In South-East Asia intensive cassava cultivation on predominantly sandy, low-fertility oxisols poses a severe risk of soil degradation through water erosion. Howeler (1991b, 1998) presented an overview of the problem and discussed options on how to control it. Under tropical Asian conditions, conventional tillage including ploughing and harrowing – although leading to greater cassava yields – poses great risks of soil erosion unless followed by contour ridging. On Hainan Island in southern China, ploughing and harrowing followed by contour-ridge preparation produced fresh root yields of 26.3 t ha^{-1}, the greatest among seven methods of soil preparation examined, with the second smallest amount of erosion.

This information corroborates the general soil-conservation effectiveness of contour ridging. Special caution is warranted, however, when slopes are too steep, no proper sideward inclination of the furrows is used to facilitate surface drainage, or rainfall is overly heavy. Under these conditions too much water may accumulate behind the ridges, causing them to break and open the way to gully erosion, which leads to even greater soil losses than without ridges. In these cases an additional soil cover or more solid structures to reduce runoff and erosion, such as planted barriers, may be needed.

Cover crops

In the context of conservation tillage, Lal (1990) reviewed information on perennial cover crops providing ample documentation on their runoff and erosion-reducing potential. He also provided a list of suitable grasses and legumes for both tropical and temperate zones. Whilst most of the listed species are suitable only for rotations because of their strong competitive effect, a few (e.g. *Mucuna utilis*) can also be used as a simultaneous cover with food crops, being termed as 'live mulches'.

In cassava a fast-establishing simultaneous cover may be advantageous as soil protection by the crop itself usually sets in too late to be effective during the critical 2–4 months after planting. Grain legumes such as common bean, cowpea, mungbean or groundnut grown as intercrops

simultaneously with cassava, can provide a rapid ground cover without being overly aggressive competitors. This might be one reason why farmers in Indonesia frequently intercrop legume food crops with cassava (Howeler, 1998). The positive effect on farm income is seen as an additional benefit of this form of soil protection. Reports from southwestern Colombia were less encouraging, where Reining (1992) recorded a cassava solecrop ground cover of only 10–35% 2 months after planting. To accelerate soil protection, he used intercropped cowpea and common bean, reaching average soil covers of 58% after 2 months at the warmer test site and 50% at the cooler one. However, the improved early ground cover reached through intercropping did not result in less erosion. Average soil loss in the cassava–legume intercrops was 26 versus 3 t ha^{-1} with sole cropped cassava planted on contour ridges. This was apparently the result of the more intense soil preparation to obtain a fine legume seed bed, together with more compaction during manual legume seeding.

As annual grain legume intercropping is not always an effective option for soil protection in cassava, the focus has frequently been on perennial legume covers. Although the maintenance of a continuous, simultaneous ground cover in cassava has definite beneficial effects in controlling runoff, erosion and nutrient leaching, root yields can be suppressed by vigorously growing or climbing legumes. Howeler (1991a) stressed the cassava–legume competition issue. When cassava was planted in an already established legume cover crop, soil protection was good, but cassava yields generally decreased because of severe competition from the legume. With less-aggressive legumes such as *Arachis pintoi*, the drop in cassava yields is slight, but with highly productive legumes such as *S. guianensis*, it is considerable. The deep-reaching legume taproots compete strongly with cassava for nutrients and for water during droughts. Leihner *et al.* (1996b) pointed to the difficulty of establishing cover legumes under cassava in the first year of cultivation and to their competitive effects at later stages. Averaged over the first 2 years after legume undersowing, cassava yield was reduced by 37% with *P. phaseoloides*, by 35% with *Z. glabra* and by 27% with *C. acutifolium*. Even when *C. rotundifolia*, a legume with a creeping, non-aggressive growth habit was used, yield reductions on the order of 10–20% were common, restricting the attractiveness of this combination to small farmers with cattle who could make use of additional forage production.

A realistic view of cover crops in cassava appears to be that despite their positive contribution to reducing soil degradation, their adoption is compromised by difficulties in obtaining seed or planting material, laborious establishment in the field, and their adverse impact on root yield. Their actual use may thus remain limited to a rather restricted set of ecologies and farming systems where their undoubted potential is not offset by existing disadvantages.

Live barriers

Engineered soil conservation options to shorten or interrupt slopes, such as contour bunds or bench terraces are prohibitively expensive for small producers in the tropics. As a result, there is interest in low-capital technologies for reaching these objectives. Live barriers formed by grass strips or barrier hedges are among these options. An overview provided by Gallacher (1990) stated that grass strips have almost become a tradition in several countries where cultivated fields would have been more severely eroded without the filtering and retarding effect of the strips on runoff. Furthermore, the author states that grass strips do not always need to be planted, they can be left to establish naturally from native grasses if unhoed or unploughed strips are left in the field. A list of suitable grasses, perennial legumes and other plants was presented by Stocking (1993), who described the typical situation in which the specific kind of material and conservation practice was used, the farming system, the current technical recommendations for implementation, possible support practices and variations in implementation.

Among the soil conservation practices tested in cassava fields in a number of Asian countries, Howeler (1991b) found that the most successful were fertilizer application, minimal tillage, contour ridging, subsoiling, closer plant spacing, intercropping, mulching and planting live barriers of grasses, legumes or hedgerow trees. On Hainan Island, for example, experiments conducted by the South China Academy of Tropical Crops showed the effect of live barriers

planted with *S. guianensi* in reducing runoff and erosion. In reports from China, Vietnam and Thailand, Howeler *et al.* (1998) demonstrated the same positive results obtained with vetiver grass (*Vetiveria zizanioides*) barriers.

Working on Andean inceptisols in southwestern Colombia, Ruppenthal (1995) and Leihner *et al.* (1996b) described long-term testing of cassava soil conservation systems, including live barriers of dwarf elephant grass (*Pennisetum purpureum* cv. Mott) and vetiver grass. Average runoff rates (% of total rainfall) on slopes with a 7–20% gradient were lowest with contour ridges (3.6%), followed by vetiver grass (4.0%) and dwarf elephant grass (4.2%) barriers. After full establishment of the grasses, cassava yield with dwarf elephant grass barriers reached 81% and with vetiver grass barriers 90% of sole cropped cassava planted on flat land. Although yield reduction was partially due to a reduced cropping surface, it also reflects the greater competitiveness of elephant grass compared to vetiver grass. Grasses vary widely with regard to the aggressiveness with which their roots expand into neighbouring crop areas, vetiver grass being among the least spreading and competing species (Tscherning *et al.*, 1995).

Results of individual research projects and on-farm testing in the Andean region in respect of the use of live barriers in cassava were summarized by Müller-Sämann and Leihner (1999). Grasses with different uses were identified as suitable for live barrier planting in cassava. Vetiver grass, a non-forage species, exhibited outstanding technical properties as a soil-conservation component. It is recommended for very critical, erosion-prone situations on already degraded land. Citronella grass (*Cymbopogon nardus*) was less dense and effective, and its root system competed slightly more with cassava; however, it is also recommended for acid, low-fertility hillside soils because of its good adaptation and ease of propagation and handling. Adoption constraints for using this non-forage grass were overcome by constructing an essential oil extraction plant, adding value to this by-product of soil conservation. Among forage species, imperial grass (*Axonopus scoparius*) and dwarf elephant grass were the most promising, although the latter competed severely with the adjacent crop 3–4 years after establishment. Based on work with farmers, it was concluded that despite their usefulness for cut-and-carry systems, the use of forage grasses was difficult to implement on a larger scale due to high establishment and maintenance costs, amounting to 8 man-days 1000 m^{-1} or approximately US$74 for imperial grass barriers. Furthermore, transporting large quantities of forage on sloping land was difficult. It was concluded that barriers on fields distant from homesteads or stables should be functional, yet produce only a modest amount of biomass with high potential value to reduce soil-conservation costs.

Agroforestry systems with cassava

Forest areas generally show less nutrient leaching and runoff, resulting in smaller soil losses and less degradation as compared with agricultural fields planted to seasonal crops. Trees appear to have a stabilizing influence, most likely related to the permanent soil cover provided by the tree canopy, the leaf litter and the undergrowth, as well as to the fine, dense root distribution in the top layers of forest soils. Agroforestry systems try to make use of some of these advantages by combining crops with a tree component. In the end, however, management practices, more than tree or crop characteristics, determine whether the land use system has

a conservation-enhancing character and is sustainable (Lal, 1990).

Agroforestry can be regarded as a special type of intercropping system. Thus observations on cassava intercropped with trees (e.g. plantation crops) may also be relevant to cassava grown in agroforestry systems. Although cassava is a frequent element in these systems, little specific information on tree–cassava interaction is available. In southern Benin, Akondé et al. (1996) conducted alley cropping research with cassava and maize (Fig. 6.8). Over a 6-year period, cassava root yields were increased by applying an average of 3 t ha^{-1} of C. cajan DM as a mulch obtained from 4-m spaced hedgerows, but increases were significant only when mulch and a mineral fertilizer were used. When Leucaena leucocephala hedgerows were grown with cassava, twice the amount of mulch was produced; but cassava yields did not increase with or without mineral fertilizer, presumably due to the much stronger competitive effect of Leucaena as compared to Cajanus. Studying light competition in the same cropping systems, Leihner et al. (1996a) concluded from light-transmission and row-position data that there were no important shading effects in tree–crop competition when trees were pruned two to three times a year. Competition effects were attributed mostly to the interference of lateral tree roots with those of cassava. Lateral root spread was

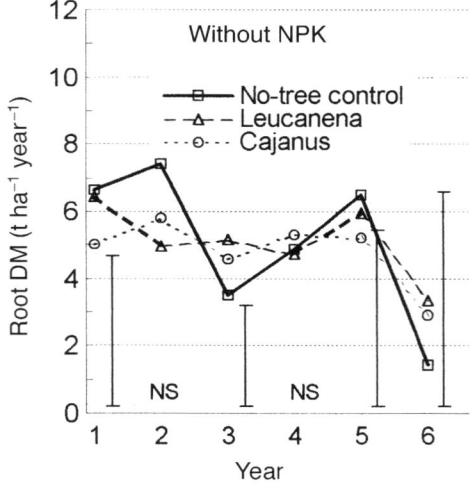
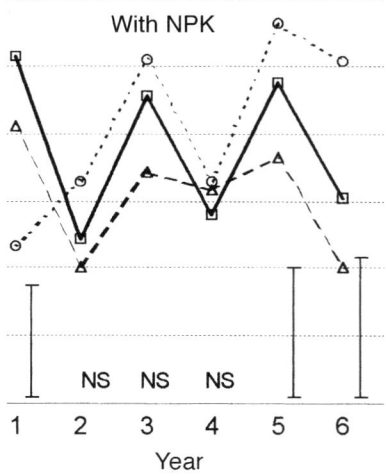

Fig. 6.8.   Effect of NPK fertilization (90 N, 39 P, 75 K) and alley cropping on cassava root yield during 6 years of crop rotation with maize on an ultisol in Benin. (Source: Akondé et al., 1996.)

much more aggressive with *Leucaena* than with *Cajanus*, which showed a characteristic taproot pattern. Considering the specific nutrient demands of cassava, *Leucaena* appeared to contribute a high amount of N whilst competing strongly for K. Given the moderate N, but high K requirement of cassava, this may have caused excessive leaf and stem growth, leading to low harvest indices and root yields.

This observation points to possible crop–tree preferences that may be founded on plant–nutrient relationships, as well as on other, yet-unexplored complementarities or antagonisms based on physiological, agronomic or microbiological interactions. In agroforestry systems with cassava, leaf cuttings from the forage legume *Flemingia macrophylla* were found to have a more positive influence on root yield than other materials, whereas *Gliricidia sepium* was reported to benefit maize in particular (Böhringer and Leihner, 1997).

Rotations

General observations on the effects of rotation as opposed to continuous cultivation systems by Gallacher (1990) and Lal (1990) coincide in that rotations benefit both soil conservation and crop productivity. Different crops deplete or recycle soil nutrients to different degrees; others may add nitrogen. Rotations can restore organic materials and promote biological soil activity with beneficial consequences for structural stability. This in turn improves soil resistance to erosion and increases infiltration capacity. Furthermore, rotations are said to slow down the build-up of pathogenic soil-borne microorganisms and noxious weed populations.

Leihner and Lopez (1988) described a cassava rotation study on a fertile inceptisol (typic eutrandept) of the Colombian central Andean mountain chain. Over a 9-year period, sole cropped, unfertilized cassava showed a fresh root yield decline from 37 to 12 t ha^{-1}. After five farmer-managed cassava sole crop cycles, the field was subdivided, and a combination of fertilization and rotation treatments were established on large, commercial-size plots. Moderate fertilization with macro- and micronutrients had no positive influence on cassava productivity. In a rotational scheme with *Crotalaria juncea* as green manure,

maize, cassava, maize, common bean, sorghum and cassava again, cassava yields returned to an average 30 t ha^{-1} level even without fertilization. It appeared that soil nutrients were not deficient, but cassava may not have been able to make use of them due to an overall biological degradation of the soil after so many years of unilateral use by a monoculture. Soil organic matter in the 'continuous cassava' system was stagnant with a tendency to decline. After 4 years of rotation cropping, however, organic matter increased by an average 1.5%. Legumes (*Crotalaria*, common bean) formed effective nodulation (suggesting efficient nitrogen fixation) and they improved P availability significantly, which was not observed under monoculture. With respect to mycorrhizal sporulation, spore counts were 36% higher at the end of the fourth year of rotation than for monoculture.

For healthy growth and good yield, cassava depends strongly on mycorrhizal symbiosis. Soil life as a whole may have been more active with rotation than under continuous cassava cultivation. Whilst there was no singular disease, insect, weed or nutritional problem responsible for the dramatic yield decline in monoculture cassava, it would appear that several minor interrelated factors added up to a substantial effect. The somewhat weaker monoculture cassava appeared to have suffered more from hornworm (*Erinnyis ello*) attack than the vigorous rotation crops. More defoliation allowed a more serious weed problem to develop in the cassava monoculture. In contrast, the greater diversity in the rotation system kept individually minor – but, in their combined action, important – phytosanitary problems below an economically relevant threshold level.

Another example of successfully controlling soil degradation and yield decline in cassava with rotations was reported by Müller-Sämann and Leihner (1999). Under cassava monoculture, the breakdown of structural stability (accompanied by an increase in erosion) occurred after just two crop cycles. Nevertheless, the factors governing erodibility such as degree of aggregation and structural stability could be restored effectively with agronomic practices including undersown cover crops, minimum tillage, weed fallow and grass–legume mixtures in rotation, the latter having the profoundest and longest lasting effect on soil strength. An analysis of hot water-extractable carbohydrates (secretions of soil microorganisms) showed that cassava–grass–legume rotations enhanced microbial soil activity, which resulted in significant levels of binding substances for soil particles, thereby increasing aggregation and decreasing erodibility (Häring, 1997). When using a modified 'productivity index' model (Flörchinger, 1999), it was possible to predict that traditional cassava monoculture would lead to complete yield loss after 25–84 years, depending upon whether a best or worst case scenario was adopted. When soil conservation practices such as contour ridges, grass barriers or rotations were used, cassava production could be expected to be feasible for another 90–100 years with only minor yield losses of 4–14%.

These examples demonstrate the importance of diversified cassava production systems. Rotations with green manure, live mulch or crops of the *Gramineae* family can potentially lead to balanced nutrient extraction and recycling whilst counteracting the build-up of cassava-specific pests. Enhancing soil life by rotating with grass–legume mixtures leads to the chemical, physical and biological restoration of degraded soils, thereby reducing vulnerability to degradation by erosion in cassava-based systems.

References

Abdurachman, A., Abujamin, S. and Kurnia, U. (1984) Soil and crop management practices for erosion control. *Pemberitaan Penelitian Tanah Dan Pupuk* 3, 7–12.

Akondé, T.P., Leihner, D.E. and Steinmüller, N. (1996) Alley cropping on an ultisol in subhumid Benin. Part 1: Long-term effect on maize, cassava and tree productivity. *Agroforestry Systems* 34, 1–12.

Andrew, C.S. and Kamprath, E.J. (eds) (1978) *Mineral Nutrition of Legumes in Tropical and Subtropical Soils*. Proceedings of a workshop held at Cunningham Laboratory, Brisbane, Australia, 16–21 January 1978. Commonwealth Scientific and Industrial Research Organisation, Melbourne, Australia.

Asher, C.J., Edwards, D.G. and Howeler, R.H. (1980) Nutritional disorders of cassava (*Manihot esculenta* Crantz). Department of Agriculture, University of Queensland, St. Lucia, Australia.

Böhringer, A. and Leihner, D.E. (1997) A comparison of alley cropping and block planting systems in subhumid Benin. *Agroforestry Systems* 35, 117–130.

Cadavid, L.F., El-Sharkawy, M.A., Acosta, A. and Sánchez, T. (1998) Long-term effects of mulch, fertilization and tillage on cassava grown in sandy soils in northern Colombia. *Field Crops Research* 57, 45–56.

Castro M., A. (1979) The new technology for cassava production. Paper presented at a Workshop on Pre-release Testing of Agricultural Technology. Centro Internacional de Agricultura Tropical, Cali, Colombia.

Castro M., A., Holguin, V.J. and Villavicencio, G.A. (1978) Efecto del ángulo de corte y posición de siembra de la estaca sobre el rendimiento de la yuca. Mimeograph. Cassava Program, Centro Internacional de Agricultura Tropical, Cali, Colombia.

Celis, E. and Toro, J.C. (1974) Métodos de semeadura e cuidados iniciais a tomar na cultura da mandioca. In: *CIAT, Curso Especial de Aperfeiçoamento para Pesquisadores de Mandioca*. CIAT, Cali, Colombia, pp. 182–186.

CIAT (1973) *Annual Report for 1972*. Cassava Program, CIAT, Cali, Colombia, pp. 69–74.

CIAT (1976) Agronomy. In: *Cassava Production Systems, Annual Report for 1975*. Cassava Program, CIAT, Cali, Colombia, pp. B47-B50.

CIAT (1977) *Annual Report for 1977*. Cassava Program, CIAT, Cali, Colombia, pp. C41–C52.

CIAT (1979) *Annual Report for 1978*. Cassava Program, CIAT, Cali, Colombia, pp. A58 & A68.

CIAT (1980) *Annual Report for 1979*. Cassava Program, CIAT, Cali, Colombia, pp. 57–60.

CIAT (1981) *Annual Report for 1980*. Cassava Program, CIAT, Cali, Colombia, pp. 74–77.

CIAT (1982) *Annual Report for 1981*. Cassava Program, CIAT, Cali, Colombia, pp. 176–182.

Cock, J.H. (1985) *Cassava – New Potential for a Neglected Crop*. Westview Press, Boulder, Colorado.

Cock, J.H., Wholey, D. and Lozano, J.C. (1976) *A Rapid Propagation System for Cassava*. Series EE-20. Centro Internacional de Agricultura Tropical, Cali, Colombia.

Cock, J.H., Castro M., A. and Toro, J.C. (1978) Agronomic implications of cassava harvesting. In: Weber, E.J., Cock, J.H. and Chouinard, A. (eds) *Cassava Harvesting and Processing*. Proceedings of a workshop held, CIAT, Cali, Colombia, 24–28 April. International Development Research Centre (IDRC), Ottawa, Canada, pp. 60–65.

Cock, J.H., Franklin, D., Sandoval, G. and Juri, P. (1979) The ideal cassava plant for maximum yield. *Crop Science* 19, 271–279.

Correa, H. and Vieira Neto, J.C. (1978) Conservação de ramas. In: *Curso de Capacitação de Técnicos para a Agricultura da Mandioca*. Ministério da Educação e

Cultura and Escola Superior de Agricultura de Lavras (ESAL), Lavras, Brazil, pp. 23–31.

Delgado, A. and Quevedo, H. (1977) Observaciones sobre el cultivo de la yuca en el sur del lago. Facultad de Agronomía, Universidad de Zulia, Maracaibo, Venezuela.

Doll, J.D. and Piedrahita, C.W. (1976) Methods of weed control in cassava. Centro Internacional de Agricultura Tropical, Series EE-21. CIAT, Cali, Colombia.

Doll, J.D., Piedrahita, C.W. and Leihner, D.E. (1982) Métodos de control de malezas en yuca (*Manihot esculenta* Crantz). In: *Yuca: Investigación, Producción y Utilización*. Referencia de los cursos de capacitación sobre yuca dictados por el Centro Internacional de Agricultura Tropical, CIAT, Cali, Colombia. CIAT and Programa de las Naciones Unidas para el Desarrollo (PNUD), Cali, Colombia, pp. 241–249.

Enjalric, F., Nguema, J., Hugot, N. and Chan Ho Tong, S. (1999) Hévéa brasiliensis et cultures vivrières associées: un système de culture pour le développement au Gabon. *Plantations, Recherche, Développement* 6(1), 5–9.

Ezumah, H.C. and Okigbo, B.N. (1980) Cassava planting systems in Africa. In: Weber, E.J., Toro, J.C. and Graham, M. (eds) *Cassava Cultural Practices*. Proceedings of a workshop held in Salvador, Bahia, Brazil, 18–21 March. IDRC Publication No. 151e, International Development Research Centre, Ottawa, Canada, pp. 44–49.

Flörchinger, F. (1999) Effects of soil erosion on crop productivity in the south-west Colombian Andes. PhD thesis, University of Hohenheim, Stuttgart, Germany.

Gallacher, R.N. (1990) The search for low-input soil and water conservation techniques. Concepts and trends. In: Baum, E., Wolff, P. and Zöbisch, M.A. (eds) *Topics in Applied Resource Management in the Tropics*, vol. 2: *Experience with Available Conservation Technologies*. Deutsches Institut für Tropische und Subtropische Landwirtschaft (GmbH), Witzenhausen, pp. 11–37.

Häring, A. (1997) Indikatoren der Bodendegradation, Strukturelle Stabilität und Bodenerosion in Traditionellen und Alternativen Maniokanbausystemen Südkolumbiens. MSc thesis, University of Hohenheim, Stuttgart, Germany.

Howeler, R.H. (1981) Nutrición mineral y fertilizacion de la yuca (*Manihot esculenta* Crantz). CIAT, Cali, Colombia.

Howeler, R.H. (1991a) Long-term effect of cassava cultivation on soil productivity. *Field Crops Research* 26, 1–18.

Howeler, R.H. (1991b) Erosion control and preservation of soil fertility: trials in Asia and Latin America. In: *CIAT Report 1991*. CIAT, Cali, Colombia, pp. 68–78.

Howeler, R.H. (1998) Cassava agronomy research in Asia – an overview 1993–1996. In: *Cassava Breeding, Agronomy and Farmer Participatory Research in Asia. Proceedings of the Fifth Regional Workshop*, Danzhou, Hainan, China. 3–8 November 1996. CIAT, Cali, Colombia, p. 335.

Howeler, R.H., Nguyen The Dang and Wilawan Vongkasem (1998) Farmer participatory selection of vetiver grass as the most effective way to control erosion in cassava-based cropping systems in Vietnam and Thailand. In: *Proceedings First International Conference on Vetiver: a Miracle Grass*. Chiang Rai, Thailand, 4–8 February 1996, Office of the Royal Development Projects Board, Bangkok, pp. 259–272.

Jacob, A. and Uexküll, H. von (1973) Fertilización, nutrición y abono de las cultivas tropicales y subtropicales, 4th edn. Ediciones Euroamericanas, Mexico D.F.

Koch, L. (1916) Het planten van de cassave volgens de methode van Heemstede Obelt vergeleken met de gewone bij de bevolking in zwang zijnde methoden. *Teysmannia* 27, 240–245.

Lal, R. (1977) Soil-conserving versus soil-degrading crops and soil management for erosion control. In: Greenland, D.J. and Lal, R. (eds) *Soil Conservation & Management in the Humid Tropics*. John Wiley & Sons, Chichester, pp. 81–86.

Lal, R. (1990) *Soil Erosion in the Tropics: Principles & Management*. McGraw-Hill, New York.

Leihner, D.E. (1979) Agronomic implications of cassava–legume intercropping systems. In: Weber, E., Nestel, B. and Campbell, M. (eds) *Intercropping with Cassava*. Proceedings of an International Workshop, Trivandrum, India, 27 November–1 December 1978, IDRC Publication No. 142e. IDRC, Ottawa, Canada, pp. 103–112.

Leihner, D.E. (1980) Cultural control of weeds in cassava. In: Weber, E.J., Toro, J.C. and Graham, M. (eds) *Cassava Cultural Practices*. Proceedings of a workshop, Salvador, Bahia, Brazil, 18–21 March. IDRC Publication No.151e, IDRC, Ottawa, Canada, pp. 107–111.

Leihner, D.E. (1983) *Management and Evaluation of Intercropping Systems with Cassava*. CIAT, Cali, Colombia.

Leihner, D.E. (1984a) The production of planting material in cassava: some agronomic implications. In: *Proceedings of the Sixth Symposium of the International Society for Tropical Root Crops*, 21–26 February 1983. Centro Internacional de la Papa, Lima, Peru, pp. 247–256.

Leihner, D.E. (1984b) Storage effects on planting material and subsequent growth and root yield of cassava (*Manihot esculenta* Crantz). In: *Proceedings of* the Sixth Symposium of the International Society for Tropical Root Crops, 21–26 February 1983. Centro Internacional de la Papa, Lima, Peru, pp. 257–266.

Leihner, D.E. (1986) Storage and regeneration of cassava (*Manihot esculenta* Crantz) planting material. In: *Proceedings of a Global Workshop on the Propagation of Root and Tuber Crops*, 12–16 September 1983. CIAT, Cali, Colombia, pp. 131–138.

Leihner, D.E. and Castro M., A. (1979) Prácticas sencillas para aumentar el rendimiento del cultivo de la yuca (*Manihot esculenta* Crantz). In: *Proceedings of the 25th Annual Assembly of 'Programa Cooperativo Centroamericano para el Mejoramiento de Cultivos Alimenticios (PCCMCA)'*, Tegucigalpa, DC, Honduras, 19–23 March.

Leihner, D.E. and Lopez, J. (1988) Effects of different cassava cropping patterns on soil fertility, crop yields and farm income. In: Degras, L. (ed.) *Proceedings of the Seventh Symposium of the International Society for Tropical Root Crops*, Gosier, Guadaloupe (FWI), 1–6 July 1985. Institut National de Recherche Agronomique (INRA), Paris, pp. 463–474.

Leihner, D.E. and Ziebell, W. (1998) Management of rubber-based intercropping systems: Optimisation of radiant energy utilisation. In: *Fonctionnement des Cultures Associées à Base d'Hévéa. Rapport Final*, Programme STD 3, Contrat No. TS3-CT 92–0148. Centre de Coopération Internationale en Recherche Agronomique pour le Développement, Département des Cultures Pérennes (CIRAD-CP), Paris, pp. 1–20.

Leihner, D.E., Ernst-Schaeben, R., Akondé, T.P. and Steinmüller, N. (1996a) Alley cropping on an ultisol in subhumid Benin. Part 2: Changes in crop physiology and tree-crop competition. *Agroforestry Systems* 34, 13–25.

Leihner, D.E., Ruppenthal, M., Hilger, T.H. and Castillo, F.J.A. (1996b) Soil conservation effectiveness and crop productivity of forage legume intercropping, contour grass barriers and contour ridging in cassava on Andean hillsides. *Experimental Agriculture* 32, 327–338.

Lozano, J.C., Toro, J.C., Castro, A. and Bellotti, A.C. (1977) *Producción de Material de Siembra de Yuca*. Serie GS-17. CIAT, Cali, Colombia.

Lozano, J.C., Bellotti, A., Reyes, J.A., Howeler, R., Leihner, D. and Doll, J. (1981) *Field Problems in Cassava*, 2nd edn, CIAT Series No. 07EC-1. CIAT, Cali, Colombia.

Mattos, P.L.P. de, Gomes, C.J. de and Matos, A.P. de (1973) Cultura da mandioca. Circular no. 27. Instituto de Pesquisas Agropecuárias do L'este, Cruz das Almas, Brazil.

Meneses, R. (1980) Efecto de diferentes poblaciones de maíz (*Zea mays* L.) en la producción de raíces de yuca (*Manihot esculenta* Crantz) al cultivarlos en asocio. MSc thesis, Universidad de Costa Rica and Centro Agronómico Tropical de Investigación y Enseñanza (CATIE), Turrialba, Costa Rica.

Montaldo, A. (1966) El cultivo de la yuca. Publicación Divulgativa No. 4. Instituto de Agronomía, Universidad Central de Venezuela, Maracay, Venezuela.

Müller-Sämann, K.M. and Leihner, D.E. (1999) Soil degradation and crop productivity research for conservation technology development in Andean hillside farming. Final Report, GTZ Project No. 95.7860.0–001.04. Institute of Plant Production and Agroecology in the Tropics and Subtropics, University of Hohenheim, Stuttgart.

Nitis, I.M. (1977) Stylosanthes as companion crop to cassava (*Manihot esculenta*). Report to International Foundation for Science (IFS), Research Grant Agreement no. 76. Faculty of Veterinary Science and Animal Husbandry, Udayana University, Denpasar (Bali), Indonesia.

Nitis, I.M. and Suarna, M. (1977) Undergrowing cassava with stylo grown under coconut. In: Cock, J.H., MacIntyre, R. and Graham, M. (eds) *Proceedings of the Fourth Symposium of the International Society of Tropical Root Crops*, CIAT, Cali, Colombia, 1–7 August 1976. IDRC-080e. IDRC, Ottawa, Canada, pp. 98–103.

Normanha, E.S. and Pereira, A.S. (1950) Aspectos agronômicos da cultura da mandioca, *Manihot utilissima* Pohl. *Bragantia* 10(7), 179–202.

Oka, M., Limsila, J., Sarakarn, S. and Sinthuprama, S. (1987) Planting materials and germination ability. In: *Eco-Physiological Studies on Cassava (Manihot esculenta Crantz) in Thailand*. Tropical Agriculture Research Centre, Ministry of Agriculture, Forestry and Fisheries, Japan and Department of Agriculture, Ministry of Agriculture and Cooperatives, Thailand, pp. 157–183.

Putthacharoen, S., Howeler, R.H., Jantawat, S. and Vichukit, V. (1998) Nutrient uptake and soil erosion losses in cassava and six other crops in a Psamment in eastern Thailand. *Field Crops Research* 57, 113–126.

Reining, L. (1992) *Erosion in Andean Hillside Farming*, Hohenheim Tropical Agricultural Series No. 1. Margraf Verlag, Weikersheim, Germany.

Rodriguez Nodals, A. (1980) Mechanical planting and other cassava cultural practices in Cuba. In: Weber, E.J., Toro, J.C. and Graham, M. (eds) *Cassava Cultural Practices*. Proceedings of a workshop, Salvador, Bahia, Brazil, 18–21 March, IDRC Publication no. 151e. IDRC, Ottawa, pp. 118–119.

Ruppenthal, M. (1995) *Soil Conservation in Andean Cropping Systems*, Hohenheim Tropical

Agricultural Series No. 3. Margraf Verlag, Weikersheim, Germany.

Sales Andrade, A. and Leihner, D. (1980) Influence of period and conditions of storage and growth and yield of cassava. In: Weber, E.J., Toro, J.C. and Graham, M. (eds) *Cassava Cultural Practices*. Proceedings of a workshop, Salvador, Bahia, Brazil, 18–21 March. IDRC Publication No. 151e. IDRC, Ottawa, pp. 33–37.

Santos, E.O., Bessa, M. and Lima, P.B. (1972) Mandioca; recomendações tecnológicas. Circular no. 8. Ministério de Agricultura, Departamento Nacional de Pesquisa Agropecuária, Instituto de Pesquisa Agropecuária do Nordeste, Brasilia, Brazil.

Silva, J.R. da (1970) O programa de investigação sobre mandioca no Brasil. In: *Primeiro Encontro de Engenheiros Agronômos Pesquisadores en Mandioca dos Países Andinos e do Estado de São Paulo*. Instituto Agronômico do Estado de São Paulo, Campinas, Brasil, pp. 61–72.

Stocking, M. (1993) Soil and water conservation for resource-poor farmers: designing acceptable technologies for rainfed conditions in eastern India. In: Baum, E., Wolff, P. and Zöbisch, M.A. (eds) *Topics in Applied Resource Management in the Tropics*, Vol. 3: *Acceptance of Soil and Water Conservation Strategies and Technologies*. Deutsches Institut für Tropische und Subtropische Landwirtschaft (GmbH), Witzenhausen, Germany, pp. 291–305.

Tan, K.H. and Bertrand, A.R. (1972) Cultivation and fertilization of cassava. In: Hendershott, C.H. (ed.) *A Literature Review and Recommendations on Cassava*. University of Georgia, Athens, Georgia, pp. 37–72.

Tardieu, M. and Fauche, J. (1961) Contribution á l'etude des techniques culturales chez le manioc. *Agronomie Tropicale* 16(4), 375–386.

Thung, M. and Cock, J.H. (1979) Multiple cropping cassava and field beans: status of present work at the International Center of Tropical Agriculture (CIAT). In: Weber, E., Nestel, B. and Campbell, M. (eds) *Intercropping with Cassava*. Proceedings of an international workshop, Trivandrum, India, 27 November–1 December 1978, IDRC Publication no. 142e. IDRC, Ottawa, Canada, pp. 7–16.

Toro, J.C. and Atlee, C. (1980) Agronomic practices for cassava production. In: Weber, C.J., Toro, M.J.C. and Graham, M. (eds) *Proceedings Workshop on Cassava Cultural Practices*, Salvador, Bahia, Brazil, Series IDRC-151e. IDRC, Ottawa, pp. 13–28, 138–152.

Tscherning, K., Leihner, D.E., Hilger, T.H., Müller-Sämann, K.M. and El-Sharkawy, M.A. (1995) Grass barriers in cassava hillside cultivation: rooting patterns and root growth dynamics. *Field Crops Research* 43, 131–140.

Wholey, D. (1977) Changes during storage of cassava planting material and their effects on regeneration. *Tropical Science* 19, 205–216.

Chapter 7
Cassava Mineral Nutrition and Fertilization

Reinhardt H. Howeler

CIAT Regional Office in Asia, Department of Agriculture, Chatuchak, Bangkok 10900, Thailand

Introduction

Cassava is generally grown by poor farmers living in marginal areas with adverse climatic and soil conditions. The crop is very suitable for these conditions because of its exceptional tolerance to drought and to acid, infertile soils. It is often grown on sloping land because of its minimal requirement for land preparation, and its ability to produce reasonably good yields on eroded and degraded soils, where other crops would fail. It has been shown (Quintiliano *et al.*, 1961; Margolis and Campos Filho, 1981; Putthacharoen *et al.*, 1998), however, that growing cassava on slopes can result in severe erosion, with high soil and nutrient losses. Thus cassava cultivation on slopes requires adequate cultural and soil conservation practices that minimize erosion (Howeler, 1994).

Cassava is well adapted to poor or degraded soils because of its tolerance to low pH, high levels of exchangeable aluminium (Al) and low concentrations of phosphorus (P) in the soil solution. Studying the effect of pH on the growth of several crops grown in flowing nutrient solution, Islam *et al.* (1980) reported that cassava and ginger (*Zingiber officinale*) were more tolerant of low pH (< 4) than tomatoes (*Lycopersicon esculentum*), wheat (*Triticum aestivum* L.) or maize (*Zea mays* L.). Centro

Internacional de Agricultura Tropical (CIAT; 1978) and Howeler (1991a) also reported that cassava and cowpeas (*Vigna unguiculata*) were more tolerant of acid soils with high levels of exchangeable Al, and were much less responsive to lime applications than common beans (*Phaseolus vulgaris*), rice (*Oryza sativa*), maize and sorghum (*Sorghum vulgaris*).

Effect of Cassava Production on Soil Fertility

Nutrient absorption, distribution and removal in harvested products

As cassava may grow well in poor and/or degraded soils and few other crops will grow well on those same soils after cassava, it is often believed that cassava is a 'scavenger crop', that is, highly efficient in nutrient absorption from a low-nutrient soil, leaving that soil even poorer than before. Thus it is often concluded that cassava extracts more nutrients from the soil than most other crops, resulting in nutrient depletion and a decline in soil fertility.

Table 7.1 shows the average removal of the major nutrients by cassava roots as compared to that for the harvested products of other crops (Howeler, 1991b). Nitrogen (N) and P removal

Table 7.1. Average nutrient removal by cassava and various other crops, expressed in both kg ha^{-1} and kg t^{-1} harvested product, as reported in the literature.

Crop/plant part	Yield (t ha^{-1})		(kg ha^{-1})			(kg t^{-1} DM produced)		
	Fresh	Dry[a]	N	P	K	N	P	K
Cassava/fresh roots	35.7	13.53	55	13.2	112	4.5	0.83	6.6
Sweet potatoes/fresh roots	25.2	5.05	61	13.3	97	12.0	2.63	19.2
Maize/dry grain	6.5	5.56	96	17.4	26	17.3	3.13	4.7
Rice/dry grain	4.6	3.97	60	7.5	13	17.1	2.40	4.1
Wheat/dry grain	2.7	2.32	56	12.0	13	24.1	5.17	5.6
Sorghum/dry grain	3.6	3.10	134	29.0	29	43.3	9.40	9.4
Common beans[b]/dry grain	1.1	0.94	37	3.6	22	39.6	3.83	23.4
Soybeans/dry grain	1.0	0.86	60	15.3	67	69.8	17.79	77.9
Groundnuts/dry pod	1.5	1.29	105	6.5	35	81.4	5.04	27.1
Sugarcane/fresh cane	75.2	19.55	43	20.2	96	2.3	0.91	4.4
Tobacco/dry leaves	2.5	2.10	52	6.1	105	24.8	2.90	50.0

Source: Howeler (1991b).
[a]Assuming cassava to have 38% DM, grains 86%, sweet potatoes 20%, sugarcane 26%, dry tobacco leaves 84%.
[b]*Phaseolus vulgaris*.
DM, dry matter.

per tonne of dry matter (DM) in cassava roots was actually much lower than that removed by most other crops, while that of potassium (K) was similar or lower than that of other crops. When cassava root yield was very high, as in Table 7.1, the N and P removal per hectare was similar to that of other crops, while K removal was indeed higher than that of any other crop. Similar results were reported by Prevot and Ollagnier (1958) and Amarasiri and Perera (1975). Putthacharoen *et al.* (1998), however, reported that with an average root yield of 11 t ha^{-1} cassava roots removed much less N and P than other crops, while K removal was similar to that of other crops and much less than that removed by pineapple.

Nutrient absorption and distribution are closely related to plant growth rate, which depends on soil fertility and climatic conditions as well as on varietal characteristics. In poor soils fertilizers can markedly increase plant growth and nutrient absorption; while in areas with a long dry season, irrigation can do the same (Fig. 7.1). At 2–3 months after planting (MAP), the tuberous roots become the major sink of DM. At harvest DM is always highest in the roots, usually followed by stems, fallen leaves, leaf blades and petioles (Fig. 7.1). Table 7.2 shows the DM and nutrient distribution at time of harvest for

fertilized and unfertilized plants in Carimagua, Colombia. Fertilization increased total DM production and root yield by about 30%, but nearly doubled the total absorption of P and K, and increased that of N by 61%. Total nutrient absorption was highest for N, followed by K, Ca, Mg, P and S; absorption of micronutrients was very low (Paula *et al.*, 1983). The roots generally accumulated more K than N, followed by P, Ca, Mg and S, while fallen leaves and stems were high in N and Ca. Similar results were reported by Nijholt (1935), Asher *et al.* (1980), CIAT (1980, 1985a,b), Howeler and Cadavid (1983), Paula *et al.* (1983) and Putthacharoen *et al.* (1998).

As indicated above, the amount of nutrients absorbed by the plant or that removed in the root harvest is highly dependent on growth rate and yield, which in turn depend on climate, soil-fertility conditions and variety. Table 7.3 shows the fresh and dry matter production, as well as the nutrient content in the roots and in the total plant from 15 experiments reported in the literature, with yields ranging from 6 to 65 t ha^{-1}. The amounts of nutrients in the roots or in the whole plant were quite variable, but tended to be very high when yields were high and quite low when yields were low. If nutrient removal were

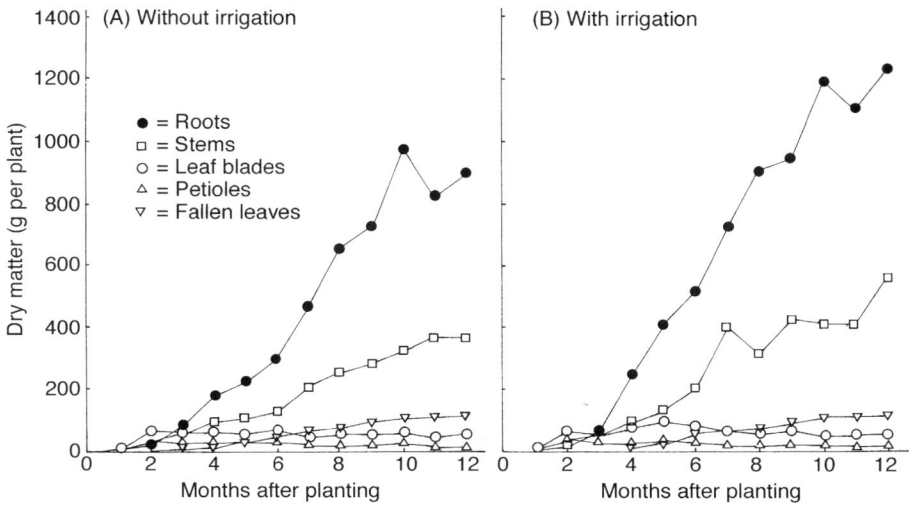

Fig. 7.1. Dry matter distribution among roots, stems, leaf blades, petioles and fallen leaves of fertilized cassava during a 12-month growth cycle in Carimagua (Colombian Eastern Plains), with (B) or without (A) irrigation. (Source: CIAT, 1985b.)

Table 7.2. Dry matter (DM) and nutrient distribution in 12-month-old cassava cv. M Ven 77, grown with and without fertilization in Carimagua, Colombia.

| | (t ha^{-1}) | (kg ha^{-1}) | | | | | | | | | | |
	DM	N	P	K	Ca	Mg	S	B	Cu	Fe	Mn	Zn
Unfertilized												
Top	5.11	69.1	7.4	33.6	37.4	16.2	8.2	0.07	0.03	0.45	0.33	0.26
Roots	10.75	30.3	7.5	54.9	5.4	6.5	3.3	0.08	0.02	0.38	0.02	0.10
Fallen leaves	1.55	23.7	1.5	4.0	24.7	4.0	2.5	0.04	0.01	–	0.37	0.18
Total	17.41	123.1	16.4	92.5	67.5	26.7	14.0	0.19	0.06	–	0.72	0.54
Fertilized												
Tops	6.91	99.9	11.7	74.3	55.0	15.3	9.6	0.08	0.03	0.78	0.57	0.30
Roots	13.97	67.3	16.8	102.1	15.5	8.4	7.0	0.07	0.03	0.90	0.06	0.17
Fallen leaves	1.86	30.5	2.0	7.1	31.9	4.7	2.6	0.05	0.02	–	0.46	0.19
Total	22.74	197.7	30.5	183.5	102.4	28.4	19.3	0.20	0.08	–	1.09	0.66

Source: Howeler (1985a).

proportional to yield, an average root yield of 15 t ha^{-1} would remove about 35 kg N, 5.8 kg P, 46 kg K, 7.0 kg Ca and 4.1 kg Mg ha^{-1}. The relationship between dry root yield and root nutrient content is, however, not linear (Fig. 7.2) because high-yielding plants generally have higher nutrient concentrations in the roots than low-yielding plants. Thus at lower yield levels, nutrient removal is considerably lower than indicated above. Figure 7.2 shows that at a fresh root yield level of 15 t ha^{-1} the nutrients removed in the harvested roots would be about 30 kg N, 3.5 kg P and 20 kg K ha^{-1}, considerably lower for P and K than previously estimated (Howeler, 1981). This may be the reason why cassava yields of less than 10 t ha^{-1} (about 3.7 t ha^{-1} of dry roots) do not seriously deplete the nutrient level of the soil. Yields can be sustained at those levels for several cropping seasons, even when no fertilizers or manures are applied, as long as plant tops are re-incorporated into the soil. The rather close fit of the experimental data reported in the literature for the relationship between P and K removal with dry root yield (Fig. 7.2B and C) indicates that yields may be more closely associated with the concentrations of P and K in the roots than with that of N.

To prevent nutrient depletion of the soil, about 60 kg N, 10–20 kg P_2O_5 and 50 kg K_2O ha^{-1} should be applied if the expected yield level is 15 t ha^{-1} and all stems and leaves are returned to the soil. If leaves and stems are also removed, at least twice this much needs to be applied. Removal of P, Ca and Mg is quite low if only roots are harvested but increases considerably (especially Ca and Mg) if plant tops are also removed.

Nutrient losses by runoff and erosion

When cassava is grown on slopes, nutrient losses in eroded sediments and runoff can be substantial. The detachment of soil particles by the impact of raindrops and the subsequent movement down slope, together with runoff water, not only physically removes part of the top soil with the associated organic matter (OM), nutrients and microorganisms, but also fertilizers, plant litter and earthworm castings. This leaves a soil that is lighter in texture and lower in OM and nutrients. Given the reduced depth of the topsoil, a highly acid, infertile subsoil may be exposed. Consequently, cassava

Table 7.3. Fresh and dry yield, as well as nutrient content in cassava roots and in the whole plant at time of harvest, as reported in the literature.

Plant part	Yield (t ha^{-1}) Fresh	Dry	Nutrient content (kg ha^{-1}) N	P	K	Ca	Mg	Source/cultivar
Roots	64.7	26.59	45	28.2	317	51	18	Nijholt (1935)
Whole plant	110.6	39.99	124	45.3	487	155	43	São Pedro Preto
Roots	59.0	21.67	152	22.0	163	20	11	Howeler and Cadavid (1983)
Whole plant	–	30.08	315	37.0	238	77	32	M Col 22, fertilized
Roots	52.7	25.21	38	27.9	268	34	19	Nijholt (1935)
Whole plant	111.1	44.65	132	48.5	476	161	52	Mangi
Roots	37.5	13.97	67	17.0	102	16	8	Howeler (1985b)
Whole plant	–	22.74	198	31.0	184	102	28	M Col 22, unfertilized
Roots	~ 33.9	12.60	161	10.0	53	16	12	Paula et al. (1983)
Whole plant	–	20.92	330	20.5	100	88	30	Branco SC, fertilized
Roots	32.3	15.39	127	19.1	71	6	5	Cadavid (1988)
Whole plant	–	25.04	243	34.4	147	56	25	CM 523–7, fertilized
Roots	~ 27.6	10.28	100	8.7	107	15	13	Paula et al. (1983)
Whole plant	–	19.56	353	24.8	174	133	37	Riqueza, fertilized
Roots	26.6	12.81	91	11.3	47	5	6	Cadavid (1988)
Whole plant	–	19.10	167	19.1	76	32	19	CM 523–7, unfertilized
Roots	26.0	10.75	30	8.0	55	5	7	Howeler (1985a)
Whole plant	–	17.41	123	16.0	92	67	27	M Ven 77, unfertilized
Roots	18.3	5.52	32	3.6	35	5	4	Sittibusaya, unpublished
Whole plant	–	9.01	95	9.9	65	37	15	Rayong 1, fertilized
Roots	16.1	3.64	30	4.7	45	9	5	Putthacharoen et al. (1998)
Whole plant	–	10.55	193	27.0	137	122	27	Rayong 1, 1990/91
Roots	~ 15.0	5.58	66	2.7	17	8	5	Paula et al. (1983)
Whole plant	–	10.62	197	8.1	61	100	20	Riqueza, unfertilized
Roots	~ 8.7	3.24	37	1.5	23	4	2	Paula et al. (1983)
Whole plant	–	6.54	93	4.0	40	30	9	Branca SC, unfertilized
Roots	8.7	2.68	13	0.9	4	3	2	Sittibusaya, unpublished
Whole plant	–	4.23	39	3.2	10	21	8	Rayong 1, unfertilized
Roots	6.0	1.52	18	2.2	15	5	2	Putthacharoen et al. (1998)
Whole plant	–	4.37	91	12.2	55	46	15	Rayong 1, 1989/90

yields on eroded soils are substantially lower than on nearby non-eroded soil (Howeler, 1986, 1987).

Research has shown (Quintiliano *et al.*, 1961; Margolis and Campos Filho, 1981; Putthacharoen *et al.*, 1998; Wargiono *et al.*, 1998)

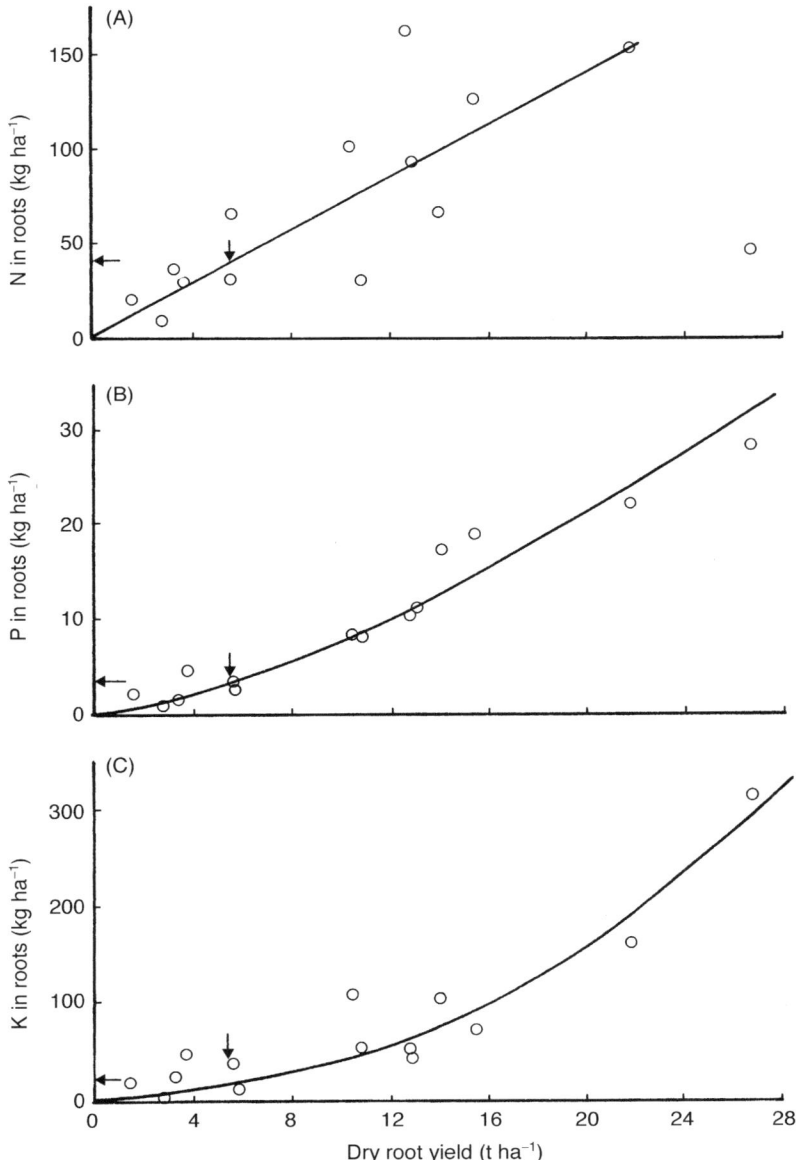

Fig. 7.2. Relation between the N, P and K contents of cassava roots and dry root yield, as reported in the literature (see Table 7.3); arrows indicate the approximate nutrient contents corresponding to a fresh root yield of 15 t ha^{-1}. (In Fig. 7.2A two points corresponding to data of Nijholt (1935) were not considered when drawing the line, as the reported levels of N removal were too low for the high yields obtained. This may have been due to faulty analytical procedures at the time. Data for P and K removal from the same author seem to be correct and are therefore included.)

that cultivating cassava on slopes results in more erosion than for most other crops. This is mainly due to the wide plant spacing used and the slow initial growth of cassava, which result in soil being exposed to the direct impact of rainfall during the first 3–4 MAP. Recent research has shown that erosion can be markedly reduced by simple agronomic practices such as selecting varieties with rapid initial growth, minimum tillage, closer plant spacing (0.8×0.8 m), vertical planting, intercropping, mulching, fertilizer application, planting at the end rather than the beginning of the rainy season, and the planting of contour hedgerows of grasses such as vetiver (*Vetiveria zizanioides*), leguminous trees or forage species (Howeler, 1998; Nguyen Huu Hy et al., 1998; Tongglum et al., 1998; Wargiono et al., 1998).

Table 7.4 shows the soil and associated nutrient losses measured in cassava trials conducted in Thailand and Colombia. Nutrient losses were much higher in Thailand than in Colombia due to the higher soil losses, even on a 'gentle' slope of only 7%. Losses of P, K and Mg are in terms of available P and exchangeable K and Mg. If the unavailable or non-exchangeable fractions had been measured, these losses would have been about ten times higher. Thus annual N losses in eroded sediments in Thailand were almost twice as high as in the harvested roots (Putthacharoen et al., 1998), while in Colombia N losses in sediments were one to two times as

high as those estimated in the harvested roots (Ruppenthal, 1995; Ruppenthal et al., 1997).

There is little quantitative information about nutrient losses from cassava fields in rainfall runoff. Naturally this depends on rainfall intensity, amount of runoff, fertilizers applied, etc. Data reported for upland rice grown on 25–35% slope in Luang Prabang, Laos, indicate that total N and P losses in runoff (3.6 kg N and 0.85 kg P ha^{-1} $year^{-1}$) were considerably lower than in the eroded sediments (35.5 kg N and 5.6 kg P ha^{-1} $year^{-1}$), but that total K losses were higher (52 kg ha^{-1} $year^{-1}$) in the runoff than in the sediment (33 kg ha^{-1} $year^{-1}$; Phommasack et al., 1995, 1996).

Long-term effect of cassava production on soil productivity

When cassava is grown continuously on the same soil without adequate fertilizer or manure inputs, soil productivity may decline due to nutrient depletion and soil loss by erosion. Sittibusaya (1993) reported that cassava yields in unfertilized plots declined from 26–30 to 10–12 t ha^{-1} after 20–30 years of cassava cultivation. Similar or even faster yield declines have been observed for other annual crops (Ofori, 1973; Nguyen Tu Siem, 1992).

Cong Doan Sat and Deturck (1998) compared the effect of long-term cultivation

Table 7.4. Nutrients in sediments eroded from cassava plots with various treatments in Thailand and Colombia.

Location and treatments	Dry soil loss (t ha^{-1} $year^{-1}$)	(kg ha^{-1} $year^{-1}$)			
		N[a]	P[b]	K[b]	Mg[b]
Cassava on 7% slope in Sriracha, Thailand[c]	71.4	37.1	2.18	5.15	5.35
Cassava on 7–13% slope in Quilichao, Colombia[d]	5.1	11.5	0.16	0.45	0.45
Cassava + leguminous cover crops in Quilichao, Colombia[d]	10.6	24.0	0.24	0.97	0.81
Cassava + grass hedgerows in Quilichao, Colombia[d]	2.7	5.8	0.06	0.22	0.24
Cassava on 12–20% slope in Mondomo, Colombia[d]	5.2	13.3	1.09	0.45	0.36
Cassava + leguminous cover crops in Mondomo, Colombia[d]	2.7	6.5	0.04	0.24	0.20
Cassava + grass hedgerows in Mondomo, Colombia[d]	1.5	3.5	0.02	0.13	0.10

[a]Total N.
[b]Available P and exchangeable K and Mg.
[c]Source: Putthacharoen et al. (1998).
[d]Source: Ruppenthal et al. (1997).

of cassava with that of natural forest, rubber, cashew and sugarcane grown on similar soils in southern Vietnam. Cassava cultivation resulted in the lowest levels of soil organic C, total N and exchangeable K and Mg, and an intermediate level of P because of some P fertilizer applications. Cassava cultivation also resulted in the lowest clay content, soil aggregate stability and water retention, as well as intermediate bulk density and infiltration rates. Compared to native forest, grasslands or perennial plantation crops, long-term cassava cultivation has a negative impact on soil productivity, especially when grown on slopes with inadequate management and no fertilizer. This is also true, however, for other annual food crops, which all require frequent land preparation, resulting in erosion and more rapid decomposition of OM. Moreover, fertilizer inputs may not compensate for nutrient removal in harvested products, or losses by leaching and erosion.

When cassava was grown for 8 consecutive years without fertilizers in Quilichao, Colombia, root yields declined from 22 t ha^{-1} in the first year to 13 t ha^{-1} in the last year (Fig. 7.3). With application of only N or P, yields declined from 27 and 29 t ha^{-1} in the first year to 20 and 15 t ha^{-1}, respectively, in the last year. This yield decline was due to the increasing intensity of K deficiency. When only K was applied (150 kg K$_2$O ha^{-1}) yields could be maintained at about 30 t ha^{-1}, while with the annual application of 100 kg N, 200 kg P$_2$O$_5$ and 150 kg K$_2$O ha^{-1}, yields actually increased from about 32 to nearly 40 t ha^{-1}. Without K application the exchangeable K content of the soil decreased in 2–3 years from 0.2 to about 0.1 meq 100 g^{-1} and remained at that level for the following five crop cycles. With application of 150 kg K$_2$O ha^{-1}, the soil K level remained constant at about 0.2 meq 100 g^{-1}, while with applications of 300 kg K$_2$O ha^{-1} it increased gradually to 0.45 meq 100 g^{-1}. Thus high yields and adequate levels of K could be maintained with annual applications of 150 kg K$_2$O ha^{-1} (Howeler and Cadavid, 1990; Howeler, 1991b).

On mineral soils in Malaysia, very high cassava yields of about 50 t ha^{-1} were maintained for 9 years with annual applications of 112 kg N, 156 kg P$_2$O$_5$ and 187 kg K$_2$O ha^{-1}, but without fertilizers yields declined from 32 t ha^{-1} in the first year to about 20 t ha^{-1} in the 9th year

(Chan, 1980; Howeler, 1992). This was also attributed mainly to increasing K deficiency. In Kerala, India, continuous cropping for 10 years also resulted in declining yields when no K was applied, while annual applications of 100 kg K$_2$O ha^{-1} maintained high yields of 20–30 t ha^{-1} (Kabeerathumma et al., 1990).

Figure 7.4 shows similar results for a long-term (> 20 year) fertility trial conducted in Khon Kaen, Thailand. Without K application yields declined from 28 t ha^{-1} in the first year to 10 t ha^{-1} in the second, and then slowly declined to about 5 t ha^{-1} during the subsequent 17 cropping cycles. With application of NK or NPK, yields could be maintained at a level of 20 t ha^{-1} for the entire period. Figure 7.4B also shows that when plant tops were re-incorporated into the soil, the yield decline without fertilizer application was much slower. After 19 years the yield was still about 10 t ha^{-1}, i.e. twice as high as when all plant parts were removed. Thus, when plant tops were returned to the soil, yields of about 10 t ha^{-1} could be maintained, even in a very poor soil and without any application of fertilizers. With adequate fertilization high yields of at least 20 t ha^{-1} could be maintained for 19 years of continuous cropping.

Diagnosis of Nutritional Disorders

If plant growth is not optimal and/or yields are low, and if other causes such as insect pests and diseases, drought, shade or cold have been ruled out, plants may be suffering from nutritional deficiencies and/or toxicities. Before effective remedial measures can be taken, it is essential to diagnose the problem correctly. This can be done in several ways, but the best diagnosis is usually obtained from a combination of different methods.

Observation of deficiency and toxicity symptoms

Cassava plants do not readily translocate nutrients from the lower to the upper leaves; instead, when certain nutrients are in deficient supply, plants respond by slowing the growth rate, producing fewer and smaller leaves and sometimes shorter internodes. Leaf life is also

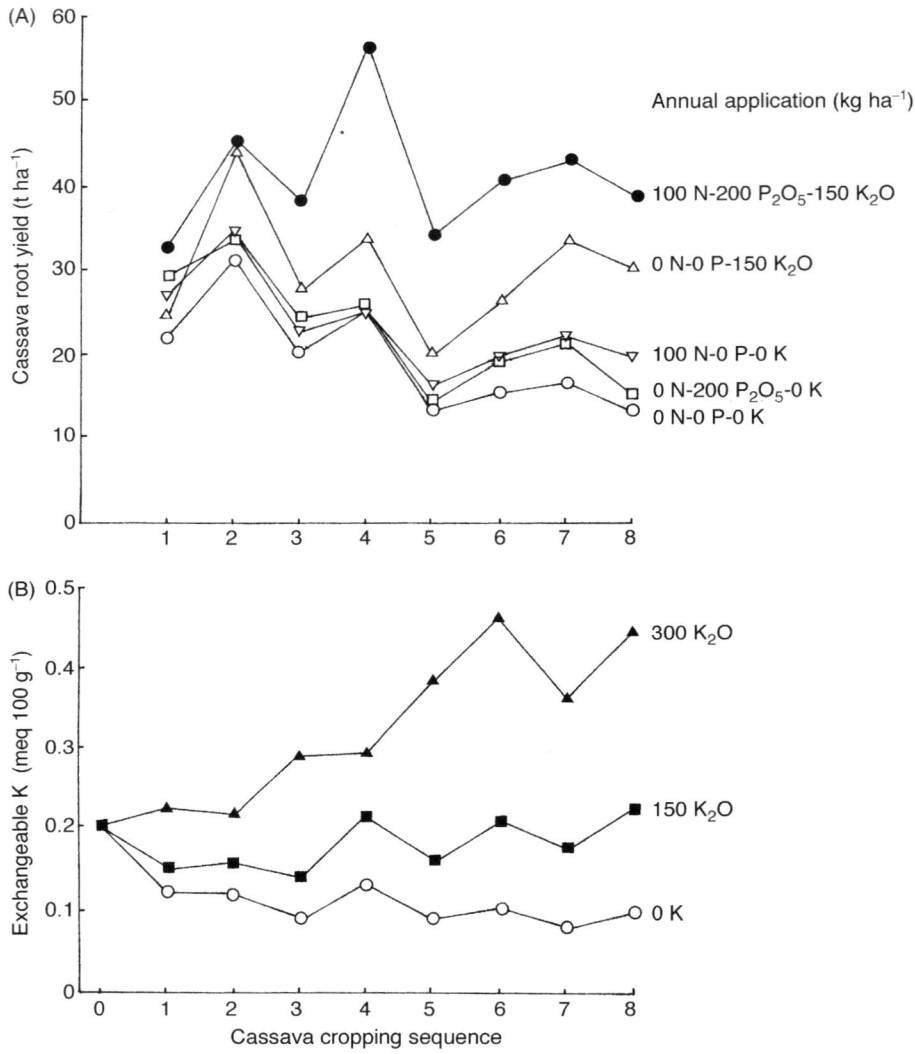

Fig. 7.3. Effect of various levels of annual applications of N, P and K on cassava root yield (A), and on the exchangeable K content of the soil (B) during eight consecutive cropping cycles in a long-term NPK trial conducted at CIAT-Quilichao, Colombia.

reduced. As nutrients are not readily mobilized to the growing point, symptoms for NPK deficiencies, normally found in the lower leaves, tend to be less pronounced in cassava than in other crops. For that reason farmers may not be aware that plant growth is reduced because of nutritional deficiencies. The initial diagnosis based on deficiency or toxicity symptoms needs to be confirmed by soil or plant tissue analyses or

from experiments. Nevertheless, visual identification is a quick, easy method to diagnose many nutritional problems. Symptoms have been described and colour photos have been included in several publications (Asher *et al.*, 1980; Howeler, 1981, 1989, 1996a,b; Lozano *et al.*, 1981; Howeler and Fernandez, 1985). The symptoms of nutrient deficiencies and toxicities are summarized in Table 7.5.

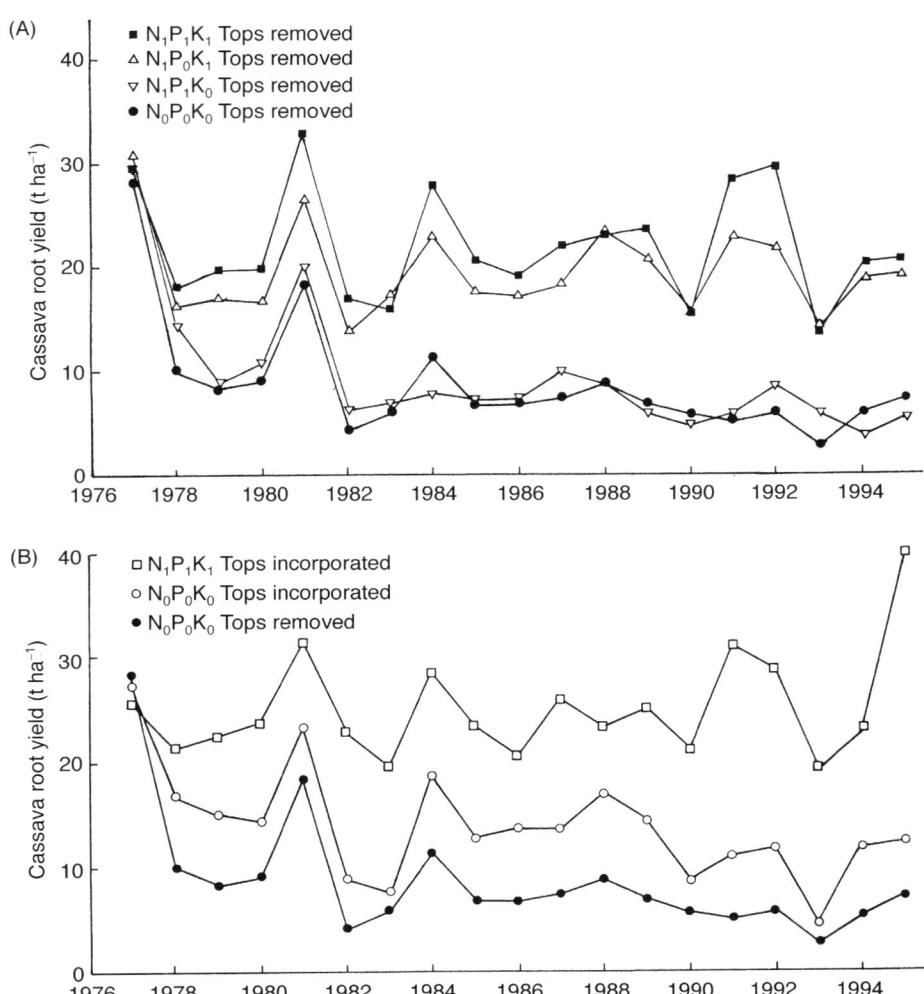

Fig. 7.4. Effect of annual applications of various combinations of N, P and K (A) and crop residue management (B) on cassava yield during 19 consecutive crops grown in Khon Kaen, Thailand, from 1977–1995. (Source: Howeler, 2000.)

Soil analysis

This method is advantageous in that problems can be detected before planting and, if necessary, lime and/or nutrients can be applied before plant growth is affected by the problem. Soil analyses are particularly useful for detecting P, K, Ca, Mg and Zn deficiencies, while soil pH will indicate whether Al and/or Mn toxicity or micronutrient deficiencies are likely to occur. Analysis for OM content is not very reliable in predicting N responses as high-OM soils may still produce a significant N response if N mineralization is slow, especially in very acid soils.

Soil analyses usually determine the amount of available or exchangeable nutrient as this part of the total soil nutrient is best correlated with plant uptake. These 'available' fractions are usually determined by shaking the soil sample with certain extracting solutions and determining the amount of nutrient in the extract. Different laboratories may use different extracting agents as there is no one method that is optimal for all soil types; thus results from one laboratory may

Table 7.5.   Symptoms of nutrient deficiencies and toxicities in cassava.

Deficiencies	Symptoms
Nitrogen (N)	Reduced plant growth
	In some cultivars, uniform chlorosis of leaves, starting with lower leaves, but soon spreading throughout the plant
Phosphorus (P)	Reduced plant growth, thin stems, short petioles; sometimes pendant leaves
	Under severe conditions 1–2 lower leaves turn yellow to orange, become flaccid and necrotic; may fall off
	In some cultivars lower leaves turn purplish/brown
Potassium (K)	Reduced plant growth with excessive branching, resulting in prostrate plant type
	Small, sometimes chlorotic upper leaves; thick stems with short internodes
	Under severe conditions premature lignification of upper stems with very short internodes, resulting in zigzag growth of upper stems
	In some cultivars purple spotting, yellowing and border necrosis of lower leaves
	In other cultivars upward curling of lower leaf borders, similar to drought stress symptoms
Calcium (Ca)	Reduced root and shoot growth
(rare in the field)	Chlorosis, deformation and border necrosis of youngest leaves with leaf tips or margins bending downwards
Magnesium (Mg)	Marked intervenal chlorosis or yellowing in lower leaves
(often seen in field)	Slight reduction in plant height
Sulphur (S) (similar to N deficiency; seldom seen in field)	Uniform chlorosis of upper leaves, which soon spreads throughout the plant
Boron (B)	Reduced plant height, short internodes, short petioles and small deformed upper leaves
	Purple–grey spotting of mature leaves in middle part of plant
	Under severe conditions gummy exudate on stem or petioles (almost never seen in field)
	Suppressed lateral development of fibrous roots
Copper (Cu)	Deformation and uniform chlorosis of upper leaves, with leaf tips and margins bending up- or downward
(mainly in peat soils)	Petioles of fully expanded leaves long and bending down
	Reduced root growth
Iron (Fe) (mainly in calcareous soils)	Uniform chlorosis of upper leaves and petioles; under severe conditions leaves turn white with border chlorosis of youngest leaves
	Reduced plant growth; young leaves small, but not deformed
Manganese (Mn) (mainly in sandy and high pH soils)	Intervenal chlorosis or yellowing of upper or middle leaves; uniform chlorosis under severe conditions
	Reduced plant growth; young leaves small, but not deformed.
Zinc (Zn) (often seen in high pH or calcareous soils; also in acid soils)	Intervenal yellow or white spots on young leaves
	Leaves become small, narrow and chlorotic in growing point; necrotic spotting on lower leaves as well
	Leaf lobes turn outward away from stem
	Reduced plant growth; under severe conditions, death of young plants

Toxicities	Symptoms
Aluminium (Al) (only in very acid mineral soils)	Reduced root and shoot growth
	Under very severe conditions yellowing of lower leaves
Boron (B) (only observed after excessive B application)	Necrotic spotting of lower leaves, especially along leaf margins
Manganese (Mn) (mainly in acid soils and when plant growth stagnates)	Yellowing or oranging of lower leaves with purple-brown spots along veins
	Leaves become flaccid and drop off
Salinity (observed only in saline/ alkaline soils)	Uniform yellowing of leaves, starting at bottom of plant but soon spreading throughout
	Symptoms very similar to Fe deficiency
	Under severe conditions border necrosis of lower leaves, poor plant growth and death of young plants

differ from those of another. In interpreting the results, therefore, it is important to consider the methodology used.

Representative soil samples should be taken in areas that appear to be uniform in terms of plant growth and previous management. About 10–20 subsamples are taken in zigzag fashion across the whole area. These subsamples are thoroughly mixed together and then about 300–500 g are air dried or dried at about 65°C in a forced-air oven. This combined sample is then finely ground, screened and sent to the lab for analysis.

Results of the soil analysis can be compared with published data obtained from correlation studies, which indicate either the 'critical level' of the nutrient, as determined with a specific extracting agent or the nutrient ranges according to the particular nutritional conditions of the crop. Table 7.6 gives the ranges corresponding to the nutritional requirements of cassava determined with particular methodologies.

These data were determined from many fertilizer experiments conducted in Colombia and in various Asian countries by relating the relative yield in the absence of N, P or K fertilizers (yield without the nutrient over the highest yield obtained with the nutrient) with the OM, available P and exchangeable K content of the soil, respectively. Figure 7.5 gives an example from nine locations in four Asian countries. A line was drawn visually through the points to show the relationship and to estimate the 'critical level' of the nutrient or soil parameter. This critical level is interpreted as the concentration of the nutrient in the soil or plant tissue above which there is no further significant response to application of the nutrient (usually defined as corresponding to 95% of maximum yield). Critical levels are 3.2% for OM, 7 μg g^{-1} for P (Bray II) and 0.14 meq 100 g^{-1} for exchangeable K. The critical levels for P and K are close to those reported earlier in the literature (Table 7.7). Those for available soil-P reported for cassava (4–10 μg g^{-1}) are much lower than for most other crops (10–18 μg g^{-1}), indicating that cassava will grow well in soils that are low in P and where other crops would suffer from P deficiency. This is due to the effective association between cassava roots and vesicular–arbuscular mycorrhizas (VAM) occurring naturally in the soil (Howeler, 1990).

The critical levels for exchangeable K for cassava (0.08–0.18 meq K 100 g^{-1}; Table 7.7) are also lower than for most other crops

Table 7.6. Approximate classification of soil chemical characteristics according to the nutritional requirements of cassava.

Soil parameter	Very low	Low	Medium	High	Very high
pH[a]	< 3.5	3.5–4.5	4.5–7	7–8	> 8
Organic matter[b] (%)	< 1.0	1.0–2.0	2.0–4.0	> 4.0	
Al saturation[c] (%)			< 75	75–85	> 85
Salinity (mS cm^{-1})			< 0.5	0.5–1.0	> 1.0
Na saturation (%)			< 2	2–10	> 10
P[d] (μg g^{-1})	< 2	2–4	4–15	> 15	
K[d] (meq 100 g^{-1})	< 0.10	0.10–0.15	0.15–0.25	> 0.25	
Ca[d] (meq 100 g^{-1})	< 0.25	0.25–1.0	1.0–5.0	> 5.0	
Mg[d] (meq 100 g^{-1})	< 0.2	0.2–0.4	0.4–1.0	> 1.0	
S[d] (μg g^{-1})	< 20	20–40	40–70	> 70	
B[e] (μg g^{-1})	< 0.2	0.2–0.5	0.5–1.0	1–2	> 2
Cu[e] (μg g^{-1})	< 0.1	0.1–0.3	0.3–1.0	1–5	> 5
Mn[e] (μg g^{-1})	< 5	5–10	10–100	100–250	> 250
Fe[e] (μg g^{-1})	< 1	1–10	10–100	> 100	
Zn[e] (μg g^{-1})	< 0.5	0.5–1.0	1.0–5.0	5–50	> 50

[a]pH in H_2O.
[b]OM = Walkley and Black method.
[c]Al saturation = 100 × Al (Al + Ca + Mg + K) in meq 100 g^{-1}.
[d]P in Bray II; K, Ca, Mg and Na in 1N NH_4-acetate; S in Ca phosphate.
[e]B in hot water; and Cu, Mn, Fe and Zn in 0.05 N HCl + 0.025 N H_2SO_4.
Source: Howeler (1996a,b).

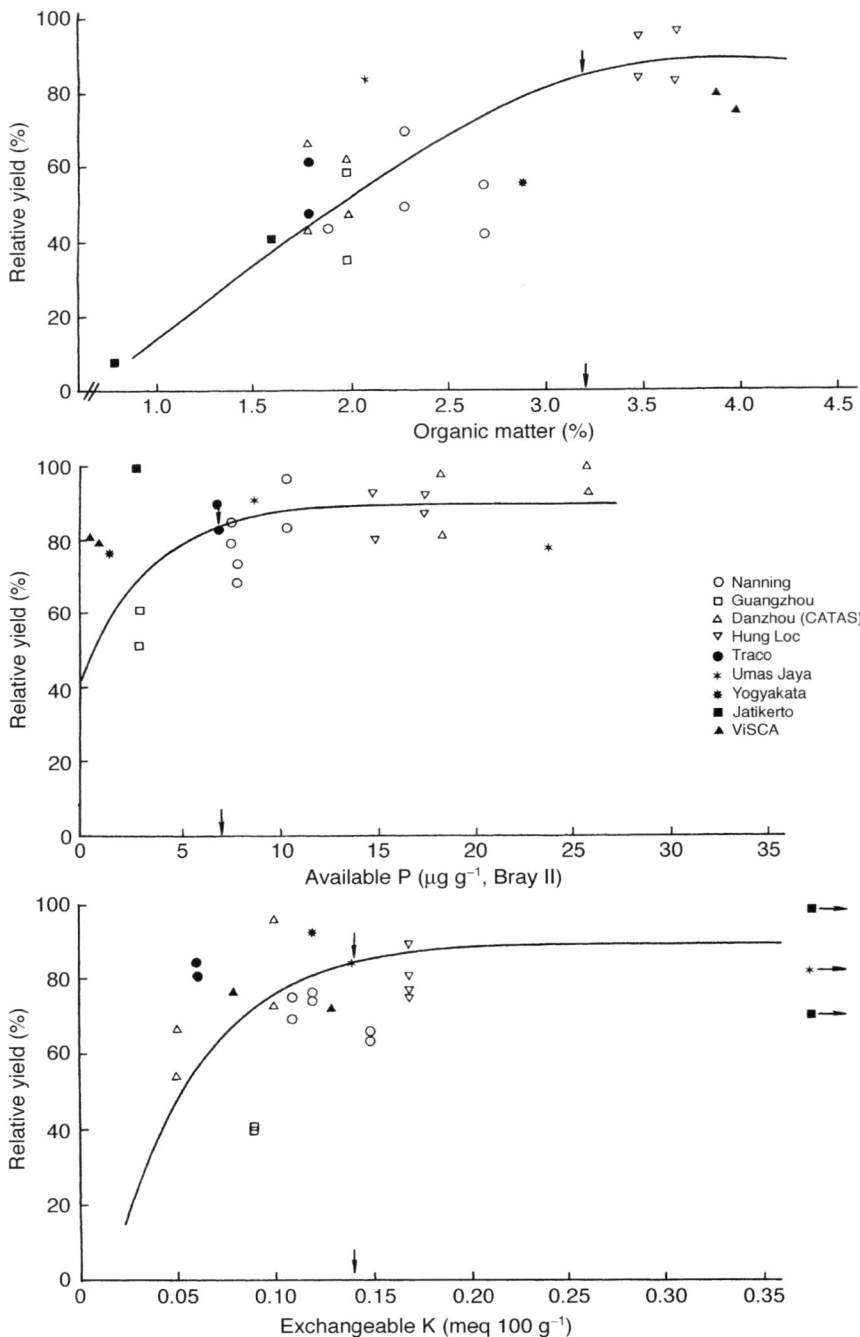

Fig. 7.5. Relationship between relative yield of cassava (i.e. without nutrient as % of highest yield with nutrient) and organic matter available P and exch. K content of the soil in nine NPK trials conducted in Asia (1993–1996). (Source: Howeler, 1998.)

Table 7.7. Critical levels[a] of nutrients for cassava and other crops according to various methods of soil analysis, as reported in the literature.

Soil parameter	Method[c]	Crop	Critical level	Source
Organic matter (%)	Walkley and Black	Cassava	3.1	Howeler (1998)
P (μg g^{-1})	Bray I	Cassava	7	Howeler (1978)
		Cassava	8[b]	Kang et al. (1980)
		Cassava	4.2	Cadavid (1988)
		Cassava	7	Howeler (1989)
		Maize	14	Kang et al. (1980)
		Soybeans	15	Kang et al. (1980)
	Bray II	Cassava	8	CIAT (1982)
		Cassava	4	Howeler (1985a)
		Cassava	6	CIAT (1985a)
		Cassava	5.8[b]	Cadavid (1988)
		Cassava	10	Howeler (1989)
		Cassava	10	Hagens and Sittibusaya (1990)
		Cassava	4	Howeler and Cadavid (1990)
		Cassava	4.5	Howeler (1995)
		Cassava	7	Howeler (1998)
		Common beans[d]	10–15	Howeler and Medina (1978)
	Olsen-EDTA	Cassava	3	van der Zaag et al. (1979)
		Cassava	7.5[b]	Cadavid (1988)
		Cassava	8	Howeler (1989)
	North Carolina	Cassava	5.0[b]	Cadavid (1988)
		Cassava	9	Howeler (1989)
		Common beans	18	Goepfert (1972)
K (meq 100 g^{-1})	NH$_4$-acetate	Cassava	0.09–0.15	Obigbesan (1977)
		Cassava	< 0.15	Kang (1984)
		Cassava	< 0.15	Kang and Okeke (1984)
		Cassava	0.18	Howeler (1985b)
		Cassava	0.175[b]	Cadavid (1988)
		Cassava	0.15	Howeler (1989)
		Cassava	0.18	Howeler and Cadavid (1990)
		Cassava	0.08–0.10	Hagens and Sittibusaya (1990)
		Rice	0.21	Jones et al. (1982)
		Potatoes	0.20–1.00	Roberts and McDole (1985)
		Sugarcane	0.16–0.51	Orlando Filho (1985)
	Bray II	Cassava	0.15	CIAT (1985a)
		Cassava	0.17	Howeler (1985b)
		Cassava	0.16	CIAT (1988b)
		Cassava	0.175[b]	Cadavid (1988)
		Cassava	0.17	Howeler and Cadavid (1990)
		Cassava	0.12	Howeler (1995)
		Cassava	0.14	Howeler (1998)
	North Carolina	Cassava	0.15	Howeler (1989)
Ca (meq 100 g^{-1})	NH$_4$-acetate	Cassava	0.25	CIAT (1979)
		Common beans	4.5	Howeler and Medina (1978)
Mg (meq 100 $^{-1}$)	NH$_4$-acetate	Cassava	< 0.20	Kang (1984)
pH	1 : 1 in water	Cassava	4.6 and 7.8	CIAT (1977, 1979)
		Common beans	4.9	Abruña et al. (1974)
Al (% sat.)	KCl	Cassava	80	CIAT (1979)
		Common beans	10–20	Abruña et al. (1974)

footnote on next page

(0.16–0.51 meq K 100 g^{-1}), indicating that despite the crop's relatively high K requirement, it will still grow well on soils with only intermediate levels of K.

As mentioned above, there is seldom a good relationship between the relative response to N and the soil OM content (Howeler, 1995). Using data from 56 NPK trials conducted in Brazil from 1950 to 1983 (Gomes, 1998), the critical level determined for OM was only 1.3%, considerably lower than the 3.1% determined in Asia (Howeler, 1998).

Using data from 20 NPK cassava trials conducted in Colombia to compare different methods of extracting available P, Cadavid (1988) reported the highest correlation between relative cassava yields and available soil P using Bray I, followed by Bray II, North Carolina and Olsen-EDTA extractants. For determining exchangeable K, Cadavid (1988) found no significant difference between the use of Bray II and NH$_4$-acetate; both resulted in a critical level of 0.175 meq K 100 g^{-1}.

Plant tissue analysis

Analysis of plant tissue indicates the actual nutritional status of the plants. The total amount of a certain nutrient is determined, resulting in data that are fairly similar among different laboratories. These analyses are particularly useful for diagnosing N and secondary or micronutrient deficiencies.

Given that nutrient concentrations vary among different tissues, it is imperative to use an 'indicator' tissue, the nutrient concentration of which is best related to plant growth or yield. For cassava, the best 'indicator' tissue is the blade of the youngest fully expanded leaf (YFEL), i.e. normally about the fourth to fifth leaf from the top. Blades without petioles are analysed as nutrient concentrations are quite different in these two tissues (Table 7.8). Nutrient concentrations also change during the growth cycle, depending on the rate of plant growth (Howeler and Cadavid, 1983; CIAT, 1985a,b). As they tend to stabilize after about 4 months, leaf samples should be taken at about 4 MAP.

About 20 leaf blades (without petioles) are collected from a plot or uniform area in the field and combined into one sample (Howeler, 1983). If leaves are dusty or have received chemical sprays, they should be washed gently and rinsed in distilled or deionized water. To prevent continued respiration with consequent loss of DM, leaves should be dried as soon as possible at 60–80°C for 24–48 h. If no oven is available, leaves should be dried as quickly as possible in the sun, preferably in a hot but well-ventilated area, and away from dust. After drying, samples are finely ground in a lab mill. For Cu analysis samples should be passed through a stainless steel sieve. For Fe analysis the dry leaves should be ground with an agate mortar and pestle. Samples are normally collected in paper bags to facilitate drying, but for analysis of B, plastic bags should be used. Once ground and sieved, samples are stored in plastic vials until analysis.

To diagnose nutritional problems, the results are compared with the nutrient ranges corresponding to the various nutritional states of the plant (Table 7.9), or with critical levels reported in the literature (Table 7.10). While the numbers may vary somewhat given the different varieties, soil and climatic conditions (Howeler, 1983), the data in these tables can be used as a general guide for interpreting plant tissue analyses.

Footnote for Table 7.7.
[a]Critical level defined as 95% of maximum yield.
[b]Critical level defined as 90% of maximum yield.
[c]Methods:
   Bray I = 0.025 N HCl + 0.03 N NH$_4$F.
   Bray II = 0.10 N HCl + 0.03 N NH$_4$F.
   Olsen-EDTA = 0.5 N NaHCO$_3$ + 0.01N Na-EDTA.
   North Carolina = 0.05 N HCl + 0.025 N H$_2$SO$_4$.
   NH$_4$-acetate = 1 N NH$_4$ -acetate at pH 7.
[d]*Phaseolus vulgaris*.

Table 7.8. Nutrient concentration in various plant parts of fertilized and unfertilized cassava cv. M Ven 77 at 3–4 MAP in Carimagua, Colombia.

	(%)						(μg g^{-1})				
	N	P	K	Ca	Mg	S	Fe	Mn	Zn	Cu	B
Unfertilized											
Leaf blades											
Upper	4.57	0.34	1.29	0.68	0.25	0.29	198	128	49	9.9	26
Middle	3.66	0.25	1.18	1.08	0.27	0.25	267	185	66	8.7	37
Lower	3.31	0.21	1.09	1.48	0.25	0.25	335	191	89	7.6	42
Fallen[a]	2.31	0.13	0.50	1.69	0.25	0.22	4850	209	121	9.4	39
Petioles											
Upper	1.50	0.17	1.60	1.32	0.37	0.10	79	172	40	4.4	16
Middle	0.70	0.10	1.32	2.20	0.43	0.10	76	304	72	2.9	15
Lower	0.63	0.09	1.35	2.69	0.45	0.13	92	361	110	2.8	15
Fallen	0.54	0.05	0.54	3.52	0.41	0.13	271	429	94	2.5	18
Stems											
Upper	1.64	0.20	1.22	1.53	0.32	0.19	133	115	36	9.7	14
Middle	1.03	0.18	0.87	1.45	0.30	0.16	74	103	39	8.9	13
Lower	0.78	0.21	0.81	1.19	0.32	0.16	184	95	54	7.9	10
Roots											
Rootlets[a]	1.52	0.15	1.02	0.77	0.38	0.16	5985	191	165	–	10
Thickened roots	0.42	0.10	0.71	0.13	0.06	0.05	127	10	16	3.0	4
Fertilized											
Leaf blades											
Upper	5.19	0.38	1.61	0.76	0.28	0.30	298	177	47	10.6	26
Middle	4.00	0.28	1.36	1.08	0.27	0.26	430	207	63	9.6	30
Lower	3.55	0.24	1.30	1.40	0.22	0.23	402	220	77	8.5	37
Fallen[a]	1.11	0.14	0.54	1.88	0.23	0.19	3333	247	120	8.9	38
Petioles											
Upper	1.49	0.17	2.18	1.58	0.36	0.10	87	238	33	4.9	17
Middle	0.84	0.09	1.84	2.58	0.41	0.07	88	359	49	3.0	14
Lower	0.78	0.09	1.69	3.54	0.42	0.07	95	417	70	3.2	15
Fallen	0.69	0.06	0.82	3.74	0.20	0.08	294	471	155	3.1	17
Stems											
Upper	2.13	0.23	2.09	2.09	0.47	0.14	94	140	37	9.8	14
Middle	1.57	0.21	1.26	1.30	0.26	0.11	110	120	46	10.8	12
Lower	1.37	0.28	1.14	1.31	0.23	0.09	210	99	36	10.0	10
Roots											
Rootlets[a]	1.71	0.19	1.03	0.71	0.33	0.20	3780	368	136	–	10
Thickened roots	0.88	0.14	1.05	0.16	0.06	0.05	127	15	15	3.9	4

[a]Fallen leaves and rootlets were probably contaminated with micronutrients from the soil.
Source: Howeler (1985a).

Greenhouse and field experiments

If analysis of soil or plant tissue is not possible, one can also diagnose nutritional problems by planting cassava in pot experiments using the soil in question, or directly in the field. To diagnose nutrient deficiencies in a particular soil in either pot or field experiments, it is recommended to use the 'missing element' technique, where all nutrients are applied to all treatments at rates that are expected to be non-limiting, while one nutrient is missing in each treatment (i.e. -N, -P, -K, etc.). Treatments with the poorest growth or yield indicate the element that is most deficient.

Table 7.9. Nutrient concentrations in youngest fully expanded leaf (YFEL) blades of cassava at 3–4 MAP, corresponding to various nutritional states of the plants; data are averages of various greenhouse and field trials.

Nutrient	Very deficient	Deficient	Low	Sufficient	High	Toxic
	Nutritional states[a]					
N (%)	< 4.0	4.1–4.8	4.8–5.1	5.1–5.8	-> 5.8	–[b]
P (%)	< 0.25	0.25–0.36	0.36–0.38	0.38–0.50	> 0.50	–
K (%)	< 0.85	0.85–1.26	1.26–1.42	1.42–1.88	1.88–2.40	> 2.40
Ca (%)	< 0.25	0.25–0.41	0.41–0.50	0.50–0.72	0.72–0.88	> 0.88
Mg (%)	< 0.15	0.15–0.22	0.22–0.24	0.24–0.29	> 0.29	–
S (%)	< 0.20	0.20–0.27	0.27–0.30	0.30–0.36	> 0.36	–
B (μg g^{-1})	< 7	7–15	15–18	18–28	28–64	> 64
Cu (μg g^{-1})	< 1.5	1.5–4.8	4.8–6.0	6–10	10–15	> 15
Fe (μg g^{-1})	< 100	100–110	110–120	120–140	140–200	> 200
Mn (μg g^{-1})	< 30	30–40	40–50	50–150	150–250	> 250
Zn (μg g^{-1})	< 25	25–32	32–35	35–57	57–120	> 120

[a]Very deficient, < 40% maximum yield.
 Deficient, 40–80% maximum yield.
 Low, 80–90% maximum yield.
 Sufficient, 90–100% maximum yield.
 High, 100–90% maximum yield.
 Toxic, < 90% maximum yield.
[b]– = no data available.
Source: Howeler (1996a,b).

For pot experiments it is recommended not to sterilize or fumigate the soil, in order not to kill the native mycorrhizas. Rooted plant shoots rather than stakes should be used as the stakes have high nutrient reserves and their use would therefore delay responses to nutrient additions. In pot experiments cassava plants are generally harvested 3–4 MAP, and dry weights of top growth are used as indicators of nutrient response.

Correcting Nutritional Disorders

Chemical fertilizers

While cassava performs better than most crops on infertile soils, the crop is highly responsive to fertilizer applications. High yields can be obtained and maintained only when adequate amounts of fertilizers and/or manures are applied. Thousands of fertilizer experiments conducted by FAO worldwide indicate that cassava is as responsive to fertilizer applications as other crops, with yield increases of 49% (West Africa) to 110% (Latin America) versus increases of 43% (yams and rice in West Africa) to 102% (rice in Latin America) for other crops. In West Africa (Ghana) cassava responded mainly to K, in Latin America (Brazil) to P, and in Asia (Indonesia and Thailand) to N (Richards, 1979; Hagens and Sittibusaya, 1990).

Cassava is sensitive to over-fertilization, especially with N, which will result in excessive leaf formation at the expense of root growth. Cock (1975) reported that cassava has an optimal leaf area index (LAI) of 2.5–3.5 and that high rates of fertilization may lead to excessive leaf growth and an LAI > 4. High N applications not only reduce the harvest index (HI) and root yield, but can also reduce the starch and increase the HCN content of the roots. Moreover, nutrients generally interact with each other, and the excessive application of one nutrient may induce a deficiency of another. Howeler *et al.* (1977) and Edwards and Kang (1978) have shown that high rates of lime application may actually reduce yields by inducing Zn deficiency. Spear *et al.* (1978b) showed that increasing the K concentration in nutrient solution decreased the

Table 7.10.   Critical nutrient concentrations for deficiencies and toxicities in cassava plant tissue.

Element	Method	Plant tissue	Critical level[a]	Source
N deficiency	Field	YFEL blades	5.1%	Fox *et al.* (1975)
	Field	YFEL blades	5.7%	Howeler (1978)
	Field	YFEL blades	4.6%	Howeler (1995)
	Field	YFEL blades	5.7%	Howeler (1998)
	Nutrient solution	Shoots	4.2%	Forno (1977)
P deficiency	Field	YFEL blades	0.41%	CIAT (1985a)
	Field	YFEL blades	0.33–0.35%	Nair *et al.* (1988)
	Nutrient solution	Shoots	0.47–0.66%	Jintakanon *et al.* (1982)
K deficiency	Nutrient solution	YFEL blades	1.1%	Spear *et al.* (1978a)
	Field	YFEL blades	1.2%	Howeler (1978)
	Field	YFEL blades	1.4%	CIAT (1982)
	Field	YFEL blades	1.5%	CIAT (1982)
	Field	YFEL blades	< 1.1%	Kang (1984)
	Field	YFEL blades	1.5%	CIAT (1985a)
	Field	YFEL blades	1.7%	Howeler (1995)
	Field	YFEL blades	1.9%	Nayar *et al.* (1995)
	Field	YFEL blades	1.9%	Howeler (1998)
	Nutrient solution	Petioles	0.8%	Spear *et al.* (1978a)
	Field	Petioles	2.5%	Howeler (1978)
	Nutrient solution	Stems	0.6%	Spear *et al.* (1978a)
	Nutrient solution	Shoots and roots	0.8%	Spear *et al.* (1978a)
Ca deficiency	Nutrient solution	YFEL blades	0.46%	CIAT (1985a)
	Field	YFEL blades	0.60–0.64%	CIAT (1985a)
	Nutrient solution	Shoots	0.4%	Forno (1977)
Mg deficiency	Nutrient solution	YFEL blades	0.29%	Edwards and Asher (unpublished)
	Field	YFEL blades	< 0.33%	Kang (1984)
	Field	YFEL blades	0.29%	Howeler (1985a)
	Nutrient solution	YFEL blades	0.24%	CIAT (1985a)
	Nutrient solution	Shoots	0.26%	Edwards and Asher (unpublished)
S deficiency	Field	YFEL blades	0.32%	Howeler (1978)
	Nutrient solution	YFEL blades	0.27%	CIAT (1982)
	Field	YFEL blades	0.27–0.33%	Howeler, unpublished
Zn deficiency	Field	YFEL blades	37–51 µg g^{-1}	CIAT (1978)
	Nutrient solution	YFEL blades	43–60 µg g^{-1}	Edwards and Asher (unpublished)
	Nutrient solution	YFEL blades	30 µg g^{-1}	Howeler *et al.* (1982c)
	Field	YFEL blades	33 µg g^{-1}	CIAT (1985a)
Zn toxicity	Nutrient solution	YFEL blades	120 µg g^{-1}	Howeler *et al.* (1982c)
B deficiency	Nutrient solution	YFEL blades	35 µg g^{-1}	Howeler *et al.* (1982c)
	Nutrient solution	Shoots	17 µg g^{-1}	Forno (1977)
B toxicity	Nutrient solution	YFEL blades	100 µg g^{-1}	Howeler *et al.* (1982c)
	Nutrient solution	Shoot	140 µg g^{-1}	Forno (1977)
Cu deficiency	Nutrient solution	YFEL blades	6 µg g^{-1}	Howeler *et al.* (1982c)
Cu toxicity	Nutrient solution	YFEL blades	15 µg g^{-1}	Howeler *et al.* (1982c)
Mn deficiency	Nutrient solution	YFEL blades	50 µg g^{-1}	Howeler *et al.* (1982c)
	Nutrient solution	Shoots	100–120 µg g^{-1}	Edwards and Asher (unpublished)
Mn toxicity	Nutrient solution	YFEL blades	250 µg g^{-1}	Howeler *et al.* (1982c)
	Nutrient solution	Shoots	250–1450 µg g^{-1}	Edwards and Asher (unpublished)
Al toxicity	Nutrient solution	Shoots	70–97 µg g^{-1}	Gunatilaka (1977)
	Nutrient solution	Roots	2000–14,000 µg g^{-1}	Gunatilaka (1977)

[a]Range corresponds to values obtained in different varieties.
YFEL, youngest fully expanded leaf.

absorption of Ca and especially Mg, leading to Mg deficiency. However, in both nutrient solution and field experiments with varying rates of applications of K, Ca and Mg, Howeler (1985b) did not find a significant effect of increasing K on the Ca concentration in the leaves. The Mg concentration decreased slightly in the field, but increased in the nutrient solution experiment. Increasing Mg supply markedly decreased the concentrations of K and Ca. Similarly, Ngongi et al. (1977) reported that high applications of KCl induced S deficiency in a low-S soil in Colombia, while Nair et al. (1988) found that high rates of P application induced Zn deficiency. Hence, it is important not only to apply the right amount of each nutrient, but also the right balance among the various nutrients.

Nitrogen

Severe N deficiency is usually observed in very sandy soils low in OM, but may also be found in high-OM, but acid soils, mainly due to a low rate of N mineralization.

Significant responses to N have been observed more frequently in Asia than in Latin America or Africa. In nearly 100 NPK trials conducted by FAO on farmers' fields in Thailand, there was mainly a response to N, followed by K and P (Hagens and Sittibusaya, 1990). Similar results were obtained in 69 trials conducted in Indonesia (FAO, 1980). In Africa relatively few fertilizer trials have been conducted with cassava, mainly because very few cassava farmers apply fertilizers. In West Africa the responses to N

were probably the most frequent (Okogun et al., 1999). In Latin America responses to N were the least frequent, with significant responses reported in only five out of 41 trials conducted in Brazil (Gomes, 1998) and in five out of 22 trials conducted in Colombia (Howeler and Cadavid, 1990).

On a very sandy soil (89% sand, 0.7% OM) in Jaguaruna, Santa Catarina, Brazil, two local varieties showed a nearly linear response up to 150 kg N ha^{-1}. For both varieties highest yields were obtained with a split application of N, with one-third applied at 30, 60 and 90 days after planting (Moraes et al., 1981). A similarly spectacular response to N was also observed in a clay soil with 1.2% OM in Jatikerto, East Java, Indonesia (Fig. 7.6). In this case, cassava was intercropped with maize, which competed strongly for the limited supply of N in the soil (Wargiono et al., 1998). In Nanning, Guangxi, China, there was also a highly significant response to N (Zhang Weite et al., 1998), up to 200 kg N ha^{-1} in one cultivar (SC205), but only up to 50 kg N ha^{-1} in the other (SC201). As the latter cultivar is extremely vigorous, high N levels produced too much top growth at the expense of root production. Similar negative responses to high N applications have been reported by Acosta and Perez (1954), Vijayan and Aiyer (1969), Fox et al. (1975) and Obigbesan and Fayemi (1976). Krochmal and Samuels (1970) reported a root yield reduction of 41% and top growth increase of 11% due to high N applications. These high rates also stimulate production of N-containing compounds, such as protein and HCN, and may result in a decrease in root starch content. High

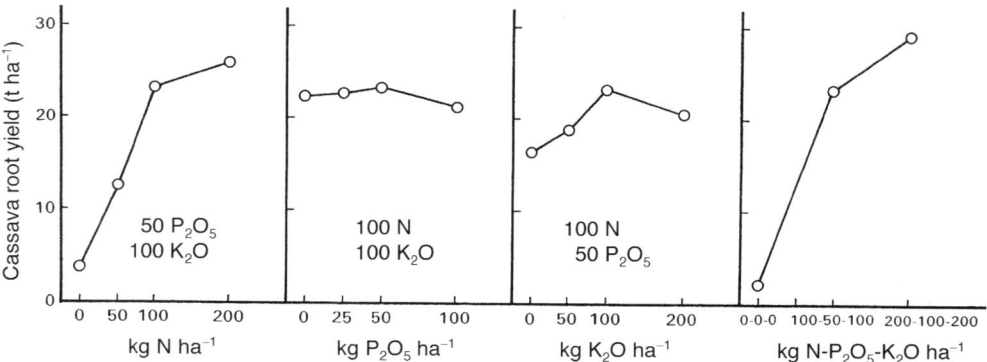

Fig. 7.6. Response of cassava cv. Faroka to the annual application of different levels of N, P and K during the seventh cycle in Jatikerto, East Java, Indonesia in 1994/95. (Source: Wargiono et al., 1998.)

rates of N application may also increase the intensity of diseases such as cassava bacterial blight (Kang and Okeke, 1984). Thus N rates must not only be adjusted to a particular soil but also tailored to the needs of a particular cultivar.

Trials on optimum time and partitioning of N applications have generally shown non-significant differences between single applications at planting, at 1 MAP or various partitions (0–3 MAP) using N rates up to 100 kg N ha^{-1} (Howeler, 1985a). At higher rates, partition of the nitrogen application was found to be better than a single application.

There are usually no significant differences among N sources such as urea, NH_4NO_3, mono- or di-ammonium phosphate. Vinod and Nair (1992) reported significantly higher yields with slow-release N sources such as neem cake-coated urea or super-granules of urea.

When cassava is grown for forage production, spacing is greatly reduced (0.6×0.6 m), and green tops are cut off at 3–4 month intervals. The offtake of N is very high, sometimes > 300 kg N ha^{-1} year^{-1} (CIAT, 1988a,b; Putthacharoen et al., 1998). High rates of N (> 200 kg ha^{-1}) need to be applied to sustain high levels of shoot and root production.

Phosphorus

Cassava's tolerance to low P concentrations in soil solution is not due to the efficient uptake of P by the root system; in fact, cassava grown in flowing nutrient solution required a much higher P concentration for maximum growth than rice, maize, cowpeas or common beans (Howeler et al., 1981; Jintakanon et al., 1982; Howeler 1990). When inoculated with endotrophic VAM, the growth of cassava in nutrient solution improved significantly (Howeler et al., 1982a). Masses of mycorrhizal hyphae growing in and around the fibrous roots of cassava markedly increased the plant's ability to absorb P from the surrounding medium (Fig. 7.7). When planted in natural soil, the crop's fibrous roots soon become infected with native soil mycorrhizas. The resulting hyphae grow into the surrounding soil and help in the uptake and transport of P to the cassava roots. Through this highly effective symbiosis, cassava is able to absorb P from soils with low levels of available

P, mainly by extending the soil volume from which P can be absorbed through the associated mycorrhizal hyphae.

Responses of cassava to P application depend on the available-P level of the soil, the mycorrhizal population and the variety used. van der Zaag et al. (1979) reported high yields of 42 t ha^{-1} in an oxisol in Hawaii with only 3 µg P g^{-1} ($NaHCO_3$-extractant) using the cultivar Ceiba. CIAT (1988a) similarly reported that some varieties produced yields of 40–50 t ha^{-1} without P application in a soil with only 4.6 µg P g^{-1} (Bray II). In other soils with equally low levels of available P but with a less-efficient mycorrhizal population, cassava responded very markedly to P applications. Thus in the oxisols of the Eastern Plains of Colombia, with only 1.0 µg P g^{-1} (Bray II), cassava responded markedly to applications of 200–400 kg P_2O_5 ha^{-1} (Fig. 7.8). Of the seven P sources tested, banding of triple superphosphate (TSP) or broadcast applications of basic slag were most effective.

Fig. 7.7. Comparison of a vesicular–arbuscular mycorrhizas (VAM)-inoculated plant (right) with a non-inoculated plant grown in a phosphorus-deficient nutrient solution.

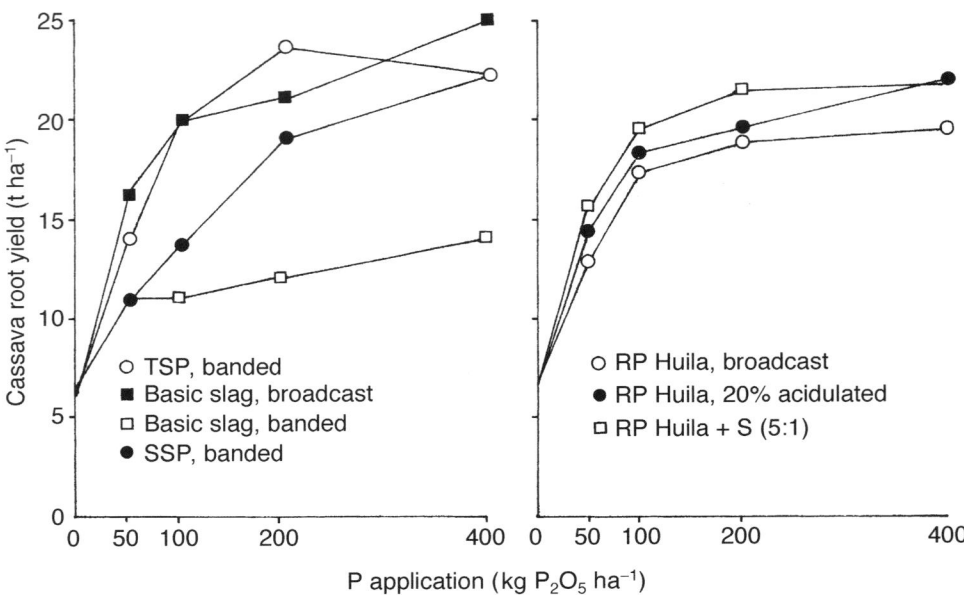

Fig. 7.8. Response of cassava cv. Llanera to application of different levels and sources of P in Carimagua, Colombia. RP, rock phosphate; TSP, triple super-phosphate; SSP, single super-phosphate. (Source: CIAT, 1977.)

Partially acidulated rock phosphate (RP) or RP mixed with elemental sulphur (S) were also quite effective in these acid soils (CIAT, 1978). Locally produced simple super phosphate (SSP) was less effective, except at high rates of application. Similarly, Santos and Tupinamba (1981) reported significant responses to 60 or 120 kg P_2O_5 ha^{-1} in three soils of Sergipe, Brazil, with TSP and hyperphosphate being more effective than two local sources of RP. Soluble-P sources like TSP, SSP, mono- or di-ammonium phosphate should be band applied near the stakes, while less-soluble sources such as basic slag and RPs should be broadcast and incorporated. All P should be applied at or shortly after planting as fractionation of P had no significant effect on yield. Alternative methods of P application, such as stake treatments or foliar sprays, are not as effective as soil application in increasing yields (Howeler, 1985a).

Severe P deficiency has been reported mainly in Latin America, particularly on oxisols, ultisols and inceptisols in Brazil and Colombia. These soils are highly P fixing and have available (Bray II or Mehlich I) P levels of only 1–2 µg g^{-1}. During the first year(s) of cropping, cassava responds markedly to P application, but with continuous cropping on the same land, responses to P become less significant as soil P levels build up from previous applications (Nair et al., 1988; Howeler and Cadavid, 1990; Kabeerathumma et al., 1990).

In Asia P deficiency is seldom the principal limiting factor for cassava production because most cassava is grown on soils with more than 4 µg g^{-1} of available P and/or that have been previously fertilized with P. Nevertheless, significant responses to P application have been observed in Guangzhou (Guangdong), Nanning (Guangxi) and on Hainan Island of China; in northern and southern Vietnam; and on Leyte Island of the Philippines. In low-P soils in Kerala State, India, significant initial responses to 100 kg P_2O_5 ha^{-1} were reported, but these declined over time. Nair et al. (1988) determined an optimum economic rate of 45 kg P_2O_5 ha^{-1}.

In Africa few P experiments have been conducted with cassava. Responses to P application have been reported mainly in Ghana (Stephens, 1960; Takyi, 1972) and Madagascar (Cours et al., 1961). Ofori (1973) reported a negative effect of P application on cassava yields on a forest ochrosol in Ghana.

Large varietal differences have been observed in cassava's ability to grow on low-P soils (CIAT, 1988a,b). Pellet and El-Sharkawy (1993a,b) found that varietal differences in response to applied P were not due to genetic differences in P-uptake efficiencies, but rather to contrasting patterns of DM distribution and P-use efficiency (root yield total per P in plant). Low-P tolerant cultivars had a high fine root-length density, moderate top growth, and a high, stable HI.

Potassium

Although K is not a basic component of protein, carbohydrates or fats, it plays an important role in their metabolism. Potassium stimulates net photosynthetic activity of a given leaf area and increases the translocation of photosynthates to the tuberous roots. This results in low carbohydrate levels in the leaves, further increasing photosynthetic activity (Kasele, 1980).

Blin (1905), Obigbesan (1973) and Howeler (1998) reported that K application not only increased root yields but also their starch content. Similar increases in starch content with increasing applications of K have been observed in Carimagua (CIAT, 1982) and Pescador, Colombia (Howeler, 1985a), as well as in southern Vietnam (Nguyen Huu Hy et al., 1998) and China (Howeler, 1998). In general root starch content increases up to 80–100 kg K_2O ha^{-1} and then decreases at higher rates of K application. Obigbesan (1973) and Kabeerathumma et al. (1990) reported that K also decreased the HCN content of roots, while Payne and Webster (1956) found highest levels of HCN in roots produced in low-K soils.

Potassium deficiency in cassava is generally found in tropical soils with low-activity clay such as oxisols, ultisols and inceptisols, as well as in alfisols derived from sandstone. After land clearing the alfisols have a reasonable level of exchangeable K, but often show a significant K response in the second year of planting because of low K reserves in the parent material (Kang and Okeke, 1984).

Long-term experiments in Asia and Colombia have shown that K deficiency invariably becomes the main limiting factor when cassava is grown continuously on the same soil without adequate K fertilization. Figure 7.9 shows that, without K application, yields declined from 22.4 to 6.3 t ha^{-1} during 10 years of continuous cropping in Trivandrum, Kerala, India (Kabeerathumma et al., 1990). During that time, exchangeable K of the soil gradually declined from 0.18 to 0.064 meq 100 g^{-1}. High yields of 20–30 t ha^{-1} could be maintained with annual applications of 100 kg N, 100 P_2O_5 and 100 K_2O ha^{-1}. The high rate of P used led to a buildup of soil P to excessive levels (~100 µg P g^{-1}). There was also a slight buildup of exchangeable K to a level of 0.25 meq 100 g^{-1}. This contrasts with reports by Howeler and Cadavid (1990), which show that 8 years of continuous cropping with annual applications of 150 kg K_2O ha^{-1} in Quilichao, Colombia, maintained the exchangeable K content at only 0.2 meq 100 g^{-1}. In a very poor sandy soil near the Atlantic Coast of Colombia, Cadavid et al. (1998) also found that annual applications of 50 kg N, 50 P_2O_5 and 50 K_2O ha^{-1} increased yields during 8 years of continuous cropping, but had no effect on soil K, which remained at a low level (0.06 meq 100 g^{-1}).

Long-term NPK trials were conducted in four Asian countries. After 4–10 years of continuous cropping, there was a significant response to N in eight out of 11 sites, to P in four and to K in seven. Responses to K increased most markedly with time (Nguyen Huu Hy et al., 1998; Howeler, 2000), but in Malang, Indonesia, responses to N increased more markedly than elsewhere (Wargiono et al., 1998).

Data from short-term NPK trials conducted in Brazil (Gomes, 1998) indicate that significant responses to K were obtained in only nine out of 48 trials; similarly in Colombia there was a significant K response in six out of 22 locations, mainly in the Eastern Plains.

In Africa, significant responses to K have been found on strongly acid soils of eastern Nigeria (Okeke, unpublished) and on slightly acid soils (0.23 meq K 100 g^{-1}) of southwestern Nigeria (Kang and Okeke, 1984). No significant response to K was observed in field experiments conducted in Nigeria (Obigbesan, 1977) nor in Ghana (Takyi, 1972). In Madagascar, however, Roche et al. (1957) and Cours et al. (1961) found that K was the main limiting nutrient, and applications of 110 kg K_2O ha^{-1} were recommended (Anon, 1952, 1953).

Experiments on different sources of K have generally shown no significant differences between the use of KCl, K_2SO_4 or Sulphomag (CIAT, 1985a), while in southern India the use of syngenite and schoenite (extracted from sea-water) were also found to be equally effective (Central Tuber Crops Research Institute; CTCRI, 1974, 1975). In low-S soils in the Eastern Plains

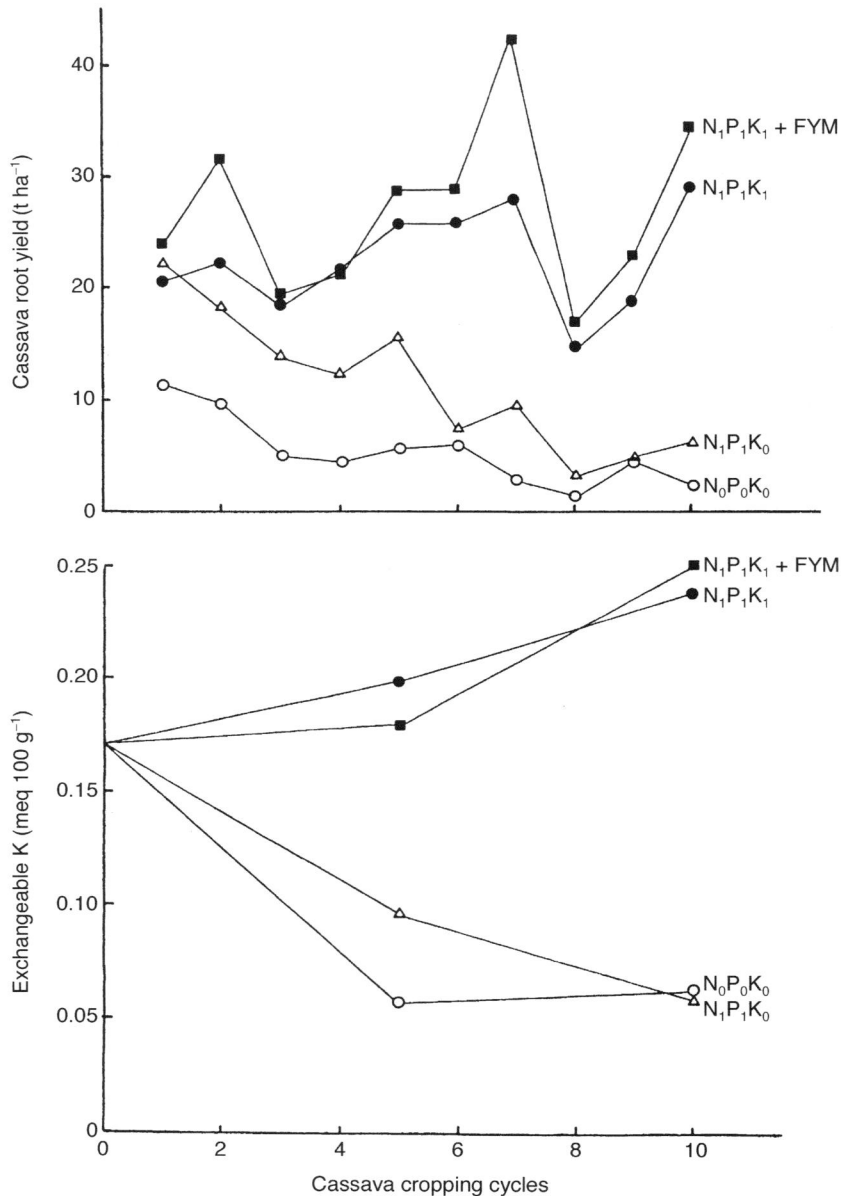

Fig. 7.9. Cassava yield (top) and the exchangeable K content of the soil (bottom) during 10 years of continuous cropping with various NPK treatments in Trivandrum, Kerala, India. (Source: Kabeerathumma *et al.*, 1990.)

of Colombia, Ngongi *et al.* (1977) obtained much better yields with application of K_2SO_4 or KCl + S than with KCl. However, this was mainly a response to S; the same experiment conducted in Jamundí, Colombia, did not show any difference between K sources.

Experiments on the optimum time of K application have produced somewhat contradictory results, but generally there were no significant differences between single or split applications, or, among different times of application (CIAT, 1982). Overall, a single application at one MAP produced the highest yield. In India, Ashokan and Shreedharan (1980) recommended split application of K only when low rates were applied, but CTCRI (1972) found no significant differences among different times of application in Trivandrum, India.

Calcium and magnesium

Calcium plays an important role in the supply and regulation of water in the plant, while Mg is a basic component of chlorophyll and as such is essential for photosynthesis.

Symptoms of Ca deficiency are seldom observed in the field; but in very acid soils with low levels of exchangeable Ca (< 0.25 meq 100 g^{-1}), the crop may respond to Ca applications. In Carimagua-Alegría, Colombia, highly significant responses to application of Ca were obtained on a sandy loam soil with a pH of 5.1 and only 0.18 meq Ca 100 g^{-1} and 0.05 meq Mg 100 g^{-1}. Highest yields were obtained with application of 200–400 kg Ca ha^{-1} as broadcast gypsum. Broadcast calcitic or dolomitic limes were less effective, while band-applied gypsum was ineffective in increasing cassava yields (CIAT, 1985a). As these Ca sources are relatively insoluble, they should all be broadcast and incorporated before planting. The good response to gypsum was not a response to S because either $MgSO_4$ or S were applied uniformly to all plots.

In the same soil in Carimagua, an experiment was conducted to determine the optimum rates and best sources of Mg (CIAT, 1985a). There was a significant response up to the highest level of 60 kg Mg ha^{-1}, but there were no overall significant differences among sources; the more soluble Sulphomag was more effective

at intermediate rates, while banded $MgSO_4$ or broadcast MgO were better at higher rates of application.

Experiments conducted in nutrient solution culture showed that increasing concentrations of Ca in solution markedly reduced the concentrations of K and Mg in YFEL blades at 2 MAP, and that increasing concentrations of Mg in solution similarly decreased the concentrations of K and Ca in the leaves. In field experiments in Carimagua, however, increasing levels of Ca or Mg had no effect on leaf concentration of K, and only slightly decreased the concentrations of Mg or Ca, respectively (Howeler, 1985b). There appears to be a strong antagonistic effect between K and Mg and between Ca and Mg, but not between K and Ca.

Sulphur

This basic component of certain amino acids is essential for producing protein. In industrial areas much of the plant's S requirements are met from S emissions into the atmosphere, but in isolated areas cassava may suffer from S deficiency. This has been reported only for Carimagua, Colombia, which is far removed from any industrial centres. Soils there contained only 23 µg of Ca phosphate-extractable S g^{-1} of soil; with application of 40 kg S ha^{-1} as elemental S this increased to 36 µg g^{-1}. There was a response to applying S up to 20–40 kg S ha^{-1}. There were no significant differences among S sources although yields were slightly higher with banded K- and Mg-sulphate than with broadcast elemental S. Clear S-deficiency symptoms were observed in the control plots. These plants had 0.20–0.25% S in YFEL blades, compared with 0.30–0.32% in plants that had received S applications. Critical levels of 0.27 and 0.33% S were estimated in two field experiments (Howeler, unpublished).

Micronutrients

Deficiencies are generally observed in high pH or calcareous soils, but deficiencies of Zn have been observed in both acid and alkaline soils. Lime application to acid soils with low levels of available Zn may induce Zn deficiency, resulting in low yields and even death of young plants (see Acid soils, below).

Zinc

Cassava is quite susceptible to Zn deficiency, especially in the early stages of growth. Plants showing early symptoms of Zn deficiency may later recuperate once the fibrous root system is well established and roots become infected with mycorrhizas. If the deficiency is severe, however, plants may either die or produce very low yields. The response of two varieties to soil application of different levels of Zn as $ZnSO_4.7H_2O$ was tested in Carimagua-Alegría after applying 2 t ha^{-1} of lime (CIAT, 1985a). Both varieties were seriously affected by Zn deficiency in the check plots, but reached maximum yields with application of 10 kg Zn ha^{-1}, band applied with NPK at planting. A critical level of 33 µg Zn g^{-1} in YFEL blades was estimated. Broadcast application of 10–20 kg ha^{-1} of Zn as ZnO was also effective in increasing yields in acid soils (CIAT, 1978).

In high-pH soils, application of $ZnSO_4.7H_2O$ to the soil is not as effective because the applied Zn is precipitated rapidly (CIAT, 1978). Foliar application or stake treatments may be more effective. When 20 cassava cultivars were planted in a high-pH (7.9), low-Zn (1.0 µg g^{-1}) soil, with or without treating stakes for 15 min in a solution of 4% $ZnSO_4.7H_2O$ before planting, yields increased from an average 11.5 to 25.0 t ha^{-1} due to the Zn treatment (CIAT, 1985a). Large varietal differences in low-Zn tolerance were observed, with some cultivars dying off completely without the Zn treatment, while others produced high yields with or without Zn. Dipping stakes for 15 min in a $ZnSO_4.7H_2O$ solution and air-drying the stakes overnight before planting is a very inexpensive yet effective way to reduce Zn deficiency in cassava when grown on low-Zn soils.

Copper

This deficiency in cassava has been reported only on peat soils in Malaysia (Chew, 1971). Chew et al. (1978) recommended a basal application of 2.5 kg Cu ha^{-1} as $CuSO_4. 5H_2O$.

Iron and manganese

Deficiency symptoms of Fe are often observed in calcareous soils such as in the Yucatan peninsula of Mexico, parts of northern Colombia, southeast Java of Indonesia and western Nakorn Rachasima province of Thailand. Mn deficiency has been observed in sandy soils along the coast in northeast Brazil, in alkaline soils at CIAT, Colombia, and in northern Vietnam, near houses where lime had been used for their construction. A practical solution is probably a stake treatment with 2–4% $FeSO_4.H_2O$ or $MnSO_4.4H_2O$ before planting, or foliar sprays with sulphates or chelates.

Boron

Symptoms of B deficiency have been observed in both acid soils of Carimagua and Quilichao, as well as in alkaline soils at CIAT, Colombia. Similar symptoms were also observed in North Vietnam and southern China, although the exact nature of that problem was never identified. In Colombia, application of 1–2 kg B ha^{-1}, band applied as borax at planting, eliminated the symptoms, increased plant height, increased B concentrations in the leaves from 3 to 40 µg g^{-1}, but had no significant effect on yield. Thus it seems that cassava is quite tolerant of low levels of available B in the soil.

Soil Amendments

Soil amendments are usually applied to increase the pH of acid soils or decrease the pH of alkaline soils, but they can also serve to improve the physical conditions of the soil or improve nutrient availability.

Acid soils

Very acid soils present a complex of problems for plant growth, including low pH, high concentrations of Al and/or Mn, low levels of Ca, Mg and K, and sometimes low P and N. Cassava as a species is particularly tolerant of soil acidity and high levels of Al (Gunatilaka, 1977; CIAT, 1979; Islam et al., 1980), but some varietal differences in acid soil tolerance have also been observed (CIAT, 1982, 1985a; Howeler, 1991a). In very acid (pH < 4.5) and high Al (> 80% Al saturation) soils, lime application may increase cassava yields, mainly by supplying Ca and Mg as nutrients. High rates of lime may, however, induce micronutrient deficiencies, particularly Zn, resulting in decreased yield (Spain et al., 1975; Edwards and Kang, 1978). Field trails (CIAT, 1976) showed that without Zn, cassava responded to lime applications only up to 2 t ha^{-1}, but with applied Zn there was a positive response up to 6 t ha^{-1} of lime. Analysis of cassava leaves confirmed that liming reduced Zn uptake and that with 6 t ha^{-1} of lime without Zn, the Zn concentrating in YFEL blades dropped below the critical level of 40–50 µg g^{-1}. Large varietal differences have been found for both high-Al and low-Zn tolerance (Spain et al., 1975).

Saline–alkaline soils

Cassava is not well adapted to saline and alkaline soils and may suffer from a combination of high pH, high Na, high salt and low uptake of micronutrients. Yields can be improved by applying 1–2 t ha^{-1} of elemental S or 1–2 t ha^{-1} of H_2SO_4 (CIAT, 1977), but this is seldom justified economically. Most cultivars will tolerate a pH up to about 8.0, Na saturation up to about 2%, and conductivity up to 0.5 mS cm^{-1} (CIAT, 1977). Large varietal differences in tolerance have been observed, and the use of tolerant varieties is probably the most practical solution.

Animal manures and compost

Many cassava farmers apply animal manures or compost to cassava with good results, but little research has been conducted to determine optimum rates and methods of application.

In Vietnam and south China, most farmers apply 5–10 t ha^{-1} of pig manure, in Indonesia up to 9 t ha^{-1} of cattle manure, and in Cauca province of Colombia, 4–5 t ha^{-1} of chicken manure. Animal manures tend to have low nutrient contents (< 10% of that in most compound fertilizers), but they do contain Ca, Mg, S and some micronutrients not found in most chemical fertilizers (Howeler, 1980b). In addition, they may improve the physical conditions of the soil, although this has not been well documented.

Silva (1970) reported good responses to applications of 6–15 t ha^{-1} of cattle manure in Rio Grande do Norte of Brazil; higher applications decreased yields. Howeler (1985a) reported that 4.3 t ha^{-1} chicken manure (corresponding to 170 kg P_2O_5 ha^{-1}) increased cassava yields in Mondomo, Colombia, from 19 to 31 t ha^{-1}. The chicken manure was about twice as effective as cattle manure or 10–30–10 compound fertilizer, applied at equivalent levels of P. The total amount of nutrients applied with the chicken manure was considerably higher than that applied in the chemical fertilizer, but the greater beneficial effect must also be due to improved soil structure, the presence of essential elements other than NPK, and the stimulation of beneficial microorganisms such as VAM. If these manures have to be transported over long distances, the higher cost of transport and application may make them more expensive than chemical fertilizers. Moreover, in many countries like Thailand, animal manures are not readily available.

In Bahia, Brazil, Gomes et al. (1983) obtained very high yields with a system called 'parcagem', which is basically in situ application of cattle manure, where a large number of cattle are enclosed overnight on a small piece of land. It was calculated that 30 animals kept overnight on 1 ha for 60 days will produce about 8 t of dry manure containing 40 kg N (plus the N in the urine). At an equal dosage of 40 kg N ha^{-1}, cassava yields with the parcagem system (combined with additional P and K) increased 30–90% as compared to application of only chemical fertilizers. This system may be economically viable in areas with large cattle populations. Good results were also obtained in Bahia when applying 5 t ha^{-1} of cattle manure combined with 10 kg P_2O_5 ha^{-1} (Diniz et al., 1994).

Little information is available on the effectiveness of applying compost, which is considerably lower in nutrients than animal manures (Howeler et al., 2002); but when applied in large quantities (10–15 t ha^{-1}), they may supply considerable amounts of nutrients as well as improving the soil's physical condition and water-holding capacity. Application of 6–10 t ha^{-1} of compost made from cassava peel residues in starch factories has given promising results in Thailand. In Brazil the application of 5 t ha^{-1} of vinhoto (a by-product of alcohol production) increased yields, especially when fortified with 60 kg P_2O_5 and 1 t lime (Souza et al., 1992).

Green manures, cover crops and mulch

Planting of green manures and their subsequent mulching or incorporation into the soil have been practised traditionally as a form of 'improved fallow' to maintain soil fertility. Research on the use of green manures and cover crops to maintain soil fertility in cassava fields was recently reviewed (Howeler *et al.*, 2002a, b). In the absence of chemical fertilizers, green manures increased yields slightly in Quilichao, Colombia, but significantly on sandy soils in Media Luna, Colombia. In the presence of fertilizers the effect was minimal. Most effective were kudzu (*Puerarea phaseoloides*), Zornia (*Zornia latifolium*) and groundnuts (*Arachis hypogea*) in Quilichao, and local weeds and *Canavalia ensiformis* in Media Luna. In Thailand, *Crotalaria juncea* was the most productive and effective in increasing yields. The growing of green manures before planting cassava was found to be impractical, however, as cassava yields were markedly reduced by planting too late in the wet season (Howeler, 1992, 1995).

Research on cover crops in both Colombia (Muhr *et al.*, 1995; Ruppenthal, 1995) and Thailand (Howeler, 1992; Tongglum *et al.*, 1992, 1998) indicate that almost all species compete too strongly with cassava for soil water and nutrients, resulting in unacceptably low cassava yields, especially in the second and third year after cover-crop establishment. It is clear that cassava is a weak competitor, and yields are seriously reduced by competition from weeds or cover crops.

Planting intercrops such as maize, groundnut, cowpea, common bean, mungbean or soybean and incorporating the residue after harvest may improve soil fertility (especially if the intercrops are fertilized), help reduce erosion and provide the farmer with additional food or income, without reducing cassava yields too seriously (Leihner, 1983).

Application of mulch of local weeds, cut-and-carry grass or crop residues such as rice straw or maize stalks can also improve soil fertility and moisture, as well as reduce surface temperature and erosion. In Africa application of mulch, especially that of leguminous species, increased cassava yields in acid sandy soils (Ofori, 1973; Hulugalle *et al.*, 1991). In sandy soils on the Atlantic Coast of Colombia, Cadavid

et al. (1998) reported that annual applications of 12 t ha^{-1} of dry *Panicum maximum* grass as mulch significantly increased cassava yields during eight consecutive years of cropping, especially in the absence of chemical fertilizers. It also increased root DM content and decreased HCN levels. Annual mulching gradually increased soil P and especially soil K, and prevented a decline in soil Ca and Mg. In addition, the mulch cover reduced soil temperatures in the top 20 cm of soil and enhanced the maintenance of soil C. Thus application of mulch, where available, can be another effective way of improving soil productivity.

Inoculation with mycorrhizas

As indicated above, cassava can grow well in low-P soils because of a highly efficient symbiosis with VAM, which occur naturally in the soil. Without VAM, cassava would require an application of at least 1–2 t ha^{-1} of P to obtain the same yield as plants with VAM but without P (Howeler, 1980a; Howeler *et al.*, 1982b). Compared with six other tropical crops and forages, cassava was found to be the most dependent on VAM (Howeler *et al.*, 1987).

Soils, however, differ in both quantity and quality of native mycorrhizae and, thus, in the crop's responses to P application (Sieverding and Howeler, 1985; Howeler *et al.*, 1987). Of many VAM species tested, *Glomus manihotis* was one of the most effective species for increasing cassava growth and yield in acid soils. *G. manihotis* was also found to compete strongly with other VAM species in the range of 50–200 kg ha^{-1} of applied P. Inoculation with *G. manihotis* markedly increased cassava growth in greenhouse experiments using sterilized soils. The effect was significant but less dramatic in non-sterilized soil. When plots in the field were sterilized with methyl bromide to kill the native VAM in the topsoil, re-inoculation of the soil with VAM markedly improved initial plant growth (Fig. 7.10). Non-inoculated plants grew poorly, showing clear symptoms of P deficiency. Once the roots of these non-inoculated plants reached the non-sterilized subsoil, however, they soon became infected with VAM and recuperated. In Quilichao, Colombia, which has a highly effective native VAM population dominated by *G.*

Fig. 7.10. Comparison of plants inoculated with vesicular–arbuscular mycorrhizas (VAM); (right) and non-inoculated plants in plots previously sterilized with methyl bromide.

manihotis, non-inoculated plants in the sterilized plots attained as high yields as those in the unsterilized soil. Inoculated plants growing in the sterilized soil produced 40% higher yields (Howeler *et al.*, 1982b). In other experiments at Quilichao, soil sterilization markedly reduced yields, but inoculation of plants growing in unsterilized soil did not increase yields because of the highly effective native VAM population.

In soils with a less effective native VAM population, such as in Carimagua, Colombia, inoculation of plants grown in sterilized soil increased yields nearly threefold without applied P and 164% with 100 kg P ha^{-1}. In unsterilized soil the effect of inoculation was not significant without applied P, but significant with 100 kg P ha^{-1}. In a similar trial in Carimagua, using different sources of P, VAM inoculation in unsterilized soil had no effect without P application but increased yields significantly (22%) at an intermediate application of 100 kg P ha^{-1} (Howeler and Sieverding, 1983).

Numerous experiments on VAM inoculation of cassava growing in natural soils in Colombia indicate that responses vary from location to location, depending on the efficiency of the native VAM population and the ability of the introduced species to compete with the native population. In areas with less-effective native populations, such Carimagua, Colombia, yields increased 23% by inoculation, but this may decrease over time once cassava cultivation has

stimulated a build-up of native mycorrhizas. In other locations the effect of inoculation was smaller and not consistent. It is clear that mycorrhizas are absolutely essential for cassava growth, but it seems difficult to improve on an already highly efficient, naturally occurring symbiosis.

References

Abruña, F., Perez-Escolar, R., Vicente-Chandler, J., Figarella, J. and Silva, S. (1974) Response of green beans to acidity factors in six tropical soils. *Journal of Agriculture* (University of Puerto Rico) 58(1), 44–58.

Acosta, J.R. and Perez, G.J. (1954) Abonamiento en yuca. *Suelo Tico (Costa Rica)* 7, 300–308.

Amarasiri, S.L. and Perera, W.R. (1975) Nutrient removal by crops growing in the dry zone of Sri Lanka. *Tropical Agriculturist* 131, 61–70.

Anon. (1952) Le manioc. *Recherche Agronomique de Madagascar* 1, 49–52.

Anon. (1953) Essais de fumure du manioc. *Recherche Agronomique de Madagascar. Compte Rendu* 2, 85–88.

Asher, C.J., Edwards, D.G. and Howeler, R.H. (1980) *Nutritional Disorders of Cassava* (Manihot esculenta *Crantz*). University of Queensland, St Lucia, Queensland, Australia.

Ashokan, P.K. and Sreedharan, C. (1980) Effect of potash on growth; yield and quality of tapioca variety H. 97. In: *Proceedings National Seminar on Tuber Crops Production Technology*, Tamil Nadu

Agricultural University, Coimbatore, Tamil Nadu, India, pp. 78–80.

Blin, H. (1905) La fumure du manioc. *Bulletin Economique de Madagascar* 3, 419–421.

Cadavid, L.F. (1988) *Respuesta de la yuca (*Manihot esculenta *Crantz) a la aplicacion de NPK en suelos con diferentes características.* Monograph, Universidad Nacional de Colombia, Palmira, Colombia.

Cadavid, L.F., El-Sharkawy, M.A., Acosta, A. and Sanchez, T. (1998) Long-term effects of mulch, fertilization and tillage on cassava grown in sandy soils of northern Colombia. *Field Crops Research* 57, 45–56.

CTCRI (1972) *Annual Report 1971.* CTCRI, Trivandrum, India.

CTCRI (1974) *Annual Report 1973.* CTCRI, Trivandrum, India.

CTCRI (1975) *Annual Report 1974.* CTCRI, Trivandrum, India.

CIAT (1976) *Annual Report for 1975.* CIAT, Cali, Colombia.

CIAT (1977) *Annual Report for 1976.* CIAT, Cali, Colombia.

CIAT (1978) *Annual Report for 1977.* CIAT, Cali, Colombia.

CIAT (1979) *Annual Report for 1978.* CIAT, Cassava program. Cali, Colombia, pp. A76–84.

CIAT (1980) Cassava program. *Annual Report for 1979.* CIAT, Cali, Colombia.

CIAT (1982) Cassava program. *Annual Report for 1981.* CIAT, Cali, Colombia.

CIAT (1985a) Cassava program. *Annual Report for 1982 and 1983.* CIAT, Cali, Colombia.

CIAT (1985b) Cassava program. *Annual Report for 1984.* Working Document No. 1. CIAT, Cali, Colombia.

CIAT (1988a) Cassava program. *Annual Report for 1985.* Working Document No. 38. CIAT, Cali, Colombia.

CIAT (1988b) Cassava program. *Annual Report for 1986.* Working Document No. 43. CIAT, Cali, Colombia.

CIAT (1998) *Improving Agricultural Sustainability in Asia. Integrated Crop-Soil Management for Sustainable Cassava-Based Production Systems.* End-of-Project Report, 1994–1998 submitted to the Nippon Foundation. CIAT, Cali, Colombia.

Chan, S.K. (1980) Long-term fertility considerations in cassava production. In: Weber, E.J., Toro, J.C. and Graham, M. (eds) *Workshop on Cassava Cultural Practices*, Salvador, Bahia, Brazil, 18–21 March, IDRC-151e, pp. 82–92.

Chew, W.Y. (1971) The performance of tapioca, sweet potato and ginger on peat at the Federal Experiment Station, Jalan Kebun, Selangor. Agronomy

Branch, Division of Agriculture, Kuala Lumpur, Malaysia.

Chew, W.Y., Ramli, K. and Joseph, K.T. (1978) Copper deficiency of cassava (*Manihot esculenta* Crantz) on Malaysian peat soil. *MARDI Research Bulletin* 6(2), 208–213.

Cock, J.H. (1975) Fisiología de la planta y desarrollo. In: *Curso sobre Produccion de Yuca.* Instituto Colombiano Agropecuario, Regional 4, Medellín, Colombia.

Cong Doan Sat and Deturck, P. (1998) Cassava soils and nutrient management in South Vietnam. In: Howeler, R.H. (ed.) *Cassava Breeding, Agronomy and Farmer Participatory Research in Asia*, Proceedings 5th Regional Workshop, Danzhou, Hainan, China, 3–8 November 1996, pp. 257–267.

Cours, G., Fritz, J. and Ramahadimby, G. (1961) El diagnóstico felodérmico de la mandioca. *Fertilité* 12, 3–20.

Diniz, M. de S., Gomes C. J. de and Caldas, R.C. (1994) Sistemas de adubação na cultura da mandioca. *Revista Brasileira de Mandioca* (Cruz das Almas, Bahia, Brazil) 13, 157–160.

Edwards, D.G. and Kang, B.T. (1978) Tolerance of cassava (*Manihot esculenta* Crantz) to high soil acidity. *Field Crops Research* 1, 337–346.

EMBRAPA, CNPMF (1981) *Relatória Técnico Annual. Projecto de Mandioca.* Cruz das Almas, Bahia, Brazil, pp. 119–167.

EMBRAPA, CNPMF (1984) *Relatória Técnico Annual do CNPMF 1983.* Cruz das Almas, Bahia, Brazil.

FAO (1980) *Review of Data on Responses of Tropical Crops to Fertilizers, 1961–1977.* FAO, Rome, Italy.

Forno, D.A. (1977) The mineral nutrition of cassava (*Manihot esculenta* Crantz) with particular reference to nitrogen. PhD thesis. University of Queensland, St Lucia, Queensland, Australia.

Fox, R.H., Talleyrand, H. and Scott, T.W. (1975) Effect of nitrogen fertilization on yields and nitrogen content of cassava, Llanera cultivar. *Journal of Agriculture* (University of Puerto Rico) 56, 115–124.

Goepfert, C.F. (1972) Experimento sobre o efeito residual da adubação fosfatada em feijoeiro (*Phaseolus vulgaris*). *Agronomia Sulriograndense* 8, 41–47.

Gomes, C.J. de (1998) *Adubação de mandioca.* Curso Internacional de Mandioca para Países Africanos de Lingua Portuguesa. Cruz das Almas, Bahia, Brazil, 13–30 April.

Gomes, C.J. de and Ezeta, F.N. (1982) Nutrição e adubação potássica da mandioca no Brasil. In: *Anais de Simpósio Sobre Potássio na Agricultura Brasileira*, Londrina, Paraná. Instituto da Potassa/IAPAR, Piracicaba, SP, Brazil, pp. 487–502.

Gomes, C.J. de, Souza, R.F., Rezende, J. de O. and Lemos, L.B. (1973) Efeitos de N, P, K, S, micronutrientes e calagem na cultura de mandioca. *Boletim Técnico 20*. Instituto de Pesquisa Agropecuária do Leste (IPEAL). Cruz das Almas, Bahia, Brazil, pp. 49–67.

Gomes, C.J. de, Carvalho, P.C.L. de, Carvalho, F.L.C. and Rodrigues, E.M. (1983) Adubação orgânico na recuperação de solos de baixa fertilidade com o cultivo da mandioca. *Revista Brasileira de Mandioca* (Cruz das Almas, Bahia, Brazil) 2(2), 63–76.

Gunatilaka, A. (1977) Effects of aluminium concentration on the growth of corn, soybean, and four tropical root crops. MSc thesis. University of Queensland, St. Lucia, Queensland, Australia.

Hagens, P. and Sittibusaya, C. (1990) Short and long term aspects of fertilizer applications on cassava in Thailand. In: Howeler, R.H. (ed.) *Proceedings 8th Symposium International Society of Tropical Root Crops*, Bangkok, Thailand, 30 October–5 November, 1988, pp. 244–259.

Howeler, R.H. (1978) The mineral nutrition and fertilization of cassava. In: *Cassava Production Course*. CIAT, Cali, Colombia, pp. 247–292.

Howeler, R.H. (1980a) The effect of mycorrhizal inoculation on the phosphorus nutrition of cassava. In: Weber, E.J., Toro, J.C. and Graham, M. (eds) *Cassava Cultural Practices*, Proceedings Workshop, Salvador, Bahia, Brazil, 18–21 March, International Development Research Centre (IDRC) 151e, Ottawa, Canada, pp. 131–137.

Howeler, R.H. (1980b) Soil-related cultural practices for cassava. In: Weber, E.J., Toro, J.C. and Graham, M. (eds) *Cassava Cultural Practices*, Proceedings Workshop, Salvador, Bahia, Brazil, 18–21 March, IDRC 151e, Ottawa, Canada, pp. 59–69.

Howeler, R.H. (1981) *Mineral Nutrition and Fertilization of Cassava*. Series 09EC-4, CIAT, Cali, Colombia.

Howeler, R.H. (1983) *Análisis del Tejido Vegetal en el Diagnóstico de Problemas Nutricionales: Algunos Cultivos Tropicales*. CIAT, Cali, Colombia.

Howeler, R.H. (1985a) Mineral nutrition and fertilization of cassava. In: *Cassava; Research, Production and Utilization*. UNDP-CIAT Cassava Program, Cali, Colombia, pp. 249–320.

Howeler, R.H. (1985b) Potassium nutrition of cassava. In: *Potassium in Agriculture*. International Symposium, Atlanta, GA, 7–10 July. ASA, CSSA, SSSA, Madison, Wisconsin, USA, pp. 819–841.

Howeler, R.H. (1986) El control de la erosión con prácticas agronómicas sencillas. *Suelos Ecuatoriales* 16, 70–84.

Howeler, R.H. (1987) Soil conservation practices in cassava-based cropping systems. In: Tay, T.H., Mokhtaruddin, A.M. and Zahari, A.B. (eds) *Proceedings International Conference Steepland Agriculture in the Humid Tropics*, Kuala Lumpur, Malaysia, 17–21 August, pp. 490–517.

Howeler, R.H. (1989) Cassava. In: Plucknett, D.L. and Sprague, H.B. (eds) *Detecting Mineral Deficiencies in Tropical and Temperate Crops*. Westview Press. Boulder, Colorado, pp. 167–177.

Howeler, R.H. (1990) Phosphorus requirements and management of tropical root and tuber crops. In: *Proceedings Symposium on Phosphorus Requirements for Sustainable Agriculture in Asia and Oceania*. IRRI, Los Baños, Philippines. 6–10 March, 1989, pp. 427–444.

Howeler, R.H. (1991a) Identifying plants adaptable to low pH conditions. In: Wright, R.J., Baligar, V.C. and Murrmann, R.P. (eds) *Plant-Soil Interactions at Low pH*. Kluwer Academic, The Netherlands, pp. 885–904.

Howeler, R.H. (1991b) Long-term effect of cassava cultivation on soil productivity. *Field Crops Research* 26, 1–18.

Howeler, R.H. (1992) Agronomic research in the Asian Cassava Network – an overview, 1987–1990. In: Howeler, R.H. (ed.) *Cassava Breeding, Agronomy and Utilization Research in Asia*, Proceedings 3rd Regional Workshop, Malang, Indonesia, 22–27 October 1990, pp. 260–285.

Howeler, R.H. (1994) Integrated soil and crop management to prevent environmental degradation in cassava-based cropping systems in Asia. In: Bottema, J.W.T. and Stoltz, D.R. (eds) *Upland Agriculture in Asia*, Proceedings Workshop, Bogor, Indonesia, 6–8 April 1993, pp. 195–224.

Howeler, R.H. (1995) Agronomy research in the Asian Cassava Network – towards better production without soil degradation. In: Howeler, R.H. (ed.) *Cassava Breeding, Agronomy Research and Technology Transfer in Asia*. Proceedings 4th Regional Workshop, Trivandrum, Kerala, India, 2–6 November 1993, pp. 368–409.

Howeler, R.H. (1996a) Diagnosis of nutritional disorders and soil fertility maintenance of cassava. In: Kurup, G.T., Polaniswami, M.S., Potty, V.P., Padmaja, G., Kabeerathumma, S. and Pillai, S.V. (eds) *Tropical Tuber Crops: Problems, Prospects and Future Strategies*. Oxford and IBH Publishing Co., New Delhi, India, pp. 181–193.

Howeler, R.H. (1996b) Mineral nutrition of cassava. In: Craswell, E.T., Asher, C.J. and O'Sullivan, J.N. (eds) *Mineral Nutrient Disorders of Root Crops in the Pacific*. Proceedings Workshop, Nuku'alofa, Kingdom of Tonga, 17–20 April 1995, ACIAR Proceedings no. 5, Canberra, Australia, pp. 110–116.

Howeler, R.H. (1998) Cassava agronomy research in Asia – an overview, 1993–1996. In: Howeler, R.H. (ed.) *Cassava Breeding, Agronomy and Farmer Participatory Research in Asia*. Proceedings 5th Regional Workshop, Danzhou, Hainan, China, 3–8 November 1996, pp. 355–375.

Howeler, R.H. (2000) Cassava production practices – can they maintain soil productivity? In: Howeler, R.H., Oates, C.G. and O'Brien, G.M. (eds) *Cassava, Starch and Starch Derivatives*, Proceedings of an International Symposium, Nanning, Guangxi, China, 11–15 November 1996. pp. 101–117.

Howeler, R.H. and Cadavid, L.F. (1983) Accumulation and distribution of dry matter and nutrients during a 12-month growth cycle of cassava. *Field Crops Research* 7, 123–139.

Howeler, R.H. and Cadavid, L.F. (1990) Short- and long-term fertility trials in Colombia to determine the nutrient requirements of cassava. *Fertilizer Research* 26, 61–80.

Howeler, R.H. and Fernandez, F. (1985) *Nutritional Disorders of the Cassava Plant. Study Guide*. CIAT, Cali, Colombia.

Howeler, R.H. and Medina, C.J. (1978) *La fertilización en el frijol* Phaseolus vulgaris: *elementos mayores y secundarios*. Revisión de la literatura para el Curso de Producción de Frijol. CIAT, Cali, Colombia.

Howeler, R.H. and Sieverding, E. (1983) Potentials and limitations of mycorrhizal inoculation illustrated by experiments with field grown cassava. *Plant and Soil* 75, 245–261.

Howeler, R.H., Cadavid, L.F. and Calvo, F.A. (1977) The interaction of lime with minor elements and phosphorus in cassava production. In: *Proceedings 4th Symposium International Society of Tropical Root Crops*, Cali, Colombia. IDRC, Ottawa, Canada, pp. 113–117.

Howeler, R.H., Edwards, D.G. and Asher, C.J. (1981) Application of the flowing solution culture techniques to studies involving mycorrhizas. *Plant and Soil* 59, 179–183.

Howeler, R.H., Asher, C.J. and Edwards, D.G. (1982a) Establishment of an effective endomycorrhizal association in cassava in flowing solution culture and its effect on phosphorus nutrition. *New Phytologist* 90, 229–238.

Howeler, R.H., Cadavid, L.F. and Burckhardt, E. (1982b) Response of cassava to VA mycorrhizal inoculation and phosphorus application in greenhouse and field experiments. *Plant and Soil* 69, 327–339.

Howeler, R.H., Edwards, D.G. and Asher, C.J. (1982c) Micronutrient deficiencies and toxicities of cassava plants grown in nutrient solutions. I. Critical tissue concentrations. *Journal of Plant Nutrition* 5, 1059–1076.

Howeler, R.H., Sieverding, E. and Saif, S. (1987) Practical aspects of mycorrhizal technology in some tropical crops and pastures. *Plant and Soil* 100, 249–283.

Howeler, R.H., El-Sharkawy, M.A. and Cadavid, L.F. (2002a) The use of grain and forage legumes for soil fertility maintenance and erosion control in cassava in Colombia (in press).

Howeler, R.H., Oates, C.G. and Costa Allem, A.C. (2002b) *Strategic Assessment on the Impact of Small-holder Cassava Production and Processing on the Environment and Biodiversity*. International Fund for Agricultural Development (IFAD), Rome, Italy (in press).

Howeler, R.H., Thai Phien and Nguyen the Dang (2002) Sustainable cassava production on sloping lands in Vietnam. *Proceedings of International Workshop on Sustainable Land Management in the Northern Mountainous Region of Vietnam*. Hanoi, Vietnam (in press).

Hulugalle, N.R., Lal, R. and Gichuru, M. (1991) Effect of five years of no-tillage and mulch on soil properties and tuber yield of cassava on an acid ultisol in southeastern Nigeria. *IITA Research* 1, 13–16.

Islam, A.K.M.S., Edwards, D.G. and Asher, C.J. (1980) pH optima for crop growth: results of flowing culture experiment with six species. *Plant and Soil* 54(3), 339–357.

Jintakanon, S., Edwards, D.G. and Asher, C.J. (1982) An anomalous, high external phosphorus requirement for young cassava plants in solution culture. In: *Proceedings 5th International Symposium Tropical Root Crops*, Manila, Philippines, 17–21 September 1979, pp. 507–518.

Jones, U.S., Katyal, J.C., Mamaril, C.P. and Park, C.S. (1982) Wetland-rice nutrient deficiencies other than nitrogen. In: *Rice Research Strategies for the Future*. IRRI, Los Baños, Philippines, pp. 327–378.

Kabeerathumma, S., Mohankumar, B., Mohankumar, C.R., Nair, G.M., Prabhakar, M. and Pillai, N.G. (1990) Long range effect of continuous cropping and manuring on cassava production and fertility status of soil. In: Howeler, R.H. (ed.) *Proceedings 8th Symposium International Society of Tropical Root Crops*, Bangkok, Thailand, 30 October–5 November 1988, pp. 259–269.

Kang, B.T. (1984) Potassium and magnesium responses of cassava grown in ultisol in southern Nigeria. *Fertilizer Research* 5, 403–410.

Kang, B.T. and Okeke, J.E. (1984) Nitrogen and potassium responses of two cassava varieties grown on an alfisol in southern Nigeria. In: *Proceedings 6th Symposium International Society of Tropical Root Crops*, Lima, Peru, 21–26 February 1983, pp. 231–237.

Kang, B.T., Islam, R., Sanders, F.E. and Ayanaba, A. (1980) Effect of phosphate fertilization and inoculation with VA-mycorrhizal fungi on performance of cassava (*Manihot esculenta*, Crantz) grown on an alfisol. *Field Crops Research* 3, 83–94.

Kasele, I.N. (1980) Investigation on the efffect of shading, potassium and nitrogen and drought on the development of cassava tuber at the early stage of plant growth. MSc thesis, University of Ibadan. Ibadan, Nigeria.

Krochmal, A. and Samuels, G. (1970) The influence of NPK levels on the growth and tuber development of cassava in tanks. *CEIBA* 16, 35–43.

Leihner, D. (1983) *Management and Evaluation of Intercropping Systems with Cassava*. CIAT, Cali, Colombia.

Lozano, J.C., Bellotti, A., Reyes, J.A., Howeler, R., Leihner, D. and Doll, J. (1981) *Field Problems in Cassava*. CIAT Series No. 07EC-1, Cali, Colombia.

Margolis, E. and Campos Filho, O.R. (1981) Determinação dos fatores da equação universal de perdas de solo num podzólico vermelho amarelo de Glória do Goitá. In: *Anais do 3rd Encontro Nacional de Pesquisa Sobre Conservação do Solo*, Rengife, Pernambuco, Brazil, 28 July–1 August 1980, pp. 239–250.

Moraes, O. de, Mondardo, E., Vizzotto, J. and Machado, M.O. (1981) *Adubação química e calagem da mandioca*. Boletim Técnico No.8. Empresa Catarinense de Pesquisa Agropecuãria. Florianopolis, Santa Catarina, Brazil.

Muhr, L., Leihner, D.E., Hilger, T.H. and Müller-Sämann, K.M. (1995) Intercropping of cassava with herbaceous legumes. II. Yields as affected by below-ground competition. *Angewandte Botanik* 69, 22–26.

Nair, P.G., Mohankumar, B., Prabhakar, M. and Kabeerathumma, S. (1988) Response of cassava to graded doses of phosphorus in acid lateritic soils of high and low P status. *Journal of Root Crops* 14(2), 1–9.

Nayar, T.V.R., Kabeerathumma, S., Potty, V.P. and Mohankumar, C.R. (1995) Recent progress in cassava agronomy in India. In: Howeler, R.H. (ed.) *Cassava Breeding, Agronomy Research and Technology Transfer in Asia*. Proceedings 4th Regional Workshop, Trivandrum, Kerala, India, 2–6 November 1993, pp. 61–83.

Ngongi, A.G.N., Howeler, R.H. and MacDonald, H.A. (1977) Effect of potassium and sulphur on growth, yield, and composition of cassava. In: *Proceedings 4th Symposium International Society of Tropical Root Crops*, Cali, Colombia, 1–7 August 1976. IDRC, Ottawa, Canada, pp. 107–113.

Nguyen Huu Hy, Pham Van Bien, Nguyen The Dang and Thai Phien (1998) Recent progress in cassava agronomy research in Vietnam. In: Howeler, R.H. (ed.) *Cassava Breeding, Agronomy and Farmer Participatory Research in Asia*. Proceedings 5th Regional Workshop, Danzhou, Hainan, China, 3–8 November 1996, pp. 235–256.

Nguyen Tu Siem (1992) Organic matter recycling for soil improvement in Vietnam. In: *Proceedings 4th Annual Meeting*, IBSRAM Asialand Network.

Nijholt, J.A. (1935) *Opname van voedingsstoffen uit den bodem bij cassave* [Absorption of nutrients from the soil by a cassava-crop]. Buitenzorg. Algemeen Proefstation voor den Landbouw. Korte Mededeelingen No. 15.

Normanha, E.S. and Pereira, A.S. (1950) Aspectos agronômicos da cultura da mandioca (*Manihot utilissima* Pohl). *Bragantia* (Campinas, SP, Brazil) 10, 179–202.

Nunes, W.O., de, Brito, D.P.P.S., de, Meneguelli, C.A., Arruda, N.B. de and Oliveira, A.B. de (1974) Resposta da mandioca a adubação mineral e métodos de aplicação do potássio em solos de baixa fertilidade. *Pesquisa Agropecuária Brasileira* (*Série Agronomia*) (Rio de Janeiro, Brazil) No. 9, pp. 1–9.

Obigbesan, G.O. (1973) The influence of potassium nutrition on the yield and chemical composition of some tropical root and tuber crops. In: *10th Coloquium International Potash Institute*, Abidjan, Ivory Coast, pp. 439–451.

Obigbesan, G.O. (1977) Investigations on Nigerian root and tuber crops: effect of potassium on starch yield, HCN content and nutrient uptake of cassava cultivars (*Manihot esculenta*). *Journal of Agricultural Science* 89, 29–34.

Obigbesan, G.O. and Fayemi, A.A.A. (1976) Investigations on Nigerian root and tuber crops: influence of nitrogen fertilization on the yield and chemical composition of two cassava cultivars (*Manihot esculenta*). *Journal of Agricultural Science* 86, 401–406.

Ofori, C.S. (1973) Decline in fertility status of a tropical forest ochrosol under continuous cropping. *Experimental Agriculture* 9, 15–22.

Okogun, J.A., Sanginga, N. and Adeola, E.O. (1999) *Soil Fertility Maintenance and Strategies for Cassava Production in West and Central Africa*. IITA, Ibadan, Nigeria (mimeograph).

Orlando Filho, J. (1985) Potassium nutrition of sugarcane. In: Bishop, W.D. *et al.* (eds) *Potassium in Agriculture*. ASA-CSSA-SSSA, Madison, Wisconsin, pp. 1045–1062.

Paula, M.B. de, Nogueira, F.D. and Tanaka, R.T. (1983) Nutrição mineral da mandioca: absorção de nutrientes e produção de materia seca por duas cultivares de mandioca. *Revista Brasileira de Mandioca*, Cruz das Almas, 31–50.

Payne, H. and Webster, D.C. (1956) The toxicity of cassava varieties on two Jamaican soil types of differing K status. Ministry of Agriculture and Fisheries. Crop Agronomy Division, Kingston, Jamaica.

Pellet, D. and El-Sharkawy, M.A. (1993a) Cassava varietal response to phosphorus fertilization. I. Yield, biomass and gas exchange. *Field Crops Research* 35, 1–11.

Pellet, D. and El-Sharkawy, M.A. (1993b) Cassava varietal response to phosphorus fertilization. II. Phosphorus uptake and use efficiency. *Field Crops Research* 35, 13–20.

Phommasack, T., Sengtaheuanghung, O. and Phanthaboon, K. (1995) The management of sloping lands for sustainable agriculture in Laos. In: Sajjapongse, A. and Elliot, C.R. (eds) *The Management of Sloping Lands for Sustainable Agriculture in Asia*, (Phase 2, 1992–1994). IBSRAM/Asialand Network Doc. no. 12. IBSRAM, Bangkok, Thailand, pp. 87–101.

Phommasack, T., Sengtaheuanghung, O. and Phanthaboon, K. (1996) The management of sloping lands for sustainable agriculture in Laos. In: Sajjapongse, A. and Leslie, R.N. (eds) *The Management of Sloping Lands in Asia*. IBSRAM/Asialand Network Doc. no. 20. IBSRAM, Bangkok, Thailand, pp. 109–136.

Prevot, P. and Ollagnier, M. (1958) La fumure potassique dans les regions tropicales et subtropicales. In: *Potassium Symposium*, Berne, Switzerland, pp. 277–318.

Putthacharoen, S., Howeler, R.H., Jantawat, S. and Vichukit, V. (1998) Nutrient uptake and soil erosion losses in cassava and six other crops in a Psamment in eastern Thailand. *Field Crops Research* 57, 113–126.

Queiroz, G.M. de and Pinho, J.L.N. de (1981) Resultados do experimento efeito da fertilização com macronutrientes NPK em mandioca no Estado do Ceará. EPACE. Pacajús, Ceará, Brazil.

Queiroz, G.M. de, Pinho, J.L.N. de, Lima, A.R.C. da and Verde, N.G.L. (1980) Efeito da fertilização com macronutrientes NPK em mandioca (*Manihot esculenta* Crantz) no Estado do Ceará. In: EPACE, *Relatório Annual de Pesquisa Fitotécnica*. Fortaleza, Ceará, Brazil, pp. 83–96.

Quintiliano, J., Margues, A., Bertoni, J. and Barreto, G.B. (1961) Perdas por erosão no estado de São Paulo. *Brigantia* 20(2), 1143–1182.

Richards, I.R. (1979) Response of tropical crops to fertilizer under farmers conditions – analysis of results of the FAO Fertilizer Programme. *Phosphorus in Agriculture* 76, 147–156.

Rio Grande do Norte (1976) Pesquisa e experimentação com culturas alimentares: mandioca, 1971–1975. *Secretaria da Agricultura*. Rio Grande do Norte, Brazil.

Roberts, S. and McDole, R.E. (1985) Potassium nutrition of potatoes. In: Bishop, W.D. *et al.* (eds) *Potassium in Agriculture*. International Symposium,

Atlanta, Georgia, 7–10 July. ASA-CSSA-SSSA, Madison, Wisconsin, pp. 800–818.

Roche, P., Velly, J. and Joliet, B. (1957) Essai de determination des seuils de carence en potasse dans le sol et dans les plantes. *Revue de la Potasse* 1957, 1–5.

Ruppenthal, M. (1995) *Soil Conservation in Andean Cropping Systems*. Hohenheim Tropical Agriculture Series no. 3, Hohenheim University, Germany.

Ruppenthal, M., Leihner, D.E., Steinmuller, N. and El-Sharkawy, M.A. (1997) Losses of organic matter and nutrients by water erosion in cassava-based cropping systems. *Experimental Agriculture* 33, 487–498.

Santos, Z.G. dos and Tupinamba, E.A. (1981) Resultados do experimento de niveis e fontes de fósforo na produção de mandioca (*Manihot esculenta* Crantz). EMBRAPA-UEPAE, Aracaju, Sergipe, Brazil.

Sieverding, E. and Howeler, R.H. (1985) Influence of species of VA mycorrhizal fungi on cassava yield response to phosphorus fertilization. *Plant and Soil* 88, 213–222.

Silva, J.R. da and Freire, E.S. (1968) Efeitos de doses crescentes de nitrogênio, fósforo e potássio sobre a produção de mandioca em solos de baixa e alta fertilidade. *Brigantia* (*Campinas, SP, Brazil*) 27 (29), 357–364.

Silva, L.G. (1970) Adubação NPK na cultura da mandioca em Tabuleiró Costeiro no estado da Paraiba. *Pesquisas Agropecuárias no Nordeste* 2(1), 73–75.

Silva, L.G., Souza, J.B. de, Silva, J.C. da and Lucas, A.P. de (1969) Ação de macronutrientes e manganês na cultura da mandioca em solos de tabuleiros costeiros do Nordeste. SUDENE, Recife. Pernambuco, Brazil.

Sittibusaya, C. (1993) [*Progress report of soil research on fertilization of field crops, 1992*]. Annual Cassava Program Review, Rayong, Thailand, 19–20 January 1993 [in Thai].

Sobral, L.F., Barreto, A.C., Siqueira, L.A., Santos, Z.G. dos, Souza, R.F., Rezende, J.O. de and Ribeiro, J.V. (1976) Efeitos de macro e micronutrientes em produção da mandioca (*Manihot esculenta* Crantz). *Comunicado Técnico, 1*. EMBRAPA-Rep. No Estado de Sergipe, Aracaju, Sergipe, Brazil.

Souza, L.D., Gomes C.J. de and Caldas, R.C. (1992) Interação vinhoto, calage, e fósforo na cultura da mandioca no Norte de Mato Grosso. *Revista Brasileira de Mandioca* 11(2), 148–155.

Spain, J.M., Francis, C.A., Howeler, R.H. and Calvo, F. (1975) Differential species and varietal tolerance to soil acidity in tropical crops and pastures. In: Bosnemisza, E. and Alvarado, A. (eds) *Soil*

Management in Tropical America. North Carolina State University, Raleigh, North Carolina, pp. 308–329.

Spear, S.N., Asher, C.J. and Edwards, D.G. (1978a) Response of cassava, sunflower, and maize to potassium concentration in solution. I. Growth and plant potassium concentration. *Field Crops Research* 1, 347–361.

Spear, S.N., Edwards, D.G. and Asher, C.J. (1978b) Response of cassava, sunflower, and maize to potassium concentration in solution. III. Interactions between potassium, calcium, and magnesium. *Field Crops Research* 1, 375–389.

Stephens, D. (1960) Fertilizer trials on peasant farms in Ghana. *Empire Journal of Experimental Agriculture* 109, 1–22.

Takyi, S.K. (1972) Effects of potassium, lime and spacing on yields of cassava (*Manihot esculenta* Crantz). *Ghana Journal of Agricultural Science* 5(1), 39–42.

Tongglum, A., Vichukit, V., Jantawat, S., Sittibusaya, C., Tiraporn, C., Sinthuprama, S. and Howeler, R.H. (1992) In: Howeler, R.H. (ed.) *Cassava Breeding, Agronomy and Utilization Research in Asia*. Proceedings 3rd Regional Workshop, Malang, Indonesia, 22–27 October 1990, pp. 199–223.

Tongglum, A., Pornpromprathan, V., Paisarncharoen, K., Wongwitchai, C., Sittibusaya, C., Jantawat, S., Nual-on, T. and Howeler, R.H. (1998) Recent progress in cassava agronomy research in Thailand. In: Howeler, R.H. (ed.) *Cassava Breeding, Agronomy and Farmer Participatory*

Research in Asia. Proceedings 5th Regional Workshop, Danzhou, Hainan, China, 3–8 November 1996, pp. 211–234.

Vijayan, M.R. and Aiyer, R.S. (1969) Effect of nitrogen and phophorus on the yield and quality of cassava. *Agricultural Research Journal of Kerala* 7, 84–90.

Vinod, G.S. and Nair, V.M. (1992) Effect of slow-release nitrogenous fertilizers on the growth and yield of cassava. *Journal of Root Crops* 18, 124–125.

Wargiono, J., Kushartoyo, Suyamto, H. and Guritno, B. (1998) Recent progress in cassava agronomy research in Indonesia. In: Howeler, R.H. (ed.) *Cassava Breeding, Agronomy and Farmer Participatory Research in Asia*. Proceedings 5th Regional Workshop, Danzhou, Hainan, China, 3–8 November 1996, pp. 307–330.

Zaag, P. van der (1979) The phosphorus requirements of root crops. PhD thesis, University of Hawaii, Hawaii.

Zangrande, M.B. (1981) Resultados do experimento estudo de níveis de NPK para a cultura da mandioca (*Manihot esculenta* Crantz) no Estado do Espírito Santo, EMCAPA. Cariacica, ES, Brazil.

Zhang Weite, Lin Xiong, Li Kaimian, Huang Jie, Tian Yinong, Lee Jun and Fu Quohui (1998) Cassava agronomy research in China. In: Howeler, R.H. (ed.) *Cassava Breeding, Agronomy and Farmer Participatory Research in Asia*. Proceedings 5th Regional Workshop, Danzhou, Hainan, China, 3–8 November 1996, pp. 191–210.

Chapter 8
Breeding for Crop Improvement

D.L. Jennings[1] and C. Iglesias[2]

[1]'Clifton', Honey Lane, Otham, Maidstone, Kent ME15 8JR, UK
(formerly of EAFRO and IITA); [2]Weaver Popcorn Co., PO Box 20,
New Richmond, IN 47967, USA (formerly of CIAT, Colombia)

Introduction

Cassava has been evolving as a food crop ever since it became important in the second and third millennium BC (Reichel-Dolmatoff, 1965; Lathrap, 1973), but its adaptation to African and Asian conditions did not begin until post-Columbian times. In the Americas, Africa and Asia, progress towards improved adaptation and quality was first through subconscious selection by farmers. A wide range of genetic diversity was generated through centuries of such farmer selection (Bonierbale *et al.*, 1995). It was not until the present century that serious attempts began by national organizations to improve the crop by plant breeding. Much of this was instigated by the colonial powers and was very successful, but progress slowed considerably when countries became independent. This trend was arrested in the 1960s, when the increasing world population and the limited supply of energy foods prompted a surge of interest in the crop.

High priority was given to cassava breeding and related research when the International Institute of Tropical Agriculture (IITA) was opened in Nigeria and the Centro Internacional de Agricultura Tropical (CIAT) was opened in Colombia. For the first time breeders and associated scientists were given resources to study the crop in depth and to assess the extensive variation available. The two International Centres collaborated with existing national programmes and instigated the initiation of new ones. In India the Central Tuber Crops Research Institute (CTCRI) took on a similar role. The objectives were to increase both the yield per unit area and the area under cultivation, and also to improve root quality.

Cytotaxonomy of the Genus *Manihot*

Manihot esculenta (cassava) is placed in the *Fruticosae* section of the genus *Manihot*, which is a member of the Euphorbiaceae. The *Fruticosae* section contains low-growing shrubs adapted to savannah, grassland or desert and is considered less primitive than the *Arboreae* section, which contains the tree species. The genus occurs naturally only in the Western hemisphere, between the southwest USA (33°N) and Argentina (33°S), and shows most diversity in two areas, one in northeastern Brazil extending towards Paraguay, and the other in western and southern Mexico.

All the species so far studied have 36 chromosomes, which show regular bivalent pairing at meiosis. However, in both cassava and *Manihot glaziovii* (sect. *Arboreae*) there is evidence of polyploidy from studies of pachytene karyology. There are three nucleolar chromosomes which is high for true diploids, and duplication for some of the chromosomes. This indicates that *Manihot* species are probably segmental allotetraploids derived from crossing between two taxa whose haploid complements had six chromosomes in common but differed in the other three. Studies with biochemical markers identified by electrophoresis support this interpretation, in that they show disomic inheritance at 12 loci, with evidence of gene duplication (Jennings and Hershey, 1985; Charrier and Lefevre, 1987).

Flower Behaviour, Hybridization Techniques and Seed Management

Cassava is monoecious. The female flowers normally open 10–14 days before the males on the same branch, but self-fertilization can occur because male and female flowers on different branches or on different plants of the same genotype open simultaneously. The proportions of self- and cross-pollinated seed produced depends on genotype, planting design and the type of pollinating insects present.

The availability of flowers is influenced by plant habit, because branching always occurs when an inflorescence is formed (Fig. 8.1). Hence tall, unbranched plants are less floriforous than highly branched, low-growing ones.

To make a controlled cross between two parents, unopened flowers are first enclosed in muslin bags and the chosen pollen is applied to the stigmas as soon as the female flowers open. The muslin bags are then replaced with netting bags to catch the seed when the ripe fruits dehisce explosively. Another system, for example where cyclic breeding methods such as those used in other out-breeding crops are followed, is to plant a set of varieties in a specially designed crossing block (Fig. 8.2), and to remove all the male flowers from the varieties to be used as females. The separation of male and female flowers makes the control of pollination easy, but it is still laborious to produce large quantities of seed (Kawano, 1980; Hahn, 1982). Polycross designs similar to the ones used for forage crops can also be used for cassava, using a random distribution of elite genotypes replicated several times. This method does not prevent self-pollination, but it produces considerably more cross-bred seeds than controlled pollination methods (Wright, 1965).

The fertility of clones is variable and can be very low. An average of one seed per fruit is commonly achieved through controlled pollination from a maximum of three from the trilocular ovary. The genotype of the female parent is more important in determining success than that of the pollen (Jennings, 1963).

Newly harvested seeds are dormant and require 3–6 months storage at ambient temperatures before they will germinate. Germination can be hastened by carefully filing the sides of the seed coat at the radicle end and by temperature management. Ellis *et al.* (1982) found that

Fig. 8.1. Flowering cassava showing the association of branching with inflorescence development.

Fig. 8.2. Cassava crossing block at IITA in which all male flowers were removed from parents being used as females and the resulting fruits from cross-pollinated flowers are being collected in muslin bags before they dehisce explosively.

few seeds germinated unless the temperature exceeded 30°C for at least a part of the day and the mean temperature exceeded 24°C; the best rates occurred at 30–35°C. A dry heat treatment of 14 days at 60°C was also beneficial for newly harvested seeds. If temperatures permit and irrigation is available the easiest method is to sow the seeds directly into the soil. This is successful at IITA because temperatures from January to March range from 30 to 35°C (Hahn *et al.*, 1973). At CIAT seeds are frequently planted in a screen-house and the emerging seedlings held until they reach 20–25 cm before being transplanted to well-prepared soil with good moisture conditions. Seeds for storage should be kept at 5°C and 60% relative humidity (IITA, 1978) because they lose viability rapidly during a year's storage at ambient temperature (Kawano, 1978).

Breeding Strategy

The worldwide emphasis of the breeding work at the international centres implied that the objectives would be broad and that a large and variable number of characteristics would be required to achieve them. To meet all the local needs, improvements at IITA and CIAT would first have to be incorporated into broadly based breeding populations which would then be

subjected to further selection at national centres. Previously, it was sufficient to make crosses between the best local varieties.

The policy adopted was to create improved populations into which exotic germplasm from several sources could be introgressed, while retaining the desirable gene complexes already present and allowing sufficient inbreeding for the expression and elimination of recessive ones (Hahn *et al.*, 1973, 1979; Hershey, 1981, 1984; Hahn, 1982). It was desirable to minimize inbreeding and to restore heterozygosity fully to avoid inbreeding depression (Kawano *et al.*, 1978a). The improved germplasm generated was distributed either in the form of elite genotypes transferred *in vitro*, or as populations of recombinant seeds (full-sibs or half-sibs) (Bonierbale *et al.*, 1995).

Breeding strategy for Africa

The germplasm of the original importations to Africa was inevitably narrowly based, but natural intercrossing among highly heterozygous varieties and subconscious selection among the resulting self-sown seedlings made possible the rapid progress towards local adaptation. Natural crossing with the introduced *M. glaziovii* (Ceara rubber) produced the 'tree cassava' and may have broadened the genetic base of the crop in Africa. All the introduced germplasm was probably as highly susceptible to cassava mosaic disease (CMD) as most of the present American germplasm, but a level of tolerance was achieved quickly, and acceptable yields were usually obtained.

Nevertheless, the effects of CMD were often so devastating that most national programmes concentrated on breeding for resistance to it. At IITA, Hahn and his co-workers set out to create base populations by cyclic selection and recombination, and to upgrade them with a range of new high-yielding germplasm which included some highly CMD-susceptible germplasm from the Americas. They used a system based upon half-sib test-crosses, in which the selections and local, virus-resistant tester varieties were grown in isolation blocks. Three-plant plots of each of the selections and the local varieties were planted in several replications to favour random crossing, and the local varieties became universal pollinators when the male flowers from the other parents were all removed. The resulting progenies were grown in replicated trials which were sometimes duplicated in contrasting environments.

Improved cultivars have been released in several African countries (Table 8.1) although their adoption has sometimes been disappointing.

Both controlled and open-pollinated methods of breeding are used currently at IITA, but the scale of the latter procedure being followed in the 1980s can be seen from the following:

- Year 1: Up to 100,000 seedlings were raised from field-sown seed and screened for resistances to CMD and cassava bacterial blight (CBB). At harvest, selection was for compact roots with short necks, stems branching at about 100 cm and for low HCN in the leaves.
- Year 2: Up to 3000 of the selections from year 1 were grown in small non-replicated plots. Further selection was for disease resistances, yield potential and dry matter content of the roots, and the HCN in the roots was assayed enzymatically.
- Year 3: Up to 100 of the selections from year 2 were tested in replicated trials at three locations and consumer acceptance was assessed.

Table 8.1.   Cassava cultivars released by National Programmes in Africa.

Country	Variety
Benin	TMS 30572, TMS 4(2) 1425, TMS 30572A, Ben 86052
Burundi	40160–1, 40160–3
Cameroon	8034, 8017, 8061, 820516, 1005, 658, 244
Ivory Coast	TMS 30572, TMS 4(2) 1425
Gabon	CIAM 76–6, CIAM 76–7, CIAM 76–13, CIAM 76–33
Gambia	TMS 60124, TMS 4(2) 1425
Ghana	TMS 30572, TMS 50395, TMS 4(2) 1425
Guinea Conakry	TMS 30572, TMS 4(2) 1425
Guinea Bissau	TMS 4(2) 1425, TMS 60142
Liberia	CARICASS 1, CARICASS 2, CARICASS 3
Malawi	Mbundumali, Gomani, Chitembwe
Mozambique	TMS 30001, TMS 30395, TMS 42025
Nigeria	N. C. idi-osi (TMS 30572), N. c. savannah (TMS 4(2) 1425), TMS 91934, TMS 90257, TMS 84537, TMS 81/00110, TMS 82/00661
Rwanda	Gakiza, Karana, TMS 30572
Sierra Leone	ROCASS 1, ROCASS 2, ROCASS 3, NUCASS 1, NUCASS 2, 80/40, 80/61
Togo	TMS 4(2) 1425, TMS 30572
Uganda	NASE 1 (TMS 60142), NASE 2(TMS 30337), MIGYERA (TMS 30572)
Zambia	LUC 133
Zaire	Kinuani, F100, 4023/3, 02864, Lwenyi/3

Source: Mahungu *et al.* (1994).

- Year 4: Selection was continued for up to 25 selections in larger trials at more locations.
- Year 5: Five of the best selections were tested on farms.
- Year 6: The final selections were multiplied and distributed.

It was found that germplasm from the Americas gave populations with large yield improvements, but that two generations of breeding with parents resistant to CMD and CBB were necessary to achieve acceptable resistance levels. Hybrids between East African selections and Nigerian varieties were the best sources of the two resistances. Seeds from the improved material were used to establish new populations in other parts of Africa. The new environments imposed new requirements, but the wide genetic base of the parental populations allowed for the selection of new traits, including resistance to diseases not prevalent at IITA. Ideally, where resources permitted, the best selections were intercrossed to produce new populations for progressive improvement in their adaptation to the local ecologies.

Breeding strategy for the Americas

Although cassava is indigenous to the Americas, the first breeding did not start there until 1948 at Campinas in Brazil, and little work elsewhere was started until the 1970s (Normanha, 1970). The main reason was that there was no single widespread and devastating disease of overriding importance as in Africa, and the crop was considered to have few problems requiring plant breeding work: diseases and pests were plentiful but the problems were not too serious and essentially local ones. Also, the crop never enjoyed a high priority within the research plans formulated by National Programmes. However, the intensification of production accentuated the need for better varieties, and, soon after its inception, CIAT began a programme which was based upon crop improvement by controlled breeding.

The objective of the breeding at CIAT was to provide germplasm for environments extending throughout both the American and Asian tropics and subtropics. The diversity in climates, soils,

pests and diseases presented such a broad array of objectives that they could not easily be achieved by selection within a single population or at a single site. The breeders (Hershey, 1984) therefore classified the areas into seven so-called edapho-climate zones, each characterized by a set of soil and climatic conditions which differentially affected the performance of cassava genotypes and determined the incidence and severity of pests and diseases. Each zone had its own adaptation requirements and the landraces grown in each of them had almost certainly persisted because they were suitably adapted to them. Any attempt to improve them, therefore, had to be through selection in the particular ecosystem and from germplasm adapted to the particular stresses present. A gene pool for each zone was therefore created by intercrossing among genotypes selected for good performance in each zone.

The zones were differentiated first by temperature and then by rainfall and soil type. They included four ecosystems in the lowland tropics: humid, subhumid, acid soil savannahs and semi-arid; areas of medium altitudes; highlands; and the subtropics. Priority for germplasm development was related to the importance of the crop in the different ecosytems worldwide. The distinct populations were created by both controlled and open-pollination methods, and were continually upgraded by recurrent selection and the introduction of new germplasm. Hershey (1984) describes the following procedure being followed in the 1980s:

- Year 1: Up to 50,000 seedlings were established at CIAT in groups based upon the adaptation of the parents to particular edapho-climatic zones. After 6 months, low selection pressure was applied for plant and root type to give about 25,000 selections; one cutting from each was used for testing in the zone for which the population was planned and another was retained at CIAT. Further selection was then made, including selection for disease and pest resistances.
- Year 2: Up to 3000 selections from year 1 were tested further in non-replicated plots for the above characteristics, plus root dry matter and HCN contents, both at CIAT and at one of the other target sites.

- Year 3: Up to 300 selections from year 2 were further tested in yield trials at several sites.
- Year 4: Up to 100 selections from year 3 were tested in larger trials at several sites.
- Year 5: Up to 20 selections were further tested in a Colombian trial network.
- Year 6: Promising selections were distributed for evaluation to national centres, usually in similar edapho-climatic zones.

From this programme a very broad range of improved genetic diversity was produced and distributed to cassava breeding programmes all over the world from CIAT in Colombia (Bonierbale *et al.*, 1995).

Breeding strategy for Asia

CIAT provides support to Asia in the form of seed for local selection, and IITA provides germplasm segregating for resistance to mosaic disease to India. India has a distinct form of this disease but is the only Asian country affected. The main cassava breeding activity in Asia was developed jointly by CIAT and the Field Crop Research Centre in Rayong, Thailand. According to Kawano *et al.* (1998) the basic scheme consisted of:

- Year 1: Up to 5000 seedlings from up to 100 crosses between Asian and Latin American parents were sown and transplanted in seedling trials in Rayong. After 8–10 months a low selection pressure was applied for plant and root type to give about 700 selections.
- Year 2: Up to 700 selections from year 1 were tested further in Rayong in non-replicated single rows of ten plants for the above characteristics, plus root dry matter and root yield. Special emphasis was given to selection for high harvest index.
- Year 3: Up to 80 selections from year 2 were further tested in a preliminary yield trial at Rayong, consisting of two replications of 50 plants each. The same traits were evaluated as in the single row trial.
- Year 4: Between 20 and 25 advanced selections were tested in larger yield trials in at least three locations: Rayong, Khon Kaen and Mahasarakarm, that represent a wide range of production conditions.

- Year 5–6: Between six and eight elite genotypes were planted in regional trials in at least seven locations for 2 years.
- Year 7: Selected genotypes were distributed for evaluation in other national programmes (Vietnam, China, Indonesia and Philippines), and were multiplied for further testing and distribution to farmers in Thailand.

Kawano *et al.* (1998) commented that over a period of 14 years, some 372,000 genotypes from 4130 crosses had been evaluated in their programme in Thailand, but only three genotypes had passed the tests for official release. These improved varieties occupied almost 400,000 ha in 1996, generating an economic impact estimated at around US$278 million.

Breeding for High Yield

High yield is achieved first by selecting plants that have both a genetic structure and a plant structure which maximizes performance, and then by bringing together resistances or tolerances to the factors which limit yield. Hybrid vigour through heterozygosity is the main requirement for the genetic structure of new varieties and this is a major objective of the strategies described above. The genetic base of the material imported into Africa was necessarily narrow; hence the considerable hybrid vigour obtained at IITA from crosses with new germplasm from the Americas. Elsewhere, vigour has been maintained by keeping the genetic base as wide as possible.

It may be beneficial to enlarge the genetic base further by making interspecific crosses with some of the many shrub species of the Fruticosae section of *Manihot*. Jennings (1959) obtained considerable hybrid vigour by crossing cassava with *Manihot melanobasis*, but this species may be a misnamed form of *M. esculenta* (Rogers and Appan, 1970, 1973). The subgenus has other as yet untried candidates that have tuberous roots and may provide new opportunities for increasing heterozygosity. These might include *Manihot aesculifloia, Manihot rubricantis, Manihot augustiloba* and *Manihot priuglei* that are mentioned by Rogers and Appen (1970), as well as the subspecies, *M. esculenta flabellifolia* and other species discussed in Chapter 4.

Models for high yield; the significance of plant habit, leaf longevity and disease resistances

Cassava plant habits are so variable that efforts have been made to discover which of them is best equipped for giving high yields: essentially the ability to convert solar energy into starch and store it in the roots. As physiological information became available, computer modelling was used to estimate the effects of the many variables, including those associated with stress and disease (Hunt et al., 1977; Cock et al., 1979).

The hypothesis followed was that crop growth rate increases at a decreasing rate as leaf area increases, whereas the dry matter required for stem and leaf production increases linearly with the leaf area index (LAI, a function of the rate of leaf formation, leaf size and longevity). Hence root growth rate, which is the difference between the total growth rate and that of the tops, increases up to a certain level of LAI and then decreases. Thus there is an optimum LAI for yield, and manipulations of the components of LAI can bring it closer to this optimum and maximize yield.

It turns out that root growth declines at values of LAI above 4, apparently because the resources required to form and maintain a higher LAI increase approximately linearly with LAI and leave less material for root growth. Leaf and stem growth have preference over root growth and the latter receives only the carbohydrate remaining after the requirements of the tops have been met (Gijzen et al., 1990). The size of the roots rarely limits yield, and it is the LAI and not the root sink that determines it. Indeed, the roots can accept much more carbohydrate than is normally available (Lian and Cock, 1979; Pellet and El-Sharkawy, 1994).

These findings have a profound effect on selection criteria, selection procedure and even selection priorities. Among existing varieties, branching habit affects LAI the most. Varieties that branch 6–8 weeks after planting and six to eight times a year with four branches formed on each occasion, allocate too little of their resources to the roots. Their total dry matter production may be high and they compete well in unimproved 'extensive' agriculture, but their distribution of dry matter to the roots is too low for high yield. The models show that branching should be delayed until 30 weeks, leaf size to 500–600 cm^2 and leaf life prolonged to 15–20 weeks (Cock et al., 1979).

Plants with delayed branching are desirable not only for high yield but also because they facilitate mixed cropping with other crops, which leads to the maximum food yield per land unit. Leaf longevity was not previously considered important, but it prolongs dry matter production without using resources (El-Sharkawy, 1993). Hence resistance to diseases that cause premature leaf fall are important too.

The models aid selection procedures because they explain why there is no correlation between the yields of plants in mixed populations and their yields in single row trials ($r = 0.068$). This is because the former is determined by competitiveness, for which the optimum LAI is higher than the optimum for the latter. Selection cannot therefore be done for yield itself in the mixed populations present in the early stages of breeding, but it can be done on harvest index, which is the root weight expressed as a proportion of total plant weight. Not only is this value correlated in the two kinds of population ($r = 0.608$), but it is highly correlated with root yield ($r = 0.763$), and has a high heritability, the harvest indices of progenies being highly correlated with the means of their parents ($r = 0.745$). Hence, Kawano and Thung (1982) and Kawano et al. (1998) reported high correlation and regression coefficients for harvest index with root yield and demonstrated the effectiveness of using the trait at all stages of selection as an indirect selection for root yield. In practice, plant competition is minimized by wide spacing (e.g. 1 m apart in rows which are 2 m apart), and the selection criterion is for plants whose main stem does not branch until it reaches about 1 m (Fig. 8.3; Kawano et al., 1978b; Hahn et al., 1979). Taller plants with higher branching are also less productive (Fig. 8.4), but are often preferred by smallholders because they facilitate mixed cropping with other food plants.

The models also explain the genotype interactions with temperatures that are encountered when breeding is for adaptation to altitudes above 2000 m, where average temperatures often fall to 17°C. Most varieties yield badly in these situations because they produce an inadequate LAI, but the LAI may become excessive if temperatures rise and the top growth of the plant

Fig. 8.3. Cassava with plant habit ideal for high root yield, showing branching at an intermediate height. Courtesy of CIAT.

Fig. 8.4. Tall cassava with high branching and relatively low root yield. Courtesy of CIAT.

increases. A common situation is for the most vigorous genotypes to yield the most at 20°C and the least at 28°C, and for the least vigorous ones to yield the least at 20°C and the most at 28°C.

The optimum phenotype (plant type) is the same, but the genotypes that achieve it change with temperature (Irikura *et al.*, 1979).

By determining the consequences of malfunction in each organ, the models help to decide breeding priorities, i.e. they show which diseases are sufficiently damaging for resistance to them to be given a place in the programme (Cock, 1978). It turns out that priority cannot be justified for diseases which cause plant death on a moderate scale, small decreases in tuber number or small decreases in leaf size. Breeding emphasis is justified for all disorders that reduce leaf life or photosynthetic efficiency, cause stem damage or high levels of early plant death. However, although traditional farming systems often require only limited disease control, the lack of resources for purchased inputs by small farmers often forces breeders to take account of host resistance. In any case, the priorities of disease control will increase as progress is made towards an optimum plant habit, when leaf area, for example, will become more important, or if a change towards more densely planted monocrops aggravates the existing disease problems.

Breeding for Root Quality: Starch and Dry Matter Content

Cassava is used for diverse purposes and so most of the criteria for quality are also diverse, but high starch content and quality is always required. Starch content is usually estimated from dry matter percentage, to which it is highly correlated ($r = 0.810$; IITA, 1974; CIAT, 1975), but a quicker method is to determine the root's specific gravity, which is related to both dry matter and starch content. A calculation can be obtained from the specific weight of a sample (3–5 kg) of unpeeled roots in air and water, or by passing samples through a series of sodium chloride solutions of increasing specific gravity to find the one with the lowest specific gravity in which the sample will float (Hershey, 1982).

A high dry matter content is not necessarily ideal because, for reasons unknown, it is associated with postharvest deterioration. This can be serious for commercial outlets, but not where roots are used immediately as in subsistence agriculture. Dry matter content is not associated with fresh root yield and it is still uncertain whether a high level can be maintained when yields are high: progress in one may require sacrifice in the other (CIAT, 1981; Iglesias et al., 1994). Similarly, substantial progress towards a capacity for prolonged postharvest storage may be difficult, but genetic differences have been identified (Kawano and Rojanaridpiched, 1983). More recently, Iglesias et al. (1996) showed that it was possible to break the association between high dry matter and high postharvest deterioration, and that the heritability of the trait is high enough for considerable progress through conventional breeding.

Starch quality is influenced by the amylose content, which for good cooking varieties is 21%, for industrial varieties (more waxy types) is 15% and for multiple-purpose varieties is 17% (IITA, 1977). Wheatley et al. (1992) found a range of 15–28% amylose in the roots of plants in the CIAT germplasm collection. No waxy (zero amylopectin) mutants have been detected, but variations in the ratio of amylose to amylopectin could open new markets for cassava starch in future. Most of the efforts in the early days of the international centres were devoted to cassava as a human staple, but nowadays, knowledge of the genetic variation available in terms of root and starch quality may provide opportunities for marginal regions to expand into global markets.

Breeding for Low Content of Cyanogenic Glucosides

Hydrocyanic acid (HCN) forms when two cyanogenic glucosides (linamarin and lutaustralin) are hydrolysed by endogenous enzymes. Mutant acyanogenic varieties which lack genes for the production of either the glucosides or the enzymes would be ideal, but no such mutants have been found in germplasm collections or segregating progenies, probably because they are likely to be recessive and difficult to discover in cassava because of its polyploid make-up. Breeders therefore select for low HCN content, which is conferred by a complex of recessive minor genes (Hahn et al., 1973). Recent studies have suggested a role of cyanide in the resistance of cassava to pests (Bellotti et al., 1999). The possibility of confining the cyanogenic glucosides to non-edible plant parts in order to maintain pest resistance therefore needs to be explored, in parallel to efforts to decrease cyanide in the edible parts.

Since the glucosides are synthesized in the leaves and translocated to the roots, a common practice was to screen leaves semi-quantitatively using a sodium picrate test. However, a more accurate enzymatic analysis (Cooke, 1978) has been automated for rapid determinations (Hahn, 1984). Root analyses are now preferred because the correlation ($r = 0.36$) between leaf and root results is very low, probably because of the high variation detected for the trait in both tissues. Reports of independent synthesis of linamarin in the roots (Makame et al., 1987) would certainly reduce the correlation between the occurrence of cyanogens in the leaves and roots. The correlation between the two tissues is better if determinations are confined to young leaves and root peel (CIAT, 1982). Most breeding programmes screen at the early stages with the sodium picrate test (Cooke et al., 1978; CIAT, 1982), sometimes using modifications suggested by O'Brien et al. (1994) and Yeoh et al. (1998), and then evaluating advanced selections using enzymatic methods.

There appears to be no obstacle to combining low HCN with the other desirable root qualities sought, but it is difficult to reduce levels to below 10–20 p.p.m. Hahn (1984) produced a low HCN population by continuous selection and recombination. Selections from such material have a special role where leaves are eaten as a vegetable. Leaves are rich in protein and provide a dietary complement to the roots. However, HCN is not the only factor responsible for bitterness in the roots, though roots with levels below 10 mg 100 g^{-1} are generally considered to be sweet.

Breeding for High Content of Protein and Other Nutritional Elements in the Root

The primary function of cassava roots is to store starch, and attempts to enhance the protein content could well have adverse effects on this function. It is probably better to obtain protein from other sources. Nevertheless, cassava germplasm with high root protein content is available, and attempts to use it in breeding have been made.

Several Indian varieties have a high protein content (Hrishi and Jos, 1977), as well as some related species, notably several from Brazil (Nassar and Costa, 1977; Nassar 1978), *Manihot saxicola* (Bolhuis, 1953) and *Manihot melanobasis* (Jennings, 1959). The protein contents of the interspecific hybrids derived from the last two species tended to decrease as backcrossing proceeded, however.

Selection for increased vitamin and mineral content was recently initiated at CIAT, targeted toward regions with severe deficiencies, mainly in vitamin A. Considerable improvement in carotene content can be achieved within the existing cassava germplasm (Iglesias *et al.*, 1996, 1997).

Breeding for Resistance to Cassava Mosaic Disease (CMD) and to Cassava Brown Streak Disease (CBSD)

CMD is caused by whiteflyborne cassava mosaic geminiviruses (CMGs) and occurs in Africa and India (Chapter 12).Variation in virus virulence occurs and an exceptionally virulent variant found in Uganda is probably a hybrid of the two African forms (Deng *et al.*, 1997). There is a wide but apparently continuous range in the expression of host resistance. There is no evidence that

the resistance is specific to particular virus forms, and the highest levels of resistance available are needed to control the most virulent virus forms encountered. Hence the resistant germplasm bred in East Africa was used to initiate resistance breeding at IITA in Nigeria (Beck, 1982; Jennings, 1994) and selections from the IITA programme were successfully used as resistance donors in India (Hahn *et al.*, 1980a).

After many years of limited progress in breeding with cassava germplasm at many African centres, a change of emphasis occurred when the programmes at Amani (Tanzania) and Lac Alaotra (Madagascar) began the task of transferring resistance from the tree species *M. glaziovii* Muell-Arg, *Manihot dichotoma* Ule, *Manihot catingae* Ule, *Manihot pringlei* Watson (which was probably misnamed; Rogers and Appan, 1970) and 'tree cassava', which was thought to be a natural hybrid of *M. glaziovii* and cassava. It took three or four backcrosses, made over 15 years, to restore tuberous roots and lose the tree-like characteristics of the donor species (Fig. 8.5). Only the hybrids with *M. glaziovii* were ultimately successful (Nichols, 1947; Cours, 1951; Jennings, 1957).

The resistance of the best backcross hybrids from Tanzania was adequate for most but not all situations. Resistance was later shown to be multi-genic and recessive in inheritance (Jennings, 1978; Hahn *et al.*, 1980b). Intercrossing among resistant selections began in 1953, and was successful probably because it concentrated the recessive genes from different sources and made them homozygous. The material obtained from this intercrossing was much more resistant than the backcross hybrids and was the origin of the resistant parents used at IITA later.

Resistance was assessed by the proportion of symptom-bearing plants or branches present, and by the symptom intensity. The two estimates were correlated ($r = 0.43$ and 0.48 in two trials) and the efficiency of assessment was improved by 38% when both aspects were considered together (Jennings, 1957, 1994).

Experiments and observations in Tanzania (Jennings, 1957, 1960a), Madagascar (Cours-Darne, 1968) and Nigeria (IITA, 1980; Rossel *et al.*, 1988) lead to the following concept of resistance (Jennings, 1994): when exposed to infection, an unknown proportion of resistant

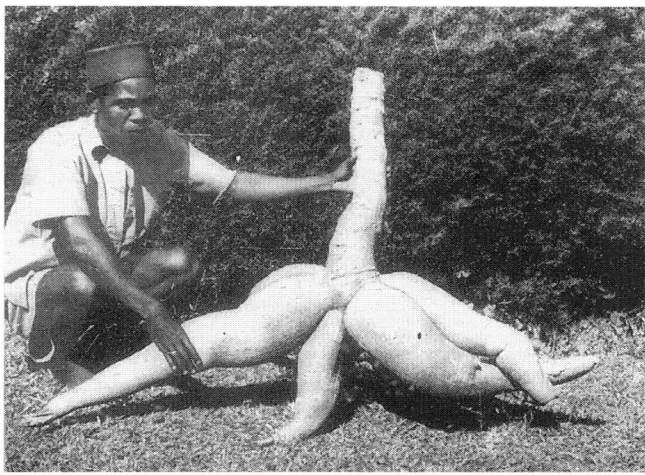

Fig. 8.5. Vigorous second backcross hybrid of *Manihot glaziovii* to cassava with large tuberous roots but of only intermediate quality produced at Amani, Tanzania.

plants (which could well be 100%) become infected at one or more of their stem apices. Some of them localize the virus at their bases and either remain symptom-free or show only transient symptoms, while others become symptom-bearing and remain so. Similarly, a proportion of plants derived from infected cuttings become symptom-free. There is thus a dynamic situation in which new infections are occurring and previously symptom-bearing plants are becoming symptom-free. A point of equilibrium is reached which depends on the resistance level of the host and the virulence and inoculum pressure of the virus, and is influenced by the management practices used.

Almost 100% of Latin American germplasm introduced into Africa has shown extreme susceptibility to CMD. Limited improvements occurred in F1 crosses and first generation backcrosses to African germplasm (Porto *et al.*, 1994). A molecular map for cassava was developed at CIAT from a cross between a Latin American susceptible variety and a genotype resistant to CMD (Angel *et al.*, 1993; Fregene *et al.*, 1997). The objective was to locate genes for resistance to CMD on the molecular map because the discovery of close linkages between genes for resistance and molecular markers would make it possible to breed for resistance to CMD in the absence of the disease.

Brown streak virus disease occurs only in coastal areas of East Africa and in Malawi,

(Chapter 12) but breeding for resistance to it has been done only in Tanzania, where the resistance was considered almost as important as resistance to CMD.

The important sources of resistance to CBSD were *M. glaziovii*, *M. melanobasis* and several cassava varieties of Brazilian origin (Jennings, 1957, 1960b). Symptoms can occur in mature leaves, leaf bases on the stem and in the roots, and are best scored in mature plants at harvest time. Symptoms can be transient in resistant plants when the old leaves are shed and the necrotic tissues of the stems and roots are occluded by new symptom-free growth. The observation that symptoms in resistant plants tend to be confined to the roots suggests that the resistance mechanism may involve a localization of the virus to the lower parts of the plant as postulated for resistance to CMD (Jennings, 1960b).

Breeding for Resistance to Cassava Bacterial Blight (CBB; *Xanthomonas campestris* pv. *manihotis*)

Resistance to CBB has always been an important requirement for several of the zones considered by CIAT, but the disease did not become prevalent in Africa until the late 1970s. American varieties show a continuous range of resistance but at CIAT only 15 out of 2800 clones tested were rated highly for resistance. The inheritance of resistance is by recessive, mainly additive genes and there is a good correlation between the mean resistance of parents and that of their hybrids ($r = 0.549$), resulting in a heritability of 48% (CIAT, 1978; IITA, 1978).

An important finding at IITA was that resistance to CBB in progenies derived from the crossing of cassava with *M. glaziovii* is associated with resistance to CMD. Hahn *et al.* (1980b) found phenotypic and genetic correlation coefficients between the two resistances in half-sib families of 0.423 and 0.899, respectively. They attributed the result to the occurrence of linked recessive gene complexes on one or more of the chromosomes inherited from *M. glaziovii*. Random transmission of some of the parental chromosomes of this species into backcross progenies with cassava has in fact been observed (Magoon *et al.*, 1969). Jennings (1978) found a discontinuity for this joint resistance both in populations of *M. glaziovii* itself and in its backcross progenies with cassava. He suggested that this discontinuity was conferred by some kind of genetic unit that was not invariably present in *M. glaziovii*, but had been retained through seven generations of breeding following the interspecific cross with this species.

These results mean that selection for one resistance should lead to an increase in the other. Studies of genetic variances and heritabilities for the two resistances suggest that genetic gain would be greater for CMD resistance in response to a first selection for CBB resistance, than in the reciprocal order, but a first screening for resistance to CMD is always preferred for practical reasons (Hahn *et al.*, 1980b). Recently, both ORSTROM in France and CIAT have studied the pathogen's genetic diversity and its implications for breeding. Restrepo and Verdier (1997) found considerable diversity of the pathogen in Latin America, but a restricted range in Africa. A set of strains representing different genetic groups is being tested to evaluate CIAT's germplasm and to select the most appropriate ones to use in screening for resistance. Clearly, the variability found in the pathogen will affect future breeding efforts in Latin America and has implications for other regions if virulent strains of the blight pathogen, presently confined to Latin America, spread to other parts of the world.

Breeding for Resistance to Fungal Diseases

Fungal diseases vary in importance in different zones, and resistance breeding for them has rarely been a primary objective of major programmes until recently. Resistances conferred by minor genes invariably occur in the landraces obtained from affected areas and are always non-race specific when physiological specialization of the pathogen occurs (CIAT, 1976). Notable examples are resistances to super-elongation disease (*Elsinoe brasiliensis*; Kawano *et al.*, 1983) and anthracnose (*Colletotrichum* species; Ezumah, 1980). For the high rainfall lowland tropics, leaf spot diseases caused by species of *Cercospora*, *Cercosporidium*, *Phaeoramularia* or *Phoma* can reduce yield by reducing the efficiency and longevity of the leaves; control of *Cercospora henningsii* in susceptible varieties, for example, improved yield by 10–23% (Teri *et al.*, 1978).

Technical problems have hindered breeding for resistance to root pathogens, but inoculation techniques under high humidity conditions have now identified moderate to strong resistance to *Diplodia* species and given reliable screening of progenies. Resistances to *Phytophthora drechsleri* and *Phytophthora nicotianae* var. *nicotianae* have been identified by a technique in which plugs of infected tissue are inserted into harvested roots which are then incubated in plastic bags for 2 weeks (CIAT, 1990). Stable, high resistance to *Phytophthora* species combined with resistance to *Fusarium* species have been identified in three Brazilian varieties which are being widely used as parents in breeding (Hershey and Jennings, 1992). Recent studies of genetic diversity for resistance to isolates from different *Phytophthora* species have revealed a number of germplasm accessions tolerant to all of them (Alvarez *et al.*, 1999). However, the strategy for breeding for resistance should still be based on an initial diagnosis of the predominant pathogen present in order to choose the most appropriate resistance genes.

Breeding for Resistance to Mites, Mealybugs and Whiteflies

Resistance to the green cassava mite (*Mononychellus tanajoa*), which caused devastation when it arrived in Africa in the 1970s, was discovered in several African varieties (Nyiira, 1975; IITA, 1980) and appeared to be associated with plant vigour and leaf pubescence (Fig. 8.6; Leuschner, 1980). In Colombia, Byrne *et al.* (1982) identified both tolerance and antibiosis resistance mechanisms, both having high heritability. Bellotti and Byrne (1979) and Kawano and Bellotti (1980) considered that prospects for resistance breeding were good following systematic surveys of American germplasm for resistances to the three important mites, namely, *M. tanajoa*, *Tetranychus urticae* and *Oligonychus peruvianus*.

Resistances to the mealybugs *Phenacoccus manihoti* in Africa and *Phenacoccus herreni* in the Americas are important where there are extended dry periods. On both continents it is closely correlated with the pubescence of leaf buds and unexpanded leaves (van Schoonhoven, 1974; Ezumah, 1980; Hahn, 1984) and is therefore easily identified in the absence of the pest and will probably be durable. It is evaluated either by scoring the density of the pubescence or by counting the hairs on the underside of an unexpanded leaf. More recent breeding at CIAT has emphasized selection for resistance of the antibiosis type and for tolerant types that recover from pest damage (Hershey and Jennings, 1992).

Cassava is one of the few crops for which high levels of resistance to whiteflies have been detected. A high frequency of resistance is found in accessions from Ecuador. The national programme in Colombia is about to release two whitefly-resistant varieties for regions where the pest causes considerable direct damage and also acts as a vector for viruses (Bellotti *et al.*, 1999).

Breeding for Efficient Use of Basic Resources

Cassava is commonly grown in marginal regions under water stress and in soils that are low in nutrients, particularly phosphates. Hence the efficient use of these resources often reduces stress and may result in production stability, even under marginal conditions. The plant characteristics which confer tolerance to prolonged water stress are complicated, but selection under natural drought conditions is effective in improving at least some of them (Hershey and Jennings, 1992; El-Sharkawy, 1993; Tafur *et al.*, 1997).

Varieties for areas with a short rainy season are specialized in that they must produce a crop in 6 months. To do this they must distribute a very high proportion of their dry matter to the roots, and consequently they produce an insufficient photosynthetic source for a longer growing season (Cock, 1976).

Breeding for tolerance of low phosphate nutrition is possible by comparing yields on high and low phosphate plots. Large differences have

Fig. 8.6.   Subglabrous (left) and pubescent (right) shoot tips and unexpanded leaves associated respectively with genotypes susceptible and resistant to mites and mealybugs. Courtesy of CIAT.

been detected, and the most tolerant types, which use the phosphates most efficiently, have been used as parents for progenies which are grown for selection on low phosphate soils (CIAT, 1981; Hershey and Jennings, 1992). Ideally, genotypes should tolerate low fertility but must also respond to improved fertility by increasing their root yield. This requires the separate evaluation of selections under high and low phosphate conditions. Plant architecture is important for efficient nutrient use. Genotypes with a short or intermediate plant habit may use 20% less nutrients than taller ones to produce a unit of yield at similar productivity levels (CIAT, 1997).

Iglesias *et al.* (1994) applied sensitivity analysis across a set of environments and showed that this method could contribute to the overall objective of improving dry matter production under poor growing conditions, while maintaining the capacity of the crop to respond to favourable environments.

Cassava Breeding in the Future: the Role of Biotechnology

The classical methods of breeding described here have produced plants capable of large advances in root yield. Hershey and Jennings (1992) reported improvements of over 200% for the period from 1976, when work began at CIAT, until 1990 for material tested at two trial sites and subjected to stress conditions. Similar advances have been made at IITA. However, the rate of improvement in average national cassava yields in the most important production countries has not paralleled the progress at experimental level, except for some Asian countries (Kawano, 1978).

Progress in the future will be aided by new biotechnology tools such as gene transfer from other species and molecular marker assisted selection (Chapter 10). Genetic engineering has a special role for improving heterozygous, clonally propagated crops such as cassava, because genes can be introduced into popular varieties without changing their positive attributes. All the quality combinations which make these varieties preferred by farmers could be maintained, allowing a higher rate of adoption of improved genotypes.

For many years this particular advantage was precluded because it was not possible to regenerate plants from transformed single cells or somatic tissues: routine regeneration was possibly only from embryogenic tissue. However, it should now be possible to achieve it by producing somatic embryos or 'artificial seeds' from somatic tissues (Stamp and Henshaw, 1982; Schoeple *et al.*, 1996; Taylor *et al.*, 1996).

Work on both *Agrobacterium tumefaciens* mediated and particle gun methods for transferring DNA to cassava is making progress (Calderon-Urrea, 1988; CIAT, 1991; Raemakers *et al.*, 1997). The DNA transfer which has the potential for the most valuable improvement is the transfer of part of the CMV genome, possibly the part which codes for coat protein, which would be expected to inhibit the synthesis of virus particles and either reduce or prevent symptom expression (Fauquet and Beachy, undated; Padidam *et al.*, 1999). In readiness for this development, the sequence has been determined of the two genomic components of the virus, which has two circular, single-stranded DNAs (Stanley and Gay, 1983). Other prospects are the transfer of a trypsin inhibitor gene from potato, which is expected to confer broad-spectrum insect resistance (Hershey and Jennings, 1992).

Research at the University of Bath and CIAT aims to identify key genes activated during the root deterioration process, with the objective of altering them in the future. Genes that have key roles have been identified by Beeching *et al.* (1994), and are now the target for genetic modification.

In the area of starch quality, the genes for anti-sense construct for granule bound starch synthase isoforms I and II, branching enzyme and ADP-glucose pyrophosphorylase isolated by Munyikwa *et al.* (1997) are being incorporated into cassava genotypes, to generate waxy genotypes and other starch variants that could open new markets for industrial uses.

The genetic control of cyanogen production is being studied at several institutions in the world and recent developments in gene identification have opened promising avenues for control of cyanogen synthesis and accumulation (Koch *et al.*, 1994).

References

Alvarez, E., Chacon, M.I. and Sanchez-Cusguen, N.J. (1999) DNA polymorphism and virulence variation of *Phytophthora* population isolated from cassava *Manihot esculenta* Crantz. In: *Proceedings of the 7th International Congress of Plant Pathology*. British Society of Plant Pathology, Birmingham, p. 30.

Angel, F., Arias, D., Tohme, J., Iglesias, C. and Roca, W. (1993) Towards the construction of a molecular map of cassava (*Manihot esculenta* Crantz): comparison of restriction enzymes and prove sources in detecting RFLPs. *Journal of Biotechnology* 31, 103–113.

Beck, B.D.A. (1982) Historical perspectives of cassava breeding in Africa. In: Hahn, S.K. and Ker, A.D.R. (eds) Root crops in Eastern Africa. *Proceedings of a Workshop at Kigali, Rwanda, 23–27 November 1980*. IDRC-177e, Ottawa, Canada.

Beeching, J.R., Dodge, A.D., Moore, K.M., Phillips, H. and Rickard, J. (1994) Physiological deterioration in cassava: possibilities for control. *Tropical Science* 34, 335–343.

Bellotti, A.C. and Byrne, D. (1979) Host plant resistance to mite pests of cassava. In: Rodriguez, J.G. (ed.) *Recent Advances in Acarology* 1, 13–21. Academic Press, New York.

Bellotti, A.C., Smith, L. and Lapointe, S.L. (1999) Recent advances in cassava pest management. *Annual Review of Entomology* 44, 343–370.

Bolhuis, G.G. (1953) A survey of attempts to breed cassava varieties with a high protein content in the roots. *Euphytica* 2, 17–112.

Bonierbale, M., Iglesias, C. and Kawano, K. (1995) Genetic resources management of cassava at CIAT. In: MAFF, *International Workshop on Genetic Resources: Root and Tuber Crops*. MAFF, Tsukuba, Japan, pp. 39–52.

Byrne, D.H., Guerrero, J.M., Bellotti, A.C. and Gracen, V.E. (1982) Yield and plant growth responses of *Mononychellus* mite resistant and susceptible cassava cultivars under protected versus infested conditions. *Crop Science* 22, 486–490.

Calderon-Urrea, A. (1988) Transformation of *Manihot utilissima* (cassava) using *Agrobacterium tumefaciens* and the expression of the introduced foreign genes in transformed cell lines. *PhD thesis, Vrijes University, Brussels, Belgium*, p. 37.

Charrier, A. and Lefevre, F. (1987) The genetic variability of cassava: origin, evaluation and utilization. In: Fauquet, C. and Fargette, D. (eds) *Proceedings of an International Seminar on African cassava mosaic disease and its control*, Côte d'Ivoire, Yamoussoukro. CTA/FAO/ORSTOM/IITA/IAPC, pp. 77–91.

CIAT (1975, 1976, 1978, 1981, 1982, 1990, 1991) *Annual Reports of the Centro Internacional de Agricultura Tropical*. Cassava Program. CIAT, Cali, Colombia.

CIAT (1997) Improved cassava genepools. *Annual Report 1997*. CIAT, Cali, Colombia.

Cock, J.H. (1976) Characteristics of high yielding cassava varieties. *Experimental Agriculture* 12, 135–143.

Cock, J.H. (1978) A physiological basis of yield loss in cassava due to pests. In: Brekelbaum, T., Bellotti, A. and Lozano, J.C. (eds) *Proceedings of the Cassava Protection Workshop*, at CIAT, Colombia, 7–12 November 1977. CIAT, Cali, Colombia.

Cock, J.H., Franklin, D., Sandoval, G. and Juri, P. (1979) The ideal cassava plant for maximum yield. *Crop Science* 19, 271–279.

Cooke, R.D. (1978) An enzymatic assay for the total cyanide content of cassava (*Manihot esculenta* Crantz). *Journal of the Science of Food and Agriculture* 29, 345–352.

Cooke, R.D., Howland, A.K. and Hahn, S.K. (1978) Screening cassava for low cyanide using an enzymatic assay. *Experimental Agriculture* 14, 367–372.

Cours, G. (1951) Le Manioc à Madagascar. *Mémoires de L'institut Scientifique de Madagascar Série B, Biologie Végétale* 3, 203–400.

Cours-Darne, G. (1968) Improving cassava in Africa. The Abidjan Conference: Agricultural Research Priorities for Economic Development in Africa. *US National Academy of Sciences* 2, 330–339.

Deng, D., Otim-Nape, W.G., Sangare, A., Ogwal, S., Beachy, R.N. and Fauquet, C.M. (1997) Presence of a new virus closely related to East African cassava mosaic geminivirus, associated with a cassava mosaic outbreak in Uganda. *African Journal of Root and Tuber Crops* 2, 23–28.

El-Sharkawy, M.A. (1993) Drought-tolerant cassava for Africa, Asia and Latin America. *BioScience* 43, 441–451.

Ellis, R.H., Hong, T.D. and Roberts, E.H. (1982) An investigation of the influence of constant and alternating temperature on the germination of cassava seed using a two-dimensional temperature gradient plate. *Annals of Botany* 49, 241–246.

Ezumah, H.C. (1980) Cassava improvement in the programme national manioc in Zaire: objectives and achievements up to 1978. In: Terry, E.R., Oduro, K.A. and Caveness, F. (eds) *Proceedings of the 1st Triennial Root Crops Symposium of the International Society for Tropical Root Crops–Africa Branch*, Ibadan, Nigeria, 8–12 September 1980. IDRC-163e, Ottawa, Canada, pp. 29–30.

Fauquet, C. and Beachy, R.N. (undated) *Cassava Viruses and Genetic Engineering*. Technical Centre for Agricultural and Rural Cooperation, Wageningen, The Netherlands, pp. 1–30.

Fregene, M.A., Angel, F., Gómez, R., Rodríguez, F., Roca, W., Tohme, J. and Bonierbale, M. (1997) A molecular genetic map of cassava (*Manihot esculenta* Crantz). *Theoretical and Applied Genetics* 95, 431–441.

Gijzen, H., Veltkamp, H.J., Govdriaan, J. and de Bruij, G.H. (1990) Simulation of dry matter production and distribution in cassava (*Manihot esculenta* Crantz). *Netherlands Journal of Agricultural Science* 38, 159–173.

Hahn, S.K. (1982) Research priorities, techniques and accomplishments in cassava breeding at IITA. In: Hahn, S.K. and Ker, A.D.R. (eds) *Root Crops in Eastern Africa, Proceedings of a Workshop*, Kigali, Rwanda, 23–27 November 1980. IDRC-179e, Ottawa, Canada, pp. 19–26.

Hahn, S.K. (1984) Progress of root and tuber improvement at IITA. In: *Proceedings of the 6th Symposium of the International Society for Tropical Root Crops*, Lima, Peru, 20–25 February 1983.

Hahn, S.K., Howland, A.K. and Terry, E.R. (1973) Cassava breeding at IITA. In: Leakey, C.L.A. (ed.) *Proceedings of the 3rd Symposium of the International Society for Tropical Root Crops*, Ibadan, Nigeria, 2–9 December 1973. IITA, Ibadan, Nigeria, pp. 4-10.

Hahn, S.K., Terry, E.R., Leuschner, K., Akobundu, I.O., Okali, C. and Lal, R. (1979) Cassava improvement in Africa. *Field Crops Research* 2, 193–226.

Hahn, S.K., Terry, E.R. and Leuschner, K. (1980a) Breeding cassava for resistance to cassava mosaic disease. *Euphytica* 29, 673–683.

Hahn, S.K., Howland, A.K. and Terry, E.R. (1980b) Correlated resistance of cassava to mosaic and bacterial blight diseases. *Euphytica* 29, 305–311.

Hershey, C.H. (1981) Germplasm flow at CIAT's Cassava Program. *CIAT Annual Review*, CIAT, Cali, Colombia, pp. 1–29.

Hershey, C.H. (1982) Quick estimation of dry matter content of cassava roots possible using rapid evaluation technique. *Cassava Newsletter 11*. CIAT, Cali, Colombia, pp. 4–5.

Hershey, C.H. (1984) Breeding cassava for adaptation to stress conditions: development of a methodology. In: *Proceedings of the 6th Symposium of the International Society for Tropical Root Crops*, Lima, Peru 20–25 February 1983, pp. 303–314.

Hershey, C.H. and Jennings, D.L. (1992) Progress in breeding cassava for adaptation to stress. *Plant Breeding Abstracts* 62, 823–831.

Hrishi, N. and Jos, J.S. (1977) Breeding for protein enhancement in cassava. In: Leakey, C.L.A. (ed.) *Proceedings of the 3rd Symposium of the International Society for Tropical Root Crops*, Ibadan,

Nigeria. 2–9 December 1973. IITA, Ibadan, Nigeria, pp. 11–13.

Hunt, L.A., Wholey, D.W. and Cock, J.H. (1977) Growth physiology of cassava. *Field Crops Abstracts* 30, 77–91.

Iglesias, C., Calle, F., Hershey, C. and Jaramillo G. (1994) Sensitivity of cassava (*Manihot esculenta* Crantz) clones to environmental changes. *Field Crops Research* 36, 213–220.

Iglesias, C., Bedoya, J., Morante, N. and Calle, F. (1996) Genetic diversity for physiological deterioration in cassava roots. In: Kurup, G.T. (ed.) *Tropical Tuber Crops: Problems, Prospects and Future Strategies*. Science Publishers Incorporated, Lebanon, New Hampshire, pp. 115–126.

Iglesias, C., Mayer, J., Chavez, L. and Calle, F. (1997) Genetic potential and stability of carotene content in cassava roots. *Euphytica* 94, 367–373.

IITA (1974, 1977, 1978, 1980) *Annual Reports of the International Institute of Tropical Agriculture*. IITA, Ibadan, Nigeria.

Irikura, Y., Cock, J.H. and Murray, B.G. (1979) The physiological basis of genotype-temperature interactions in cassava. *Field Crops Research* 2, 227–239.

Jennings, D.L. (1957) Further studies in breeding cassava for virus resistance. *East African Agricultural Journal* 22, 213–219.

Jennings, D.L. (1959) *Manihot melanobasis* Muell. Agr, – a useful parent for cassava breeding. *Euphytica* 8, 157–162.

Jennings, D.L. (1960a) Observations on virus diseases of cassava in resistant and susceptible varieties. I. Mosaic disease. *Empire Journal of Experimental Agriculture* 28, 23–34.

Jennings, D.L. (1960b) Observations on virus diseases of cassava in resistant and susceptible varieties. II. Brown Streak. *Empire Journal of Experimental Agriculture* 28, 261–270.

Jennings, D.L. (1963) Variation in pollen and ovule fertility in varieties of cassava, and the effect of interspecific crossing on fertility. *Euphytica* 12, 69–76.

Jennings, D.L. (1978) Inheritance of linked resistances to African cassava mosaic and bacterial blight diseases. In: Brekelbaum, T., Bellotti, A. and Lozano, J.C. (eds) *Proceedings of a Cassava Protection Workshop*. CIAT, Cali, Colombia, pp. 45–49.

Jennings, D.L. (1994) Breeding for resistance to African cassava mosaic geminivirus in East Africa. *Tropical Science* 34, 110–122.

Jennings, D.L. and Hershey, C.H. (1985) Cassava breeding: a decade of progress from international programmes. In: Russell G.E. (ed.) *Progress in Plant Breeding 1*. Butterworths, London, 89–116.

Kawano, K. (1978) Genetic improvement of cassava (*Manihot esculenta* Crantz) for productivity.

Tropical Agriculture Research, Series 11, Ministry of Agriculture and Forestry, Japan, p. 21.

Kawano, K. (1980) Cassava. In: Fehr, W.R. and Hadley, H.H. (eds) *Hybridisation of Crop Plants*. ASA, CSSA, Madison, Wisconsin, pp. 225–233.

Kawano, K. and Bellotti, A. (1980) Breeding approaches in cassava. In: Maxwell, F.G. and Jennings, P.R. (eds) *Breeding Plants Resistant to Insects*. John Wiley & Sons, New York, pp. 313–315.

Kawano, K. and Rojanaridpiched, C. (1983) Genetic study on post-harvest deterioration in cassava. *Kasetsart Journal* 17, 14–25.

Kawano, K. and Thung, M. (1982) Intergenotypic competition with associated crops in cassava. *Crop Science* 22, 59–63.

Kawano, K., Amaya, A., Daza, P. and Rios, M. (1978a) Factors affecting efficiency of hybridisation and selection in cassava. *Crop Science* 18, 373–376.

Kawano, K., Daza, P., Amaya, A., Rios, M. and Goncalves, M.F. (1978b) Evaluation of cassava germplasm for productivity. *Crop Science* 18, 377–380.

Kawano, K., Umemura, Y. and Kano, Y. (1983) Field assessment and inheritance of cassava resistance to superelongation disease. *Crop Science* 23, 201–205.

Kawano, K., Narintaraporn, K., Narintaraporn, P., Sarakarn S., Limsila, A., Limsila, J., Suparhan, D., Sarawat, V. and Watananonta, W. (1998) Yield improvement in a multistage breeding program for cassava. *Crop Science* 38, 325–332.

Koch, B.M., Sibbesen, O., Swain, E., Kahn, A., Liangcheng, D., Bak, S., Halkier, A. and Moller, B.M. (1994) Possible use of biotechnological approach to optimize and regulate the content and distribution of cyanogenic glucosides in cassava to increase food safety. *Acta Horticulturae* 315, 45–60.

Lathrap, D.W. (1973) The antiquity and importance of long-distance trade relationships in the moist tropics of Pre-Columbian South America. *World Archaeology* 5, 170–186.

Lian, T.S. and Cock, J.H. (1979) Branching habit as a yield determinant in cassava. *Field Crops Research* 2, 281–290.

Leuschner, K. (1980) Screening for resistance against green spider mite. In: Terry, E.R., Oduro, K.A. and Caveness, F. (eds) *Proceedings of the 1st Triennial Root Crops Symposium of the International Society for Tropical Root Crops – Africa Branch*, Ibadan, Nigeria, 8–12 September 1980. IDRC-163e Ottawa, Canada, pp. 75–78.

Magoon, M.L., Krishnan, R. and Vijaya Bai, K. (1969) Cytogenetics of F1 hybrid between cassava and ceara rubber and its backcrosses. *Genetica* 41, 425–436.

Mahungu, N.M., Dixon, A.G.O. and Mkumbua, J. (1994) Breeding cassava for multiple pest

resistance in Africa. *African Crop Journal* 2, 539–552.

Makame, M., Akoroda, M.O. and Hahn, S.K. (1987) Effects of reciprocal stem grafts on translocation in cassava. *Journal of Agricultural Sciences* 109, 605–608.

Munyikwa, T.R.I., Langeveld, S., Salehuzzeman, S.N.I.M., Jacobson, E. and Visser, R.G.F. (1997) Cassava starch biosynthesis: new avenues for modifying starch quantity and quality. *Euphytica* 96, 65–75

Nassar, N.A. (1978) Wild *Manihot* species of central Brazil for cassava breeding. *Canadian Journal of Plant Science* 58, 257–261.

Nassar, N.A. and Costa, C.P. (1977) Tuber formation and protein content in some wild (Mandioca) species native of central Brazil. *Experientia* 33, 1304–1306.

Nichols, R.F.W. (1947) Breeding cassava for virus resistance. *East African Agricultural Journal* 12, 184–194.

Normanha, E.S. (1970) [Cassava breeding work at the São Paulo State Agronomic Institute, Campinas, Brazil.] *Trabalhos do I. encoutro de engenheiros agronômicos pesquisadores em mandioca dos paises andinos e do Estado de São Paulo*, pp. 40–47.

Nyiira, Z.M. (1975) Advances in research on the economic significance of the green cassava mite, *Mononychellus tanajoa* Border in Uganda. International exchange and testing of cassava germplasm in Africa. In: Terry, E.R. and MacIntyre, R. (eds) *Proceedings of an Interdisciplinary Workshop*, Ibadan, Nigeria, 17–21 November 1975. IDRC-063e, Ottawa, Canada, pp. 22–29

O'Brien, G., Wheatley, C., Iglesias, C. and Poulter, N. (1994) Evaluation, modification and comparison of two rapid assays for cyanogens in cassava. *Journal of Science of Food and Agriculture* 59, 391–399.

Padidam, M., Beachy R.N. and Fauquet, C.M. (1999) A phage ssDNA binding protein complements ssDNA accumulation of geminivirus and interferes with viral movement. *Journal of Virology* 73, 1609–1616.

Pellet, D. and El-Sharkawy, M.S. (1994) Sink-source relations in cassava: effects of reciprocal grafting and leaf photosynthesis. *Experimental Agriculture* 30, 359–367.

Porto, M.C.M., Asiedu, R., Dixon, A. and Hahn S.K. (1994) An agro-ecological oriented introduction of cassava germplasm from Latin America into Africa. In: Ofori, F. and Hahn, S.K. (eds) *Tropical Root Crops in a Developing Economy*. ISTRC/ISHS, Wageningen, The Netherlands, pp. 118–129.

Raemakers, C.J.J.M., Sofiari, E., Jacobsen, E. and Visser, R.S.F. (1997) Regeneration and transformation of cassava. *Euphytica* 96, 153–161.

Reichel-Dolmatoff, G. (1965) Colombia. In: *Ancient Peoples and Places*. Thames and Hudson, London, pp. 62–75.

Restrepo, S. and Verdier, V. (1997) Geographical differentiation of the population of *Xanthomonas axonopodis* pv. *manihotis* in Colombia. *Applied and Environmental Microbiology* 63, 4427–4434.

Roa, A.C., Maya, M.M., Duque, M.C., Tohme, J., Llem, A.M. and Bonierbale, M.W. (1997) AFLP analysis of relationships among cassava and other *Manihot* species. *Theoretical Applied Genetics* 95, 741–750.

Rogers, D.J. and Appan, S.G. (1970) Untapped genetic resources for cassava improvement. In: Plucknett, D.L. (ed.) *Proceedings of the Second International Symposium on Tropical Root and Tuber Crops*. College of Tropical Agriculture, Hawaii, pp. 72–75.

Rogers, D.J. and Appan, S.G. (1973) Manihot, Manihotoides (Euphorbiaceae). *Flora Neotropica*, Mongraph 13. Hafner Press, New York.

Rossel, H.W., Thottappilly, G., Van Lent, J.M.W. and Huttinga, H. (1988) The etiology of cassava mosaic in Nigeria. In: African cassava mosaic disease and its control. *Proceedings of the International Seminar on African Cassava Mosaic Disease 4–8 May 1987*, Yamoussoukro, Côte d'Ivoire. CTA, Wageningen, pp. 57–63.

Schoeple, C., Taylor, N., Caramo, R., Konan, K.N., Marmey, Y., Henshaw, G.G., Beachy, R.N. and Fauquet, C. (1996) Regeneration of transgenic cassava plants (*Manihot esculenta*). *Nature Biotechnology* 14, 731–735.

Stamp, J. and Henshaw, G. (1982) Somatic embryogenesis in cassava. *Zeitschrift für Pflanzenphysiology* 105, 97–102.

Stanley, J. and Gay, M.R. (1983) Nucleotide sequence of cassava latent virus DNA. *Nature* 301, 260–262.

Tafur, S.M., El-Sharkawy, M.A. and Calle, F. (1997) Photosynthesis and yield performance of cassava in seasonally dry and semi-arid environments. *Photosynthetica* 33, 249–257.

Taylor, N.J., Edwards, M., Kiernan, R.J., Davey, C., Blakesley, D. and Henshaw, G.G. (1996) Development of friable embryonic callus and suspension culture system in cassava (*Manihot esculenta*). *Nature Biotechnology* 14, 726–730.

Teri, J.M., Thurston, H.D. and Lozano, J.C. (1978) The *Cereospora* leaf diseases of cassava. In: Brekelbaum, T., Bellotti, A. and Lozano, J.C. (eds) *Proceedings of a Cassava Protection Workshop*, CIAT, Cali, Colombia, 7–12 November 1977. CIAT, Cali, Colombia, pp. 101–116.

van Schoonhoven, A. (1974) Resistance to thrips damage in cassava. *Journal of Economic Entomology* 67, 728–730.

Wheatley, C.C., Orrego, J.I., Sanchez, T. and Granados, E. (1992) Quality evaluation of the cassava core collection at CIAT. In: Thro, A.M. and Roca, W. (eds) *Proceedings of the First International Scientific Meeting*. CIAT, Cali, Colombia, pp. 255–267.

Wright, C.E. (1965) Field plans for a systematically designed polycross. *Record of Agricultural Research* 14, 31–41.

Yeoh, H.-K., Sanchez, T. and Iglesias, C. (1998) Large-scale screening of cyanogenic potential in cassava roots using the enzyme-based dip-sticks. *Journal of Food Comparison and Analysis* 11, 2–10.

Chapter 9
Genetic Resources and Conservation

N.Q. Ng and S.Y.C. Ng

International Institute of Tropical Agriculture (IITA), Oyo Road, Ibadan, Nigeria

Taxonomy, Origin and Distribution

Cassava belongs to the botanical species *Manihot esculenta* Crantz of the family Euphorbiaceae, subfamily Crotonoideae, and tribe Manihotae. The genus *Manihot* contains about 100 species of herbs, shrubs and trees among which the production of latex and cyanogenic glucosides is common (Rogers and Fleming, 1973; Bailey, 1976). These species are grouped into 19 taxonomic sections and cassava is classified under the section *Manihot* (Rogers and Appan, 1973). All *Manihot* species originated in the tropical Americas.

Vavilov (1951) placed the origin of cassava in Brazil. In addition, other lowland areas of tropical Americas have been considered as places where cassava could have originated (Smith, 1968). Rogers (1963) identified two geographic centres of speciation of cassava: (i) the drier areas of western and southern Mexico and portions of Guatemala; and (ii) the dry northeastern portions of Brazil. Nassar (1978a,b) identified four areas of diversity of the wild species: (i) central Brazil; (ii) northeastern Brazil; (iii) southwestern Mexico; and (iv) western Mato Grosso (Brazil) and Bolivia (chapter 1).

The cassava crop may have been cultivated in Colombia and Venezuela from 3000 to 7000 years ago (Rouse and Cruxent, 1963, cited in Hershey, 1985). Ugent *et al.* (1986) cited evidence for domestication on the Peruvian coast before 4000 BC. Sauer (1952, quoted by Smith,

1968) proposed the heart of domestication as northwestern South America. Vavilov (1939) suggested early cultivation of cassava in the equatorial region of South America. It is believed that cassava was carried by the Arawak tribes of Central Brazil to the Caribbean Islands and Central America in the 11th century (Brucher, 1989), by the Portuguese to the west coast of Africa, via the Gulf of Benin and the Congo River at the end of the 16th century (Jones, 1959), and to the east coast via the islands of Reunion, Madagascar, and Zanzibar at the end of the 18th century (Barnes, 1975; Jennings, 1976). The crop arrived in India about 1800. The Spaniards took it into the Pacific, but it was not widely used as a food crop there until the 1960s (Jennings, 1976). Cassava is now widely cultivated in the tropics. The tuberous roots provide a major food source for more than 500 million people in Africa, Latin America and Asia. Its leaves are also used for human consumption in parts of Africa and Asia.

The Mesoamerican region extending from the northwestern coast of Mexico and covering parts of Guatemala, El Salvador and Nicaragua is also a potential area for early domestication (Rogers, 1965), where the wild species *Manihot aesculifolia*, *Manihot pringlei* and *Manihot isoloba* may have contributed to the cultigen through extensive hybridization, although this theory has been questioned (Brucher, 1989). Rogers and Fleming (1973) believed that cassava is a complex species with multiple sites of initial cultivation. Allem (1994) believed that *M. esculenta* was derived from two wild ancestor species, *Manihot flabellifolia* Pohl and *Manihot peruviana* Mueller. He placed the two wild species in the primary genepool of the cultivated cassava, while *Manihot glaziovii*, *Manihot dichotoma*, *Manihot pringlei*, *Manihot aesculifolia*, *Manihot pilosa*, *Manihot triphylla* and *Manihot pruinosa* he placed in the secondary genepool.

Cassava is monoecious and predominantly outcrossing, which leads to a very high degree of heterozygosity in plants and among populations produced from true seeds. Cassava varieties are heterozygous individuals which are propagated vegetatively to maintain the desired genotypes. Though cultivated germplasm has erratic flowering habits, it produces seed readily in many environments. Its seeds disperse naturally, as the fruit capsules dehisce at maturity. Many cassava plants growing from naturally dispersed seeds occur in farmers' fields in Africa. Through outcrossing among heterogeneous plant populations and subsequent selection by nature and by farmers, specific recombinants or new variants are created under the African agroecosystem. Specialized cultivars, such as those with a high level of resistance to cassava mosaic disease, could have been selected in this way.

Some wild *Manihot* species, such as *Manihot glaziovii* and *Manihot tristis* were introduced to many parts of Africa and Asia, at least since the early 20th century. They have now become naturalized in some parts of Africa and Asia (Rogers and Appan, 1973). These wild species were introduced to Africa and Asia initially for use as a source of rubber and later as shade trees in cocoa plantations and home compounds, and in Africa as a tree for fencing (Allem and Hahn, 1991). Introgression from *Manihot glaziovii* into cassava seems to be occurring under natural conditions in Africa. This phenomenon could have generated additional new variability in cassava in Africa, outside the place where it was first domesticated. Indeed, it has been widely recognized that an important secondary centre of diversity of cassava has become established in Africa (Gulick *et al.*, 1983).

The native range of *Manihot* species is from southern Arizona (*Manihot davisiae* and *Manihot angustiloba*) to Argentina (*Manihot grahami* and *Manihot anisophylla*). Only one wild species, *Manihot brachyloba*, is native to the West

Indies. The species of *Manihot* are all rather sporadic in their distribution. However, there are two major concentrations of species, one in Mexico, the other in Brazil (Rogers, 1963). The species found in the two areas are remarkably disjunct; with the exception of cultivated species, none of the North American species is found in South America. The preponderant number of species are South American; the greatest number are found in eastern central Brazil (Rogers and Appan, 1973). Most *Manihot* species are found in relatively dry areas, and only a few are found typically in rainforest regions. The North American species are mostly found on limestone-derived soils. All of the species in the genus are sensitive to frost. There are only two species, *M. grahami* and *M. anisophylla* whose native distributions are in regions with occasional frost.

Germplasm Conservation

The importance of plant genetic resources to global food security as well as to the security of the livelihood of millions of rural families was underlined at the first United Nations Conference on the Human Environment in Stockholm, 1972. A decade before that conference, widespread loss of diversity at different levels of biological organization was recognized (TAC/FAO, 1972; Frankel, 1973). The Stockholm conference called for concerted efforts to conserve and utilize naturally occurring genetic variability in all plants, whilst considering the interests of both present and future generations.

Several international research institutes of the Consultative Group for International Agricultural Research (CGIAR), such as Centro Internacional de Agricultura Tropical (CIAT) and International Institute of Tropical Agriculture (IITA), were already established at that time. They had started assembling genetic resources of their mandated crops, consisting of local or introduced landraces, improved cultivars and related wild species for use in their breeding programmes. This initial effort of collecting plant genetic resources was formalized when many CGIAR centres established *ex situ* conservation programmes to collect and conserve the genetic resources of their own mandated crops, as an integral part of their activities. CIAT and IITA share the CGIAR mandate for cassava. They collaborated with many national programmes with active programmes in cassava breeding or genetic resources conservation in collection, utilization and conservation of the genetic resources of this crop. The 1992 United Nations Conference on Environment and Development was held in Rio de Janeiro, where the Convention on Biological Diversity (CBD) was signed by over 150 heads of state. This marked a historic commitment by the nations of the world to conserve biodiversity and to ensure that biological resources are used sustainably and that the benefits of such use are shared equitably.

Germplasm collections

Recent reports indicated that some 20,000 accessions of cassava germplasm and its wild relatives are being preserved *ex situ* in CIAT, IITA, and national programmes in more than 45 countries throughout the world (Bonierbale *et al.*, 1997; IITA, 1997, 1998). Table 9.1 lists the number of accessions held in the two CGIAR centres and national programmes. The figures may include duplicates, especially among those that are held by CIAT and national programmes in the Americas, and by IITA and national programmes in West and Central Africa. The collection at CIAT includes germplasm from Argentina (72), Brazil (1334), Colombia (2001), Costa Rica (148), Cuba (77), Ecuador (117), Guatemala (91), Indonesia (51), Malaysia (67), Mexico (102), Nigeria (19), Panama (43), Paraguay (231), Peru (405), Puerto Rico (15), Thailand (31), USA (10), Venezuela (249), and five other countries (Bonierbale *et al.*, 1997). The IITA collection includes materials from Cameroon (247), Togo, (289), Ghana (81), Nigeria (363), Republic of Benin (378), Congo (64), Kenya (12), East Africa (29), Brazil (49), and five other countries in Africa. The total number of unique accessions is likely to be much smaller than 20,000. South America has the largest collection, followed by West and Central Africa. The collection held at CIAT represents the largest diversity of cassava from the Americas and Asia, and from West and Central Africa at IITA. The IITA-coordinated research networks, East Africa Root Crops Research Network (EARRNET) based in Uganda, and the Southern Africa Root Crops Research Network

(SARRNET) based in Malawi, initiated activities in network member countries to collect and evaluate the local cassava germplasm. A total of 1245 accessions were assembled and maintained in EARRNET member countries, including Burundi, Kenya, Madagascar, Rwanda and Uganda (IITA, 1997). Similarly, a total of 797 accessions of local germplasm were collected and maintained in SARRNET member countries, including Angola, Botswana, Malawi, Mozambique, Namibia, Tanzania, Swaziland, South Africa, Zambia and Zimbabwe (IITA, 1998). Cassava germplasm accessions preserved at CIAT and IITA have been designated as 'in trust' collections under the auspices of the Food and Agriculture Organization of the United Nations (FAO) for public access. They are freely available to researchers worldwide.

Ex situ conservation methods

There are two basic approaches to the conservation of plant genetic resources, *ex situ* and *in situ*, and they are complementary. There are various methods for the conservation of genetic resources. The best way is to adopt a combination of cost-effective and practical methods, to conserve the targeted genepool of a species. The cassava crop is an outcrossing species and produces botanical seed in many environments, but is mainly propagated vegetatively using stem cuttings with two to six nodes, or by shoot tips in *in vitro* cultures to maintain genotypes. Its wild relatives are also predominantly outcrossing and are propagated mainly by botanical seed, and in some species, by stem cuttings. Many wild species could also be propagated in *in vitro* cultures (Iwanaga and Iglesias, 1994; Ng and Ng, 1997). Various *ex situ* techniques and options are available for the conservation of cassava genetic resources. These are the field genebank, seed storage, *in vitro* reduced growth storage and cryopreservation of shoot tips and pollen. DNA storage could also eventually become one of the options for cassava germplasm conservation. However, strategies and procedures for the utilization of the stored DNA have yet to be devised. Recent advances in cassava biotechnology which facilitate the selective transfer of genes, stored DNA from outside the genepool of cassava is likely to become more relevant in breeding (Withers, 1994).

Table 9.1. Cassava germplasm collection in some national and international research centres (maintained as field genebanks).

Region/country	No. accessions	Institute/programme
South America		
Colombia	4695	CIAT
Brazil	4132	CNPMF/CENARGEN
Paraguay	360	IAN
Ecuador	101	INIAP
Argentina	177	INTA
Bolivia	18	IIA
Central America		
Costa Rica	154	CATIE
Mexico	225	INIFAP
Panama	50	IIA
Nicaragua	37	UNA
Caribbean		
Cuba	495	INIVIT
Dominican Rep.	46	–
Eastern and Southern Africa		
Angola	13	
Botswana	11	
Tanzania	254	RTCP
Malawi	170	RTCP
Uganda	413	RTCP
Kenya	250	RTCP
Mozambique	81	INIA
Zambia	96	
Rwanda	280	
Zimbabwe	6	
South Africa	100	
West and Central Africa		
Benin	340	SRCV
Cameroon	250	
Côte d'Ivoire	300	
Gabon	42	
Ghana	2000	PGRC/CRI
Burkina Faso	14	
Nigeria	435	NRCRI
	2861	IITA
Guinea, Conakry	168	
Senegal	57	ISRA/CDH
Sierra Leone	134	IAR
Togo	734	
D.R. Congo	250	
Asia – Oceania		
China	86	SCATC/UCRI/GAAS
India	1507	CTCRI
Indonesia	251	CRIFC/MARIF
Israel	5	Israel Genebank for Agric. Crop
Malaysia	92	MARDI
Myanmar	21	ARI
Pakistan	3	Plant Introduction Center
Philippines	384	PRCRTC/IPB
Sri Lanka	112	CARI/PGRC
Thailand	250	RFCRC
Vietnam	36	Hung Loc. Agric. Centre

Footnote on next page

Field genebanks

Cassava germplasm can be maintained in the field, as living collections which are relatively easy to establish and maintain with little need of sophisticated equipment. This method has an advantage over other conservation methods in that it provides plant materials readily for evaluation/characterization and for cross-pollination. A major disadvantage is the requirement for large fields to maintain germplasm collections. Germplasm materials in a field genebank are under pressure of constant exposure to diseases and pests, which could lead to a loss of genetic materials or genetic drift. Proper maintenance of a field genebank in a cost-effective way is important to ensure the survival of germplasm in a sustainable manner.

The planting distance of cassava for food production is usually a minimum of 1 m × 1 m between plants. Cassava stakes (stems with two to six nodes cut from mature stems) are used as propagules, and planted directly in the ground or in soil ridges/mounds, with at least one node buried under the surface of the soil. To achieve greater root yields, cassava is more frequently planted on soil ridges/mounds. For the purpose of maintaining a germplasm collection, it is less important to produce high root yields or improved root quality, than to maintain the clonal materials in perpetuity. A common practice used by IITA for cassava field genebank maintenance is to plant cassava stakes on flat ground in rows. Each accession is planted on a 2.5-m row plot, at a distance of 25 cm between hills within the row, and 50 cm between rows, with a total of 11 plants. The close spacing between plants suppresses weed growth and minimizes the area of land required. Nine months after planting, the plants are pruned; after 18–24 months the materials are planted in a new field. Where the activities for germplasm characterization/preliminary evaluation at IITA are combined with the maintenance of germplasm, cassava stakes are planted on ridges at a normal planting distance of 1 m × 1 m within and between rows, with 10 plants per accession. The materials are planted in new land every year. IITA maintains a collection of 24 wild *Manihot* species and CIAT has 26 species. Both the landraces and wild species are important sources of resistance to pests and diseases as well as quality characters. Work at IITA and CIAT has confirmed previous reports on useful traits in several wild *Manihot* species, such as high protein content, insect resistance and high levels of carotene (Asiedu *et al.*, 1992). IITA has used several wild *Manihot* species in interspecific hybridization with cassava to transfer their desirable genes into cassava (Hahn *et al.*, 1980, 1990).

Footnote for Table 9.1.
Acronyms: ARI, Agricultural Research Institute, Yezin, Myanmar; CARI/PGRC, Central Agricultural Research Institute, Gannoruwa, Peradeniya, Sri Lanka/Plant Genetic Resources Centre, Peradeniya, Sri Lanka; CATIE, Centro Agronomico Tropical de Investigacion y Ensenanza, Turrialba, Costa Rica; CENARGEN, Centro Nacional de Recursos Geneticos e Biotecnologia (of EMBRAPA), Brasilia, Brazil; CNPMF, Centro Nacional de Pesquisa de Mandioca e Fruticultura Tropical, Brazil; CRIFC/MARIF, Central Research for Food Crops, Indonesia/Malang Research Institute for Food Crop, Indonesia; CTCRI, Central Tuber Crops Research Institute, India; IAN, Instituto Agronomico Nacional, Paraguay; IAR, Institute of Agronomic Research, Sierra Leone; IIA, Instituto de Investigacion Agricola, Bolivia; IIA, Institute de Investigaciones Agropecuarias, Panama; INIA, Instituto Nacional de Investigacao Agronomica, Mozambique; INIAP, Instituto Nacional de Investigacion Agropecuarias, Ecuador; INIFAP, Instituto Nacional de Investigaciones Forestales y Agropecuarias, Mexico; INIVIT, Instituto Nacional de Investigacion de Viandas Tropicales, Cuba; INTA, Instituto Nacional de Tecnologia Agropecuaria, Argentina; ISRA, Institut Senegalais de Recherches Agricole/Centro pour le Developement de l'horticulture, Senegal; MARDI, Malaysian Agricultural Research and Development Institute, Malaysia; NRCRI, National Root Crop Research Institute, Nigeria; PGRC/CRI, Plant Genetic Resources Centre/Crop Research Institute, Ghana; PRCRTC/IPB, Philippine Root Crop Research and Training Centre/Institute of Plant Breeding, Philippines; RFCRC, Rayong Crop Research Centre, Thailand; RTCP, Root and Tuber Crop Research Programme; SCATC/UCRI/GAAS, South China Academy of Tropical Crops/ Upland Crop Research Institute/Guang Dong Academy of Agricultural Science, China; UNA, Universidad Nacional Agraria, Facultad de Agronomia, Nicaragua.

Seed genebanks

It was reported that cassava seeds could tolerate desiccation to 3.2% moisture content (MC), and there was no loss in seed viability following 4 months hermetic storage at −20°C with 6% MC (Ellis *et al.*, 1981), and after 14 years hermetic storage at −20°C with 6% MC (unpublished results, Seed Science Laboratory, Department of Agriculture, the University of Reading, cited by Hong *et al.*, 1996). Cassava seeds lost their viability after 2 years of storage under ambient conditions in Ibadan, Nigeria. Thus, cassava seeds are orthodox: they store best in cool and dry conditions.

At IITA, cassava seeds are harvested in bulk from all plants of each individual clonal accession and kept separately as a population of the clonal accession. Freshly harvested seeds are dried to 5–7% seed MC, sealed in aluminium foil envelopes, and stored at −20°C for long-term conservation. Seeds for use in breeding programmes and for distribution are stored in paper bags in a cold store maintained at 5°C and *c.* 30% relative humidity.

In vitro genebanks

In vitro propagation and conservation techniques for cassava have been well developed and have been applied routinely in many genebanks, particularly in CIAT and IITA, and in national programmes in Brazil, Argentina, Paraguay and Cuba (Bonierbale *et al.*, 1997; Ng and Ng, 1997; Ng *et al.*, 1999). Table 9.2 summarizes information on the cassava germplasm collections that are maintained in an *in vitro* genebank, either by reduced growth storage or under normal culture conditions. Shoot-tip cultures of cassava clonal accessions are usually conserved under reduced-growth or slow-growth culture media,

Table 9.2. Cassava germplasm collections maintained *in vitro* in some national and international research institutes.

Country/institution	No. accessions (Reference)	Culture condition	Storage duration (months)
Argentina	120 (Bonierbale *et al.*, 1997)	–	–
Brazil	1307 (De Goes *et al.*, 1999)	–	3–14
Caribbean	28 (Bateson, 1999)	18–22°C	–
CIAT	5714 (Guevara and Mafla, 1999)	23–24°C, 1000 lux Low sucrose medium	10–18
Cuba	– (Morales *et al.*, 1999)	20–22°C, 500 lux Normal culture medium	–
Ghana	9 (Acheampong, 1999)	28°C	–
IITA	727 (Ng and Ng, 1997)	18–22°C	8–12
India	30 (Bonierbale *et al.*, 1997)	–	–
Paraguay	101 (Bonierbale *et al.*, 1997)	–	–
Philippines	13 (Zamora and Paet, 1999)	8 h light, normal culture medium	2
South Pacific	19 (Taylor, 1999)	20°C	9
Sri Lanka	56 (Bonierbale *et al.*, 1997)	–	–
Vietnam	10 (Bonierbale *et al.*, 1997)	–	–

or reduced incubation conditions (temperature and light intensity). Cassava clones in the *in vitro* genebank at CIAT are conserved under the following conditions: (i) constant temperature of 23–25°C for 24 h through the day and night, with 12 h light illumination of 1000–1500 lux; (ii) a slightly modified Murashige–Skoog (MS) culture medium (Murashige and Skoog, 1962). Three to five tubes per accession are maintained (Iwanaga and Iglesias, 1994). At IITA, the cultures are maintained under lower incubation temperatures, and lower light intensity than at CIAT, with five to ten tubes per accession (Ng and Ng, 1997; Ng *et al.*, 1999). Shoot-tip and nodal cultures have also been used at IITA to transfer cassava germplasm collections from collaborating National Agricultural Research Systems in Cameroon, Ghana and the Republic of Benin.

The source of materials for *in vitro* propagation are *apical* buds or nodes collected at their active growth stage, from cassava plants growing in a field genebank or screen house. Buds collected are surface disinfected with 70% ethanol, followed by 7% sodium hypochlorite solution with Tween 20 for 20 min. They are then rinsed with three changes of sterile distilled water. Meristems with one to two leaf primordia are excised from the buds and inoculated in MS basal medium supplemented with 3% sugar, 80 mg l^{-1} adenine sulphate, 0.15 mg l^{-1} benzyl amino purine (BAP), 0.2 mg l^{-1} naphthalene acetic acids (NAA), 0.04 mg l^{-1} gibberellic acid (GA$_3$) and 0.6% agar. For nodal cutting culture, nodes from young green shoots are surface disinfected with 70% ethanol for 5 min, followed by 10% sodium hypochlorite solution with Tween 20 for 20 min and then by 5% sodium hypochlorite solution for 10 min. The nodes are rinsed three times with sterile distilled water, then placed on MS medium supplemented with 3% sugar, 0.01 mg l^{-1} NAA, 0.05 mg l^{-1} BAP and 0.7% agar. Ten tubes (16×125 mm) per accession are cultured. Cultures are incubated in a culture room maintained at 28–30°C with 12 h light illumination of 1000–1500 lux intensity, for 3–4 weeks. Regenerated plantlets are then micropropagated for initial increase and transferred to stores at 18–24°C, and low light intensity. Cultures are checked regularly. Those that are contaminated with fungi or bacteria are discarded and those that show deterioration

are subcultured. Cultures can be kept for 8–12 months before subcultures are made. Theoretically, this cycle can be repeated indefinitely and plantlets can also be transplanted to the isolation room for virus and other disease indexing. Through disease or virus indexing, healthy germplasm accessions certified by Plant Quarantine Services are then multiplied for use in exchange with research partners.

Cryopreservation

Cryopreservation enables long-term conservation of germplasm. Recent developments in cryopreservation, especially those reducing cryodamage, offer an improvement of survival after freezing and the list of successfully cryopreserved species is increasing.

Early research by Bajaj (1983) showed that cassava shoot tips frozen directly in liquid nitrogen (LN) with cryoprotectants resumed growth and formed plantlets after thawing. A survival percentage of 26 was obtained after 3 years storage in LN. Other successful approaches involving controlled cooling regimes have been described for shoot tips (Escobar *et al.*, 1993), seeds and zygotic embryos (Marin *et al.*, 1990) and somatic embryos (Sudarmonowati and Henshaw, 1990). Pollen/anthers are also a good source of plant material for long-term conservation. Of the different explants used, the shoot tip is the most suitable plant material for use in the cryopreservation of cassava clonal germplasm. It is most amenable to tissue culture and can be available at any time of the year in any quantity.

The classical cryopreservation protocols have normally been by slow cooling with the use of cryoprotectants and these require sophisticated and expensive programmable freezing equipment. The new techniques recently developed offer an opportunity to use fast freezing, and direct immersion in LN. Pregrowth of shoot tips in proliferation medium, followed by treatment with cryoprotectants and dehydration before direct immersion of shoot tips in LN resulted in a plant recovery rate as high as that with programmable freezing (Escobar *et al.*, 1995). Shoot tips were also successfully cryopreserved by encapsulation–dehydration followed by direct immersion in LN (Escobar *et al.*, 1998). The vitrification technique has been employed to simplify handling of explants and has secured a high level of recovery. Shoot tips were cryopreserved successfully using this technique on cassava germplasm from Thailand (Charoensub *et al.*, 2000) and from Africa (Ng and Ng, 2000). Some important factors influencing the recovery include pre-culture, treatment with vitrification solution, thawing, size of shoot tips and genotype. The procedure involves pre-culture of shoot tips in a high sucrose proliferation medium, treatment with vitrification solution, followed by direct plunging into LN. Frozen plant materials after retrieval from LN are thawed either by fast or slow thawing, washed with an appropriate solution and transferred to culture medium for re-growth. Up to 60% recovery was recorded in cassava using this procedure (Ng and Ng, unpublished).

In situ and on-farm conservation

In situ conservation of genetic resources is to maintain genetic material, usually the wild relatives of crop plants or trees, in their natural ecosystems. On-farm conservation is to conserve local landraces/cultivars in farmers' fields. *In situ* or on-farm conservation of genetic resources maintains the evolutionary processes of the targeted plant species. The process can make a direct contribution to the wellbeing of farmers and communities by ensuring that adapted plant types remain directly available to them for their own continuing use.

Maintenance of landraces or traditional crop varieties on-farm and in home gardens or wild relatives of crop plants *in situ* has gained recognition and been widely promoted as an effective way of conserving traditional crop varieties and wild relatives of crop species. For thousands of years, farmers selected and managed their plant genetic resources through gathering and domestication, continuous selection and cultivation of cultivars most suitable to the surrounding environments in which the farmers work and live. Farmers are creators and conservators of crop genetic resources, while at the same time they abandon what they do not need or cannot maintain because of social factors or biological constraints. It is impossible at this stage to rely solely on farmers to save all traditional cassava varieties they themselves have created or

selected, unless there are some incentives to do so and support from institutions and governments. Development of on-farm or *in situ* conservation strategies require very broad knowledge, ranging from social, biological and environmental factors. Such information required for the development of effective *in situ* conservation strategies for cassava species and their wild relatives is not well documented or understood. It requires a multidisciplinary team approach to gather data and analyse the information for the design of an appropriate *in situ* conservation strategy for cassava.

Before a well-designed strategy for *in situ* or on-farm conservation of cassava and their wild relatives is in place, scientific communities, social workers and policy makers should encourage/ support farmers or volunteers and institutions to conserve as many of their traditional varieties on-farm as possible. *Ex situ* conservation of genetic resources for the foreseeable future still remains the most effective and reliable way of conserving cassava genetic resources.

Conclusions

With the increasing importance of cassava around the world for human consumption, animal feed and industrial uses, there will be an increasing need for a wide range of genetic diversity to develop cultivars having specific characteristics and for adaptation to different ecologies. It is important that the existing cassava genetic resources held in various institutes be well maintained, and safely duplicated using a combination of available conservation methods.

There are some gaps in the existing *ex situ* collection, with respect to the representation of genetic diversity from many geographical areas (Bonierbale *et al.*, 1997). Few collections of wild relatives of cassava have been assembled and conserved in *ex situ* genebanks. Because of the destruction of natural habitats where wild relatives of cassava are growing and also abandonment of old traditional cassava cultivars by farmers, there is some urgent need to collect the cassava diversity not represented in the existing collections for *ex situ* conservation. Concurrently, germplasm characterization using agrobotanical descriptors and molecular markers should be intensified and accelerated. Information obtained from the characterization will assist in the selection of core collection(s) and the elimination of duplicates, thus increasing the efficiency of germplasm management and use. Recent progress in cryopreservation research in cassava is very encouraging, especially the vitrification and fast freezing protocols. This technique when fully developed offers a relatively simple and inexpensive method for long-term conservation of cassava germplasm. This technique will be affordable for cassava conservation in national programmes. Research in this area should focus on the optimization of the protocol to achieve high recovery rate and application to a wide range of cassava germplasm.

Investigation of *in situ* or on-farm conservation of cassava genetic resources by national programmes and international research institutes should be encouraged. Appropriate strategies for *in situ* or on-farm conservation should be developed to complement *ex situ* conservation. Ideal locations for *in situ* or on-farm conservation should be identified. National/regional/international policies governing the use of the areas for the conservation of cassava genetic resources and their access should be well articulated to ensure that the conserved genetic resources would be accessible to researchers worldwide.

References

Acheampong, E. (1999) *In vitro* genebank management of clonally propagated crops under minimal conditions. In: Engelmann, F. (ed.) *Management of Field and* in vitro *Germplasm Collections*. IPGRI, Rome, Italy, pp. 73–75.

Allem, A.C. (1994) The origin of *Manihot esculenta* Crantz (Euphorbiaceae). *Genetic Resources & Crop Evolution* 41(3), 133–150.

Allem, A.C. and Hahn, S.K. (1991) Cassava germplasm strategies for Africa. In: Ng, N.Q., Perrino, P., Attere, F. and Zedan, H. (eds) *Crop Genetic Resources of Africa*, Vol. II. Ibadan, Nigeria, pp. 127–149.

Asiedu, R., Hahn, S.K., Bai, K.V. and Dixon, A.G.O. (1992) Introgression of genes from wild relatives into cassava. In: Akoroda, M.O. and Arene, O.B. (eds) *Proceedings of the Fourth Triennial Symposium of the International Society of Tropical Root Crops – Africa Branch*. ISTRC-AB/IDRC/CTA/IITA, Nigeria, pp. 89–91.

Bailey, H. (1976) *Hortus Third. A Concise Dictionary of Plants Cultivated in the United States and Canada*. McMillan, New York.

Bajaj, Y.P.S. (1983) Cassava plant from meristem cultures freeze-preserved for three years. *Field Crops Research* 7, 161–167.

Barnes, H. (1975) The diffusion of the manioc plant from South America to Africa: an essay in ethnobotanical culture history. Dissertation, Colombia University.

Bateson, J.M. (1999) *In vitro* genebank management at CARDI tissue laboratory. In: Engelmann, F. (ed.) *Management of Field and* in vitro *Germplasm Collections*. IPGRI, Rome, Italy, pp. 90–92.

Bonierbale, M., Guevara, C., Dixon, A.G.O., Ng, N.Q., Asiedu, R. and Ng, S.Y.C. (1997) Chapter 1. Cassava. In: Fuccillo, D., Sears, L. and Stapleton, P. (eds) *Biodiversity in Trust*. Cambridge University Press, Cambridge, pp. 1–20.

Brucher, H. (1989) *Useful Plants of Neotropical Origin and their Wild Relatives*. Springer-Verlag, Berlin.

Charoensub, R., Phansiri, S., Sakai, A. and Yongmanitchai, W. (2000) Cryopreservation of *in vitro* grown shoot-tips of cassava cooled to −196°C by vitrification. In: Engelmann, F. and Takagi, H. (eds) *Cryopreservation of Tropical Plant Germplasm – Current Research Progress and Applications*. Japan International Research Center for Agricultural Sciences, Tsukuba, Japan and International Plant Genetic Resources Institute, Rome, Italy, pp. 401–403.

De Goes, M., Mendes, R.A., Labuto, L.B.D. and Cardoso, L.D. (1999) Present status of germplasm conservation *in vitro* at CENARGEN/EMBRAPA, Brazil. In: Engelmann, F. (ed.) *Management of Field and* in vitro *Germplasm Collections*. IPGRI, Rome, Italy, pp. 86–87.

Ellis, R.H., Hong, T.D. and Roberts, E.H. (1981) The influence of desiccation on cassava seed germination and longevity. *Annals of Botany* 47, 173–175.

Escobar, R.H., Roca, W.M. and Mafla, G. (1993) Cryopreservation of cassava shoot tips. In: Roca, W.M. and Thro, A.M. (eds) *Proceedings of the 1st International Scientific Meeting of the Cassava Biotechnology Network*. Colombia, pp. 116–119.

Escobar, R.H., Mafla, G. and Roca W.M. (1995) Cryopreservation for long-term conservation of cassava genetic resources. In: *Proceedings of the 2nd Scientific Meeting of the Cassava Biotechnology Network*, Vol. 1, Colombia, pp. 190–193.

Escobar, R.H., Palacio, J.D., Rangel, M.P. and Roca, W.M. (1998) Cassava cryopreservation II. In: *Abstract of JIRCAS/IPGRI Joint International Workshop on Cryopreservation of Tropical Plant Germplasm – Current Research Progress and*

Applications, 20–23 October 1998, Tsukuba, Japan, p. 25.

Frankel, O.H. (1973) *Survey of Crop Genetic Resources in Their Centers of Diversity*. First Report. FAO/IBP, Rome, Italy.

Guevara, C.L. and Mafla, G. (1999) *Manihot* collections held at CIAT. In: Engelmann, F. (ed.) *Management of Field and* in vitro *Germplasm Collections*. IPGRI, Rome, Italy, pp. 109–112.

Gulick, P., Hershey, C. and Esquinas-Alcazar, J. (1983) *Genetic Resources of Cassava and Wild Relatives*. International Board for Plant Genetic Resources, Rome, Italy.

Hahn, S.K., Terry, E. and Leuschner, R. (1980) Cassava breeding for resistance to cassava mosaic disease. *Euphytica* 29, 673–683.

Hahn, S.K., Bai, K.V. and Asiedu, R. (1990) Tetraploids, triploids and $2n$ pollen from diploid interspecific crosses with cassava. *Theoretical and Applied Genetics* 79, 433–439.

Hershey, C. (1985) Cassava germplasm resources. Presentation at a Workshop on Cassava Breeding: a Multidisciplinary Review, 4–7 March 1985, Philippines.

Hong, T.D., Linington, S. and Ellis, R.H. (1996) *Seed Storage Behaviour: a Compendium*. Handbooks for Genebanks: No.4. International Plant Genetic Resources Institute, Rome, p. 287

IITA (1997) *Project 16. Conservation and Genetic Enhancement of Plant Biodiversity Annual Report 1996*. IITA, Ibadan, Nigeria, p. 2.

IITA (1998) *Project 16. Conservation and Genetic Enhancement of Plant Biodiversity Annual Report 1997*. IITA, Ibadan, Nigeria, p. 4.

Iwanaga, M. and Iglesias, C. (1994) Cassava genetic resources management at CIAT. In: *Report of the First Meeting of the International Network for Cassava Genetic Resources. International Crop Network Series No. 10*. IPGRI, Rome, Italy, pp. 77–86.

Jennings, D.L. (1976) Cassava, *Manihot esculenta* (Euphorbiaceae). In: Simmonds, N. (ed.) *Evolution of Crop Plants*. Longman, London, pp. 81–84.

Jones, W.O. (1959) *Manioc in Africa*. Stanford University Press, Stanford, Connecticut.

Marin, M.L., Mafla, G., Roca, W.M. and Withers, L.A. (1990) Cryopreservation of cassava zygotic embryos and whole seeds in liquid nitrogen. *Cryo-Letters* 1, 251–264.

Morales, S.R., Garcia, M.G., Torres, J.L., Vega, V.M. and Martin, J.C.V. (1999) *In vitro* germplasm storage of tropical root and tuber crops, and banana and plantain in the Cuban Republic. In: Engelmann, F. (ed.) *Management of Field and* in vitro *Germplasm Collections*. IPGRI, Rome, Italy, pp. 84–85.

Murashige, T. and Skoog, F. (1962) A revised medium for rapid growth and bioassay with tobacco tissue culture. *Physiologia Plantarum* 15, 473–497.

Nassar, N. (1978a) Conservation of the genetic resources of cassava (*Manihot esculenta*) – determination of wild species localities with emphasis on probable origin. *Economic Botany* 32(3), 311–320.

Nassar, N. (1978b) Microcenters of wild cassava, *Manihot* spp. diversity in central Brazil. *Turrialba* 28(4), 345–347.

Ng, S.Y.C and Ng, N.Q. (1997) Cassava *in vitro* germplasm management at the International Institute of Tropical Agriculture. *African Journal of Root Crops* 2(1&2), 232–233.

Ng, S.Y.C. and Ng, N.Q. (2000) Cryopreservation of cassava and yam by fast freezing. In: Engelmann, F. and Takagi, H. (eds) *Cryopreservation of Tropical Plant Germplasm – Current Research Progress and Applications*. Japan International Research Center for Agricultural Sciences, Tsukuba, Japan and International Plant Genetic Resources Institute, Rome, Italy, pp. 418–420.

Ng, S.Y.C., Mantell, S.H. and Ng, N.Q. (1999) Biotechnology in germplasm management of cassava and yams. In: Benson, E.E. (ed.) *Plant Conservation Biotechnology*. Taylor and Francis, London, pp. 179–202.

Rogers, D. (1963) Studies of *Manihot esculenta* Crantz and related species. Bulletin Torrey Botany Club 90(1), 43–54.

Rogers, D.J. (1965) Some botanical and ethnological considerations of *Manihot esculenta*. *Economic Botany* 19(4), 369–377.

Rogers, D.J. and Appan, S.G. (1973) *Flora Neotropica. Monograph No. 13 Manihot Manihotoides (Euphorbiaceae)*. Hafner Press, New York.

Rogers, D.J. and Fleming, H.S. (1973) A monograph of *Manihot esculenta* with an explanation of the taximetrics methods used. *Economic Botany* 27, 1–113.

Smith, C.E. (1968) The New World centers of origin of cultivated plants and the archeological evidence. *Economic Botany* 22(3), 253–266.

Sudarmonowati, E. and Henshaw, G.G. (1990) Cryopreservation of cassava somatic embryos and embryogenic tissue. In: *Book of Abstract of International Congress for Plant Tissue and Cell Culture*. Amsterdam, The Netherlands, p. 140.

TAC/FAO (1972) The collection, evaluation and conservation of plant genetic resources. *Report of the TAC ad hoc Working Group*. Beltsville, TAC Secretariat/FAO, Rome, Italy.

Taylor, M. (1999) *In vitro* conservation of root and tuber crops in the South Pacific. In: Engelmann, F. (ed.) *Management of Field and* in vitro *Germplasm Collections*. IPGRI, Rome, Italy, pp. 93–95.

Ugent, D., Pozorski, S. and Pozorki, T. (1986) Archeological manioc (*Manihot*) from coastal Peru. *Economic Botany* 40, 78–102.

Vavilov, N.I. (1939) The important agricultural crops of pre-Columbian America and their mutual relationship. *Izd. Gos. Georgr. O-va* [Public. Nat. Dept. Geogr. USSR], 71(10), 1–25.

Vavilov, N.I. (1951) The origin, variation, immunity and breeding of cultivated plants. Translated by Starr, C.K. *Chronica Botanica* 13, 1–366.

Withers, L.A. (1994) Methods and strategies for cassava germplasm. In: *Report of the First Meeting of the International Network for Cassava Genetic Resources. International Crop Network Series No. 10.* IPGRI, Rome, Italy, pp. 135–140.

Zamora, A.B. and Paet, C.N. (1999) *In vitro* gene-banking activities at the Institute of Plant Breeding, College of Agriculture, University of the Philippines at Los Banos. In: Engelmann, F. (ed.) *Management of Field and* in vitro *Germplasm Collections.* IPGRI, Rome, Italy, pp. 76–83.

Chapter 10
Cassava Biotechnology

M. Fregene[1] and J. Puonti-Kaerlas[2]

[1]Centro Internacional de Agricultura Tropical (CIAT), A.A. 6713, Cali, Colombia;
[2]Institute for Plant Sciences, ETH-Zentrum/LFW E 17, CH-8092 Zürich, Switzerland

Introduction

Cassava ranks second to sugarcane and is better than both maize and sorghum as an efficient producer of carbohydrate under optimal growing conditions. It is the most efficient producer under suboptimal conditions of uncertain rainfall, infertile soils and limited inputs encountered in the tropics (Loomis and Gerakis, 1975; El-Sharkawy, 1990). This makes cassava an attractive source of food, feed and renewable industrial raw material in under-developed regions of the world. However, biological constraints of a long growth cycle (8–24 months), vegetative propagation and perishability of the bulky roots lessen the crop's potential as an engine of rural development. Securing sufficient clean planting material for the production of a healthy crop can be an ordeal for many small farmers, and the relatively low inputs required for cassava production contrasts sharply with the high inputs/risks involved in processing, transporting and marketing the highly perishable roots (Fresco, 1993).

Biotechnology can contribute to the solution of these problems and realize great benefits for cassava farmers. Since the late 1970s, CIAT has worked and engaged in partnerships with Advanced Research Institutes (ARIs) in Europe and the USA to bring biotechnology to bear on the problems of cassava production, especially those that cannot be dealt with effectively through conventional methods. The Cassava Biotechnology Network (CBN), funded by the Netherlands Development Assistance (NEDA), is an example of an interdisciplinary forum of cassava researchers, farmers, end users and scientists in ARIs, hosted by CIAT with an agenda to find solutions to the problems encountered in the crop (Thro and Fregene, 1999). Cassava biotechnology falls into the following broad areas: molecular genetic markers for germplasm assessment, gene cloning and cassava breeding, genetic engineering for root quality, pest and disease resistance, and tissue culture for rapid multiplication of healthy planting material and cryopreservation. This chapter considers recent advances in molecular genetic markers and genetic engineering.

Molecular markers

Genetic markers of cassava

Genetic markers have become fundamental tools for understanding the inheritance and diversity of natural variation. They provided Gregor Mendel, the father of modern-day genetics, the tools for his ground-breaking experiment on heredity and more recently markers have made possible the construction of genetic maps, the cloning of genes known only by their phenotypes and position on a genetic map, and whole genome sequencing. The earliest genetic markers in cassava were morphological. Graner (1942) described the inheritance of two markers, leaf shape morphology and root colour, more recently, eight morphological markers, located on the stem, leaves and roots have been described (Hershey and Ocampo, 1989). These markers are governed by alleles with a major phenotypic effect with little or no environmental effects. The second generation of markers were biochemical, such as isozymes, and they provided a useful tool for genetic 'fingerprinting' and the study of genetic diversity in cassava (Hussain *et al.*, 1987; Ramirez *et al.*, 1987; Ocampo *et al.*, 1992; Lefevre and Charrier, 1993a). Isozymes have been applied to characterizing relationships among accessions of African cassava germplasm (Lefevre and Charrier, 1993b; Wanyera *et al.*, 1994) and fingerprinting of the international cassava collection held at CIAT (Ocampo *et al.*, 1992). The alpha beta esterase system was found to be most informative, providing 22 alleles, which have complemented morphological descriptors for the identification of duplicates in the collection at CIAT (Ocampo *et al.*, 1995).

With the discovery of the molecular basis of natural variation, molecular or DNA markers have rapidly gained importance in the study of genes, genomes and genetic diversity. They represent a limitless source of neutral markers for the quantitative assessment of genetic diversity and 'signposts' in gene and genome mapping. Their abundance in any organism facilitates the resolution of genetic relationships and genome/gene mapping unknown until the coming of molecular markers. They represent differences in the nucleotide sequences of either nuclear or organellar genomes and can be uncovered using diverse methods based upon PCR (Mullis, 1990), DNA–DNA hybridizations (Botstein *et al.*, 1980; Fodor *et al.*, 1993), or both. The most prominent molecular marker systems include minisatellites (Jeffreys *et al.*, 1985), restriction fragment length polymorphisms (RFLPs; Botstein *et al.*, 1980) and randomly amplified polymorphic DNAs (RAPDs; Williams *et al.*, 1990). Others are microsatellites, also known as simple sequence repeat markers (SSRs; Litt and Lutty, 1989a,b), amplified fragment length polymorphisms (AFLPs; Vos *et al.*, 1995) and DNA sequencing of the internal transcribed spacer (ITS) of ribosomal

DNA (Baldwin, 1992) and of single-copy nuclear genes. More recently DNA chips or oligonucleotide arrays have been added to an ever growing list of molecular markers (Fodor *et al.*, 1993). All these marker systems, other than DNA chips, have been used in cassava to assess genetic diversity, genome mapping and gene tagging.

Molecular marker assessment of genetic diversity

The genetic resources of cassava and its wild relatives represent a critical resource for the future of the crop. It is therefore understandable that germplasm collections and the study of genetic relationships between accessions have been made using virtually every available molecular marker. Minisatellite markers, fingerprints of highly variable tandemly repeated arrays of nuclear DNA (Jeffreys *et al.*, 1985), were the first markers used to study relationships among cassava accessions and their wild relatives (Bertram, 1993; Ocampo *et al.*, 1995). The 'fingerprints' were obtained by Southern hybridization of M13 phage DNA sequences to cassava genomic DNA digested with a panel of restriction enzymes (Rogstad *et al.*, 1988). They were used to derive relationships amongst cassava accessions and *Manihot* spp. from Central America and to identify 29 possible duplicate accessions in a subset of the international cassava collection held at CIAT. RFLP analysis of chloroplast DNA and nuclear ribosomal (rDNA) sequences, using heterologous probes, were also applied to assessing phylogenetic relationships among *Manihot* species from South and Central America and cassava (Bertram, 1993). The evidence from molecular analysis contradicted the widely held view, based on morphological characters, of an origin of cassava in a Meso-American *Manihot* species, *Manihot aesculifolia* and suggested a possible domestication from some close wild relatives from Brazil, including *Manihot tristis* and *Manihot esculenta* sub-spp. *flabellifolia* (Chapter 1). AFLP markers were employed to obtain a quantitative assessment of genetic relationships in a representative sample of the crop's diversity and six wild taxa (Roa *et al.*, 1997). The study again demonstrated that the Brazilian *Manihot* species *Manihot esculenta* sub-spp. *flabellifolia*, *M. tristis* and *Manihot peruviana*, are more similar to cassava than its Mexican relative *M. aesculifolia*, and that cassava might have its origin in these close relatives (Roa *et al.*, 1997). Conclusive evidence on the origins of cassava came from a phylogeographic study based on sequencing the single-copy nuclear gene glyceraldehyde 3-phosphate dehydrogenase (*G3pdh*; Olsen and Schaal, 1999). They demonstrated that cassava originated from natural populations of *M. esculenta* sub-spp. *flabellifolia* from the southern border of the Amazon basin in Brazil. Markers have also been used to obtain a quantitative assessment of genetic similarity in cassava (Beeching *et al.*, 1993; Second *et al.*, 1997; Elias *et al.*, 2000), and to study the genetic structure of germplasm resistant to disease (Sanchez *et al.*, 1999; Fregene *et al.*, 2000). Other studies have sought to determine the genetic structure and the basis of genetic differentiation of cassava landraces in Africa (Mkumbira *et al.*, 2002; Fregene *et al.*, unpublished).

Beeching *et al.* (1993) employed cloned cassava genes coding for enzymes involved in cyanogenesis as RFLP markers to obtain genetic similarity estimates in a small collection of cassava and some wild relatives. They were able to identify possible duplicates, confirm the intermediate position of interspecific hybrids between individuals of the parent material and postulated undocumented interspecific crossing with *Manihot glaziovii*, to explain the distant position of certain cassava accessions in respect of the majority of samples analysed. Evidence of introgression into cassava from *M. glaziovii* was also observed in an AFLP evaluation of genetic diversity in a large collection of cassava from the South American centre of diversity (Second *et al.*, 1997). Bonierbale *et al.* (1994) reported a comparison among elite cassava germplasm held at CIAT, adapted to five edaphoclimatic production conditions. They found that germplasm from certain edaphoclimatic zones (ECZ) showed a broader genetic base than others, but the accessions could not generally be assigned to a particular ECZ pool based on molecular patterns, due to considerable overlap of allele frequencies. Other approaches to the assessment of genetic diversity involved analysis of the structure of genotypes resistant to disease. Multiple

correspondence analysis (MCA) of AFLP data, using two primer combinations, of cassava genotypes resistant and susceptible to two strains of *Xanthomonas axonopodis*, permitted an elucidation of the genetic structure of cassava germplasm resistant to cassava bacterial blight (CBB; Sanchez *et al.*, 1999a,b). Results revealed a random distribution of resistance/susceptibility, suggesting that resistance to CBB has arisen independently many times in cassava germplasm. In contrast, AFLP assessment of 29 landraces and improved varieties from Africa resistant and susceptible to the sometimes devastating cassava mosaic disease (CMD) revealed a non-random distribution of resistant/susceptible varieties (Fregene *et al.*, 2000). African landraces resistant to CMD were also found to be genetically differentiated from resistant elite lines and from susceptible landraces, suggesting two different sources of CMD resistance and that the landrace source may have arisen recently as a single mutational event.

In other studies of cassava landraces, the genetic variability of 31 varieties of cassava traditionally grown by Makushi Amerindians from Guyana, and a representative sample of 38 varieties from an *ex situ* world collection held at CIAT, Cali, Colombia, was assessed by AFLP markers (Elias *et al.*, 2000). Twenty-one varieties presented intravarietal polymorphism, suggesting a variety could be made up of more than one genotype. The amount of diversity found in the cassava cultivars from a single site in Guyana was equal to that in the group representative of the CIAT collection and non-correspondence was found between the structure of molecular diversity and variation observed for agronomic traits that are targets for selection by cultivators. In contrast, genetic diversity was found to be structured according to taste, bitter as against sweet varieties in a study of cassava varieties from Northern Malawi (Mkumbira *et al.*, 2002). In another study of genetic relationships in 96 landraces collected from ten villages in southern Tanzania 68 SSR markers were employed in a principal component analysis to reveal genetic differentiation amongst the landraces not based on taste or location (Fregene *et al.*, unpublished). Although the basis of clustering is unclear, it is thought to represent different introduction events. Like maize, cassava appears to have highly differentiated gene pools and a large percentage of dominant/recessive gene action loci, which are two key characteristics required for heterosis. Once the wealth of data has been analysed and crosses between clusters tested, hopefully molecular markers can be used to predict heterosis. SSR markers have also been used to identify duplicates in the CIAT core collection (Chavarriaga-Aguirre et al., 1999) and to analyse germplasm from the littoral and Amazonian regions of Brazil (Mueller et al., unpublished).

Genome mapping in cassava

Markers have also been used to generate a molecular genetic map of cassava. The map was constructed from the segregation of RFLP, SSR, RAPD and isozyme markers in an intraspecific cross between TMS 30572, an improved line from IITA, Ibadan, Nigeria, and CM2177–2, an elite line from CIAT, Cali, Colombia (Fregene et al., 1997). The F_1 mapping progeny consists of 150 individuals. The RFLP markers were derived from several genomic and cDNA libraries and cloned genes of known functions, while the SSR markers were generated from one small-fragment genomic library (Chavarriaga-Aguirre et al., 1998). A total of 150 RFLP, 30 RAPD, five microsatellite and three isoenzyme loci segregating as single-dose restriction fragments (SDFs; Wu et al., 1992) in the gametes of the female parent, define 20 linkage groups and span 950 cM with an average marker density of one per 6 cM. From another 120 RFLP, 50 RAPD, four microsatellite and one isoenzyme single-dose markers in the gametes of the male parent, 24 linkage groups were drawn with a total distance of 1220 cM and average marker density of one marker every 8 cM. Thirty RFLP and two SSR markers detected a unique segregating fragment in each parent and a common allele in both parents and were mapped to similar positions on the male/female-derived linkage group. Such allelic bridges (Ritter et al., 1991) are crucial for identifying the analogous linkage groups in the male- and female-derived maps, as they detect the same locus on both parental chromosomes, except when they represent dupicated sequences. Comparison of intervals in the male- and female-derived maps, bounded by markers heterozygous in the allelic bridges of both parents (Ritter et al., 1991), revealed

significantly less meiotic recombination in the gametes of the female compared to the male parent.

Cassava is thought to be an allopolyploid (Magoon et al., 1969) and a segmental allopolyploid (Umanah and Hartmann, 1973). Genetic mapping of its genome is expected to provide conclusive evidence of this. In well-known allopolyploids such as maize, wheat and cotton, blocks of duplicated loci have been clearly identified by RFLP mapping (Helentjaris et al., 1988; Devos et al., 1993; Reinisch et al., 1994). Results from the genetic mapping of the cassava genome revealed only a few randomly distributed duplicated loci, less than 5% of the total number of markers, a number corresponding roughly to that reported in many diploids (Causse et al., 1994; Tanksley et al., 1995). Moreover, more than 90% of the markers detected only one locus compared to more than 50% in maize and cotton (Helentjaris et al., 1988; Reinisch et al., 1994). A similar low percentage of duplicated loci was found during RFLP mapping of Hevea, a member of the family Euphorbiaceae which has the same number of chromosomes as cassava ($2n = 36$; Lespinasse et al., 2000). Evidence of the allopolyploid origin of cassava ($2n = 36$) is cytogenetic and relies heavily on the identification of two sets of dissimilar nucleolar-organizing regions, on the repetition of chromosome types (Magoon et al., 1969; Umanah and Hartman, 1973), and on the basic chromosome numbers of other genera in the Euphorbiaceae, which range from six to 11 (Perry, 1943). No evidence of tetrasomic inheritance or of wild Manihot relatives with chromosome numbers of $2n = 18$ has been found to support the allopolyploid theory in cassava. The karyology of the 18 haploid chromosomes of cassava reveals six identical pairs and three different pairs of homologous chromosomes (Magoon et al., 1969). Assuming random assortment between homologous chromosomes of the six pairs of identical chromosomes, a reduction of 67% would be expected for all markers linked in repulsion (17% as against 50%). The percentage of markers linked in the repulsion phase in the genetic map of cassava reported here, 30%, is significantly higher than this although lower than the 50% expected for full diploids. This suggests that a small amount of random pairing may occur. Consequently, a purely disomic mode of inheritance has been concluded for cassava, although it is not currently clear if this represents the vestiges of an ancient allopolyploid or a true diploid.

Together the female- and male-derived maps have more than 300 markers and are estimated to cover 80% of the cassava genome. The cassava map therefore requires completion. However, the majority of markers on the map are RFLP markers and do not lend themselves easily to large scale, high-throughput marker-assisted analysis of plant populations, the principal application of the map. In an attempt to make marker technology more widely applicable in cassava, work began on a second generation map made up of highly polymorphic PCR-based markers, such as SSRs and sequence-tagged sites (STSs). Two SSR-enriched genomic DNA libraries were constructed and about 6000 clones were screened for the presence of the following SSR motifs: TC, GT, CAA, CAG, ACG, AAT, and CAGA and GATA (Mba *et al.*, 2000). A cDNA library constructed from leaf and root mRNA isolated from the elite cassava clone TMS 30572 was also screened for the SSR motifs mentioned above. More than 87,000 clones were screened. A total of 322 SSR markers were obtained from the enriched libraries, of which 92 have currently been mapped to the existing map of cassava (Mba *et al.*, 2000; Fregene *et al.*, unpublished data). Another 200 SSR markers were obtained from the cDNA libraries of which 10 have been mapped (R.E.C. Mba, unpublished data). The level of polymorphism of the SSR markers derived from the cDNA library in the parents of the cassava map population was 40%, considerably less than that found in the enriched genomic DNA libraries (60%).

The genetic map of cassava comprises of 250 RFLP markers on the male- and female-derived maps. It is an important resource that can be converted into PCR-based co-dominant markers known as STSs. The RFLP probes used in generating the markers have therefore been sequenced and primers designed to amplify them. The primers will be used in PCR reaction and gel electrophoresis without or with restriction enzyme digestion (cleaved amplified polymorphisms; CAPs) to discover polymorphisms which can serve as markers. This will ensure the rapid completion of a PCR-based map of cassava.

Another initiative toward completing the cassava map is the generation of expressed sequence tags (ESTs). Isolation of genes that are differentially expressed in the parents of a mapping population, has been proposed as a way of developing ESTs around specific traits for the candidate locus approach to increase the accuracy of mapping quantitative traits (Boventius and Weller, 1994; Suarez *et al.*, 2000). The cDNA/AFLP technique (Bachem *et al.*, 1996) was applied to mRNA from the parents of the cassava genetic map population and more than 500 transcript-derived fragments (TDFs) were obtained that were unique in either parent (Suarez *et al.*, 2000). A subset of 50 TDFs were cloned and sequenced. Sequence alignment of the expressed sequence tags (ESTs) revealed mostly genes of unknown function. Six of the TDFs have been mapped as RFLP markers to the existing molecular genetic map of cassava; the TDFs as RFLP markers were more polymorphic than random cDNAs. A number of cloned genes of known function have also been included on the molecular genetic map of cassava. They include two cytochrome P-450 genes that convert the amino acids L-valine and L-isoleucine during the biosynthesis of cyanogenic glucosides linamarin and lotaustralin in cassava (Andersen *et al.*, 2000), the AGPase phosphorylase, and the granule-bound starch synthase (GBSSII) gene involved in the biosynthesis of starch (Munyikwa *et al.*, 1997).

Genetic mapping of genes controlling agronomic traits

Molecular genetic maps provide a set of 'landmarks' for the complete genome and consequently a high probability of detecting linkage with any gene(s) of interest in genetics or breeding. Practical applications have been seen for many crops in 'tagging' genes of agronomic interest, usually as a more efficient selection parameter in breeding schemes, and for dissecting quantitative genetic variation into simpler Mendelian components (Tanksley *et al.*, 1989; Lee, 1995; Mohan *et al.*, 1997). One of the primary objectives of genetic mapping and gene tagging efforts in cassava is to provide tools that can increase the cost-effectiveness and efficiency of cassava breeding. Desirable characters that are difficult to evaluate using traditional methods are logical targets for 'molecular breeding' of cassava. These include resistance to pests and pathogens (especially those that are subject to quarantine exclusion), traits expressed only at the end of the crop's growing cycle, and traits for which the phenotype is difficult to measure. Prerequisites for the molecular mapping of agronomic traits generally include importance of the trait in question, the difficulty of screening by direct methods and a large variable population with a simple pedigree. Other requirements include a source of genome-wide markers and a reliable phenotypic screening method. The population selected for the construction of a genetic map of cassava was designed to segregate for traits regarded as priorities for the development of molecular markers and marker-assisted selection (MAS; Fregene *et al.*, 1997). They include resistance to CMD, bacterial blight, early bulking and root quality characters such as cyanogenesis, postharvest deterioration, culinary quality and starch content.

Cassava mosaic disease (CMD)

CMD is the most important disease of cassava in Africa (Chapter 12), and a potential threat to the crop in Latin America where the disease is still not known, although the whitefly vector has recently been found on cassava. Host plant resistance is the principal method of control, and was first identified in third backcross derivatives of an interspecific cross between cassava and *M. glaziovii*. Resistance is thought to be polygenic with a recessive component. The female parent TMS 30572 of the cassava map population has this source of resistance. Recently, several Nigerian cassava landraces have been identified that show strong resistance to CMD. Several mapping populations were developed that segregate for the old and new landrace sources of resistance. They include a half-sib backcross population, derived from crossing five F_1 progeny of the mapping population to the CMD-resistant parent and F_1 of the resistant landraces to susceptible varieties. The crosses were evaluated at two sites with high disease pressure over 2 years in Nigeria. Classical genetic analysis confirmed the polygenic nature of the *M. glaziovii* source of resistance and a major dominant gene control for the new resistance source (Akano *et al.*, 2002).

A bulk segregant analysis (BSA) approach was used to quickly identify markers linked to both sources of resistance. An SSR marker, SSRY40, on linkage group D of the genetic map of cassava, was found to be associated with CMD resistance and explains 48% of the phenotypic variance of CMD resistance ($P < 0.001$). This gene(s) has been designated *CMD1*. An SSR marker, SSRY28, and a RFLP marker, GY1, located on linkage group R explain 68% and 70%, respectively, of phenotypic variance of the new source of resistance (Akano *et al.*, 2002). The dominant CMD resistance gene has been designated *CMD2* and it is flanked by SSRY28 and GY1 at 9 and 8 cM, respectively. *CMD2* and markers associated to it are particularly valuable tools for breeding resistance to CMD in Latin America where the absence of the disease makes it impossible to select for resistance to CMD. In Africa, where a rapid deployment of strong resistance into cassava gene pools is required to protect cassava from CMD, selecting for the high level of resistance with a marker maybe more efficient compared to conventional breeding. The advantage of MAS is that it enables the breeder to eliminate at an early stage CMD-susceptible genotypes, which in the case of the heterozygous CMD-resistant landraces, is 50%. This halves the cost of disease evaluation and increases selection efficiency. The elimination of inferior genotypes at an early stage increases the efficiency of selection by allowing the breeder to concentrate on fewer genotypes at the seedling and crucial single-row trial stages, where progenies are reduced by up to 95%.

Cassava bacterial blight (CBB)

CBB, caused by *X. axonopodis* pv. *manihotis* (*Xam*), is a major disease of cassava in Africa and South America. Resistance to CBB was evaluated in individuals of the F_1 cross by controlled greenhouse inoculations and symptoms were assessed visually at 7, 15 and 30 days after inoculation, using a scale where 0 = no disease and 5 = maximum susceptibility (Jorge *et al.*, 2000). Five *Xam* strains, CIO-84, CIO-1, CIO-136, CIO-295 and ORST X-27 were used. Area under the disease progress curve (AUDPC) was used as a quantitative measure of resistance in quantitative trait loci (QTL) analysis by single-marker regression. Based on the AUDPC values, 12

QTLs, located on linkage groups B, C, D, G, L, N and X of the female-derived framework map, were found to explain 9–27% of the phenotypic variance of response to the five *Xam* strains. A scheme to confirm the usefulness of these markers in evaluating segregating populations for resistance to CBB has also been proposed (Jorge *et al.*, 2000).

Early bulking

Early bulking is another trait evaluated in the F_1 mapping progeny. A preliminary assessment of early bulking was conducted in 1998 by harvesting the F_1 mapping population at 7 months after planting (MAP) in the CIAT Palmira site (Fregene *et al.*, 2001). Dry matter yield was determined on three plants per genotype. Based on results from this evaluation, 40 early-bulking genotypes and 40 late-bulking groups were selected. Carefully picked healthy cuttings of the 80 selected genotypes were planted in a new experiment in December 1998 at Palmira. An early-bulking cassava landrace (*Mandioca de tres meses*) introduced from Brazil was used as control. Nine harvests were done within a 7-month period after which the experiment was terminated (July 1999). At each harvest, four plants in a row within a plot, per genotype, were evaluated for root yield and other traits assumed related to bulking. The traits evaluated were plant height, plant vigour, leaf area index (LAI), fresh root yield, fresh foliage, number of roots per plant, root diameter of the biggest five storage roots. Others were harvest index (HI), measured as the ratio of root yield to total harvested biomass, root dry matter and dry foliage. Multiple regression analyses of the evaluated traits (independent variable) and dry matter root yield (dependent variable) revealed that early bulking is affected mostly by HI and dry foliage. QTL analysis was done for traits significantly linked to early bulking using QGENE (Nelson, 1997) and markers linked to the traits associated with early bulking were significant at $P < 0.005$. Three QTLs each were found for dry foliage weight that explain 25–33% of phenotypic variance while five QTLs were found for harvest index that explain 18–27% of phenotypic variance (Okogbenin, unpublished). Kawano *et al.* (1998) showed that selection of HI in breeding scheme is an efficient indirect

selection parameter for root yield. Based on the marker analysis of the early-bulking study, it is apparent that early bulking (and by extension yield), can be increased more effectively by using a selection criteria based on markers for foliage (or total plant biomass) and HI.

Other traits

Several other gene-tagging projects ongoing at CIAT include resistance to the cassava whitefly (*Aleurotrachelus socialis*), CBB, using different crosses, the cassava root rot (*Phytophthora* spp.) and root qualities such as postharvest deterioration (PHD), starch quality and culinary quality.

Cloning of genes of agronomic interest

The heterozygous nature of cassava implies that attempts to introduce any trait, even when it is controlled by a single gene, may lead to the loss of a favoured variety. A more efficient way to introduce traits controlled by a single gene, such as CMD- resistance, is through genetic engineering. However, it is necessary first to clone the genes controlling the trait of interest. There are several approaches to cloning known only by its phenotype, or by its biochemical role in a biosynthetic pathway. The first is that of positional cloning (Martin *et al.*, 1993; Tanksley *et al.*, 1995) and cloning of genes via heterologous genes (Bothwell *et al.*, 1990). Three important criteria for positional cloning are a fine map based on a large mapping population of the appropriate genome region, a bacterial artificial chromosome (BAC) library and an efficient transformation protocol for complementation analysis.

A BAC library has been constructed for cassava, for positional cloning of genes identified during genetic mapping of traits of agronomic interest (Fregene *et al.*, 2000, 2001). DNA was isolated from the cassava variety TMS 30001 and embedded in agarose plugs as described by Zhang *et al.* (1995). TMS 30001, developed at the International Institute for Tropical Agriculture (IITA), shows strong resistance to CMD and resistance to some strains of CBB. Large genomic DNA, in one-third of an agarose plug, was partially digested with *Hin*d III, 1.5 units (U) for 20 min at 37°C, and DNA fragments of

100–300 kb, size-selected by pulse field gel electrophoresis (CHEF MAPPER, Bio-Rad Corp.). Size-selected DNA was ligated into the Hind III cloning site of pBeloBAC11 in a vector:insert ratio of 10 : 1, using 14 U of ligase, in a final volume of 100 µl. Twenty microlitres of DH10B competent cells (GIBCO BRL) were transformed, with 2 µl of the ligation reaction, by electroporation, and white colonies picked for DNA insert sizing. Colonies were grown for 14 h in LB+ 30 mg ml^{-1} Chloramphenicol and plasmid DNA isolated by the Autogen automatic plasmid isolation robot (Kurabo Inc.). Plasmid DNA was digested with 10 U of Not I to liberate inserts and separated on a 1% agarose gel by pulse field gel electrophoresis. The rest of the ligation was transformed, plated out and 55,000 clones with average size of 80,000 base pairs picked with the Q-bot robot (Genetix PLC). The library has a 5× coverage of the cassava genome.

Discovery of genetic markers linked to the gene controlling the new landrace source of resistance to CMD and construction of a BAC library, facilitates positional cloning of the CMD resistance gene. The nearest marker to the dominant CMD resistance gene is 8 cM and it is not appropriate for the construction of an array of overlapping BAC clones (contigs) across the genome intervals; contigs across > 1 cM is not feasible. Therefore, fine or dense maps of the genomic regions identified carrying the resistance gene are needed. The most efficient way of fine mapping in cassava is by employing a variant of the bulk segregant analysis and AFLP markers (Giovannoni et al., 1991). The method utilizes the unique ability of the RAPD and AFLP technique to sample many loci throughout the genome, using many primer combinations, and the ability to find additional linked markers in any region, by screening bulks of genotypic classes of markers adjoining that region. This method is currently being used with a large mapping population of 700 genotypes to identify more markers in the region of the cassava genome carrying the CMD2 resistance gene. Once fine maps have been obtained, a relationship between genetic distances and physical distances in the relevant regions will be estimated. Based upon the estimated physical distance required to transverse the region bearing resistance genes, a BAC contig will be constructed by BAC clone digestion and fingerprinting.

Finally candidate BAC clones will be introduced into cassava genotypes susceptible to CMD via genetic transformation using the BIBAC system.

Other genes of agronomic interest have also been cloned by the use of heterologous probes. One of the most important is the biosythesis gene for the generation of cyanogenic glucosides. Cassava produces the cyanogenic glucosides linamarin and lotaustralin in the roots, that can be toxic to human and animal health if not removed by processing. To block the production of these glucosides it is necessary to identify the key genes in the biosynthesis of linamarin and lotaustralin. The first steps in the biosynthesis of the two cyanogenic glucosides are the conversion of L-valine and L-isoleucine, respectively, to the corresponding oximes. Two full-length cDNA clones that encode cytochromes P-450 catalysing these reactions have been isolated using a heterologous probe (Anderson et al., 2000). The two cassava cytochromes P-450 are 85% identical and share 54% sequence identity to CYP79A1 from sorghum. They have been designated CYP79D1 and CYP79D2. Functional expression in the methylotrophic yeast, Pichia pastoris, reveal that each cytochrome P-450 metabolizes L-valine as well as L-isoleucine, consistent with the co-occurrence of linamarin and lotaustralin in cassava. Both CYP79D1 and CYP79D2 are actively transcribed in the cassava genome and production of acyanogenic cassava plants would therefore require down-regulation of both genes. Cassava trangenics with CYP79 anti-sense constructs have been generated and are currently being tested for the production of linamarin and lotaustralin (Moller, 2000, personal communication).

Genes involved in the in situ breakdown of the cyanogenic glucosides of cassava following tissue damage leading to the production of hydrocyanic acid, have also been cloned. A linamarase cDNA clone (pCAS5) was isolated from a cotyledon cDNA library using a white clover beta-glucosidase heterologous probe (Hughes et al., 1992). Concanavalin A affinity chromatography and endoglycosidase H digestion demonstrate that linamarase from cassava is glycosylated, having high-mannose-type N-asparagine-linked oligosaccharides. Consistent with this structure and the extracellular location of the active enzyme is the identification of an N-terminal signal peptide on the deduced amino acid sequence of pCAS5. Several genes controlling starch biosynthesis in cassava are also included in the list of cloned cassava genes (Munyikwa et al., 1997). They include the ADP glucose pyrophosphorylase (AGPase) B and S gene that catalyses the synthesis of ADP glucose and the granule-bound synthetase (GBSSII) gene, the predominant starch synthase gene that catalyses the conversion of ADP-glucose to amylose. They were cloned using homologous genes from potato. Other starch biosynthesis genes cloned include the starch branching enzyme that gives rise to amylopectin, the branched starch polymer. Potato transgenics expressing the cassava AGPase B gene in anti-sense orientation had reduced levels of AGPase B mRNA and 1.5–3 times less starch and more than five times more soluble sugars than the controls (Munyikwa et al., 1997). Similarly, cassava transgenics expressing GBSSII in the anti-sense orientation, produced little or no GBSSII transcripts and almost 100% amylopectin starch (T.R.I. Munyikwa, personal communication). These starch biosynthesis genes open up an opportunity to produce a range of cassava roots with different quantities of soluble sugars and starches with different amylose/ amylopectin ratios.

Another trait of economic interest for which cloned genes are needed to block its biosynthetic pathway, is PHD. In cassava, this is thought to be a physiological wound response mechanism that undermines the nutritional integrity of tuberous roots (Wheatley, 1982; Wheatley and Schwabe, 1985). Several genes known to be involved in wound healing in plants have been cloned and characterized for their expression during PHD. They include the ACC oxidase that catalyses the last reaction of ethylene biosynthesis in plants, phenylalanine ammonia-lyase (PAL), a key enzyme of the phenyl propanoid metabolism pathway, and catalase, involved in the breakdown of hydrogen peroxide (Li et al., 2002a,b; Reilly et al., 2002). A cDNA, mSOD1, encoding cytosolic copper/zinc superoxide dismutase (CuZnSOD) has also been cloned and characterized from cell cultures of cassava which produce a high yield of SOD (Lee et al., 1999). The mSOD1 gene is highly expressed in cultured cells, as well as in intact stems and tuberous roots. Levels of mSOD1 transcript increased dramatically a few hours after heat stress at 37°C and showed a synergistic effect with wounding stress.

In conclusion, genes, molecular tags and the knowledge accumulated can be used by plant breeders and genetic engineers to address the constraints cassava faces worldwide that have no ready solution using conventional tools. This will require sustained support from both public and private institutions and has the potential to produce a highly productive and profitable crop.

Genetic Engineering

Genetic engineering is a powerful tool that complements traditional breeding and can extend the genetic pool of useful gene sources beyond the species. Transgene technology also offers the advantage of transferring single or even quantitative traits, without the problems of linkage encountered in traditional breeding. The areas where genetic engineering can have an impact in cassava include yield and root quality improvement, pest and disease resistance, and the production of novel compounds for value-added products from cassava. Other important unsolved problems are the low protein content of the roots, the poor storability of freshly harvested roots and the cyanogenic nature of cassava.

Production of stably transformed plants requires an efficient in vitro culture system that allows regeneration of plants. Plant cells are generally considered to be totipotent, thus being able to regenerate whole plants from single cells in vitro. The ability to regenerate in vitro is, however, often limited to certain tissues and developmental stages, and the requirements for transformation and regeneration competence may not always be compatible. Furthermore, a method for efficient transfer and stable integration of the transgenes into the plant genomic DNA is essential for transformation, as well as a means for identifying and selecting transformed cells. The main constraint is usually not the delivery of foreign DNA to the regenerable cells, but the recovery of the transformed cells. Finally, the introduced genes must be correctly expressed in the primary transgenic plants and transmitted stably to their progeny. As cassava is vegetatively propagated, the transgenes can be fixed already at the level of the primary transgenic plants, and stable inheritance is of concern only when the transgenic plants are to be incorporated in breeding programmes.

Cassava transformation and regeneration methods

Of the several methods for delivering foreign DNA into plant cells (Potrykus and Spangenberg, 1995), the most commonly used are *Agrobacterium*-mediated gene transfer and particle bombardment. As stable transformation frequencies are low, the use of different marker genes is necessary to allow the identification and to be susceptible to *Agrobacterium* (Calderon-Urrea, 1988), but the pathogenicity of different strains is highly variable and genotype-dependent (Chavarriaga-Aguirre *et al.*, 1993; Sarria *et al.*, 1993; Li *et al.*, 1996; Puonti-Kaerlas *et al.*, 1997b). Until recently, mainly due to selection problems, cassava was considered recalcitrant to genetic engineering. The first reports on successful regeneration of transgenic cassava plants have been published only in the second half of the 1990s (Li *et al.*, 1996; Raemakers *et al.*, 1996; Schöpke *et al.*, 1996). Table 10.1 shows the current status of cassava transformation.

The most commonly used visual markers are GUS-encoded by the *uidA* gene (Jefferson *et al.*, 1986; Jefferson 1987), the luciferase genes from the firefly *Photinus pyralis* (Ow *et al.*, 1986) and soft coral *Renilla reniformis* (Mayerhofer *et al.*, 1995), and the green fluorescent protein (GFP; Chalfie *et al.*, 1994). Selectable marker genes can render transformed plant cells resistant to an antibiotic, a metabolic analogue, or a herbicide, thus allowing cells containing a transgene to survive and proliferate, while the wild-type cells are either arrested in their growth or killed. The most commonly used selectable marker genes encode resistance to aminoglycoside antibiotics (*npt*II) (Bevan *et al.*, 1983; Fraley *et al.*, 1983; Herrera-Estrella *et al.*, 1983), hygromycin (*hpt*; van den Elzen *et al.*, 1985; Waldron *et al.*, 1985), and phosphinotricin, the active ingredient in many herbicides including Basta (*pat* and *bar*; Murakami *et al.*, 1986; De Block *et al.*, 1987; Thompson *et al.*, 1987; Wohllenben *et al.*, 1988). As public concern regarding the use of antibiotic-resistance genes in transgenic plants has become an important factor, development is moving towards other selection systems. An example is the use of antibiotic-resistance genes whose expression can be inhibited by the introduction of introns into the coding region (Wang *et al.*, 1997). A selectable marker system based on the *ipt* gene from the T-DNA of *Agrobacterium tumefaciens*, either linked to the *Ac* transposable element, to allow its removal from regenerating shoots (Ebinuma *et al.*, 1997), or under the control of an inducible promoter (Kunkel *et al.*, 1999) is another example with potential. New non-antibiotic methods have been developed based on positive selection, which favours the regeneration and growth of transgenic cells while suppressing the growth and proliferation of non-transgenic ones. The use of a glucuronide derivative of the cytokinin benzyladenine (Joersbo and Okkels, 1996), xylose (Haldrup *et al.*, 1998a,b) and mannose (Joersbo *et al.*, 1998), have been shown to improve significantly the frequencies of transgenic plant regeneration in tobacco, potato and sugarbeet when compared to antibiotic selection schemes.

Of the different explants used for regeneration, meristems are the tissue of choice as they represent 'growth centres' of plants. In *Arabidopsis thaliana* and *Medicago truncatula* (Trieu *et al.*, 2000), developing flower meristems can be transformed efficiently using *Agrobacterium*. In cassava, meristems can be induced to form multiple shoots on cytokinin-containing medium. Most of the shoots are derived from pre-existing axillary meristems, but also *de novo* formation of new meristems and shoots occurs (Konan *et al.*, 1994a, 1995, 1997). Transient and stable expression of both GUS and luciferase have been demonstrated in meristems and meristem-derived somatic embryos and multiple shoot clusters after particle bombardment (Puonti-Kaerlas *et al.*, 1997a) or co-cultivation with *Agrobacterium* (Konan *et al.*, 1995; Puonti-Kaerlas *et al.*, 1997a). Transgenic sectors were detected in the developing shoots, but no fully transgenic plants have been regenerated. Somatic embryogenesis is now the most commonly used regeneration method for cassava. Somatic embryogenesis in cassava is restricted to meristematic and embryonic tissues. Somatic embryos can only be induced on a limited number of explants such as cotyledons or embryonic axes from zygotic embryos (Stamp and Henshaw,

Table 10.1. Methods used for the genetic transformation of cassava.

Target tissue	Regeneration mode	Gene transfer system	Selection	Transgenic tissues	Analysis	Reference
Somatic embryos	Somatic embryogenesis	Electroporation	–	Chimeric embryos	Transient GUS expression	Luong et al. (1995)
Somatic cotyledons	Shoot organogenesis	Agrobacterium	Hygromycin geneticin	Transgenic plants	Southern, Northern	Li et al. (1996)
Embryogenic suspension	Somatic embryogenesis	Particle bombardment	Paromomycin	Transgenic plants	Southern	Schöpke et al. (1996)
Embryogenic suspension	Somatic embryogenesis	Particle bombardment	Luciferase	Transgenic plants	Southern	Raemakers et al. (1996) Schöpke et al. (1997a)
Embryogenic suspension	Somatic embryogenesis	Particle bombardment	–	Chimeric suspensions	Transient gene expression	Munyikwa et al. (1998a)
Embryogenic suspension	Somatic embryogenesis	Particle bombardment	Luciferase and phosphinotricin	Transgenic plants	Southern, Northern	González et al. (1998)
Embryogenic suspension	Somatic embryogenesis	Agrobacterium	Paromomycin	Transgenic plants	Southern	Sarria et al. (2000)
Somatic cotyledons	Somatic embryogenesis	Agrobacterium	Basta	Transgenic plants	Southern	Zhang et al. (2000)
Somatic cotyledons	Shoot organogenesis	Particle bombardment	Hygromycin	Transgenic plants	Southern, RT-PCR	Zhang and Puonti-Kaerlas (2002)
Somatic cotyledons, Embryogenic suspension	Shoot organogenesis, Somatic embryogenesis	Particle bombardment	Mannose, hygromycin	Transgenic plants	Southern, Northern, RT-PCR	Zhang et al. (2001)
Embryogenic suspension	Shoot organogenesis, Somatic embryogenesis	Agrobacterium	Mannose, hygromycin	Transgenic plants	Southern, Northern, RT-PCR	

1982; Konan *et al.*, 1994a,b) and immature leaf lobes (Stamp and Henshaw, 1987a; Szabados *et al.*, 1987; Matthews *et al.*, 1993; Raemakers, 1993; Raemakers *et al.*, 1993a; Li *et al.*, 1995, 1996, 1998a; Puonti-Kaerlas *et al.*, 1997a,b). Other tissues include meristems and shoot tips (Szabados *et al.*, 1987; Narayanaswami *et al.*, 1995; Frey, 1996; Puonti-Kaerlas *et al.*, 1998), anthers (Mukherjee, 1995) and immature inflorescences on auxin-containing media (Woodward and Puonti-Kaerlas, 1998). Primary somatic embryos can be induced to produce secondary somatic embryos by further subculturing on auxin-containing medium (Stamp and Henshaw, 1987b). By constant subculturing of somatic embryos, a cyclic embryogenesis system can be established either in liquid or solid medium, where the embryos rarely pass the 'torpedo' stage, until transferred to germination medium.

In cassava, both primary and secondary somatic embryos develop from groups of cells, usually located at or near the vascular tissue (Stamp, 1987; Raemakers *et al.*, 1995b). The multicellular origin of cassava somatic embryos makes them poorly suited for genetic engineering. Moreover, their location under the plant epidermis also limits their accessibility to *Agrobacterium*. Transgenic sectors were detected in somatic embryos after electroporation, but no transgenic plants were regenerated from these embryos (Luong *et al.*, 1995, 1997). Particle bombardment of embryogenic clusters led to high transient expression of visible marker genes in several laboratories, but only sectorial transgenic embryos could be regenerated from bombarded embryos. Transgenic callus can be obtained readily from explants from cycling somatic embryos, but the competence for embryogenesis is lost when the cultures are treated with antibiotics to select for transformed cells (Chavarriaga-Aguirre *et al.*, 1993; Schöpke *et al.* 1993; Puonti-Kaerlas, unpublished). There is only one report on successful regeneration of transgenic cassava plants by secondary somatic embryogenesis, supported by molecular data to show the presence of transgenes in one of the 15 regenerated Basta-resistant plant lines (Sarria *et al.*, 1995, 2000). In this case, secondary somatic embryos were induced on cotyledon explants from primary somatic embryos of the cassava landrace MPeru183, after

co-cultivation with the wild-type *Agrobacterium* strain CIAT1182, carrying the genes encoding GUS and phosphinotricin resistance. Selection on 8–32 mg l^{-1} Basta allowed the regeneration of putative transgenic secondary embryos that could be regenerated to mature plants. MPeru183 and the *Agrobacterium* strain CIAT1182 were selected after a screen for the most efficient combination to ensure high transformation frequency (Sarria *et al.*, 1993). The reproducibility of this method has not been assessed so far, and the CIAT1182 strain is oncogenic, thus its use for production of transgenic plants on routine basis is still limited. Disarming the vector should offer great potential for further improvement of *Agrobacterium*-mediated transformation methods. The use of immature cassava leaves to regenerate transgenic plants after co-cultivation with *Agrobacterium* was reported (Arias-Garzon and Sayre, 1998) but needs verification at the molecular level.

As transformation of somatic embryos can lead to the formation of chimerics (Sarria *et al.*, 1995), a more promising option is the use of friable embryogenic callus (FEC). A fraction of the cycling somatic embryos maintained on a MS medium (Murashige and Skoog, 1962), or even more efficiently, on a GD medium (Gresshoff and Doy, 1974) supplemented with picloram, produces a highly FEC (Taylor *et al.*, 1996). Pure FEC can be transferred easily to liquid culture to establish a rapidly proliferating embryogenic suspension in SH (Schenk and Hildebrandt, 1972) medium supplemented with 6% sucrose and 10–12 mg l^{-1} picloram. Maturing embryos develop on transfer to hormone-free medium. In contrast to the primary or secondary somatic embryos, the new embryogenic units in FECs develop from the surface cells of the globular embryo clusters, and appear to be of single cell origin, which makes them good targets for transformation (Taylor *et al.*, 1996). Particle bombardment of embryogenic suspensions of 'TMS 60444' allowed regeneration of transgenic cassava plants using three different approaches. The first is based on antibiotic selection using paromomycin (Schöpke *et al.*, 1996; González *et al.*, 1998) or hygromycin (Zhang and Puonti-Kaerlas, 2002). The second method employs visual selection using firefly luciferase as a screenable marker gene (Raemakers *et al.*, 1996), or on a combination of antibiotic selection and luciferase screening (Munyikwa *et al.*, 1998a,b) and the third on positive selection (Zhang and Puonti-Kaerlas, 2002). With paromomycin selection the bombarded tissues were first grown for 2 weeks after transformation in liquid medium containing 12 mg l^{-1} picloram, after which, paromomycin was added to the liquid culture at 15 mg l^{-1}. After 4–5 weeks of liquid selection, the developing embryogenic units were transferred to solid culture medium under the same selective conditions and cultured for another 4 weeks. Regeneration of plants was only possible when no selection was applied.

After the transgenic units were multiplied as FECs, shoot regeneration was initiated by sequential transfer and culture on a series of media to induce the differentiation of globular and torpedo-stage embryos (1.2 mg l^{-1} picloram), development of cotyledons (0.93 mg l^{-1} NAA) and maturation of embryos (0.5% activated charcoal, no growth regulators). Before attempting to root the regenerants, a multiplication step to induce multiple shoot formation from the apical meristem of the germinating embryos was applied (1 mg l^{-1} BA). The capacity for regeneration differed greatly between the selected lines, and sometimes no plants regenerated. Southern data were published to prove the presence of the *uidA* gene in one regenerated shoot. When hygromycin was used, selection was started 3 days after bombardment. The cultures were grown in a liquid medium with 12 mg l^{-1} picloram and 50 mg l^{-1} hygromycin for 2–4 weeks after which they were transferred to a solid medium supplemented with 1 mg l^{-1} NAA and 25 mg l^{-1} hygromycin for 2–8 weeks to allow FEC formation and embryo emergence. The hygromycin-resistant embryos developed into shoots on transfer to an elongation medium containing 0.4 mg l^{-1} BA and no hygromycin. GUS assays, PCR, reverse transcribed (RT)-PCR as well as Southern and Northern analyses confirmed the transgenic nature and stable expression of the transferred genes in two regenerated plant lines. This protocol allows transgenic plants to be regenerated in 15 weeks, thus reducing considerably the time required for *in vitro* culture. Selection systems based on paromomycin (Schöpke *et al.*, 1997a; González *et al.*, 1998) and hygromycin have also been adapted to *Agrobacterium*-mediated transformation of embryogenic suspension cultures.

After a 2-day co-cultivation period of the suspension cells with *Agrobacterium* strain ABI containing a plasmid carrying the *nptII* and intron-interrupted *uidA* genes, the cultures were grown for 8–10 days without selection, followed by culture in liquid SH medium containing paromomycin for 5–6 weeks and on solidified selective medium for 4 weeks. They were then transferred to regeneration medium without selection for embryo differentiation (4 weeks), cotyledon development (4 weeks), maturation (2 weeks), shoot development (4 weeks) and rooting (4–8 weeks) using the media described for bombarded suspensions. Southern blot analysis demonstrated stable integration of the *uidA* gene into the cassava genome in two plant lines. After 3–4 days co-cultivation with *Agrobacterium* strain LBA4404 containing a plasmid carrying intron-interrupted *hpt* (Wang *et al.*, 1997) and *uidA* genes, the tissue was cultured for 3 days without selection and then for 15 days in liquid medium containing 12 mg l^{-1} picloram and 50 mg l^{-1} hygromycin. After another 15 days' liquid culture using 25 mg l^{-1} hygromycin, the suspensions were transferred to solid medium with 1 mg l^{-1} NAA and 25 mg l^{-1} hygromycin. Embryos developed on this medium in 4–8 weeks, and shoots were regenerated from the cotyledonary stage embryos on a medium with 0.4 mg l^{-1} BA without hygromycin in 4 weeks. Molecular analyses confirmed the transgenic nature of 12 regenerated lines. Raemakers *et al.* (2000) have further modified the protocols based on antibiotic selection.

When firefly luciferase was used to select for transgenic tissues, the embryogenic suspensions were cultured for 1 day after bombardment on solid medium containing 6% sucrose. They were then transferred either to liquid medium containing 6% sucrose, or, to solid medium with reduced sucrose concentration of 6% to 2% in two 3-day subculture steps. Two weeks after bombardment, the cultures were monitored for luciferase activity, and clusters of embryogenic units at and around the luciferase positive spots, were isolated and cultured further. The luciferase screen and tissue selection was repeated at 2-week intervals, until 2 months after bombardment the friable embryogenic calli clusters containing at least 1% luciferase-positive units were transferred to maturation medium containing 1 mg l^{-1} picloram for development of somatic embryos. The maturing luciferase-positive embryos were further multiplied by secondary somatic embryogenesis on media containing either 10 mg l^{-1} NAA or 8 mg l^{-1} dinitrophenol (2,4-D), and in subsequent steps of cyclic somatic embryogenesis 10 mg l^{-1} NAA. After desiccation, the somatic embryos were induced to grow on a medium containing 1 mg l^{-1} BA, and then multiplied as shoot cultures by nodal cuttings. The efficiency of the secondary embryo formation from selected embryos was 83%, but only 1–15% of the embryos grow into transplantable shoots. Southern hybridization data were presented to support the transgenic nature of three plants.

A method combining antibiotic selection, using phosphinotricin, 20 mg l^{-1}, and luciferase, allowed the development of both transformed and non-transformed maturing embryos from bombarded FECs, but the luciferase screening could be used to exclude escapes (Snepvangers et al., 1997). The inclusion of 20 mg l^{-1} phosphinotricin in the maturation medium could not block the maturation of non-transgenic embryos completely. Later studies demonstrated that strict visual selection using luciferase was more efficient and led to the production of transgenic lines faster than antibiotic selection. Southern analysis confirmed the transgenic nature of 18 plant lines produced by combining antibiotic and visual selection (Munyikwa et al., 1998a). The selection protocol using luciferase has been further refined by Raemakers et al. (2000). A positive selection system was developed for embryogenic suspensions (Zhang and Puonti-Kaerlas, 2002; Zhang et al., 2002). After transformation, the cells were grown for 4 weeks in liquid culture using a modified SH medium containing 12 mg l^{-1} picloram, 4% mannose and 1% sucrose, before transfer to solid medium containing 2% sucrose, 2% mannose and 1 mg l^{-1} NAA for 2 weeks. After another 4 weeks culture on the same medium without mannose, shoots were regenerated in 4 weeks from embryos cultured on a medium containing 2% sucrose and 0.1 mg l^{-1} BA. GUS assays and molecular analyses confirmed the transgenic nature of 14 regenerated plant lines.

The use of the protocols described above depends on the availability of embryogenic suspensions, the establishment of which is still genotype-dependent and labour-intensive. The low regeneration rates and the regeneration of possibly abnormal plants reduce the transformation efficiencies obtained after transformation of embryogenic suspensions (Taylor et al., 1996; Raemakers et al., 1997b, 2000; Snepvangers et al. 1997; Schöpke et al., 1997b). The use of paromomycin selection reduces the regenerative potential of the transgenic material and relatively complicated, time-consuming and labour-intensive regeneration schemes must be followed. This has led to somaclonal variation and reduced growth rates in up to 50% of the regenerants, and to lower survival rate (40% versus 90% in controls) of the plantlets upon transplanting to the greenhouse (Raemakers et al., 1997b; Munyikwa et al., 1998a). Luciferase, though a non-invasive detection method, requires access to costly equipment including a coupled device camera for detection and localization of the bioluminescence. In contrast, combining GUS assays with hygromycin or mannose selection, should allow rapid and easy selection of cultures for regeneration, resulting in 100% selection efficiency (Zhang and Puonti-Kaerlas, 2002) and possibly reducing the length of time required for tissue culture before plant regeneration.

An efficient alternative regeneration system using organogenesis was developed in order to circumvent the problems encountered when regenerating plants by germinating somatic embryos (Li et al., 1995, 1996, 1998a). Shoot primordia were induced directly on cotyledon explants from germinating cycling somatic embryos on a medium containing cytokinins. A cycling system where the secondary somatic embryos were induced on cotyledon explants from maturing somatic embryos was established. To induce shoot organogenesis, cycling somatic embryos were transferred to maturation medium containing 1.0 mg l^{-1} BA and 0.5 mg l^{-1} IBA. After a passage on elongation medium containing 0.4 mg l^{-1} BA, the regenerating shoots were easily rooted on hormone-free medium and transplanted into soil in the greenhouse. Shoot induction frequency of different cassava cultivars was 42–67% and shoot primordia could be induced on cotyledons from cycling embryos maintained either on 2,4-D or picloram. Cotyledon explants derived from cycling somatic embryos showed the highest competence for organogenesis, while those from primary somatic embryos responded very poorly. The organogenesis frequencies have improved further by using silver nitrate in the medium (Puonti-Kaerlas, unpublished). The transferability of this protocol has been demonstrated with ten different Latin American, African and Asian cultivars, and the system is currently being implemented at IITA, Ibadan.

Compared to regeneration via germination of embryos derived from suspensions, shoot regeneration via organogenesis is faster, and requires less time in tissue culture. Using organogenesis, transplantable shoots can be regenerated from cotyledon explants within 60–65 days. In addition, the germination/maturation steps in the protocol ensure the selection for highly regeneration competent embryos, hence minimizing the risk of producing embryogenic cultures that will be arrested in their development. The shoots regenerated via organogenesis develop from cells at or close to the cut edges of the cotyledon explants, which makes them good targets for *Agrobacterium*-mediated gene transfer. In contrast to the protocols based on somatic embryo production, both callus and shoot development were inhibited by similar amounts of antibiotics, and both geneticin, hygromycin and phosphinotricin could be used as selective agents (Li *et al.*, 1996, 1998b; Puonti-Kaerlas *et al.*, 1997a,b). The choice of an appropriate *Agrobacterium* strain and pre-induction of the bacteria with acetosyringone were essential for high transformation frequency. Extending the co-cultivation time to 4 days resulted in the highest transient transformation rates, without excessive bacterial contamination. The developmental state of the explants was also found to be a critical factor in the transformation procedure. Cotyledons from newly germinated embryos were very sensitive to *Agrobacterium*, and survived the co-cultivation procedure poorly, which resulted in low transformation rates. Cotyledon explants from older embryos survived better, but explants from germinating embryos older than 20 days regenerated less efficiently. The highest regeneration and transformation frequencies could be obtained by using cotyledon explants from somatic embryos of 'M Col 22' cultured for 15 days on maturation medium.

Following co-cultivation, callus and small resistant shoot primordia developed on selection medium containing 15 mg l^{-1} hygromycin or 20 mg l^{-1} geneticin from the cotyledon explants co-cultured with LBA4404 (pBin9GusInt) or LBA4404 (pTOK233), respectively. In GUS assays, three of 27 regenerated geneticin-resistant shoot primordia and six of 30 hygromycin-resistant shoots stained blue. After rooting, the putative transgenic shoots were transferred to soil in the greenhouse (Li *et al.*, 1996). Cloned plant material was stained for GUS activity in order to assess the expression of the 35S promoter in different cassava tissues. In contrast to earlier reports based on transient assays after particle bombardment (Arias-Garzón and Sayre, 1993), the 35S promoter was shown to be highly expressed in all cassava tissues, including all parts of the roots. The highest expression levels, as determined by the intensity of the blue colour, were in the youngest tissues, including apical and axillary meristems and root tips (Puonti-Kaerlas *et al.*, 1997a,b). The stable integration of the transgenes into cassava nuclear DNA was demonstrated in five transgenic plants and Northern data to prove the transcriptional activity of the transgenes were presented. The selection system needs further improvement, as many escapes occurred. Southern analysis of 18 lines selected on geneticin verified their transgenic nature (Li *et al.*, 1998b, Puonti-Kaerlas, unpublished). Recently, the organogenesis protocol was adapted to use with the Particle Inflow Gun to allow the widest possible range of cultivars to be transformed (Zhang and Puonti-Kaerlas, 2002; Zhang *et al.*, 2000). The bombardment parameters were partially optimized and the selection procedure was modified. Using 7.5 mg l^{-1} hygromycin for the first 10 days after bombardment, and then 15 mg l^{-1} hygromycin for 2 weeks, followed by culture of the developing shoot primordia on 10 mg l^{-1} hygromycin, increased the stringency of the selection. Molecular analyses confirmed the transgenic nature of nine regenerated plant lines.

The positive selection system using mannose was shown to be compatible also with the organogenesis-based regeneration mode of cassava (Zhang and Puonti-Kaerlas, 2002). Bombarded cotyledon pieces were first cultured for 3–4 weeks on a medium containing 1.0 mg l^{-1} BA, 0.5 mg l^{-1} IBA, 1% mannose and 0.5% sucrose, then the developing shoot primordia were transferred for 3 weeks to elongation medium containing 0.4 mg l⁻¹ BA, 2% sucrose and 1% mannose. The regenerated shoots were maintained on a medium supplemented with 1% mannose. Molecular analyses confirmed the transgenic nature of two plant lines. In order to improve the selection efficiency of both antibiotic and positive selection, rooting assays were developed. The rooting of axillary shoots of non-transgenic cassava was inhibited by 5.5 mg l⁻¹ hygromycin and 8 mg l⁻¹ geneticin (Zhang et al., 2000) and 1% mannose (Zhang and Puonti-Kaerlas, 2002), while the transgenic plants could root normally. The rooting tests allow an efficient secondary screening for elimination of escapes from primary selection.

Prospects for genetic engineering

Disease resistance

The use of agrochemicals to protect plants against most fungal and bacterial diseases is in principle possible, even if not economically viable or environmentally sustainable, but in the case of viral diseases this option does not exist, as suitable 'viricides' are not available for routine use. Resistance strategies are therefore of the highest priority. Cassava common mosaic virus (CsCMV), which causes up to 20% yield losses (CIAT, 1991), is currently the most important viral disease of cassava in Latin America. No vector is known for this virus, and it is spread mainly by mechanical transmission and the use of infected cuttings. The coat protein gene of CcSMV has been transferred to cassava as a potential source of resistance (Schöpke et al., 2000). CMD causes losses of up to 20–80% of total yields throughout the African continent, and can locally result in complete crop failure (Lozano and Booth 1974; Thresh et al., 1994a,b; Otim-Nape, 1995). It has been ranked as the most important vector-borne disease of all African food crops (Geddes, 1990). CMD is caused by a number of viruses transmitted by the whitefly (Bemisia tabaci), and disseminated in vegetative planting material, but not via true seed (Swanson and Harrison, 1994). The most recent pandemic caused by a new recombinant strain has spread in eastern Africa from north to south through Uganda and western Kenya at

15–20 km year⁻¹ (Otim-Nape et al., 1994a,b, 1997a,b; Otim-Nape, 1995; Gibson et al., 1996). In Uganda the use of new cultivars and disease-free planting material have met with some success in managing the disease (Otim-Nape et al., 1997a,b). The heterozygous nature of cassava implies that breeding of new CMD-resistance cultivars using traditional methods can lead to the loss of favoured local landraces and improved lines, and genetic transformation technology may be necessary to transfer the desired traits to preferred varieties. So far CMD has not reached Latin America, but a new biotype of the vector that feeds on cassava has been found in the neotropics (CIAT, 1990; Franca et al., 1996). This now makes CMD a serious threat to cassava production in Latin America, as the germplasm in America is highly susceptible to at least one of the causal viruses referred to as the cassava mosaic geminiviruses (CMGs).

African cassava mosaic virus (ACMV) is one of at least four CMGs having a genome consisting of two covalently closed circular single-stranded DNA (ssDNA) molecules known as DNA A and DNA B (Stanley, 1983). DNA A is responsible for the replication of both DNA components (Townsend et al., 1986; Etessami et al., 1991) and virus proliferation rates (Haley et al., 1992; Hong and Stanley, 1995) and, together with DNA B, for vector transmission and virus spread (von Arnim et al., 1993; Haley et al., 1995; Liu et al., 1997; Briddon et al., 1998). Both genomic components are also necessary for the systemic infection of susceptible host plants (Stanley, 1983). DNA B is involved in cell-to-cell and long-distance virus spread and production of disease symptoms (von Arnim et al., 1993; Haley et al., 1995). The promoter activity of ACMV is regulated by its own gene products (Haley et al., 1992; Hong and Stanley, 1995). Understanding the regulation of ACMV replication and gene expression will allow the development of new resistance strategies using transgene technology. Until recently all studies were conducted with herbaceous model species, but now studies on the simultaneous regulation of both sides of the bi-directional promoter of ACMV using dual luciferase assays, have shown differences between cassava and model plants (Frey et al., 2002). A virus replication assay has been developed for cassava, which should allow more detailed experiments using cassava instead of model systems (Puonti-Kaerlas, unpublished).

There are indications from heterologous species that resistance against ACMV could also be transferred to cassava. Expression of the defective interfering DNA (Stanley *et al.*, 1990; Frischmuth and Stanley, 1991) and the movement protein of *Tomato golden mosaic virus* (von Arnim and Stanley, 1992) have been shown to increase their resistance to the virus in variable degrees. So too has the expression of the AC1 gene of ACMV (Hong and Stanley, 1996), dianthin (Hong *et al.*, 1996, 1997) and a mutated replicase (Sängare *et al.*, 1999) in transgenic model plants. Different constructs carrying the AC1 gene have been used to transform cassava (Schöpke *et al.*, 2000). In addition to these, novel strategies are being designed for engineering sustainable resistance to ACMV (Berrie *et al.*, 1998; Schärer-Hernandez *et al.*, 1998). Preliminary results using a virus-induced cell death system show that it is possible to reduce ACMV replication by 37–99% in transgenic model plants (Frey, 2000), and transgenic cassava plants containing this construct are being developed.

Resistance to insect pests

Due to its long growth period, cassava is subject to prolonged and repeated attacks from numerous insect pests (Bellotti, 1979; Bellotti *et al.*, 1994, 1999; Chapter 11). In Latin America, lepidopteran insects are currently the main cassava pests. These include stem borer (*Chilomina clarkei*) and hornworm (*Erinnyis ello*). Hornworm causes variable yield losses along the northern coast of South America and in the Caribbean. Stem borer causes considerable damage to cassava in Colombia (Bellotti *et al.*, 1999). It not only causes considerable yield loss, but as the larvae live and feed inside cassava stems, they are protected from external applications of insecticides, and a severe attack can lead to reduced quality or even complete loss of planting material. The soil bacterium *Bacillus thuringiensis* (Bt) carries a set of *cry* genes encoding insect-specific endotoxins (Bt toxins), which are efficient in combating a variety of insects. Spraying Bt was shown to be efficient in biological control of cassava hornworm (Bellotti and Arias, 1979). Expression of the *cry* genes in transgenic cassava would complement the available methods for pest control in an environmentally and economically sustainable way. Several groups are currently working on this topic.

Root quality and yield

CYANOGENESIS. Cassava is cyanogenic, i.e. hydrogen cyanide (HCN) is produced in all parts of the plant when the tissues are damaged. To prevent cyanide poisoning, linamarin and lotaustralin have to be removed by labour-intensive processing, and shortcuts in processing can lead to fatal consequences (Akintonwa and Tunwashe, 1992; Mlingi *et al.*, 1992; Akintonwa *et al.*, 1994). All known cassava cultivars contain cyanogenic glucosides, and despite considerable efforts, no acyanogenic cultivar has been found or produced (Jennings, 1976; Bokanga, 1994; Dixon *et al.*, 1994). High cyanide cultivars are favoured in many areas as they are considered tolerant to environmental stress and also relatively safe from theft by humans and mammals (Rosling *et al.*, 1993; Kapinga *et al.*, 1997). The occurrence of severe neurological disorders is closely linked to prolonged exposure to cyanide (Osuntokun and Monekosso, 1969; Tylleskär *et al.*, 1991, 1992, 1995; Tylleskär, 1994). Recent studies also seem to indicate a neurotoxic effect of linamarin itself (Banea-Mayambu *et al.*, 1997). In addition to the health risks of cassava-based food, the effluent from cassava processing plants often contains toxic amounts of cyanide, and consequently can be a serious pollutant if not properly managed (Manilal *et al.*, 1983).

By manipulating the key enzymes in linamarin synthesis, the cytochrome P-450 oxidases, through down-regulation using anti-sense technology, the levels of cyanogenic glucosides in cassava roots could be reduced. Acyanogenic cassava could considerably increase household income by eliminating the fear of buying fresh cassava roots in some regions and reduce the emission of pollutants from starch factories. Of the genes in the biosynthesis pathway, the key enzymes, cytochrome P-450 oxidases (Koch *et al.*, 1994) were recently isolated and characterized (Andersen *et al.*, 2000), and anti-sense constructs to down-regulate the expression of these genes are currently being transformed into cassava (B.L. Møller, personal communication). Alternatively, increased expression of linamarase and α-hydroxynitrilase, the enzymes breaking down linamarin, would offer a way to enhance the rate of HCN release during cassava processing. The residual cyanohydrins in processed cassava are the main source of dietary cyanide (Tylleskär *et al.*, 1992), and thus ways of maintaining a high activity of the α-hydroxynitrilase during cassava processing would be of great interest. To enhance the breakdown of cyanohydrins during cassava root processing, production of transgenic cassava over-expressing the α-hydroxynitrilase gene is in progress (Arias-Garzon and Sayre, 1998).

PROTEIN CONTENT.   One of the disadvantages of cassava is the low protein content of the roots, which can lead to qualitative protein malnutrition in areas where the diet is based predominantly on cassava. Also, protein deficiency has been shown to aggravate symptoms related to cassava toxicity (Rosling, 1988). Synthetic storage proteins designed to improve the protein quality and quantity have been introduced to potato (Yang *et al.*, 1989) and to sweetpotato (Prakash and Egnin, 1997; Demgen, 2000). A novel synthetic storage protein gene (Kim *et al.*, 1992) has been transferred to cassava to improve the nutrient balance of cassava, so as to allow its use as a cheap protein source. The stable expression of the storage protein gene has been verified on transcriptional and translational levels and accumulation of the protein in cassava tissues has been shown in both primary roots and leaves of the regenerated transgenic plants (Puonti-Kaerlas, unpublished). Cassava leaves contain valuable high quality protein and provide a reliable, low cost source of vitamins, minerals and proteins (Balagopalan *et al.*, 1988; Bokanga, 1994; Dahniya, 1994). Excessive leaf harvesting, however, reduces storage root production, and therefore leaves can only be harvested every 2 months in order to minimize losses in root yields (Bokanga, 1994).

Leaf retention

Leaf longevity has been shown to be one of the main traits associated with high yields, together with a leaf area index of 3.0–3.5 (Hunt *et al.*, 1977; Cock, 1979). Furthermore, high drought tolerance and productivity were associated with leaf retention during drought (El-Sharkawy *et al.*, 1992; Osiru *et al.*, 1994). Leaf retention capacity during periodic drought was also positively correlated with root quality. Prolonging the life of individual leaves could help to produce cultivars with improved yield and root quality and to permit more frequent harvesting of leaves while maintaining a satisfactory photosynthetic area to ensure storage root production. As the market value of leaves in the areas where they are consumed is often higher than that of the roots (Lutaladio and Ezumah, 1981), this could also contribute to household economies. Prolongation of photosynthetically active leaf life has already been achieved in transgenic tobacco carrying the *ipt* gene encoding cytokinin production under the control of a senescence-regulated promoter from *Arabidopsis thaliana* (Gan and Amasino, 1995). Transgenic cassava plants containing the same construct have been produced and are currently being analysed (Li *et al.*, 1998b).

Future prospects for genetic engineering

The development of new techniques allowing more efficient improvement of cassava has proceeded rapidly in the past few years. Isolation of promoters and genes as cDNAs and genomic clones from cassava will contribute to the increased knowledge of metabolic pathways, their regulation, and eventually, their genetic engineering, allowing the production of new cassava cultivars resistant to pests and diseases with improved nutritional quality or altered starch composition. Vitamin A is vital to the normal development in humans, and the consequences of vitamin A deficiency range from night blindness through total blindness, to reduced resistance to various terminal diseases (Sommer, 1988; West *et al.*, 1989). According to UNICEF, *c.* 124 million children in the world suffer from vitamin A deficiency (Sommer, 1988; Humphrey *et al.*, 1992). Expressing the corresponding genes from plants producing β-carotene, the precursor for the synthesis of vitamin A, in cassava roots would help to alleviate this problem. Cassava roots have a basic capacity for β-carotene synthesis, as shown by the identification of cassava cultivars with yellow roots containing carotenoids (Moorthy *et al.*, 1990; Adewusi and Bradbury, 1993a,b). The carotenoid content of cassava roots can be as high as 2 mg per 10 g fresh weight (Iglesias *et al.*, 1997), but a large part of this may exist in non-provitamin A like forms, e.g. as luteolin (Adewusi and Bradbury, 1993a,b). Breeding for high carotenoid content was reported to have produced cassava lines with up to sevenfold higher carotene contents in the roots (Nair and Pillai, 1996). Thus the expression of the first enzyme in the pathway, phytoene synthase, might suffice to produce cassava roots with high β-carotene content in lines favoured by farmers, without changing their other characteristics. Should this prove insufficient, it is now also possible to transfer the whole pathway required for β-carotene production, as recently demonstrated with transgenic rice (Burkhardt *et al.*, 1997).

Similarly, iron deficiency anaemia, one of the most serious deficiencies in the developing countries affecting over 1 billion people, could be combated by increasing iron content and improving its availability in cassava roots using transgene technology (Anon., 1999; Gura, 1999; Lucca, 1999).

The development of cassava cultivars with different starch composition would increase the value of cassava as an industrial crop. Amylose-free cassava starch and cassava roots with increased levels of sugars would create special market niches. The expression of ADP glucose pyrophosphorylase B gene in anti-sense in transgenic cassava has already shown to result in plants with little or no transcript and increased soluble sugars in the storage roots (Munyikwa *et al.*, 1998a,b). Means to control physiological postharvest deterioration of cassava roots may be developed once the process is better understood, possibly by manipulating the regulation of the key enzymes involved in the wound response reaction in roots.

Future possibilities of adding value to cassava as an income-generating crop include also the production of biodegradable plastics. As a first step towards production of biodegradable plastics in plants, accumulation of polyhydroxy-alkanoates has been demonstrated in transgenic *Arabidopsis* plants expressing bacterial PHA genes (Poirier *et al.*, 1995), and an *ex ante* study indicated that production of biodegradable plastic in cassava may well be a viable option (Stoeckli, 1998).

Other possible strategies for value-added cassava could include the improvement of iodine content and availability, production of cyclodextrins, improved baking quality of cassava flour, nematode resistance and herbicide resistance.

It is now possible to regenerate transgenic cassava plants, which is the prerequisite for genetic improvement of cassava, but so far there is little information on the transferability of the current protocols to other cultivars beyond the model cultivars, or to other laboratories. The genetic improvement of cassava via biotechnology is constrained by the lack of routine, inexpensive, efficient and genotype-independent transformation methods. Despite the recent breakthroughs, the absence of efficient and reliable technology will be one of the greatest constraints in applying genetic engineering to cassava improvement. Different transformation systems may be needed for different cultivars, and thus both direct gene transfer methods, as well as those based on *Agrobacterium*, should be developed further. The high proliferation rate of embryogenic suspensions will make the multiplication of transgenic tissues efficient. In contrast, organogenesis may be a less genotype-dependent regeneration mode for cassava than germination of somatic embryos, and may also allow more flexibility in the choice of selectable marker genes.

The main limitation to the rapid development of new techniques for cassava remains the lack of funding for research. It is to be hoped that the progress achieved with limited resources will help to increase the interest in this important crop. Cassava biotechnology has potential to alleviate poverty, increase food security and promote the efficient use and conservation of genetic resources, provided that its results are available to cassava-growing countries. Access to this technology should be equitable, and it should not create new dependencies, nor lead to raw material substitution in industrial countries. As cassava remains irreplaceable in marginal environments, the people living in these areas should be the first to benefit from the new technology.

References

Adewusi, S.R.A. and Bradbury, J.H. (1993a) Carotenoids in cassava: comparison of open-column and HPLC methods of analysis. *Journal of Science of Food and Agriculture* 62, 375–383.

Adewusi, S.R.A. and Bradbury, J.H. (1993b) Carotenoid profile and tannin content of some cassava cultivars. In: Roca, W.M. and Thro, A.M. (eds) *Proceedings of the First International Scientific Meeting of the Cassava Biotechnology Network.* Cartagena de Indias, Colombia, CIAT Working Document 123, pp. 270–276.

Akano, A., Barrera, E., Dixon, A.G.O. Mba C. and Fregene, M. (2002) Molecular genetic mapping of resistance to the African cassava mosaic disease. *Theoretical and Applied Genetics* (in press).

Akintonwa, A. and Tunwashe, O.L. (1992) Fatal cyanide poisoning from cassava based meal. *Human Experimental Toxicology* 11, 4–49.

Akintonwa, A., Tunwashe, O. and Onifade, A. (1994) Fatal and non-fatal acute poisoning attributed to cassava-based meal. *Acta Horticulturae* 375, 285–288.

Andersen, M.D., Busk, P.K., Svendsen, I. and Møller, B.L. (2000) Cytochromes P-450 from bcassava (*Manihot esculenta* Crantz) catalyzing the first steps in the biosynthesis of the cyanogenic glucosides linamarin and lotaustralin. Cloning, functional expression in *Pichia pastoris*, and substrate specificity of the isolated recombinant enzymes. *Journal of Biological Chemistry* 275, 1966–1975.

Anon. (1978) Pest control in tropical root crops Pans manual. Center for Overseas Pests Research, Ministry of Overseas Development UK.

Arias-Garzon, D.I. and Sayre, R.T. (1998) Genetic engineering approaches to reducing the cyanide toxicity in cassava (*Manihot esculenta* Crantz). In: Carvalho, L.J.C.B., Thro, A.-M. and Vilarinhos, A.D. (eds) *Cassava Biotechnology.* IV Int. Scientific Meeting CBN, pp. 231–221.

Bachem, C.W.B., Van der Hoeven, R.S., de Brujin, S.M., Vreugdenhill, D., Zabeau, M. and Visser, R.G.F. (1996) Visualization of differential gene expression using a novel method of RNA fingerprinting based on AFLP: analysis of gene expression during potato tuber development. *The Plant Journal* 9, 745–753.

Balagopalan, C., Padmaja, G., Nanda, S.K. and Moorthy, S.N. (1988) *Cassava Food Feed and Industry.* CRC Press, Boca Raton, USA.

Baldwin, B.G. (1992) Phylogenetic utility of the internal transcribed spacers of nuclear ribosomal DNA in plants: an example from the compositae. *Molecular Phylogenetic Evolution.*

Banea-Mayambu, J.P., Tylleskär, T., Gitebo, N., Matadi, N., Gebre-Medhin, M. and Rosling, H. (1997) Geographical and seasonal association between linamarin and cyanide exposure from cassava and the upper motor neurone disease konzo in former Zaire. *Tropical Medicine and International Health* 2, 1143–1151.

Beeching, J.R., Marmey, P., Gavalda, M.C., Noirot, M., Haysom, H.R., Hughes, M.A. and Charrier, A. (1993) An assessment of genetic diversity within a collection of cassava (*Manihot esculenta* Crantz) germplasm using molecular markers. *Annals of Botany* 72, 515–520.

Bellotti, A.C. (1979) An overview on cassava entomology. In: Brekelbaum, T., Bellotti, A. and Lozano, J. (eds) *Proceedings of Cassava Protection Workshop.* CIAT, Cali, Colombia, 1977, pp. 17–28.

Bellotti, A.C. and Arias, B. (1979) Biology, ecology and biological control of cassava hornworm *Erinnyis ello.* In: Brekelbaum, T., Bellotti, A. and Lozano, J. (eds) *Proceedings of Cassava Protection Workshop,* CIAT, Cali, Colombia, 1977, pp 227–232.

Bellotti, A.C., Smith, L. and Lapointe, S.L. (1999) Recent advances in cassava pest management. *Annual Review of Entomology* 44, 343–370.

Berrie, L.C., Rybicki, E.P. and Rey, M.E.C. (1998) Antisense RNA-mediated virus resistance in tobacco plants transformed with the antisense Rep gene of South African cassava mosaic virus. In: Pires de Matos, A. and Vilarinhos, A. (eds) *Proceedings IV International Scientific Meeting of the Cassava Biotechnology Network.* Rev. Brasil. Mandioca 17, p. 37.

Bertram, R.B. (1993) Application of molecular techniques resources of cassava (*Manihot esculenta* Crantz, Euphorbiaceae): interspecific evolutionary relationships and intraspecific characterization. PhD dissertation, University of Maryland.

Bevan, M.W., Flavell, R.B. and Chilton, M.D. (1983) A chimaeric antibiotic resistance gene as a selectable marker for plant cell transformation. *Nature* 304, 184–187.

Bokanga, M. (1994) Processing of cassava leaves for human consumption. *Acta Horticulturae* 375, 203–207.

Bothwell, A.L., Yancopoulis, G.D. and Alt, F.W. (1990) *Methods for Cloning and Analysis of Eukaryotic Genes.* Jones and Bartlett, Boston, Massachusetts, pp. 8–19.

Botstein, D., White, R.L., Skolnick, M.H. and Davis, R.W. (1980) Construction of a genetic map in man using restriction fragment length polymorphisms. *American Journal of Human Genetics* 32, 314–331.

Boventius, H. and Weller, J.I. (1994) Mapping and analysis of dairy cattle quantitative trait loci by maximum likelihood methodology using milk protein genes as genetic markers. *Genetics* 137, 267–280.

Briddon, R.W., Liu, S., Pinner, M.S. and Markham, P.G. (1998) Infectivity of African cassava mosaic virus clones to cassava by biolistic inoculation. *Archives of Virology* 143, 2487–2492.

Burkhardt, P.K., Beyer, P. and Wünn (1997) Transgenic rice (*Oryza sativa*) endosperm expressing daffodil (*Narcissus pseudonarcissus*) phytoene synthase accumulates phytoene, a key intermediate of provitamin A biosynthesis. *Plant Journal* 11, 1071–1078.

Calderon-Urrea, A. (1988) Transformation of *Manihot esculenta* (cassava) using *Agrobacterium tumefaciens* and expression of the introduced foreign genes in transformed cell lines. MSc thesis, Vrije University, Brussels, Belgium.

Causse, M.A., Fulton, T.M., Cho, Y.G., Ahn, S.N., Chunwongse, J., Wu, K., Xiao, J., Yu, Z., Ronald, P.C., Harrington, S.E., Second, G., McCouch, S.R. and Tanksley, S.D. (1994) Saturated molecular map of the rice genome based on an interspecific backcross population. *Genetics* 138, 1251–1274.

Chalfie, M., Tu, Y., Euskirchen, G., Ward, W.W. and Prasher, D.C. (1994) Green fluorescent protein as a marker for gene expression. *Science* 263, 802–805.

Chavarriaga-Aguirre, P., Schöpke, C., Sangare, A., Fauquet, C.M. and Beachy, R.M. (1993) Transformation of cassava (*Manihot esculenta* Crantz) embryogenic tissues using *Agrobacterium tumefaciens*. In: Roca, W.M. and Thro, A.M. (eds) *Proceedings of the First International Scientific Meeting of the Cassava Biotechnology Network*. Cartagena de Indias, Colombia, CIAT Working Document 123, pp. 222–228.

Chavarriaga-Aguirre, P., Maya, M.M., Bonierbale, M.W., Kresovich, S., Fregene, M.A., Tohme, J. and Kochert, G. (1998) Microsatellites in cassava (*Manihot esculenta* Crantz): discovery, inheritance and variability. *Theoretical and Applied Genetics* 97, 493–501.

Chavarriaga-Aguirre, P., Maya, M.M., Tohme, J., Duque, M.C., Iglesias, C., Bonierbale, M.W., Kresovich, S. and Kochert, G. (1999) Using microsatellites, isozymes and AFLPs to evaluate genetic diversity and redundancy in the cassava core collection and to assess the usefulness of of DNA-based markers to maintain germplasm collections. *Molecular Breeding* 5, 263–273.

CIAT (1990) *Annual Report, Cassava Program 1989*. Centro Internacional de Agricultura tropical (CIAT), Cali, Colombia.

CIAT (1991) *Annual Report, Cassava Program 1987–1991*. Centro Internacional de Agricultura tropical (CIAT), Cali, Colombia.

Cock, J.H. (1979) Physiological basis of yield losses in cassava due to pests. In: Brekelbaum, T., Bellotti, A. and Lozano, J. (eds) *Proceedings of the cassava*

Protection Workshop. CIAT, Cali, Colombia, pp. 9–16.

Cock, J.H. (1985) *Cassava: New Potential for a Neglected Crop*. Westview, London.

Dahniya, M.T. (1994) An overview of cassava in Africa. *African Crop Science Journal* 2, 337–343.

De Block, M., Botterman, J., Vandewiele, M., Dockx, J., Thoen, C., Gossele, V., Movva, N.R., Thompson, C., van Montagu, M. and Leemans, M.J. (1987) Engineering herbicide resistance in plants by expression of a detoxifying enzyme. *EMBO Journal* 6, 2513–2518.

Demgen (2000) www.novatero.org/rdresult.htm

Devos, K.M., Millan, T. and Gale, M.D. (1993) Comparative RFLP maps of the homeologous chromosomes of wheat, rye, and barley. *Theoretical and Applied Genetics* 85, 784–792

Dixon, A.G.O., Asiedu, R. and Bokanga, M. (1994) Breeding of cassava for low cyanogenic potential: problems, progress and perspectives. *Acta Horticulturae* 375, 153–161.

Ebinuma, H., Sugita, K., Matsunaga, E. and Yamakado, Y. (1997) Selection of marker free transgenic plants using the isopentenyl transferase gene. *Proceedings of National Academy of Science USA* 94, 2117–2121

Elias, M., Panaud, O. and Robert (2000) Assessment of genetic variability in a traditional cassava (*Manihot esculenta* Crantz) farming system, using AFLP markers. *Heredity* 85, 219–230.

El-Sharkawy, M., Cock, J., Lynam, J., Hernandez, A. and Cadavid, L. (1990) Relationship between biomass, root yield and single-leaf photosynthesis in field grown cassava. *Field Crops Research* 25, 183–201.

El-Sharkawy, M.A., Hernandez, D. and Hershey, C. (1992) Yield stability of cassava during prolonged mid-season water stress. *Experimental Agriculture* 28, 165–174.

Etessami, P., Saunders, K., Watts, J. and Stanley, J. (1991) Mutational analysis of complementary-sense genes of African cassava mosaic virus DNA A. *Journal of General Virology* 72, 1005–1012.

Fodor, S.P., Rava, R.P., Huang, X.C., Pease, A.C., Holmes, C.P. and Adams, C.L. (1993) Multiplexed biochemical assays with biological chips. *Nature* 364, 555–556.

Fraley, R.T., Rogers, S.G., Horsch, R.B., Sanders, P.R., Flick, S., Adams, S.P., Bittner, M. L., Brand, L.A., Fink, C.L., Fry, J.S., Galluppi, G.R., Goldberg, S.B., Hoffman, N.L. and Woo, S.C. (1983) Expression of bacterial genes in plant cells. *Proceedings of the National Academy of Sciences USA* 80, 4803–4807.

Franca, F.H., Villas-Boos, G.L. and Branco, M.C. (1996) Occurrence of *Bemisia argentifolii* Below & Perring (Homoptera: Aleyrodidae) in the Federal District.

Annals of the Entomological Society of Brasil 25, 369–372.

Fregene, M.A., Vargas, J., Ikea, J., Angel, F., Tohme, J., Asiedu, R.A., Akorada, M.O. and Roca, W.M. (1994) Variability of chloroplast DNA and nuclear ribosomal DNA in cassava (*Manihot esculenta* Crantz) and its wild relatives. *Theoretical and Applied Genetics* 89, 719–727.

Fregene, M., Angel, F., Gomez, R., Rodriguez, F., Chavarriaga, P., Roca, W., Tohme, J. and Bonierbale, M. (1997) A molecular genetic map of cassava (*Manihot esculenta* Crantz). *Theoretical and Applied Genetics* 95, 431–441.

Fregene, M., Bernal, A., Duque, M., Dixon, A. and Tohme, J. (2000) AFLP analysis of African cassava (*Manihot esculenta* Crantz) germplasm resistant to the cassava mosaic disease (CMD). *Theoretical and Applied Genetics* 100, 678–685.

Fregene, M., Okogbenin, E., Mba, C., Angel, F., Suarez, M.C., Guitierez, J., Chavarriaga, P., Roca, W., Bonierbale, M. and Tohme, J. (2001) Genome mapping in cassava improvement: challenges, achievements and opportunities. *Euphytica* 120, 159–165.

Fresco, L.O. (1993) The dynamics of cassava in Africa: an outline of research issues. COSCA Working Paper No. 9. *Collaborative Study of Cassava in Africa*. International Institute of Tropical Agriculture, Ibadan, Nigeria.

Frey, P. (1996) Towards regeneration and transformation of cassava meristems. MSc thesis, Swiss Federal Institute of Technology, Zürich, Switzerland.

Frey, P. (2000) A study of ACMV gene expression for the development of virus resistance. PhD thesis. Swiss Federal Institute of Technology Zurich, Switzerland.

Frey, P.M., Schärer-Hernandez, N.G., Fütterer, J., Potrykus, I. and Puonti-Kaerlas, J. (2002) Simultaneous analysis of the bidirectional African cassava mosaic virus promoter activity using two different luciferase genes. *Virus Genes* (in press).

Frischmuth, T. and Stanley, J. (1991) African cassava mosaic virus DI DNA interferes with the replication of both genomic components. *Virology* 183, 539–544.

Gan, S. and Amasino, R.M. (1995) Inhibition of leaf senescence by autoregulated production of cytokinin. *Science* 270, 1986–1988.

Geddes, A.M.V. (1990) The relative importance of crop pests in sub-Saharan Africa. *NRI Bulletin No. 6*. Natural Resources Institute, Chatham, UK.

Gibson, R.W., Legg, J.P. and Otim-Nape, G.W. (1996) Unusually severe symptoms are a characteristic of the current epidemic of mosaic virus disease of cassava in Uganda. *Annals of Applied Biology* 128, 479–490.

Giovannoni, J.J., Wing, R.A., Ganal, M.W. and Tanksley, S.D. (1991) Isolation of molecular markers from specific chromosomal intervals using DNA pools from existing mapping populations. *Nucleic Acids Research* 11, 6553–6558.

González, A.E., Schöpke, C., Taylor, N.J., Beachy, R.N. and Fauquet, C.M. (1998) Regeneration of transgenic cassava plants (*Manihot esculenta* Crantz) through *Agrobacterium*-mediated transformation of embryogenic suspension cultures. *Plant Cell Reports* 17, 827–831.

Graner, E.A. (1942) Genetica de Manihot. I. Heriteriadade da formafolha e da coloracao da pelicula externa das raices en Manihotutilissima Pohl. *Bragantia* 2, 13–22.

Gresshoff, P. and Doy, C. (1974) Derivation of a haploid cell line from *Vitis vinifera* and the importance of the stage of meiotic development of the anthers for haploid culture of this and other genera. *Zeitschrift für Pflanzenphysiol* 73, 132–141.

Haldrup, A., Petersen, S.G. and Okkels, F.T. (1998a) Positive selection: a plant selection principle based on xylose isomerase, an enzyme used in the food industry. *Plant Cell Reports* 18, 76–81.

Haldrup, A., Petersen, S.G. and Okkels, F.T. (1998b) The xylose isomerase gene from *Thermoanaerobacterium thermosulfurogenes* allows effective selection of transgenic plant cells using D-xylose as the selection agent. *Plant Molecular Biology* 37, 287–296.

Haley, A., Zhan, X.G., Richardson, K., Head, K. and Morris, B. (1992) Regulation of the activities of African cassava mosaic virus promoters by the AC1, AC2, and AC3 gene products. *Virology* 188, 905–909.

Haley, A., Richardson, K., Zhan, X. and Morris, B. (1995) Mutagenesis of the BC1 and BV1 genes of African cassava mosaic virus identifies conserved amino acids that are essential for spread. *Journal of General Virology* 76, 1291–1298.

Helentjaris, T., Weber, D. and Wright, S. (1988) Identification of the genome locations of duplicate nucleotide sequences in maize by analysis of restriction fragments length polymorphisms. *Genetics* 118, 353–363.

Herrera-Estrella, L., De Block, M., Messens, E., Hernalsteens, J.P., Van Montagu, M. and Schell, J. (1983) Chimaeric genes as dominant selectable markers in plants. *EMBO Journal* 2, 987–995.

Hershey, C.H. and Ocampo, C. (1989) New marker genes found in cassava. *Cassava Newsletter* 13(1), 1–5.

Hong, Y. and Stanley, J. (1995) Regulation of African cassava mosaic virus complementary-sense gene expression by N-terminal sequences of the replication-associated protein AC1. *Journal of General Virology* 76, 2415–2422.

Hong, Y. and Stanley, J. (1996) Virus resistance in *Nicotiana benthamiana* conferred by African cassava mosaic virus replication-associated protein (AC1) transgene. *Molecular Plant Microbe Interactions* 9, 219–225.

Hong, Y., Saunders, K., Hartley, M.R. and Stanley, J. (1996) Resistance to gemini virus infection by virus-induced expression of dianthin in transgenic plants. *Virology* 220, 119–127.

Hong, Y., Saunders, K. and Stanley, J. (1997) Transaction of dianthin transgene expression by African cassava mosaic virus AC2. *Virology* 228, 383–387.

Hughes, M.A., Brown, K., Pancoro, A., Murray, B.S., Oxtoby, E. and Hughes, J. (1992) A molecular and biochemical analysis of the structure of the cyanogenic α-glucosidase (linamarase) from cassava (*Manihot esculenta* Crantz). *Archives of Biochemistry and Biophysics* 295, 273–279.

Humphrey, J.H., West, K.P. and Sommer, A. (1992) Vitamin A deficiency and attributable mortality among under-5-year-olds. *Bulletin of the World Health Organisation* 70, 225–232.

Hunt, A., Wholey, D.W. and Cook, J.H. (1977) Growth physiology of cassava (*Manihot esculenta* Crantz). *Field Crops Abstracts* 30, 77–91.

Hussain, A., Bushuk, W., Ramirez, H., and Roca, W.M. (1987) Identification of cassava (*Manihot esculenta* Crantz) cultivars by electrophoretic patterns of esterase isozyme. *Seed Science Technology* 15, 19–22.

Iglesias, C., Mayer, J., Chavez, L. and Calle, F. (1997) Genetic potential and stability of carotene content in cassava roots. *Euphytica* 94, 367–373.

Jefferson, R.A. (1987) Assaying chimeric genes in plants: the GUS gene fusion system. *Plant Molecular Biology Reports* 5, 387–405.

Jefferson, R.A., Burgess, S.M. and Hirsh, D. (1986) β-Glucuronidase from *Escherichia coli* as a gene-fusion marker. *Proceedings of the National Academy of Science USA* 83, 8447–8451.

Jeffreys, A., Wilson, J.V. and Thein, L. (1985) Hypervariable 'minisatellite' regions in human DNA. *Nature* 314, 67–73.

Jennings, D.L. (1976) Breeding for resistance to African cassava mosaic. African cassava mosaic, report of an interdisciplinary Workshop held at Muguga, Kenya. IDRC 071e, pp. 39–44.

Joersbo, M. and Okkels, F.T. (1996) A novel principle for selection of transgenic plant cells: positive selection. *Plant Cell Reports* 16, 219–221.

Joersbo, M., Donaldson, I., Kreiberg, K., Guldager Petersen, S., Brunstedt, J. and Okkels, F.T. (1998) Analysis of mannose selection used for transformation of sugar beet. *Molecular Breeding* 4, 111–117.

Jorge, M.A., Fregene, M.C., Duque, M., Bonierbale, W., Tohme, J. and Verdier, V. (2000) Genetic mapping of resistance to bacterial blight disease in cassava (*Manihot esculenta* Crantz). *Theoretical and Applied Genetics* 101, 865–872.

Kapinga, R., Mlingi, N. and Rosling, H. (1997) Reasons for use of bitter cassava in Southern Tanzania. *African Journal of Root Tuber Crops* 1–2, 81–84.

Kawano, K., Narintaraporn, K., Narintaraporn P., Sarakarn, S., Limsila, A., Limsila J., Suparhan, D., Sarawat, V. and Watananonta, W. (1998) Yield improvement in a multistage breeding programme for cassava. *Crop Science* 38, 325–332.

Kim, J.H., Cetiner, S. and Jaynes, J.M. (1992) Enhancing the nutritional quality of crop plants: design, construction and expression of an artificial plant storage protein gene. In: Bhatnagar, D. and Cleveland, T.E. (eds) *Molecular Approaches to Improving Food Quality and Safety*. An Avi Book, Van Nostrand Reinhold, New York, pp. 1–36.

Koch, B.M., Sibbesen, O., Swain, E., Kahn, R.A., Liangcheng, D., Bak, S., Halkier, B.A. and Møller, B.L. (1994) Possible use of a biotechnological approach to optimize and regulate the content and distribution of cyanogenic glucosides in cassava to increase food safety. *Acta Horticulturae* 343, 45–60.

Konan, N.K., Sangwan, R.S. and Sangwan-Norreel, B.S. (1994a) Efficient *in vitro* shoot-regeneration systems in cassava (*Manihot esculenta* Crantz). *Plant Breeding* 113, 227–236.

Konan, N.K., Sangwan, R.S. and Sangwan, B.S. (1994b) Somatic embryogenesis from cultured mature cotyledons of cassava (*Manihot esculenta* Crantz): identification of parameters influencing the frequency of embryogenesis. *Plant Cell Tissue Organ Culture* 37, 91–102.

Konan, N.K., Sangwan, R.S. and Sangwan-Norreel, B.S. (1995) Nodal axillary meristems as target tissue for shoot production and genetic transformation in cassava *Manihot esculenta* Crantz. In: *Cassava Biotechnology Network* (ed.) *Proceedings of the Second International Scientific Meeting*, Bogor, Indonesia, CIAT Working Document 150, pp. 276–288.

Konan, N.K., Schöpke, C., Carcamo, R., Beachy, R.N. and Fauquet, C. (1997) An efficient mass propagation system for cassava (*Manihot esculenta* Crantz) based on nodal explants and axillary bud-derived meristems. *Plant Cell Reports* 16, 444–449.

Kunkel, T., Niu, Q.-W., Chan, Y.-S. and Chua, N.-H. (1999) Inducible isopentenyl transferase as a high-efficiency marker for plant transformation. *Nature Biotechnology* 17, 916–919.

Lee, H.S, Kim, K.Y. You, S.H., Kwon, S.Y. and Kwak, S.S. (1999) Molecular characterization and expression of a cDNA encoding copper/zinc superoxide dismutase from cultured cells of cassava (*Manihot esculenta* Crantz). *Molecular and General Genetics* 262, 807–814.

Lee, M. (1995) DNA markers and plant breeding programs. *Advances in Agronomy* 55, 265–344.

Lefevre, F. and Charrier, A. (1993a) Heredity of seventeen isozyme loci in cassava (*Manihot esculenta* Crantz). *Euphytica* 66, 171–178.

Lefevre, F. and Charrier, A. (1993b) Isozyme diversity within African Manihot germplasm. *Euphytica* 66, 73–80.

Lespinasse, A. (2000) A saturated molecular genetic map of rubber (*Hevea brasilensis*). *Theoretical and Applied Genetics* 100, 139–146.

Li, H., Han, Y. and Beeching, J. (2002a) Isolation and characterization of an ACC oxidase gene in cassava. *Euphytica* (in press).

Li, H., Han, Y. and Beeching, J. (2002b) Phenyl alanine ammonia-lyase gene organization, structure and activity in cassava. *Euphytica* (in press).

Li, H.Q., Huang, Y.W., Liang, C.Y. and Guo, Y. (1995) Improvement of plant regeneration from cyclic somatic embryos in cassava. In: Cassava Biotechnology Network (ed.) *Proceedings of Second International Scientific Meeting*, Bogor, Indonesia. CIAT Working Document 150, pp. 289–302.

Li, H.-Q., Sautter, C., Potrykus, I. and Puonti-Kaerlas, J. (1996) Genetic transformation of cassava (*Manihot esculenta* Crantz). *Nature Biotechnology* 14, 736–740.

Li, H.Q., Guo, J.Y., Huang, Y.W., Liang, C.Y., Liu, H.X., Potrykus, I. and Puonti-Kaerlas, J. (1998a) Regeneration of cassava plants via shoot organogenesis. *Plant Cell Reports* 17, 410–414.

Li, H.-Q., Potrykus, I. and Puonti-Kaerlas, J. (1998b) Engineering leaf life length in cassava. In: Pires de Matos, A. and Vilarinhos, A. (eds) *Proceedings of IV International Scientific Meeting of the Cassava Biotechnology Network*. *Revista Brasileira de Mandioca* 17, 31.

Litt, M. and Lutty, J.A. (1989) A hypervariable microsatellite revealed by *in vitro* amplification of dinucleotide repeat within the cardiac muscle actin gene. *American Journal of Human Genetics* 44, 397–401.

Liu, S.J., Bedford, I.D., Briddon, R.W. and Markham, P.G. (1997) Efficient whitefly transmission of African cassava mosaic geminivirus requires sequences from both genomic components. *Journal of General Virology* 7, 1791–1794.

Loomis, R. and Gerakis, P. (1975) Productivity of agricultural ecosystems. In: Cooper, J. (ed.) *Photosynthesis and Productivity in Different Environments*. Cambridge University Press, New York, pp. 145–172.

Lozano, J.C. and Booth, R.H. (1974) Diseases of cassava. *PANS* 20, 30–54.

Lucca, P. (1999) Development of iron-rich rice and ways to improve its bioavailability by genetic engineering. PhD thesis, Swiss Federal Institute of Technology, Zürich, Switzerland.

Luong, H.T., Shewry, P.R. and Lazzeri, P.A. (1995) Transient gene expression in cassava somatic embryos by tissue electroporation. *Plant Science* 107, 105–115.

Luong, H.T., Shewry, P.R. and Lazzeri, P.A. (1997) Transformation and gene expression in cassava via tissue electroporation and particle bombardment. *African Journal of Root Tuber Crops* 2, 163–167.

Lutaladio, N.B. and Ezumah, H.C. (1981) Cassava leaf harvesting in Zaire. In: Terry, E., Oduro, K. and Caveness, F. (eds) *Tropical Root Crops: Research Strategies for the 1980s*. IITA, Ibadan Nigeria, pp. 134–136.

Magoon, M.L., Krishnan, R. and Bai, K.B. (1969) Morphology of the pachytene chromosomes and meiosis in *Manihot esculenta* Crantz. *Cytology* 34, 612–626.

Manilal, V.B., Balagopalan, C. and Narayanan, C. (1983) Physico-chemical and microbiological characteristics of cassava starch factory effluents. *Journal of Root Crops* 8, 27–41.

Martin, G.B., Brommonschenkel, S.H., Chunwongse, J., Frary, A., Ganal, M.W., Spivey, R., Wu, T., Earle, E.D. and Tanksley, S.D. (1993) Map-based cloning of a protein kinase gene conferring disease resistance in tomato. *Science* 262, 1432–1436.

Matthews, H., Schpke, C., Carcamo, R., Chavarriaga, P., Fauquet, C. and Beachy, R.N. (1993) Improvement of somatic embryogenesis and plant recovery in cassava. *Plant Cell Repair* 12, 328–333.

Mayerhofer, R., Langridge, W.H.R., Cormier, M.J. and Szalay, A.A. (1995) Expression of recombinant *Renilla* luciferase in transgenic plants results in high levels of light emission. *Plant Journal* 7, 1031–1038.

Mba, R.E.C., Stephenson, P., Edwards, K., Melzer, S., Mkumbira, J., Gullberg, U., Apel, K., Gale, M., Tohme, J. and Fregene, M. (2000) Simple sequence repeat (SSR) markers survey of the cassava (*Manihot esculenta* Crantz) genome: towards a SSR-based molecular genetic map of cassava. *Theoretical and Applied Genetics* 102, 21–31.

Mkumbira, J., Lagercrantz, U., Mahungu, N.M., Chiwona Karltun, L., Saka, J., Mhone, A., Bokanga, M., Brimer, L., Gullberg, U. and Rosling, H. (2002) *Euphytica* (in press).

Mlingi, N., Poulter, N.H. and Rosling, H. (1992) An outbreak of acute intoxications from consumption of insufficiently processed cassava in Tanzania. *Nutrition Research* 12, 677–687.

Mohan, M., Nair, S., Bhagwat, A., Krishna, T., Yano, M., Bhatia, C. and Sasaki, T. (1997) Genome mapping, molecular markers and marker-assisted selection in crop plants. *Molecular Breeding* 3, 87–103.

Moorthy, S.N., Jos, J.S., Nair, R.B. and Sreekumari, M.T. (1990) Variability of beta-carotene content in cassava germplasm. *Food Chemistry* 36, 233–236.

Mukherjee, A. (1995) Embryogenesis and regeneration from cassava calli of anther and leaf. In: Cassava Biotechnology Network (ed.) *Proceedings Second International Scientific Meeting*, Bogor, Indonesia, CIAT Working Document 150, pp. 375–381.

Mullis, K.B. (1990) The unusual origin of the polymerase chain reaction. *Scientific American* 262(4), 56–61.

Munyikwa, T.R.I., Langeveld, S., Salehuzzaman, S.N.I.M., Jacobsen, E. and Visser, R.G.F. (1997) Cassava starch biosynthesis: new avenues for modifying starch quantity and quality. *Euphytica* 96, 65–75.

Munyikwa, T.R.I., Raemakers, K.C.J.M., Schreuder, M., Kok, R., Schippers, M., Jacobsen, E. and Visser, R.G.F. (1998a) Pinpointing towards improved transformation and regeneration of cassava (*Manihot esculenta* Crantz). *Plant Science* 135, 87–101.

Munyikwa, T.R.I., Raemakers, C.C.J.M., Schreuder, M., Kreuze, J., Suurs, L., Kok, R., Rozeboom, M., Jacobsen, E. and Visser, R.G.F. (1998b) Introduction and expression of antisense ADPG-pyrophosphorylase of cassava leads to decreased levels of starch and increased levels of sugars. In: Pires de Matos, A. and Vilarinhos, V. (eds) *Proceedings of the IV International Scientific Meeting of the Cassava Biotechnology Network. Revista Brasileira de Mandioca* 17, 62.

Murakami, T., Anzai, H., Imai, S., Satoh, A., Nagaoka, K. and Thompson, C. (1986) The bialaphos biosynthetic genes of *Streptomyces hygroscopicus*: molecular cloning and characterization of the gene cluster. *Molecular and General Genetics* 205, 42–50.

Murashige, T. and Skoog, F. (1962) A revised medium for rapid growth and bioassays with tobacco tissue cultures. *Physiological Plantarum* 15, 473–497.

Nair, S.G. and Pillai, S.V. (1996) Production of carotene rich cassava through gene pool development. In: *Proceedings International Meeting on Tropical Tuber Crops* (IMOTUC) – *Tuber Crops in Food Security and Nutrition*. Indian Council of Agricultural Research, Trivandrum, Kerala, India, p. 19.

Narayanaswami, T.C., Ramanswami, N.M. and Rangaswami, S.R. (1995) Somatic embryogenesis and plant regeneration in cassava. In: Thro, A. and Roca, W. (eds) *Proceedings of Second International Cassava Biotechnology Network Meeting*. Centro Internacional de Agricultural Tropical (CIAT), Cali, Colombia, pp. 324–335.

Nelson, J.C. (1997) QGENE: software for marker based genome analysis and breeding. *Molecular Breeding* 3, 229–235.

Ocampo, C., Hershey, C., Iglesias, C. and Iwanaga, M. (1992) Esterase isozyme fingerprinting of the cassava germplasm collection held at CIAT. In: Roca, W. and Thro, A.M. (eds) *Proceedings of the 1st International Scientific Meeting of the Cassava Biotech Network*, CIAT, Cali, Colombia, pp. 81–89.

Ocampo, C., Angel, F., JimÇnez, A., Jaramillo, G., Hershey, C., Granados, E. and Iglesias, C. (1995) DNA fingerprinting to confirm possible genetic duplicates in cassava germplasm. The Cassava Biotechnology Network. *Proceedings of the Second International Scientific Meeting*, Bogor, Indonesia, 22–26 August 1994. Centro Internacional de Agricultura Tropical (CIAT), Cali, Colombia. pp. 145–147.

Okogbenin, E., Porto, M.C.M. and Dixon, A.G.O. (1998) Influence of planting season on incidence and severity of African cassava mosaic disease in the sub humic zone of Nigeria.

Olsen, K.M. and Schaal, B.A. (1999) Evidence on the origin of cassava: phylogeography of *Manihot esculenta*. *Proceedings of the National Academy of Science* 96, 5586–5591.

Oshuntokun, B. and Monekosso, G. (1969) Degenerative tropical neuropathy and diet. *British Medical Journal* 3, 178–183.

Osiru, D.S.O., Hahn, S.K. and Osonubi, O. (1994) Mechanisms of drought tolerance in cassava. *African Crop Science Journal* 2, 233–246.

Otim-Nape, G.W. (1993) Epidemiology of the African cassava mosaic geminivirus disease (ACMD) in Uganda. PhD thesis, University of Reading, Reading.

Otim-Nape, G.W. (1995) The African cassava mosaic virus ACMV: a threat to food security in Africa. In: Cassava Biotechnology Network (ed.) *Proceedings Second International Scientific Meeting*, Bogor, Indonesia, CIAT Working Document 150, pp. 519–527.

Otim-Nape, G.W., Shaw, M.W. and Thresh, J.M. (1994a) The effects of African cassava mosaic geminivirus on the growth and yield of cassava in Uganda. *Tropical Science* 34, 43–54.

Otim-Nape, G.W., Bua, A. and Baguma, Y. (1994b) Accelerating the transfer of improved production technologies: controlling the African cassava mosaic virus disease epidemics in Uganda. *African Crop Science Journal* 2, 479–495.

Otim-Nape, G.W., Thresh, J.M. and Shaw, M.W. (1997a) The effects of cassava mosaic virus disease on yield and compensation in mixed stands of healthy and infected cassava. *Annals of Applied Biology* 130, 503–521.

Otim-Nape, G.W., Bua, A., Thresh, J.M., Baguma, Y., Ogwal, S., Semakula, G.N., Acola, G., Baybakama, B. and Martin, A. (1997b) *Cassava Mosaic Disease in Uganda: the Current Pandemic and Approaches to Control*. Natural Resources Institute, Chatham.

Ow, D.W., Wood, K.V., DeLuca, M., de Wet, J.R., Helinski, D.R. and Howell, S.H. (1986) Transient and stable expression of the firefly luciferase gene in plant cells and transgenic plants. *Science* 234, 856–859.

Perry, B.A. (1943) Chromosome number and phylogenetic relationships in the Euphorbiaceae. *American Journal of Botany* 30, 527–543.

Poirier, Y., Nawrath, C. and Somerville, C. (1995) Production of polyhydroxyalkanoates a family of biodegradable plastics and elastomers in bacteria and plants. *Biotechnology* 13, 142–150.

Potrykus, I. and Spangenberg, G. (eds) (1995) *Gene Transfer to Plants*. Springer-Verlag, Berlin.

Prakash, C.S. and Egnin, M. (1997) Engineered sweetpotato (*Ipomoea batatas*) plants with a synthetic storage protein gene show high protein and essential amino acid levels. In: Dean, J.F.D. (ed.) *Proceedings 5th International Congress of Plant Molecular Biology. Plant Molecular Biology Reporter*, 15 (Suppl.), 335.

Puonti-Kaerlas, J., Frey, P. and Potrykus, I. (1997a) Development of meristem gene transfer techniques for cassava. *African Journal of Root Tuber Crops* 2, 175–180.

Puonti-Kaerlas, J., Li, H.Q., Sautter, C. and Potrykus, I. (1997b) Production of transgenic cassava (*Manihot esculenta* Crantz) via organogenesis and *Agrobacterium*-mediated transformation. *African Journal of Root and Tuber Crops* 2, 181–186.

Puonti-Kaerlas, J., Li, H.Q., Wohlwend, H. and Potrykus, I. (1998) Competence for embryogenesis and organogenesis in cassava. In: Pires de Matos, A. and Vilarinhos, A. (eds) *Proceedings IV International Scientific Meeting of the Cassava Biotechnology Network. Revista Brasileira de Mandioca* 17, 32.

Raemakers, C.J.J.M. (1993) Primary and cyclic somatic embryogenesis in cassava (*Manihot esculenta* Crantz). PhD thesis, Wageningen Agricultural University, The Netherlands.

Raemakers, C.J.J.M., Bessembinder, J.J.E., Staritsky, G., Jacobsen, E. and Visser, R.G.F. (1993a) Induction germination and shoot development of somatic embryos in cassava. *Plant Cell Tissue Organ Culture* 33, 151–156.

Raemakers, C.J.J.M., Amati, M., Staritsky, G., Jacobsen, E. and Visser, R.G.F. (1993b) Cyclic somatic embryogenesis and plant regeneration in cassava. *Annals of Botany* 71, 289–294.

Raemakers, C.J.J.M., Schavemaker, C.M., Jacobsen, E. and Visser, R.G.F. (1993c) Improvements of cyclic somatic embryogenesis of cassava (*Manihot esculenta* Crantz). *Plant Cell Reports* 12, 226–229.

Raemakers, C.J.J.M., Sofiari, E., Kanju, E., Jacobsen, E. and Visser, R.G.F. (1995a) NAA-induced somatic embryogenesis in cassava. In: Cassava Biotechnology Network (ed.) *Proceedings Second International Scientific Meeting*, Bogor, Indonesia, CIAT Working Document 150, pp. 355–363.

Raemakers, C.J.J.M., Jacobsen, E. and Visser, R.G.F. (1995b) Histology of somatic embryogenesis and evaluation of somaclonal variation. *Proceedings Second International Scientific Meeting*, Bogor, Indonesia, CIAT Working Document 150, pp. 336–354.

Raemakers, C.J.J.M., Sofiari, E., Taylor, N., Henshaw, G., Jacobsen, E. and Visser, R.G.F. (1996) Production of transgenic cassava (*Manihot esculenta* Crantz) plants by particle bombardment using luciferase activity as selection marker. *Molecular Breeding* 2, 339–349.

Raemakers, C.J.J.M., Jacobsen, E. and Visser, R.G.F. (1997a) Micropropagation of *Manihot esculenta* Crantz Cassava. In: Bajaj, Y. (ed.) *Biotechnology in Agriculture and Forestry*. Springer-Verlag, Berlin, pp. 77–102.

Raemakers, C.J.J.M., Rozenboom, M.G.M., Danso, K., Jacobsen, E. and Visser, R.G.F. (1997b) Regeneration of plants from somatic embryos and friable embryogenic callus of cassava (*Manihot esculenta* Crantz). *African Journal of Root Tuber Crops* 2, 238–242.

Raemakers, C.C.J.M., Rozeboom, M., Jacobsen, E. and Visser, R.G.F. (1998) Production of transgenic plants by particle bombardment, electroporation and *Agrobacterium tumefaciens*. In: Pires de Matos, A. and Vilarinhos, A. (eds) *Proceedings IV International Scientific Meeting of the Cassava Biotechnology Network. Revista Brasileira de Mandioca* 17, 33.

Ramirez, H., Hussain, A., Roca, W.A., Bushiuk, W. (1987) Isozyme electro-phenograms of sixteen enzymes in five tissues of cassava (*Manihot esculenta* Crantz) varieties. *Euphytica* 36, 39–48.

Reilly, K., Han, J., Iglesias, C. and Beeching, J. (2002) Oxidative stress related genes on cassava post-harvest physiological deterioration. *Euphytica* (in press).

Reinisch, A.J., Dong, J., Brubaker, C.L., Stelly, D.M., Wendel, J.F. and Paterson, A.H. (1994) A detailed RFLP map of cotton *Gossypium hirsutum* × *Gossypium barbadense*: chromosome organization and evolution in a disomic polyploid genome. *Genetics* 138, 829–847.

Ritter, E., Debener, T., Barone, A., Salamini, F. and Gebhardt, C. (1991) RFLP mapping on potato chromosomes of two genes controlling extreme resistance to potato virus X (PVX). *Molecular and General Genetics* 227, 81–88.

Roa, A.C., Maya, M.M., Duque, M., Allem, C., Tohme, J. and Bonierbale, M.W. (1997) AFLP analysis of relationships among cassava and other *Manihot* species. *Theoretical and Applied Genetics* 95, 741–750.

Roca, W.M. (1984) Cassava. In: Sharp, W.R., Evans, D.A., Ammirato, P.V. and Yamada, Y. (eds) *Handbook of Plant Cell Culture*, Vol. 2. *Crop Species*. MacMillan, New York, pp. 269–301.

Rogers, D.J. and Fleming, H.S. (1973) A monograph of *Manihot esculenta* with an explanation of the taximetric methods used. *Economic Botany* 27, 1–113.

Rogstad, S.H., Patton, J.C. and Schaal, B.A. (1988) M13 repeat probe detects DNA minisatellite-like sequences in gymnosperms and angiosperms. *Proceedings of the National Academy of Science USA* 85, 9176–9178.

Rosling, H. (1988) *Cassava Toxicity and Food Security*. Unicef African Household Security Programme Tryck Kontakt, Uppsala Sweden.

Rosling, H., Mlingi, N., Tylleskär, T. and Banea, M. (1993) Causal mechanisms behind human diseases induced by cyanide exposure from cassava. In: Roca, W.M. and Thro, A.M. (eds) *Proceedings First International Scientific Meeting of the Cassava Biotechnology Network*. Cartagena de Indias, Colombia, CIAT Working Document 123, pp. 336–375.

Sanchez, G., Restrepo, S., Duque, M., Fregene, M., Bonierbale, M. and Verdier, V. (1999a) AFLP assessment of genetic variability in cassava accessions (*Manihot esculenta*) resistant and susceptible to the cassava bacterial blight (CBB). *Genome* 42, 163–172.

Sangare, A., Deng, D., Fauquet, C.M. and Beachy, R.M. (1999) Resistance to African cassava mosaic virus conferred by a mutant of the putative NTP-binding domain of the rep gene (AC1) in *Nicotiana benthamiana*. *Molecular Biology Reports* 5, 95–102.

Sarria, R., Ocamp, C., Ramirez, H., Hershey, C. and Roca, W.M. (1993) Genetics of esterase and glutamate oxaloacetate transaminase isozymes in cassava (*Manihot esculenta* Crantz). In: Roca, W.M. and Thro, A.M. (eds) *Proceedings of the First Scientific Meeting of the Cassava Biotechnology Network, Cartagena, Colombia, 1992*. Working Document no. 123. Centro Internacional de Agricultura Tropical (CIAT). Cali, Colombia, pp. 75–80.

Sarria, R., Torres, E., Balcazar, M., Destafano-Beltran, L. and Roca, W.M. (1995) Progress in Agrobacterium-mediated transformation of cassava (*Manihot esculenta* Crantz). In: *The Cassava Biotechnology Network: Proceedings of the Second International Scientific Meeting, Bogor, Indonesia, 22–26 August 1994*. Working document no. 150. Centro Internacional de Agricultura Tropical. Cali, Colombia, pp. 241–244.

Sarria, R., Torres, E., Angel, F., Chavarriaga, P. and Roca, W.M. (2000) Transgenic plants of cassava (*Manihot esculenta*) with resistance to Basta obtained by *Agrobacterium*-mediated transformation. *Plant Cell Reports* 19, 339–344.

Schärer-Hernández, N., Frey, P., Potrykus, I. and Puonti-Kaerlas, J. (1998) Towards African cassava mosaic virus resistance in cassava. In: Pires de Matos, A. and Vilarinhos, A. (eds) *Proceedings IV International Scientific Meeting of the Cassava Biotechnology Network*. *Revista Brasileira de Mandioca* 17, 43.

Schenk, R.U. and Hildebrandt, A.C. (1972) Medium and techniques for induction and growth of monocotyledonous and dicotyledonous plant cell cultures. *Canadian Journal of Botany* 50, 199–204.

Schöpke, C., Franche, C., Bogusz, D., Chavarriaga, P., Fauquet, C. and Beachy, R.N. (1993) Transformation in Cassava (*Manihot esculenta* Crantz). In: Bajaj, Y.P.S. (ed.) *Biotechnology in Agriculture and Forestry*, Vol. 23. *Plant Protoplasts and Genetic Engineering*. Springer Verlag, Berlin, pp. 273–289.

Schöpke, C., Taylor, N., Carcamo, R., Konan, N.K., Marmey, P., Henshaw, G.G., Beachy, R.N. and Fauquet, C. (1996) Regeneration of transgenic cassava plants (*Manihot esculenta* Crantz) from microbombarded embryogenic suspension cultures. *Nature Biotechnology* 14, 731–735.

Schöpke, C., Taylor, N., Carcamo, R., Gonzalez de Schöpke, A.E., Konan, N., Marmey, P., Henshaw, G.G., Beachy, R.N. and Fauquet, C. (1997a) Stable transformation of cassava (*Manihot esculenta* Crantz) by particle bombardment and by *Agrobacterium*. *African Journal of Root and Tuber Crops* 2, 187–193.

Schöpke, C., Carcamo, R., Beachy, R.N. and Fauquet, C. (1997b) Plant regeneration from transgenic and non-transgenic embryogenic suspension cultures of cassava (*Manihot esculenta* Crantz). *African Journal of Root and Tuber Crops* 2, 194–195.

Schöpke, C., Taylor, N.J., Masona, M.V., Carcamo, R., Ho, T., Beachy, R.N. and Fauquet, C.M. (2000) Production and characterisation of transgenic cassava plants containing the coat protein gene of cassava common mosaic virus. In: Carvalho, L.J.C.B., Thro, A.-M. and Vilarinhos, A.D. (eds) *Cassava Biotechnology. Proceedings of IV International Scientific Meeting of the Cassava Biotechnology Network*, pp. 236–243.

Second, G., Allem, A., Emperaire, L., Ingram, C., Colombo, C., Mendes, R. and Carvalho, L. (1997) AFLP based *Manihot* and cassava numerical taxonomy and genetic structure analysis in progress: implications for dynamic conservation and genetic mapping. *African Journal of Root and Tuber Crops* 2, 140–147.

Snepvangers, S.C.H.J., Raemakers, C.J.J.M., Jacobsen, E. and Visser, R.G.F. (1997) Optimization of chemical selection of transgenic friable embryogenic callus of cassava using the luciferase reporter gene system. *African Journal of Root and Tuber Crops* 2, 196–200.

Sommer, A. (1988) New imperatives for an old vitamin A. *Journal of Nutrition* 119, 96–100.

Stamp, J.A. (1987) Somatic embryogenesis in cassava: the anatomy and morphology of the regeneration process. *Annals of Botany* 59, 451–459.

Stamp, J.A. and Henshaw, G.G. (1982) Somatic embryogenesis in cassava. *Zeitschrift für Pflanzenphysiologie* 105, 183–187.

Stamp, J.A. and Henshaw, G.G. (1987a) Somatic embryogenesis from clonal leaf tissues of cassava. *Annals of Botany* 59, 445–450.

Stamp, J.A. and Henshaw, G.G. (1987b) Secondary somatic embryogenesis and plant regeneration in cassava. *Plant Cell Tissue Organ Culture* 10, 227–233.

Stanley, J. (1983) Infectivity of the cloned geminivirus genome requires sequences from both DNAs. *Nature* 305, 643–645.

Stanley, J., Frischmuth, T. and Ellwood, S. (1990) Defective viral DNA ameliorates symptoms of geminivirus infection in transgenic plants. *Proceedings of the National Academy of Sciences USA* 87, 6291–6295.

Stoeckli, B. (1998) Bioplastics from transgenic cassava: a pipedream, or a promising subject for research cooperation? *Agricultural Rural Development* 5, 57–59.

Suarez, M.C., Bernal, A., Gutierrez, J., Tohme, J. and Fregene, M. (2000) Development of expressed sequence tags (ESTs) from polymorphic transcription derived fragments (TDFs) in cassava (*Manihot esculenta* Crantz). *Genome* 43, 62–67

Swanson, M.M. and Harrison, B.D. (1994) Properties, relationships and distribution of cassava mosaic geminiviruses. *Tropical Science* 34, 15–25.

Szabados, L., Hoyos, R. and Roca, W. (1987) *In vitro* somatic embryogenesis and plant regeneration of cassava. *Plant Cell Reports* 6, 248–251.

Tanksley, S.D., Young, N.D., Paterson, A.H. and Bonierbale, M.W. (1989) RFLP mapping in plant breeding: new tools for an old science. *Biotechnology* 7, 257–263.

Tanksley, S.D., Ganal, M.W. and Martin, G.B. (1995) Chromosome landing: a paradigm for map-based gene cloning in plants with large genomes. *Trends in Genetics* 11, 63–68.

Taylor, N.J., Edwards, M., Kiernan, R.J., Davey, C.D.M., Blakesley, D. and Henshaw, G.G. (1996) Development of friable embryogenic callus and embryogenic suspension culture systems in cassava (*Manihot esculenta* Crantz). *Nature Biotechnology* 14, 726–730.

Thompson, G.A., Hiatt, R.W., Facciotti, D., Stalker, D.M. and Comai, L. (1987) Expression in plants of a bacterial gene coding for glyphosate resistance. *Weed Science* 35 (Suppl.), 19–23.

Thresh, J.M., Fargette, D. and Otim-Nape, G.W. (1994a) Effects of African cassava mosaic geminivirus on the yield of cassava. *Tropical Science* 34, 26–42.

Thresh, J.M., Fishpool, L.D.C., Otim-Nape, G.W. and Fargette, D. (1994b) African cassava mosaic virus disease: an under-estimated and unsolved problem. *Tropical Science* 34, 3–14.

Thresh, J.M., Otim-Nape, G.W. and Jennings, D.L. (1994c) Exploiting resistance to African cassava mosaic virus. *Aspects of Applied Biology* 39, 51–60.

Thro, A.M. and Fregene, M. (1999) Network impact and scientific advances in cassava biotechnology. *Tropical Agriculture* 75, 230–233.

Townsend, R., Watts, J. and Stanley, J. (1986) Synthesis of viral DNA in *N. plumbaginifolia* protoplasts inoculated with Cassava latent virus (CLV); evidence for the independent replication of one component of the CLV genome. *Nucleic Acids Research* 14, 1253–1265.

Trieu, A.N., Burleigh, S.H., Kardalilsky, I.V., Maldonado-Mendoza, I.E., Versaw, W.K., Blaylock, L.A., Shin, H., Chiou, T.Z., Kataki, H., Dewbre, G.R., Weigel, D. and Harrison, M.J. (2000) Transformation of *Medicago truncatula* via infiltration of seedlings or flowering plants with *Agrobacterium*. *The Plant Journal* 22, 531–541.

Tylleskär, T. (1994) The association between cassava and the paralytic disease konzo. *Acta Horticulturae* 375, 331–339.

Tylleskär, T., Banea, M., Bikangi, N., Fresco, L., Persson, L.A. and Rosling, H. (1991) Epidemiological evidence from Zaire for a dietary etiology of konzo, an upper motor neuron disease. *Bulletin of the World Health Organisation* 69, 581–590.

Tylleskär, T., Banea, M., Bikangi, N., Cooke, R.D., Poulter, N.H. and Rosling, H. (1992) Cassava cyanogens and konzo, an upper motorneuron disease found in Africa. *Lancet* 339, 208–211.

Tylleskär, T., Banea, M., Bikangi, N., Nahimana, G., Persson, L.A. and Rosling, H. (1995) Dietary determinants of a non-progressive spastic paraparesis (Konzo): a case-referent study in a high incidence area of Zaire. *International Journal of Epidemiology* 24, 949–956.

Umanah, E.E. and Hartman, R.W. (1973) Chromosome numbers and karyotypes of some Manihot species. *Journal of the American Society of Horticultural Science* 98, 272–274.

van den Elzen, P., Townsend, J., Lee, K.Y. and Bedbrook, J. (1985) A chimeric hygromycin resistance gene as a selectable marker in plant cells. *Plant Molecular Biology* 5, 299–302.

von Arnim, A. and Stanley, J. (1992) Inhibition of African cassava mosaic virus systemic infection by a movement protein from the related geminivirus tomato golden mosaic virus. *Virology* 187, 555–564.

von Arnim, A., Frischmuth, T. and Stanley, J. (1993) Detection and possible functions of African cassava mosaic virus DNA B gene products. *Virology* 192, 264–272.

Vos, P., Hogers, R., Bleeker, M., Reijans, M., van der Lee, T., Hornes, M., Fritjers, A., Pot, J., Peleman, J., Kuiper, M. and Zabeau, M. (1995) AFLP: a new technique for DNA fingerprinting. *Nucleic Acid Research* 23, 4407–4414.

Waldron, C., Murphy, E.B., Roberts, J.L., Gustafson, G.D., Armour, S.L. and Malcolm, S.K. (1985) Resistance to hygromycin B: a new marker for plant transformation studies. *Plant Molecular Biology* 5, 103–108.

Wang, M.B., Upadhaya, N.M., Brettell, R.I.S. and Waterhouse, P.M. (1997) Intron-mediated improvement of a selectable marker gene for plant transformation using *Agrobacterium tumefaciens*. *Journal of Genetics and Breeding* 51, 325–334.

Wanyera, N.M.W., Hahn, S.K. and Aken'Ova, M.E. (1994) Introgression of ceara rubber (*Manihot glaziovii* Muell-Arg) into cassava (*M. esculenta* Crantz): a morphological and electrophoretic evidence. In: *Proceedings of the Fifth Triennial Symposium of the International Society for Tropical Root Crops – Africa Branch*. Kampala, Uganda, 22–28 November 1992, pp. 125–130.

West, K.P., Howard, G.R. and Sommer, A. (1989) Vitamin A and infection: public health implications. *Annual Review of Nutrition* 9, 63–86.

Wheatley, C.C. (1982) Studies on cassava root physiological deterioration. PhD thesis, University of London.

Wheatley, C.C. and Schwabe, W.W. (1985) Scopoletin involvement in post-harvest physiological deterioration of cassava roots (*Manihot esculenta* Crantz). *Journal of Experimental Botany* 36, 783–791.

Williams, J.G.K., Kubelik, A.R., Livak, K.J., Rafalski, J.A. and Tingey, S.V. (1990) DNA polymorphisms amplified by arbitrary primers are useful as genetic markers. *Nucleic Acids Research* 18, 6531–6535.

Wohllenben, W., Arnold, W., Broer, I., Hillemann, D., Strauch, E. and Puehler, A. (1988) Nucleotide sequence of the phosphinothricin N-acetyltransferase gene from *Streptomyces viridochromogenes* Tü494 and its expression in *Nicotiana tabacum*. *Gene* 70, 25–38.

Woodward, B. and Puonti-Kaerlas, J. (1998) Somatic embryogenesis from floral tissue of cassava (*Manihot esculenta* Crantz). In: Carvalho L.J.C.B., Thro, A.M. and Vilarinhos, A.D. (eds) *Cassava Biotechnology. Proceedings of IV International Scientific Meeting of the Cassava Biotechnology Network*, pp. 431–438.

Wu, K.K., Burnquist, W., Sorrells, M.E., Tew, T.L., Moore, P.H. and Tanksley, S.D. (1992) The detection and estimation of linkage in polyploids using single-dose restriction fragments. *Theoretical and Applied Genetics* 83, 294–300.

Yang, M.S., Espinoza, N.O., Dodds, J.H. and Jaynes, J.M. (1989) Expression of a synthetic gene for improved protein quality in transformed potato plants. *Plant Science* 64, 99–111.

Zhang, H., Zhao, X., Ding, X., Paterson, A. and Wing, R. (1995) Preparation of megabase-size DNA from plant nuclei. *The Plant Journal* 7(1), 175–184.

Zhang, P. and Puonti-Kaerlas, J. (2002) PIG-mediated cassava transformation using positive and negative selection. *Plant Cell Reports* (in press).

Zhang, P., Legris, G., Coulin and Puonti-Kaerlas, J. (2000) Production of stably transformed cassava plants via particle bombardment. *Plant Cell Reports* 19, 939–945.

Zhang, P., Potrykus, I. and Puonti-Kaerlas, J. (2002) Efficient production of transgenic cassava using negative and positive selection. *Transgenic Research* (in press).

Chapter 11
Arthropod Pests

Anthony C. Bellotti

Pest and Disease Management Project, Centro Internacional de Agricultura Tropical (CIAT), A.A. 6713, Cali, Colombia

Introduction

The dynamics of cassava production, utilization and marketing vary considerably from one continent and country to another. In recent years there has been a shift in demand for particular cassava products in Latin America with a noticeable trend towards the production of chips and pellets for animal feed, as well as high-quality flours for human consumption (Henry and Gottret, 1995). This could induce a trend towards plantation-size production units.

World cassava production, which is concentrated in Latin America, Asia and Africa, can be divided into five principal agroecosystems (Henry and Gottret, 1995):

- the lowland humid tropics;
- the lowland subhumid tropics;
- the lowland semiarid tropics;
- the highland tropics; and
- the subtropics.

Approximately 65% of the area under cassava production occurs in the lowland humid and subhumid agroecosystems (Henry and Gottret, 1995). Arthropod pests do not appear to cause significant damage in the lowland humid or the highland tropical agroecosystems, where there is considerable and consistent rainfall (Montagnini and Jordan, 1983; Bellotti *et al.*, 1999). Moderate levels of crop damage can occur in the mid-altitude (1000–1400 m) tropics,

especially if there is a dry season of 3 months or more. When the crop is grown under conditions of irregular, limited rainfall as in the lowland subhumid and semiarid tropics, however, arthropod pest populations increase, causing considerable yield reduction (Bellotti *et al.*, 1999). Arthropod pests can cause low-to-moderate levels of crop damage in the subtropical regions of Latin America and Africa and in the highland regions of Africa (Henry and Gottret, 1995).

Most cassava is grown by small-scale farmers, often on marginal or fragile soils under rainfed conditions, using few purchased inputs such as fertilizers or pesticides. In these traditional farming systems where cassava is usually one of many crops being grown, pest control is often a low priority and so cassava receives minimal pesticide applications. Under such conditions, yields are often low, with a large gap between potential yield (21.3 t ha^{-1}) and that achieved by producers: a mean of 11.2 t ha^{-1} in Latin America and 9.0 in Africa (Henry, 1995). Arthropod pests and diseases are major factors in causing this yield reduction (Bellotti *et al.*, 1999); moreover, the vegetative nature of cassava propagation contributes to pest build-up and dissemination across regions and among countries and continents.

In Latin America there are indications of a shift towards large-scale production units, where cassava is grown as a plantation crop. This may result in new or worse pest problems than those found in small-scale production. Such a scenario is especially feasible in the Neotropics, where a large complex of arthropod pests has co-evolved with the crop. For example, increased frequency and severity of cassava hornworm (*Erinnyis ello* L.) attacks have been recorded, resulting in increased pesticide applications (LaBerry, 1997). In Asia, where there are extensive areas planted with cassava, major pest problems have been avoided as the most damaging pest species from the Neotropics have not been introduced.

The Cassava Arthropod Complex

Cassava (Euphorbiaceae: *Manihot esculenta* Crantz) is a perennial shrub grown commercially as an annual or biennial, throughout tropical and subtropical regions of the world. This vegetatively propagated crop has a long growth cycle and is drought-tolerant. As it is often intercropped and planting dates are staggered, cassava is almost always present in farmers' fields. These agronomic characteristics undoubtedly contribute to the considerable diversity of arthropod pests that feed on the crop.

Cassava originated in the Neotropics, although its exact centre of origin is equivocal (Renvoize, 1973; Allem, 1994; see Chapter 4); consequently, the greatest diversity of arthropods reported attacking the crop (Table 11.1) is from the region (Bellotti *et al.*, 1994). An estimated 200 species have been reported (Bellotti and Van Schoonhoven, 1978a,b), many of which are specific to cassava and have adapted in varying degrees to an array of natural biochemical defences in the host that include laticifers and cyanogenic compounds (Bellotti and Riis, 1994). The pest complex varies greatly between the main cassava-growing areas, indicating that careful quarantine measures could prevent pest introduction into uninfested areas (Frison and Feliu, 1991). The accidental introduction of the cassava green mite (CGM; *Mononychellus tanajoa* Bonder) and the cassava mealybug (*Phenacoccus manihoti* Mat. Ferr.) from the Americas into Africa has caused considerable crop loss throughout the cassava belt and has been the object of massive biological control efforts (Herren and Neuenschwander, 1991; Neuenschwander, 1994a). In Asia none of the major neotropical cassava pests has become

established and native arthropod pests that have adapted to cassava have not been reported as causing serious yield losses (Maddison, 1979).

Recent explorations in cassava-growing regions of the Neotropics indicate that the arthropod pest complex is not geographically uniform. The evidence suggests that the mealybug *Phenacoccus herreni*, which has caused considerable damage in northeast Brazil, was probably introduced from northern South America (Venezuela or Colombia), where mealybug populations are controlled by natural enemies not found in Brazil (Bellotti *et al.*, 1994; Smith and Bellotti, 1996). *P. manihoti*, which has caused severe crop damage in Africa, occurs only in Paraguay, the Mato Grosso area of Brazil and the Santa Cruz area of Bolivia (Lohr and Varela, 1990).

Studies on the CGM demonstrate a higher degree of morphological polymorphism and a greater *Mononychellus* species complex in northern South America than elsewhere in the Neotropics (Bellotti *et al.*, 1994). This diversity is associated with greater species richness within the phytoseiid complex that preys upon *Mononychellus* spp. in cassava (Bellotti *et al.*, 1987, 1999). The cassava pest complex can be divided into two groups:

- those that appear to have co-evolved with cassava, which is their primary or only host; and
- generalist feeders that may attack the cassava crop opportunistically, especially during seasonally dry periods when cassava is one of the limited food sources available.

The former group includes the *Mononychellus* mite complex, mealybugs (*P. herreni* and *P. manihoti*), the hornworm (*E. ello*), lacebugs (*Vatiga illudens*, *Vatiga manihotae*, *Amblystira machalana*), whiteflies (*Aleurotrachelus socialis* and *Aleurothrixus aepim*), the stemborer (*Chilomima clarkei* and those of the genus *Coelosternus*), fruitflies (*Anastrepha pickeli* and *Anastrepha manihoti*), the shootfly (*Neosilba perezi*), the white scale (*Aonidomytilus albus*), thrips (*Frankliniella williamsi*) and the gallmidge (*Jatrophobia brasiliensis*; Bellotti *et al.*, 1999).

The generalist feeders primarily consist of whitegrubs (*Phyllophaga* spp. and several others), termites, cutworms, grasshoppers, leaf-cutting ants, burrower bugs (*Cyrtomenus bergi*), crickets, *Tetranychus* mite species and stemborers (*Lagochirus* spp.) (Bellotti and van Schoonhoven, 1978a; Bellotti *et al.*, 1999).

Several pests found in the Neotropics could potentially cause severe crop losses if introduced inadvertently into Asian or African cassava-growing areas. They include the cassava hornworm, several mite species, lacebugs, whiteflies and stemborers. Moreover, a species which is considered to be a secondary or minor pest in the Neotropics (e.g. the cassava mealybug, which is found only in limited sites, as mentioned) could become a major pest if introduced into areas where native natural enemies and/or adapted or resistant germplasm are unavailable. *P. manihoti* has not yet reached other regions, although there are no evident natural barriers to prevent its movement, especially in Brazil, where cassava is grown extensively throughout most of the country. The Andean mountain range in western South America has undoubtedly affected the movement of cassava pests, although this is not well documented. For example, the lepidopteran stemborer *C. clarkei*, which has good flight

Table 11.1. Global distribution of important arthropod pests of cassava.

Pest	Major species	Americas	Africa	Asia
Mites	*Mononychellus tanajoa*	X	X	
	Tetranychus urticae	X		X
Mealybugs	*Phenacoccus manihoti*	X	X	
	Phenacoccus herreni	X		
Whiteflies	*Aleurotrachelus socialis*	X		
	Aleurodicus dispersus	X	X	X
	Aleurothrixus aepim	X		
	Bemisia tabaci	X	X	X
	Bemisia afer	X	X	
Hornworm	*Erinnyis ello*	X		
	Erinnyis alope	X		
Lacebugs	*Vatiga illudens*	X		
	Vatiga manihotae	X		
Burrower bugs	*Cyrtomenus bergi*	X		
Thrips	*Frankliniella williamsi*	X	X	
	Scirtothrips manihoti	X		
Scales	*Aonidomytilus albus*	X	X	X
Fruitflies	*Anastrepha pickeli*	X		
	Anastrepha manihoti	X		
Shootflies	*Neosilba perezi*	X		
	Silba pendula	X		
Gallmidge	*Jatrophobia* (*Eudiplosis*) *brasiliensis*	X		
White grubs	*Leucopholis rorida*	X	X	X
	Phyllophaga spp.	X	X	X
	Several others	X	X	X
Termites	*Coptotermes* spp.	X	X	X
	Heterotermes tenuis	X		
Stemborers	*Chilomima* spp.	X		
	Coelosternus spp.	X		
	Lagochirus spp.	X	X	X
Leaf-cutter ants	*Atta* spp.	X		
	Acromyrmex spp.	X		
Root mealybugs	*Pseudococcus mandioca*	X		
	Stictococcus vayssierei		X	
Grasshoppers	*Zonocerus elegans*	X	X	
	Zonocerus variegatus	X	X	

Source: Adapted from Bellotti *et al.* (1999).

capabilities and has spread through many regions of Colombia and Venezuela, is not reported west of the westernmost range of the Andean mountains, but has been reported from Argentina (B. Lohr, personal communication).

Crop damage and yield loss

Arthropod damage to cassava is often indirect because most pests are foliage or stem feeders, reducing leaf area, leaf life or photosynthetic rate. Field studies indicate that pests that attack the crop over prolonged periods (3–6 months) – such as mites, mealybugs, thrips, whiteflies and lacebugs – can cause severe root yield reductions (Table 11.2) as a result of their feeding on leaf cell fluids and the consequent decrease in photosynthesis. Severe attacks can induce premature leaf drop and death of the apical meristem. The potential for yield reduction by these pests is greater than that by cyclical pests such as hornworm and leaf-cutter ants, which cause sporadic defoliation. Nevertheless, these highly visible pests often cause farmers to apply insecticides (Braun et al., 1993).

The burrower bug (C. bergi; Hemiptera: Cydnidae) is one of the few pests that damage cassava roots directly. Root punctures during feeding can introduce fungal pathogens that reduces root yield and quality (García and Bellotti, 1980). Grubs, millipedes and termites are reported occasionally as feeding on tuberous roots; however these may be secondary feeders, attacking already damaged and decaying roots.

In general, arthropod pests are most damaging to cassava during the dry season and do not appear to cause significant damage in areas of considerable and consistent rainfall (Bellotti et al., 1999). The cassava plant is well adapted to long periods of limited water, responding to water shortage by reducing its evaporative (leaf) surface rapidly and efficiently and by closing the stomata partially, thereby increasing water use efficiency (Cock et al., 1985; El-Sharkawy et al., 1992). In water-deprived plants, both the accelerated shedding of old leaves and the pronounced decrease in their photosynthetic activity means that the younger leaves play a key role in the plant's carbon nutrition. Given that pests prefer the younger canopy leaves, dry-season feeding tends to cause the greatest yield losses in cassava. Once the crop enters into a wet cycle (following

Table 11.2. Yield losses due to major cassava pests.

Pest	Yield losses	References
Hornworm (*Erinnyis ello*)	In farmers' fields, natural attack resulted in 18% yield loss; simulated damage studies resulted in 0–64% loss, depending on no. of attacks, plant age and soil fertility.	Arias and Bellotti (1984); Bellotti et al. (1992)
Mites (*Mononychellus tanajoa*)	21, 25 and 53% yield loss during a 3-, 4- and 6-month attack, resp.; 73% for susceptible cultivars compared with 15% for resistant cultivars; 13–80% in Africa.	Bellotti et al. (1983b); Byrne et al. (1982); Herren and Neuenschwander (1991); Yaninek and Herren (1988)
Whiteflies (*Aleurotrachelus socialis*)	1-, 6-, 11-month attacks resulted in 5, 42 and 79% yield loss, respectively.	Bellotti et al. (1983b, 1999); Vargas and Bellotti (1981)
Mealybugs (*Phenacoccus herreni, Phenacoccus manihoti*)	68–88% loss depending on cultivar susceptibility (in Colombia); in Africa yield losses of ≈ 80% are reported.	Herren and Neuenschwander (1991); Schulthess (1987); Vargas and Bellotti (1984)
Burrower bugs (*Cyrtomenus bergi*)	Brown-to-black lesions render roots commercially unacceptable; > 50% reduction in starch content.	Arias and Bellotti (1985); Bellotti et al. (1999); Castaño et al. (1985)
Lacebugs (*Vatiga manihoti, Amblystira machalana*)	Field trials with *A. machalana* and *V. manihoti* resulted in 39% yield loss.	CIAT (1990)
Stemborers (*Chilomima clarkei*)	In Colombia, 45–62% loss when stem breakage exceeds 35%.	Lohr (1983)

rain or irrigation), it has the potential to recover and compensate for yield losses from severe drought, as well as from pest attack, because of the formation of a new leaf canopy and the higher photosynthetic rate in newly formed leaves (El-Sharkawy, 1993).

Major Pests; Bioecology Damage and Management

Cassava mites

Mites are a universal pest of cassava, causing serious yield losses in the Americas and Africa (Herren and Neuenschwander, 1991; Bellotti *et al.*, 1999). Of the > 40 species reported feeding on cassava (Byrne *et al.*, 1983), the most frequent are *Mononychellus tanajoa* (syn = *Mononychellus progresivus*), *Mononychellus carribbeanae*, *Tetranychus cinnabarinus* and *Tetranychus urticae* (also reported as *Tetranychus bimaculatus* and *Tetranychus telarius*). Cassava is the major host for the *Mononychellus* species, whereas the *Tetranychus* species have a wide host range. Other mite species (e.g. *Oligonychus peruvianus*, *Oligonychus biharensis*, *Eutetranychus banksi* or *Mononychellus mcgregori*) are not important economically and feed on cassava only sporadically (Byrne *et al.*, 1983).

CGM, the most important species, is reported causing crop losses in the Americas and Africa (Herren and Neuenschwander, 1991; Bellotti *et al.*, 1999), especially in seasonally dry regions of the lowland tropics (Yaninek and Animashaun, 1987; Braun *et al.*, 1989). In experimental trials fresh root yields were reduced by 21, 25 and 53% following 3-, 4- and 6-month attacks, respectively (Bellotti *et al.*, 1983b). Under field conditions with higher mite populations, there was a 15% yield reduction in resistant cultivars compared with a ≤ 73% loss in susceptible cultivars; 67% of the stem-cutting planting material was damaged (Byrne *et al.*, 1982, 1983).

M. tanajoa is native to the Neotropics and it was first reported from northeast Brazil in 1938. It first appeared in Africa (Uganda) in 1971 and by 1985 it had spread across most of the cassava belt, occurring in 27 countries (Yaninek, 1988) and causing estimated root yield losses of 13–80% (Yaninek and Herren, 1988; Herren and Neuenschwander, 1991; Skovgård *et al.*, 1993).

CGM populations feed preferentially on the undersides of young emerging leaves, which develop a mottled whitish-to-yellow appearance; and they may become deformed or reduced in size (Byrne *et al.*, 1983). The CGM is a serious problem only in dry regions, where heavy infestations cause defoliation which begins at the top of the plant, often killing apical buds and shoots. Regrowth may occur but if the rains are scarce, this new flush of leaves will also be attacked (Yaninek and Animashaun, 1987).

Control

Research into the control of *M. tanajoa* has had two main thrusts: host plant resistance (HPR) and biological control (Table 11.3). These two complementary strategies promise to reduce CGM populations below economic injury levels. The continual use of acaricides is not feasible for low-income farmers. Moreover, such use is not recommended because of adverse effects on natural enemies.

HPR. Substantial efforts have been made by both international research centres which have a mandate for cassava [Centro Internacional de Agricultura Tropical (CIAT) and International Institute for Tropical Agriculture (IITA)] and by national research programmes [e.g. Centro Nacional de Pesquisa en Mandioca y Fruticultura (CNPMF)/Empresa Brasileira de Pesquisa Agropecuaria (EMBRAPA)] to identify cassava cultivars and develop hybrids with resistance to CGM (Byrne *et al.*, 1983; Bellotti *et al.*, 1987; Hershey, 1987). Of nearly 5000 landrace cultivars in the CIAT cassava germplasm bank that were evaluated for CGM resistance, only *c.* 6% (300 cultivars) have been identified as having low-to-moderate levels of resistance (CIAT, 1999). After substantial effort, cultivars with moderate levels of resistance have been developed and released to farmers.

CIAT's mite-resistance research has traditionally been carried out at two sites: (i) CIAT, Palmira, located in the mid-altitude (1000 m) Andean highlands, where mite populations are moderate; and (ii) Pivijay, Magdalena, on the Colombian Atlantic Coast, in the lowland tropics with a prolonged (4–6 months) dry season and

Table 11.3.   Control options for major cassava pests.

Pest/species	Control options	References
Hornworm (*Erinnyis ello*)	Biological control. Baculovirus biopesticide; monitoring of hornworm populations with light traps and field scouting.	Bellotti *et al.* (1992, 1999); Braun *et al.* (1993); Schmitt (1988)
Mites (*Mononychellus tanajoa*, *Mononychellus caribbeanae*)	Host plant resistance (HPR). Moderate levels of resistance available in cassava clones; effective breeding programme needed to incorporate resistance into commercial cultivars.	Bellotti *et al.* (1994); Braun *et al.* (1989); Byrne *et al.* (1982, 1983); CIAT (1999)
	Biological control. Large Phytoseiidae predator complex associated with mite complex that reduces mite populations effectively; fungal pathogen (*Neozygites*) and virus disease identified.	Bellotti *et al.* (1999); Yaninek and Herren (1988)
Whitefly (*Aleurotrachelus socialis*)	HPR. Several highly resistant varieties and hybrids available; natural enemies, especially parasitoids being surveyed and evaluated.	Arias (1995); Bellotti *et al.* (1994, 1999); Castillo (1996); CIAT (1999)
Mealybugs (*Phenacoccus herreni*)	Adequate HPR not available in cassava germplasm. Several natural enemies identified: 3 parasitoid species (*Acerophagus cocois*, *Aenasius vexans* and *Apoanagyrus diversicornis*) give good control.	Bellotti *et al.* (1983a, 1999); Van Driesche *et al.* (1990)
(*Phenacoccus manihoti*)	Parasite species *Apoanagyrus lopezi* provides good control in most regions of Africa.	Herren and Neuenschwander (1991); Neuenschwander (1994a)
Thrips (*Frankliniella williamsi*)	HPR. Pubescent cultivars have effective resistance and are available to farmers.	Bellotti and Kawano (1980); Bellotti and van Schoonhoven (1978a,b); van Schoonhoven (1974)
Burrower bug (*Cyrtomenus bergi*)	High-HCN cultivars less damaged than others. Several biological control agents including entomopathogenic nematodes and fungal pathogens give promising results in lab studies. Intercropping with *Crotalaria* sp. reduces damage.	Barberena and Bellotti (1998); Bellotti and Riis (1994); Bellotti *et al.* (1999); Caicedo and Bellotti (1994); Riis (1997)
Stemborers (*Chilomima clarkei*)	Cultural practices; maintain clean fields, destroy infested stems. HPR being investigated. Possible use of transgenic cultivars (Bt) being evaluated.	Bellotti and van Schoonhoven (1978a,b); Gold *et al.* (1990); Lohr (1983)
Lacebugs (*Vatiga manihotae*, *Vatiga illuden*, *Amblystira machalana*)	HPR evaluations have given some promising results. Natural enemies identified but not thoroughly investigated.	Bellotti *et al.* (1987, 1999); Calvacante and Ciociola (1993); CIAT (1990); Farias (1985)

high mite populations. Low-to-moderate levels of resistance are indicated by a 0–3.5 damage rating on a 0–6 evaluation scale.

Of the 300 cultivars selected as promising for mite resistance over several years (two to seven field cycles), 72 have consistently had damage ratings < 3.0 (CIAT, 1999). Most of these cultivars were collected from Brazil, Colombia, Venezuela, Peru and Ecuador, but there are also several hybrids.

Mechanisms of mite resistance have usually been expressed as antixenosis (preference versus non-preference) or antibiosis (Byrne *et al.*, 1982). Mites feeding on susceptible cultivars had greater fecundity, greater acceptability,

a shorter development time, a longer adult life span and lower larval and nymphal mortality than those feeding on resistant cultivars (Byrne *et al.*, 1983). In more recent laboratory studies, *M. tanajoa* displayed a strong ovipositional preference for susceptible cultivars. When paired with the resistant cultivars M Ecu 72, M Per 611 and M Ecu 64 in free-choice tests, 95, 91 and 88%, respectively, of the eggs were oviposited on CMC 40, the susceptible cultivar (CIAT, 1999).

BIOLOGICAL CONTROL.   Extensive surveys of cassava fields and experimental data indicate that although CGM is present throughout much of the lowland Neotropics, severe outbreaks causing significant yield losses are rare, except in parts of Brazil. From 1983–1990 extensive evaluations of the natural enemy complex associated with cassava mites were carried out at 2400 sites in 14 countries of the Americas (Byrne *et al.*, 1983; Bellotti *et al.*, 1987). This led to the identification of the phytoseiid predator complex associated with the mites. The current predator mite reference collection held at CIAT conserves primarily those predators related to phytophagous mites found on cassava. Collecting zones were usually chosen for their similarity to ecological homologues in Africa and Brazil. Of the 87 species collected and stored, 25 are new or unrecorded species; 76% (66 species) were collected from cassava. A taxonomic key on phytoseiid species associated with cassava is being prepared as part of a collaborative project with Brazilian colleagues. The CIAT-Brazil collection is now organized into a true reference collection with accompanying database and can be used readily for species description or redescription, where types and paratypes may be found.

Of the 66 phytoseiid species collected on cassava, 13 occur frequently. *Typhlodromalus manihoti* was collected most frequently, being found in > 50% of the fields surveyed. This was followed by *Neoseiulus idaeus*, *Typhlodromalus aripo*, *Galendromus annectens*, *Euseius concordis* and *Euseius ho*. *T. aripo* and *N. idaeus* have played an important role in the successful control of *M. tanajoa* in Africa (Yaninek *et al.*, 1992, 1993).

Explorations also revealed several insect predators of CGM, especially the staphylinid *Oligota minuta* and the coccinellid *Stethorus* sp. These phytoseiids and insect predators have been studied extensively in the laboratory and field (Table 11.4). It is generally agreed that the phytoseiid predators are more efficient than the insect predators at controlling mites occurring in low densities (Byrne *et al.*, 1983).

Survey data also showed that CGM densities were much higher in northeast Brazil than in Colombia and that the richness of phytoseiid species was considerably higher in Colombia than in Brazil. Of the fields surveyed in Colombia, 92% were uninfested or had low CGM densities (< 25 mites per leaf) whereas in Brazilian fields only 12% were uninfested and 25% had intermediate or high densities (Bellotti *et al.*, 1994).

Data from field experiments in Colombia (Braun *et al.*, 1989) demonstrated the value and effect of the richness of phytoseiid species associated with CGM. Fresh and dry root yields in Colombia were reduced by 33% when natural enemies were eliminated, whereas applications of acaricides did not increase yields, indicating good natural biological control.

Since 1984 numerous species of phytoseiids have been shipped from Colombia and Brazil to Africa. Despite massive releases, none of the Colombian species became established, but three of the Brazilian species did (*T. manihoti*, *T. aripo* and *N. idaeus*; Yaninek *et al.*, 1992, 1993; Bellotti *et al.*, 1999). *T. aripo* appears to be the most successful of the three species. It has spread rapidly and is now in more than 14 countries. On-farm field trials indicate that *T. aripo* reduces CGM populations by 35–60% and increases fresh root yields by 30–37%.

Neozygites cf. *floridana*, a fungal pathogen (Zygomycetes: Entomophthorales), causes irregular or periodic mortality of mite populations in Colombia and northeast Brazil (Delalibera *et al.*, 1992). The pathogen has been found on mites throughout many cassava-growing regions of the Neotropics. Some strains appear to be specific to the genus *Mononychellus* (de Moraes *et al.*, 1990). The pathogen has also been found in CGM in Africa, but epizootics have not been observed (Yaninek *et al.*, 1996), indicating that Brazilian strains may be more virulent than the African strains. Molecular techniques are being used to determine taxonomic identification of fungal strains, and *in vitro* methodologies for rearing the pathogen are being developed. This fungus, which shows considerable promise for biological control of CGM, is being evaluated in Africa.

Cassava mealybugs

More than 15 species of mealybugs are reported feeding on cassava in Africa and South America. *P. herreni*, *P. manihoti*, *Phenococcus madeirensis*, *Ferrisia virgata* and *Pseudococcus mandio* are all reported from the Americas (Bellotti *et al.*, 1983a; Williams and Granara de Willink, 1992). Only *P. herreni* and *P. manihoti*, both of neotropical origin, are important economically (Table 11.2). *P. manihoti* was introduced inadvertently

Table 11.4. Life table data on selected Phytoseiidae predators of the cassava mites *Mononychellus tanajoa* (*Mt*), *Mononychellus caribbeanae* (*Mc*) and *Tetranychus urticae* (*Tu*).

Phytoseiidae species	No. colony strains/species[a] (1986–1999)	Tolerance to RH[b] (Egg eclosion)	Egg consumption[b] (24 h)	Development time (days)[c]			Fecundity[c]			Longevity (days)[c]		% Females[c]	
				Mt	*Tu*	*Mc*	*Mt*	*Tu*	*Mc*	*Tu*	*Mc*	*Mt*	*Tu*
Typhlodromalus manihoti	31	+	68	4.9	4.1	5.5	14.2	–	3.5			74	88
Typhlodromalus aripo	9	+				6.8		13.0[d]	13.0[d]	14.0	20.9		81
Typhlodromalus tenuiscutus	7	+	45.4	5.8	5.8	5.7	32.0	2.5	16.1	6.6	16.1	75	84
Neoseiulus idaeus	20	+++	26.8	4.6	4.6	5.1	13.8	32.3	12.5	21.6	27.8	73	79
Neoseiulus californicus	5	++	26.5	4.7	4.4	7.7	34.8	43.7	23.4			70	62
Typhlodromalus rapax	1			5.0	5.4	5.8	6.0	12.0	19.4			78	58
Neoseiulus anonymus	4			4.7	5.1	5.2	14.5	34.4	27.7	39.1	12.0	73	66
Galendromus helveolus	5	+		7.4	7.0		18.7[d]	8.0[d]	23.0[d]	14.2	19.0	64	85
Galendromus annectens	6	++	17.8	5.7	6.1		22.4[d]	19.0[d]	31.0[d]	23.0	27.7	74	70
Euseius concordis	1			5.7	5.0		12.7[d]					75	

RH, relative humidity; + = 75%; ++ = 60%; +++ = 40–50%.
[a]CIAT (1994).
[b]CIAT (1999).
[c]CIAT (1990).
[d]Cuéllar *et al.* (1996).

into Africa in the early 1970s, where it spread rapidly, causing considerable yield loss. It has been the object of a successful biological control programme (Herren and Neuenschwander, 1991). In the Americas, *P. manihoti* is confined to Paraguay, certain areas of Bolivia and Mato Grosso do Sul State of Brazil, causing no economic damage (Lohr and Varela, 1990).

P. herreni is distributed throughout northern South America and northeast Brazil, where high populations can cause considerable yield losses (Table 11.2). Damage caused by both species is similar: feeding by nymphs and adults causes leaf yellowing, curling and cabbage-like malformation of the growing points. High densities lead to leaf necrosis, defoliation, stem distortion and shoot death. Reductions in photosynthetic rate, transpiration and mesophyll efficiency – together with moderate increases in water pressure deficit, internal CO_2 and leaf temperature – were found in infested plants (CIAT, 1992).

P. manihoti is parthenogenic, whereas males are required for reproduction of *P. herreni*. *P. herreni* females deposit ovisacs containing several hundred eggs on the underside of the leaves, or around the apical bud. Eggs hatch in 6–8 days and there are four nymphal instars; the fourth instar is the adult stage. Males have four nymphal instars plus the adult stage. The third and fourth instars occur in a cocoon, from which the winged adult emerges. Adult males live only 2–4 days; the average life cycle of the female is 49.5 days, that of the male, 29.5. The optimal temperature for female development is 25–30°C (Herrera *et al.*, 1989).

P. herreni populations peak during the dry season. The onset of rains reduces pest populations and permits crop recovery. (Herrera *et al.*, 1989). Recent research shows that when the water supply is limited, the cassava leaves produce increased amounts of metabolites, which could favour mealybug growth and decrease parasitoid efficacy (CIAT, 1999; Polanía *et al.*, 1999; Calatayud *et al.*, 2002). These results could help explain the rapid mealybug population increases during the dry season.

Control

Considerable effort has been made to identify resistance to cassava mealybugs. Of > 3000 cultivars from the CIAT cassava germplasm bank screened for resistance to *P. herreni*, only low levels of resistance or tolerance were identified (Porter, 1988). Resistance studies at IITA in Africa and at ORSTOM have given similar results. Partial or low-to-weak levels of resistance have been reported in germplasm evaluations with *P. manihoti* (Le Ru and Calatayud, 1994; Neuenschwander, 1994a). It is suggested, however, that even low levels of plant resistance could enhance the impact of natural enemies in a biological control programme.

BIOLOGICAL CONTROL. Management of cassava mealybugs is a well-documented example of classical biological control, especially in Africa where *P. manihoti* is being controlled successfully through the introduction of the parasitoid *Apoanagyrus lopezi* from the Neotropics. Although *P. herreni* is distributed throughout northern South America, it causes serious yield losses only in northeast Brazil. Thus *P. herreni* may be an exotic pest in this region, probably coming from northern South America (Williams and Granara de Willink, 1992).

Approximately 70 species of parasites, predators and entomopathogens of *P. herreni* have been identified in the Neotropics. Many of these are generalist predators that feed upon numerous mealybug species. Nevertheless, several parasitoids show a specificity or preference for *P. herreni*. Parasitoids identified from northern South America include *Acerophagus coccois*, *Apoanagyrus diversicornis*, *Anagyrus putonophilu*, *Anagyrus insolitus*, *Apoanagyrus elegeri* and *Aenasius vexans*. The three encyrtid parasitoids (*Ap. diversicornis*, *Ac. coccois* and *Ae. vexans*) have been identified as effective parasitoids of *P. herreni* (Van Driesche *et al.*, 1988, 1990).

Ae. vexans and *Ap. diversicornis* display a marked preference for parasitizing *P. herreni*, although laboratory studies show they will also parasitize other mealybug species (Bellotti *et al.*, 1983a, 1994; Bertschy *et al.*, 1997). *Ac. coccois* showed equal preference for both *P. herreni* and *P. madeirensis*. All three parasitoids were attracted to *P. herreni* infestations (Bertschy *et al.*, 1997). Comparative studies on the life cycles of the three parasitoid species show that each would complete two cycles for each cycle of *P. herreni*, which is a favourable ratio for biological control.

Ap. diversicornis prefers third instar nymphs, whereas the much smaller *Ac. coccois* can parasitize male cocoons, adult females and second instar nymphs with equal frequency. Oviposition by *Ap. diversicornis* caused 13% mortality of first nymphal instars (Van Driesche *et al.*, 1990). *Ae. vexans* prefers second and third instar nymphs and adult females with equal frequency (CIAT, 1990). Field studies with natural populations of *Ap. diversicornis* and *Ac. coccois* determined percentage parasitism by using trap plants with mealybug hosts set out in cassava fields (Van Driesche *et al.*, 1988). *P. herreni* mortality was estimated at 55% for the combined action of the two parasitoids (Van Driesche *et al.*, 1990).

Through the combined efforts of CIAT and EMBRAPA, *Ap. diversicornis, Ac. coccois* and *Ae. vexans* were exported from CIAT and released in northeast Brazil, primarily in the states of Bahia and Pernambuco from 1994 to 1996. Prior to introduction, EMBRAPA scientists had conducted field surveys to measure damage and collect natural enemies. By the end of 1996, > 35,000 individuals of the three parasitoid species had been released. In Bahia *Ap. diversicornis* dispersed 130 km in 6 months, 234 km in 14 months and 304 km in 21 months after release. *Ac. coccois* also became established and was recovered in high numbers at ≤ 180 km from its release site 9 months later. *Ae. vexans*, although being consistently recaptured at its release site in Pernambuco, dispersed only 40 km in 5 months (Bellotti *et al.*, 1999; Bento *et al.*, 1999). Subsequently, personal observations indicate that mealybug populations have been reduced considerably and that cassava cultivation has returned to areas where it had been previously abandoned due to *P. herreni* outbreaks.

Whiteflies

Whiteflies cause major damage in cassava-based agroecosystems in the Americas, Africa and, to a lesser degree, in Asia as direct feeding pests and virus vectors. There is a large complex in the Neotropics, where 11 species are reported on cassava: *Al. socialis, Trialeurodes variabilis, Al. aepim, Bemisia tuberculata, Bemisia tabaci, Bemisia argentifolii, Trialeurodes abutiloneus, Aleurodicus dispersus, Paraleyrodes* sp., *Aleuronudus* sp. and *Tetraleurodes* sp. (Bellotti *et al.*, 1994, 1999; Castillo, 1996; França *et al.*, 1996). *A. socialis* is the predominant species in northern South America, where it causes considerable crop damage, but it is also found, to a lesser extent, in Brazil (Farias, 1994). *B. tuberculata* and *T. variabilis* are reported in low populations from Brazil, Colombia, Venezuela and several other countries (Farias, 1990a; Bellotti *et al.*, 1999). The spiralling whitefly *A. dispersus* is reported causing conspicuous damage on cassava in West Africa (Neuenschwander, 1994b; D'Almeida *et al.*, 1998); *Bemisia afer* occurs in Kenya (Munthali, 1992) and the Côte d'Ivoire and in other countries of sub-Saharan Africa (J. Legg, personal communication).

B. tabaci has a pantropical distribution, feeding on cassava throughout Africa and several countries in Asia including India (Lal and Pillai, 1981) and Malaysia. Before 1990, the *B. tabaci* biotypes found in the Americas did not feed on cassava. Whiteflies are known to transmit the viruses causing two diseases of cassava:

- cassava mosaic disease (CMD) in Africa, India and Sri Lanka, caused by geminiviruses that are transmitted by *B. tabaci* (Thresh *et al.*, 1998; see Chapter 11).
- *B. tuberculata* is the reported vector of cassava frog skin disease in the Neotropics (Angel *et al.*, 1990; see Chapter 11).

It has been speculated that the absence of CMD in the Americas may be related to the inability of its vector, *B. tabaci*, to colonize cassava. Since the early 1990s a new biotype (B) of *B. tabaci*, regarded by some as a separate species (*B. argentifolii*), has been found feeding on cassava in the Neotropics. It is considered that CMD now poses a more serious threat to cassava production given that most traditional cultivars in the Neotropics are highly susceptible to the disease. In addition the *B. tabaci* biotype complex is the vector of several viruses of crops often grown in association with or near cassava. The possibility of viruses moving between these crops or the appearance of new viruses presents a potential threat.

Whiteflies cause direct damage to cassava by feeding on the phloem of leaves, inducing leaf chlorosis and abscission, which results in considerable reduction in root yield if feeding is prolonged. Yield losses of this type are common due to *Al. socialis* and *At. aepim*. There is a correlation between duration of whitefly attack and root yield loss (Table 11.2).

Research efforts in the Neotropics have concentrated on *Al. socialis* and *At. aepim*. Populations of both species are highest during the rainy season, but may occur throughout the crop cycle (Farias *et al.*, 1991; Gold *et al.*, 1991). *Al. socialis* females oviposit individual banana-shaped eggs on the undersides of apical leaves. Eggs hatch in *c.* 10 days and pass through three feeding nymphal instars and a pupal stage (fourth instar) before reaching the winged adult stage. During the third instar, the body colour changes from beige to black, surrounded by abundant waxy white cerosine. The black pupal stage makes this species easy to distinguish from other whitefly species feeding on cassava. Egg-to-adult development time of *Al. socialis* in the growth chamber ($28 \pm 1°C$, 70% relative humidity (RH)) was 32 days (Arias, 1995).

Control

HPR and biological control agents are accepted increasingly as complementary pest-control tactics that reduce environmental contamination and other disadvantages that arise from the excessive use of chemical pesticides. Research on cassava whitefly control in the Neotropics initially emphasized activities in HPR and cultural practices. More recently, a concentrated effort is being made to identify and evaluate the use of natural enemies in an IPM context.

CULTURAL CONTROL. In traditional cropping systems, cassava is often intercropped, a practice that has been shown to reduce populations of many pests (Leihner, 1983). Intercropping cassava with cowpea reduced egg populations of *Al. socialis* and *T. variabilis*, compared to those in monoculture (Gold *et al.*, 1990). These effects were residual and persisted up to 6 months after harvest. Yield losses in cassava/maize, cassava monoculture and mixed cultivar systems were *c.* 60%; whereas in cassava/cowpea intercrops, yield losses were only 12% (Gold *et al.*, 1989a). Intercropping with maize did not reduce egg

populations (Gold, 1993), indicating that this technique can depend on the intercropped species for success, thereby limiting its effectiveness and acceptance by farmers. However, it is a promising means of reducing pest populations for small-scale farmers.

HPR. Stable HPR offers a practical, low-cost, long-term solution for maintaining reduced whitefly populations. Whitefly resistance in agricultural crops is rare, although several good sources of resistance have been identified and high-yielding, whitefly-resistant cassava hybrids are being developed. HPR studies initiated at CIAT > 15 years ago are evaluating systematically the 6000 cultivars in the germplasm bank for whitefly resistance (CIAT, 1999), especially to *Al. socialis*. In Brazil some HPR research has been done with *At. aepim* (Farias, 1990a).

Several sources of resistance to *Al. socialis* have been identified. The clone M Ecu 72 has consistently expressed the highest level of resistance. Additional cultivars expressing moderate-to-high levels of resistance include M Ecu 64, M Per 335, M Per 415, M Per 317, M Per 216, M Per 221, M Per 265, M Per 266 and M Per 365. Based on these results, *Al. socialis* resistance appears to be concentrated in germplasm originating from Ecuador and Peru, but this phenomenon needs to be investigated further.

M Ecu 72 and M Bra 12 (an agronomically desirable clone with field tolerance to whiteflies) were used in a crossing programme to provide high-yielding, whitefly-resistant clones that showed no significant differences in yield between insecticide-treated and untreated plots (CIAT, 1992; Bellotti *et al.*, 1999). Greenhouse and field studies showed that *Al. socialis* feeding on resistant clones had less oviposition, longer development periods, reduced size and higher mortality than those feeding on susceptible ones. *Al. socialis* nymphal instars feeding on M Ecu 72 suffered 73% mortality, mostly in the early instars (CIAT, 1994; Arias, 1995; Fig. 11.1). The progeny (CG 489–34, CG 489–4, CG 489–31 and CG 489–23) selected from the M Ecu 72 × M Bra 12 cross have consistently displayed moderate levels of whitefly resistance. Three of these hybrids are currently being evaluated for release to producers in Colombia.

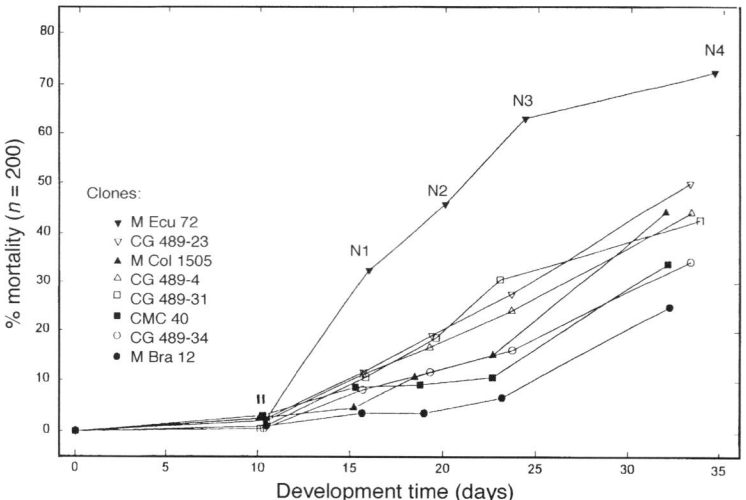

Fig. 11.1. Whitefly (*Aleurotrachelus socialis*) nymphal development and mortality on resistant and susceptible cassava clones. Source: Arias, 1995.

Resistance screening using natural *Al. socialis* populations is done in the field at two sites in Colombia:

- Nataima, Tolima, in cooperation with CORPOICA, the Colombian Agricultural Research Corporation. *Al. socialis* populations at Nataima have consistently been at moderate to high levels for nearly 15 years, offering the opportunity for sustained research over a long period.
- CIAT, Palmira, Valle del Cauca. Initially, *Al. socialis* populations at CIAT were low; however, since 1994, populations have increased dramatically and are presently higher than in Tolima. The reason for this sudden increase in *Al. socialis* populations is not understood, but is evidence of the dynamics of cassava pest eruptions, emphasizing the present and potential severity of whiteflies as cassava pests.

Research is also being done at CIAT to identify molecular markers linked to genes conferring resistance to *Al. socialis* in order to evaluate further and understand the genetics of whitefly resistance in cassava. Different breeding populations have been obtained from crosses between resistant (CG 489–34) and susceptible (M Col 2026) genotypes. Amplified fragment length polymorphism (AFLP) and simple sequence repeat (SSR) markers with bulk segregant analysis (BSA), are being used to locate markers linked with resistance for mapping and ultimately cloning the resistant genes. Co-segregating AFLP bands with resistance to *Al. socialis* have been identified and are being sequenced to generate sequence-characterized amplified region (SCARs) markers. The PCR-based marker will be the basis for the molecular mapping and used in breeding.

BIOLOGICAL CONTROL. Surveys in recent years in the Neotropics – especially in Colombia, Venezuela, Ecuador and Brazil – have identified a considerable number of natural enemies associated with the cassava whitefly complex. Gaps in knowledge about the natural enemy complex associated with the different whitefly species have limited their effectiveness in biological control programmes. Although a large complex of parasitoids has been identified, there is limited knowledge about levels of parasitism, parasitism rates for individual species, host specificity and their overall effect on regulating whitefly populations.

Since 1994, CIAT has carried out surveys in northern South America for natural enemies. The most representative group is the micro-hymenopteran parasitoids (Gold *et al.*, 1989c; Castillo, 1996; Evans and Castillo, 1998). In

Colombia, species richness – primarily from the genera *Encarsia*, *Eretmocerus* and *Amitus* – was most frequently associated with *Al. socialis*, *B. tuberculata* and *T. variabilis* (Castillo, 1996). More than ten species – several unrecorded – were collected. Three of the *Encarsia* spp. were identified as *Encarsia hispida*, *Encarsia pergandiella* and *Encarsia bellottii* (Evans and Castillo, 1998). None of the *Eretmocerus* and only one *Amitus* (*Amitus macgowni*) have been identified. The predominant species were *E. hispida*, *Amitus* sp. and *Eretmocerus* sp. The highest levels of parasitism observed for *Al. socialis*, *B. tuberculata* and *T. variabilis* were 15, 14 and 12%, respectively, although this varied according to geographic region (Castillo, 1996). Parasitism was higher in the Andean zone than in the coastal and eastern plains regions of Colombia. In surveys conducted between 1997 and 1999 in Colombia, *Encarsia* was the genus most frequently collected from the Andean highlands, while *Eretmocerus* predominated at lower altitudes on the Caribbean coast (CIAT, 1999). Moreover, the parasitoid species complex associated with each whitefly species can be influenced by geographic area. On the Caribbean coast, *Al. socialis* was most frequently parasitized by *Eretmocerus*, while in the Andean highlands it was *Encarsia*. In Valle del Cauca (1000 m altitude), 99.6% of the parasitism of *Al. socialis* was by *Encarsia* and 0.4% by *Eretmocerus*. The most numerous complex of parasitoids species was found associated with *B. tuberculata*.

Greenhouse studies with *E. hispida* parasitizing *Al. socialis* show that the third whitefly instar is preferred. Parasitism rates reached 75% in the third instar and 16, 45 and 43% in the first, second and fourth instars, respectively. The average parasitism rate was 45%, and peak parasitism occurred 72–96 h after exposure (CIAT, 1999). *E. hispida* is the most frequent parasitoid observed when there are high populations of *Al. socialis*, but its effectiveness in regulating the populations in the field is not known.

The influence of *Al. socialis*-resistant cassava cultivars on parasitoid behaviour has also been evaluated. Survival of *E. hispida* was not adversely affected by resistant genotypes. Parasitoid emergence was, however, considerably lower from pupae of *Al. socialis* that fed on the resistant cultivar M Ecu 72 than on the susceptible CMC 40 (CIAT, 1999).

Lepidoptera

Several species of Lepidoptera feed on cassava, including *E. ello* (L), *Erinnyis alope*, *Chilomima clarkei* (Amsel), *Chilozela bifilalis* (Hampson), *Phyctaenodes fibilialis*, *Agrotis ipsilon*, *Prodenia eridania* and *Phoenicoprocta sanguinea* (Bellotti and van Schoonhoven, 1978a,b; Bellotti *et al.*, 1999). The stemborer *Cm. clarkei* causes breakage of cassava stems. *A. ipsilon* and *Pr. eridania* attack recently planted stem cuttings, resulting in poor establishment; *Ph. sanguinea* is a leaf feeder. Yield reductions attributable to these last three species have not been documented. Attacks by *E. ello* and *Cm. clarkei* can reduce yields and are discussed in detail.

Cassava hornworms

E. ello (Sphingidae) is one of the most serious pests of cassava in the Neotropics (Bellotti *et al.*, 1992, 1999). It has a broad geographic range, extending from southern Brazil, Argentina and Paraguay to the Caribbean basin and the southern USA. The migratory flight capacity of *E. ello*, its broad climatic adaptation and wide host range probably account for its wide distribution and sporadic attacks (Janzen, 1987). Several other species of *Erinnyis* feed on cassava. The subspecies *E. ello ello* and *E. ello encantado* and the closely related species *E. alope* are all reported from the Neotropics.

Hornworm larvae feed on cassava leaves of all ages as well as on young, tender stems and leaf buds. Severe attacks cause complete plant defoliation, bulk root loss and poor root quality (Table 11.2). Although yield losses may be severe, complete defoliation due to hornworm attack, or even repeated attacks, do not kill cassava. The carbohydrates stored in the roots enable the plant to recover, especially under the more favourable conditions of the tropical rainy season. Repeated attacks are most common when ill-timed pesticide applications destroy natural enemies, but not fifth instar larvae or prepupae of the pest (Braun *et al.*, 1993). Moreover, large plantations of cassava are more prone to frequent and repeated hornworm attacks.

E. ello adults are grey nocturnal moths that oviposit small, round, light green-to-yellow eggs individually on the upper surface of cassava leaves. In field cage studies as many as 1850 eggs were laid per female. This high oviposition, combined with the migratory behaviour of adults, helps explain the rapid build-up of hornworm populations and their sporadic occurrence (Janzen, 1987; Bellotti *et al.*, 1992). During the larval period, each hornworm consumes *c.* 1100 cm^2 of leaf foliage; *c.* 75% of this during the fifth instar. At 15, 20, 25 and 30°C the mean duration of the larval stage is 105, 52, 29 and 23 days, respectively, indicating that peak hornworm activity may occur at lower altitudes (< 1200 m), or during the summer in the subtropics (Bellotti and Arias, 1988).

CONTROL. *E. ello*'s strong flight abilities and migratory flight capacity, combined with its broad climatic adaptation and wide host range (Janzen, 1986, 1987), often make effective control difficult to achieve. Pesticides give adequate control if hornworm populations are detected and treated while at the first three instar stages. However, farmers often react to attacks with excessive, ill-timed applications of pesticides, leading to repeated and more severe attacks (LaBerry, 1997). Larval populations in the fourth and fifth instar stages are not only difficult to control but also uneconomical, because considerable defoliation has already occurred. Pesticide use also disrupts natural enemy populations, leading to more frequent attacks (Urias López *et al.*, 1987).

An extensive complex of natural enemies is associated with *E. ello*. Nevertheless, their effectiveness is limited, most likely because of the migratory behaviour of the hornworm adults. Adults migrating *en masse* will oviposit a considerable number of eggs in cassava fields (up to 600 per plant), where natural enemy populations are too low to prevent an explosion of hornworm larvae, which can cause severe crop defoliation. Because their rate of reproduction is limited, predators and parasites cannot usually compensate sufficiently quickly to suppress dramatic hornworm eruptions (Bellotti *et al.*, 1992).

Approximately 35 species of parasites, predators and pathogens of the egg, larval and pupal stages have been recorded and reviewed extensively (Bellotti and van Schoonhoven, 1978a,b; Schmitt, 1983; Farias, 1990b; Bellotti *et al.*, 1992, 1999). Eight microhymenopteran species of the families Trichogrammatidae, Scelionidae and Encyrtidae are egg parasites, of which

Trichogramma and *Telenomus* are the most important. Tachinid flies are the most important group among the dipteran larvae parasitoids; the Braconidae, particularly *Cotesia* spp., are the most important hymenopteran (Bellotti *et al.*, 1992, 1994). The commonest egg predators are *Chrysopa* spp. Important larval predators include *Polistes* spp. (Hymenoptera: Vespidae), *Podisus* spp. (Hemiptera: Pentatomidae) and several spider species (Bellotti *et al.*, 1992). Important entomopathogens include *Cordyceps* sp. (Aconycetes: Clavicipitaceae), a soil-borne fungus that invades hornworm pupae, causing mortality. More recently, isolates of *Beauveria* sp. and *Metarhizium* sp. were found to cause high larval mortality in laboratory studies (Múnera *et al.*, 1999).

The key to the effective use of biological control agents is the ability to synchronize the release of large numbers of predators or parasites during the early stages, preferably the egg or the first to third larval instars. Predator and parasite effectiveness is limited by poor functional response during hornworm outbreaks, which are of short duration (15 days). Successful control, therefore, requires monitoring of field populations to detect immigrant adults or larvae in the early instars. This can be done with black lights (Type T20T12BLT), which trap flying adults, or by scouting for the presence of eggs or larvae (Braun *et al.*, 1993). The complexities of synchronizing inundative releases of parasites and predators with peak pest populations, suggests the need for a cheap, storable biological pesticide.

A granulosis virus species of the family *Baculoviridae* was found attacking *E. ello* in cassava fields at CIAT in the early 1970s. Pathogenicity studies in the laboratory and field resulted in nearly 100% mortality of hornworm larvae (Bellotti *et al.*, 1992; Braun *et al.*, 1993; Table 11.3). Infested larvae are collected from the field, macerated in a blender, filtered through cheesecloth, mixed with water and applied to hornworm-infested fields. Studies on the effect of virus concentration on mortality of larval instars showed a sigmoidal relationship for the first, second and fourth instars. LC$_{50}$ studies indicate that progressively higher concentrations are needed for adequate control of each succeeding larval instar. Most fifth instar larvae reached the prepupal state, but few female adults emerged and those that did had wing deformities and died without producing progeny (Bellotti *et al.*, 1992).

The hornworm baculovirus can be managed by cassava growers themselves. They can collect and macerate diseased larvae and apply the virus suspension to their fields. The virus can be stored at low cost by refrigeration and wettable powder formulations of the virus are being developed. Hornworm management with the virus was first implemented in southern Brazil, where light traps were used to detect adult movement and invasions. Virus applications, made when populations were in their early instars, resulted in almost complete control (Schmitt, 1988) and pesticide applications were reduced by 60%.

The hornworm virus is an especially attractive option for use on large cassava plantations, where pesticide applications have proved ineffective. In Venezuela, where the hornworm is endemic, the virus preparation was applied (70 ml ha^{-1}) to 7000 ha via overhead sprinkler irrigation systems when larvae were in the first and second instars, resulting in 100% control. The virus has replaced pesticides and the cost of gathering, processing, storing and applying it is only US$4 ha^{-1} (LaBerry, 1997).

Stemborers

A complex of arthropod stemborers, which includes both coleopteran and lepidopteran species, feed on and damage cassava. Long-horned beetles (*Lagochirus* spp.) are distributed worldwide, but do not appear to cause yield losses. Stemborers are most important in the Neotropics, especially in Colombia, Venezuela and Brazil. Seven species of *Coelosternus* (Coleoptera: Curculionidae) are reported reducing cassava yields and quality of planting material in Brazil. However, the damage is generally sporadic or localized and does not significantly reduce yield (Bellotti and van Schoonhoven, 1978a,b).

Populations of the stemborer *C. clarkei* (Lepidoptera: Pyralidae) have increased dramatically in Colombia and Venezuela in recent years and the species is now a major pest of cassava (Vides *et al.*, 1996). The nocturnal adult females oviposit in cassava stems, usually around the bud or node. The egg stage is ~6 days (28°C). Upon hatching, first instar larvae feed on the outer bark or stem epidermis. They are very mobile and search for an appropriate feeding site, generally around axillary buds. They form a protective web, under which the first four instars feed, enlarging the web with each instar. In the fifth instar the larvae penetrate the stem, where they complete their cycle (6–12 instars), pupate and emerge as winged adults (Lohr, 1983). The larval stage lasts 32–64 days; the pupal stage within the stem is 12–17 days. Female adults live 5–6 days (males 4–5 days), each ovipositing an average of 229 eggs.

Ch. clarkei populations occur throughout the year but are highest during the rainy season. From 4–6 overlapping cycles can occur during the 1-year crop cycle, increasing potential damage and making control more difficult. Extensive tunnelling by the larvae (> 20 can be found in one stem) cause stem breakage, leading to stem rot and a reduction in the quality and quantity of planting material (Table 11.3). Field studies show that when > 35% of the plants suffer stem breakage, significant yield loss can occur (45–62%) (Lohr, 1983). On the Colombia Caribbean coast, 85% of the cassava fields had *Ch. clarkei* damage (Vides *et al.*, 1996).

CONTROL. Once larvae enter the stem, control is very difficult. Moreover the web that covers the early larval instars offers protection from both natural enemies and pesticide applications. The highly mobile early-instar larvae are more vulnerable and can be controlled by *Bacillus thuringiensis* [Bt]. Given the overlapping generations, however, several applications would be necessary and they would be too costly for producers to adopt. Field research by Gold *et al.* (1990) showed that intercropping with maize reduced stemborer populations until the intercrop was harvested.

Several natural enemies have been identified, including the hymenopteran parasitoids *Bracon* sp., *Apanteles* sp. and *Brachymeria* sp. (Lohr, 1983). More recently emphasis has been placed on identifying resistant cassava germplasm. Approximately 1000 clones have been evaluated on the Colombia Caribbean coast, where *Ch. clarkei* populations are high. Evaluations are based on the number of *Ch. clarkei* holes and tunnels in the stems and percentage stem breakage. A number of clones with only 0–1 hole per stem have been identified, indicating varietal influence (CIAT, 1999). Nevertheless, field evaluations of germplasm using natural populations of a highly mobile pest can often give misleading results due to 'escapes' (plants that have avoided damage by chance). Thus these cultivars will need to be evaluated over several cycles.

CIAT has initiated research based on introducing insect-resistant Bt genes through *Agrobacterium*-mediated transformation into cassava embryonic tissue to develop *Ch. clarkei*-resistant cultivars. Initial results are promising (CIAT, 1999).

Cassava burrower bug: C. bergi

C. bergi is one of the few arthropod pests that feed directly on the tuberous roots of cassava, but this polyphagous species has not co-evolved with cassava. It was first recorded as a cassava pest in Colombia in 1980 (García and Bellotti, 1980) and more recently it has been reported causing commercial losses in Panama, Costa Rica and Venezuela (Riis, 1997). *C. bergi* is probably present in many other areas of the Neotropics, but feeding on other hosts including onion, groundnut, maize, potato, *Arachis pintoi* (forage groundnut), sorghum, sugarcane, coffee, coriander, asparagus, beans, pea, pastures and numerous weeds (Riis, 1997; Bellotti *et al.*, 1999).

Some hosts are strongly preferred over others. Laboratory experiments indicate that cassava is not the optimal host. *C. bergi* develops faster on maize and groundnut than on cassava and prefers maize to cassava (78 versus 22%) in free-choice feeding tests. The LD$_{50}$ for adults was 95 days on maize, 69 on onion and 66 and 64 days, respectively, on sweet (CMC 40) and bitter (M Col 1684) cassava (Riis, 1990). Optimal fecundity, survival and intrinsic rate of population increase occurred on groundnut and peanut, not on maize. Sweet cassava, sorghum and onions were not favourable hosts and *C. bergi* could not complete its life cycle on bitter cassava (Riis, 1997).

C. bergi nymphs and adults feed on cassava roots by penetrating the peel and parenchyma with a thin strong stylet. This feeding action can introduce several soil-borne pathogens (e.g. *Aspergillus*, *Diplodia*, *Fusarium*, *Genicularia*, *Phytophthora* and *Pythium* spp.) into the root parenchyma (Bellotti and Riis, 1994). Brown-to-black lesions begin to develop on the roots within 24 h after feeding is initiated, resulting in starch reduction and a serious loss in commercial value (Table 11.2). As damage is not detected until roots are harvested and peeled, producers can lose the value of the crop and also labour, time and land use.

C. bergi has five nymphal instars; nymphs and adults can live for more than 1 year feeding on cassava roots (García and Bellotti, 1980). *C. bergi* had a lifespan of 286–523 days when fed on slices of low-HCN cassava roots in the laboratory (23°C, 65 ± 5% RH). Egg eclosion averaged 13.5 days; mean development time of the five nymphal stages was 111 days; mean longevity for adults was 293 days.

C. bergi populations occur in the soil throughout the crop cycle and root damage starts in the first month of plant growth. Feeding can continue throughout the crop cycle and can result in 70–80% total root damage and > 50% reduction in total starch content. *C. bergi* can cause serious economic damage, even if populations are not high (Arias and Bellotti, 1985). Riis (1990) showed that even when populations were very low (~0), 22% of the roots were damaged. The economic injury threshold – the point where cassava root purchasers will reject a consignment – is when 20–30% of the roots are damaged, given that the 'cosmetic' damage of rot spots is not acceptable for the fresh food market.

C. bergi is strongly attracted to moist soil; it will migrate when soil moisture content is below 22% and is most persistent when it exceeds 31%. The rainy season greatly favours adult and nymphal survival, behaviour and dispersal; whereas low soil water content during the dry season restricts adult burrowing and migration and increases nymphal mortality (Riis, 1997).

Field trials and laboratory studies strongly suggest that feeding preferences of *C. bergi* may be related to the levels of cyanogenic glucosides in cassava roots. Adults and nymphs that fed on a high-HCN (> 100 mg kg^{-1} HCN) cultivar had longer nymphal development, reduced egg production and increased mortality. Oviposition on CMC 40 (43 mg kg^{-1} HCN) was 51 eggs per female versus only 1.3 on M Col 1684 (627 mg HCN equivalent kg^{-1}). Adult longevity on CMC 40 (235 days) was more than twice that on M Col 1684 (112 days; Bellotti and Riis, 1994). Riis (1997) showed that oviposition on clones with a cyanogenic potential (CNP) of < 45 p.p.m. (fresh weight) was significantly higher than on clones with a CNP > 150 p.p.m., while oviposition varied considerably on clones with CNPs between 45–150 p.p.m. Additional studies indicate that the earliest instars are most susceptible to root CNP. Due to the short length of the stylet, feeding during the first two nymphal instars is confined mainly to the root peel (Riis, 1990; Riis *et al.*, 1995), whereas third to fifth instars can feed on the root parenchyma. CMC 40 has a low cyanogen level in the root parenchyma, but a high level in the root peel (707 mg kg^{-1} HCN). Laboratory feeding experiments resulted in 56% mortality of first and second instar nymphs feeding on CMC 40 and 82% for those feeding on M Col 1684. The high cyanogen level in the peel of CMC 40 is probably responsible for the high mortality rate (Bellotti and Riis, 1994).

Feeding preference studies carried out in cassava fields in Colombia resulted in considerably more damage to CMC 40 (the low-HCN clone) than to M Col 1684. M Mex 59, with an intermediate cyanogen content (106 mg kg^{-1} HCN) suffered moderate damage. These data indicate that CNP may act as a feeding deterrent and the *C. bergi* damage should not be a problem where cassava clones with high CNP are cultivated (i.e. northeast Brazil and many parts of Africa; Bellotti and Riis, 1994).

CONTROL. Control of *C. bergi* is difficult due to the polyphagous nature of the pest and its adaptation to the soil environment. Measures should be taken early in the crop growth cycle, either during planting or within the first 2 months, given the fact that initial damage can occur during this period. Pesticide applications can reduce pest populations and damage. However, frequent applications may be required, and these are costly, environmentally hazardous, and often fail to reduce damage below economic injury levels (Castaño *et al.*, 1985). Intercropping cassava with *Crotalaria* sp. (sunn hemp) reduced root damage to ~4% compared to 61% damage in cassava monoculture (Table 11.3). However, cassava yields were reduced by 22% when intercropped and as *Crotalaria* has little commercial value, producers are reluctant to adopt this technology.

Experimental data and field observations show that high CNP cultivars are resistant to

C. bergi feeding and damage. However, in many cassava-producing regions, low CNP or 'sweet' cultivars are preferred, especially for fresh consumption. Recent studies indicate a potential resistance/tolerance to *C. bergi* in 15 low-CNP cultivars (Riis, 1997). The potential employment of this resistance justifies research on pest behaviour, resistance mechanisms, biochemistry and genetics.

The potential for biological control of *C. bergi* is being investigated and recent studies with entomopathogenic nematodes and fungi indicate a possible solution. This research, however, has been done only in the laboratory/greenhouse and field studies are required before acceptable technologies can be recommended.

The nematode *Steinernema carpocapsae* successfully parasitized *C. bergi* in the laboratory, infecting it 5–8 days after exposure. The adult was most sensitive to infection (59% parasitism after 10 days); the least sensitive were the first and second instars, with 17 and 31%, respectively (Caicedo and Bellotti, 1994). A native species, *Heterorhabditis bacteriophora*, found parasitizing *C. bergi* in the field in Colombia, resulted in 84% parasitism of the instars (Barberena and Bellotti, 1998). Isolates of the entomopathogenic fungus *Metarhizium anisoplae* have been recovered parasitizing *C. bergi* in the field. In laboratory studies mortality was highest (61%) during the fifth instar, overall average mortality was 33% (CIAT, 1994).

Cassava lacebugs

Lacebugs (Hemiptera: Tingidae) are neotropical cassava pests. Froeschner (1993) identified five *Vatiga* species that show a decided preference for cassava: *Vatiga illudens*, *Vatiga manihotae*, *Vatiga pauxilla*, *Vatiga varianta* and *Vatiga cassiae*. The first two are the most widely distributed. *V. illudens* predominates in Brazil, but also occurs throughout the Caribbean area. *V. manihotae* predominates in Colombia and Venezuela, but is also reported from Cuba, Trinidad, Peru, Ecuador, Paraguay, Argentina and Brazil. *V. varianta* is reported from Brazil and Colombia, *V. cassiae* from Brazil, and *V. pauxilla* from Argentina. Moreover, *A. machalana*, referred to as the black lacebug, damages cassava in Colombia, Venezuela and Ecuador (CIAT, 1990).

A prolonged dry period is favourable for increasing lacebug populations (Salick, 1983). Adults and nymphs feed on the undersurface of lower leaves. Initially, white feeding spots appear, increasing in number and area until the leaf centres turn white and eventually darken. High lacebug populations will cause leaves to curl and die. Younger plants (4–5 months old) attract higher populations, which tend to decline on older plants (Salick, 1983).

The relationship between damage, population density and duration is unknown. A field trial at CIAT with natural populations of *A. machalana* resulted in 39% yield reduction compared with pesticide-treated plots (CIAT, 1990).

Field observations in Colombia indicate a shift in populations of lacebug species. In CIAT cassava fields *V. manihotae* predominated until the mid-1980s. By 1990, *A. machalana* populations were considerably higher than *V. manihotae* and remained so for several years. Currently, *V. manihotae* is again the predominant species and it is difficult to find *A. machalana* in Colombian cassava fields (B. Arias, personal observation). The cause of this shift in populations is not understood. In Ecuador populations of *A. machalana* remain high.

Lacebugs are the least studied of the important cassava pests. Populations of *V. illudens* in Brazil are endemic and appear to be causing yield losses, especially in the central Campo Cerrado region and more recently in the south. Nevertheless, a sustained research effort has not been attempted in order to understand fully the dynamics of lacebug populations, yield losses and control. Considering the present and potential importance of lacebugs, there is a dearth of published information.

BIOLOGY. The egg stage of *V. manihotae* is 8–15 days, followed by five nymphal stages averaging 16–17 days. Adult longevity under field conditions averages 40 days (Borrero and Bellotti, 1983). Laboratory studies with *V. illudens* in Brazil show a nymphal duration of 13.5 days and an average adult longevity of 27 days (Farias, 1987). In laboratory studies with *A. machalana*, the egg stage averaged 8.2 days, the five nymphal instars 14 days and adult longevity was 18 and 22 days for females and males, respectively (CIAT, 1990).

CONTROL. Lacebug control appears to be difficult as few natural enemies have been observed (Borrero and Bellotti, 1983; Salick, 1983; Farias, 1985). Preliminary screening of cassava germplasm indicates that HPR may be available (CIAT, 1990; Calvacante and Ciociola, 1993), but considerable research is still required before implementation is possible.

Secondary pests

The cassava arthropod pest complex includes numerous species that feed on cassava but they do not generally cause major economic damage to the crop. These 'occasional' or 'incidental' pests either occur sporadically or at such low population levels that yield is unaffected. Outbreaks may occur in localized areas, and populations may increase to the point where they cause yield reductions. Moreover, changes in agronomic practices or varietal selection could influence pest populations, causing more crop damage (Table 11.5).

The thrips *F. williamsi* can reduce crop yields of cassava by 5–28%, depending on varietal susceptibility (Bellotti and van Schoonhoven, 1978a,b), especially in the seasonally dry tropics having a dry season of at least 3 months. *F. williamsi* can be controlled easily by using the resistant pubescent cultivars that are readily available and commonly grown in these areas (van Schoonhoven, 1974). If non-pubescent cultivars are introduced, thrips populations and damage increases, resulting in yield losses.

The fruitflies *A. pickeli* and *A. manihoti* normally attack cassava fruit. This is a problem for plant breeders, but of no concern to producers. In certain areas during the rainy season, however, females will oviposit in the tender upper portion of the cassava stem. The larvae tunnel into the stem, providing an entrance for soft rot bacteria such as *Erwinia carotovora*. The bacteria can cause severe rotting of the stem tissue, resulting in apical dieback. Yield losses due to this damage have not been reported, although there is a reduction in the quality of planting material, which may result in decreased establishment and yield in the subsequent crop cycle (Bellotti and van Schoonhoven, 1978a,b).

Shootfly (*N. perezi*) damage to the growing points of cassava breaks apical dominance,

Table 11.5 Occasional and incidental pests of cassava.

Common name	Important species	Region	Type of damage	Reported yield loss	Control strategy	References
Scales	*Aonidomytilus albus, Saissetia miranda*	Americas, Africa, Asia	Attack stems causing leaf fall; use of infested stems reduces establishment	< 20% fresh root yields; 50–60% loss in stem cutting	Destroy infested stems; use scale-free planting material	Bellotti and van Schoonhoven (1978a,b); Frison and Feliu (1991); Lozano et al. (1981); Bellotti and Peña (1978); Bellotti and van Schoonhoven (1978a,b); Lozano et al. (1981)
Fruitflies	*Anastrepha pickeli, Anastrepha manihoti*	Americas, Costa Rica, Panama, Venezuela, Colombia, Brazil, Peru	Bore fruit (seed) and stems, causing rotting of pith area	0–30% when infested stems used as planting material	Use of damage-free planting material	
Shootflies	*Neosilba perezi*	Americas	Larvae kill apical buds, retarding plant growth, inducing lateral branching	Not reported; reduced quality of planting material	None required	Bellotti and van Schoonhoven (1978a,b); Cock (1978); Lozano et al. (1981); Peña and Waddill (1982)
Gallmidges	*Jatrophobia (Eudiplosis) brasiliensis*	Americas	Yellowish green to red galls on upper leaf surface	Not reported	None required	Bellotti and van Schoonhoven (1978a,b); Lozano et al. (1981); Samways (1980)
White grubs	*Leucopholis rorida, Phyllophaga* spp., several others	All regions	Feed on planting material, roots	95% loss in germination	Soil pesticide treatment at planting	Bellotti and van Schoonhoven (1978a,b); Lal and Pillai (1981); Peña and Waddill (1982)
Termites	*Coptotermes voltkevi, Coptotermes paradoxis*	All regions	Tunnel in planting material, roots, stems, swollen roots	46–100% loss of planting material	Dusting of planting material with pesticide	Bellotti and van Schoonhoven (1978a,b); Lal and Pillai (1981); Lozano et al. (1981)
Stemborers	*Coelosternus* spp.	Americas, Brazil	Tunnel stems, stem breakage	Unknown	Cultural practices to maintain clean fields; destroy infested stems	Bellotti and van Schoonhoven (1978a,b); Lozano et al. (1986); Samways (1980); Villegas and Bellotti (1985)
	Lagochirus sp.	All regions	Tunnel stems, stem breakage	Unknown		
Leaf-cutting ants	*Atta* spp., *Acromyrmex* spp.	Americas	Foliage removed	Unknown	Toxic baits	Bellotti and van Schoonhoven (1978a,b); Diehl-Fleig et al. (1988); Samways (1980)
Grasshoppers	*Zonocerus elegans, Zonocerus variegatus*	Mainly Africa, occasionally Americas	Defoliation, stripping of bark	Unknown	Entomopathogens being investigated	Bellotti and Riis (1994); Bellotti and van Schoonhoven (1978a,b); Lomer et al. (1990); Modder (1994)
Thrips	*Frankliniella williamsi*	Americas, Africa	Leaf distortion, bud reduction	17–25%	Host plant resistance	Bellotti and Kawano (1980); Bellotti and van Schoonhoven (1978a,b); van Schoonhoven (1974)

retards plant growth and causes excessive branching. When younger plants are attacked, there is a reduction in the growth of the stems used as planting material, but seldom in yield. If plants are being grown for quality cutting production, then the crop needs to be protected only during the first 3 months of growth. Usually one timely pesticide application suffices to protect the crop.

Mealybugs feeding on and damaging cassava roots have been reported in recent years from two continents. In South America, *P. mandio* has been reported from southern Brazil, Paraguay and Bolivia (Williams, 1985; Pegoraro and Bellotti, 1994), and *Stictococcus vayssierei* is reported from the Cameroon, Africa (Ngeve, 1995). Mealybug feeding can result in reduced quality of tuberous roots and some plant defoliation. *P. mandio* females have three nymphal instars and adults oviposit an average of 300 eggs, accounting for rapid population build-ups. The life cycle from oviposition to adult was 25 days for females and 30 for males (Pegoraro and Bellotti, 1994). Yield losses of 17% have been reported; control through crop rotation is suggested.

Severe attacks of white grubs (Scarabaeidae) and termites can destroy the cuttings used to establish new plantations (Bellotti and van Schoonhoven, 1978a,b). Populations of scale insects, normally under natural biological control, can occasionally increase rapidly, especially if infested stakes are planted, causing yield losses and reduced availability of stems for use as planting material. Leaf-cutting ants can defoliate part or all of small plantations, possibly reducing yields.

Grasshopper (*Zonocerus elegans*, *Zonocerus variegatus*) attacks in Africa are reported as causing severe crop damage, but reliable data on actual root yield loss are scarce. Several countries including Nigeria, Congo, Benin, Uganda, Côte d'Ivoire and Ghana report thousands of hectares of cassava defoliated in some years, probably causing yield reductions (Modder, 1994). Simulated damage studies indicate a reduction in root yield (Toye, 1982) and in Nigeria, > 50% of the cassava crop is estimated to be lost in years of high grasshopper populations (Modder, 1994).

In Nigeria grasshopper oviposition usually occurs at the onset of the wet season and eggs hatch 6–7 months later, at the start of the dry season. This population will attack cassava toward the latter part of the dry season when preferred herbaceous food plants become scarce (Modder, 1994). Experiments show that grasshoppers are deterred from feeding on cassava by the production of large amounts of HCN. The early instars (I-IV) will not consume growing cassava and instars V and VI will eat growing cassava only if they have been deprived of food for a considerable period. Wilted cassava leaves are readily eaten by all stages and result in high grasshopper growth rate and high oviposition (Bernays *et al.*, 1977; Bellotti and Riis, 1994).

Chemical control of grasshoppers has been attempted in Africa. Although feasible, it does not appear to be financially or ecologically sustainable, and in the mid-to-long term is not effective (Modder, 1994).

Biological control using entomopathogens offers a more practical, safer and effective long-term solution for grasshopper control. The entomopathogens *Metarhizium flavoviride*, *Beauveria bassiana* and *Entomophaga grylli* have been identified infecting *Z. variegatus*. Currently, a considerable effort is being made to develop effective biopesticides and application technologies for grasshopper control. Results to date are encouraging, especially with *M. flavoviride* (Lomer *et al.*, 1990; Modder, 1994; Langewald *et al.*, 1997).

Trends in Pest Management

A successful integrated pest management (IPM) programme in cassava will depend on having effective, environmentally sound, low-cost pest management technologies available to cassava farmers in developing countries. Currently available biotechnology tools offer the potential to develop improved pest-resistant cultivars and to enhance the effectiveness of natural control organisms, including parasitoids and entomopathogens. The new generation of genetic pest management technologies presently being integrated with traditional IPM offers alternative technologies for controlling stemborers, leaf-cutting ants, grasshoppers, white grubs and other pests that are difficult to control. Research activities in these areas are already under way and recommendations should soon be available to farmers.

Pesticides

Pesticide use in traditional cassava agroecosystems is minimal due to their prohibitive costs and the long crop cycle of cassava, which may necessitate several applications. Farmers in the Neotropics, however, may respond to pest outbreaks with pesticides. As cassava production shifts to larger plantations, there may be a greater tendency to apply pesticides to control outbreaks, as already occurring in certain areas of Colombia, Venezuela and Brazil (Bellotti *et al.*, 1990).

There is considerable potential for the use of biopesticides to replace chemical pesticides in cassava pest management. The effectiveness of the hornworm baculovirus and its successful implementation, especially on large plantations (LaBerry, 1997), exemplifies this possible trend. Entomopathogens have been identified for mites, mealybugs, whiteflies, hornworms, burrower bugs, white grubs, grasshoppers and others. Further research is required to develop biopesticides and methodologies for their effective implementation. This will probably require a link to and collaboration with the biopesticide industry. This has been initiated already in Colombia.

Cultural practices

Traditional farmers in most cassava-growing regions have relied on an array of cultural practices that have often been effective in reducing pest populations (Lozano and Bellotti, 1985). Intercropping, as commonly practised by small farmers, has been shown to reduce populations and damage of whiteflies, hornworm and burrower bugs (Castaño *et al.*, 1985; Gold *et al.*, 1989b, 1990). Farmers may, however, be reluctant to adopt this practice if the intercrop species has little or no commercial value, or if cassava yields are reduced considerably (see 'cassava burrower bug'). On larger plantations where mechanization is a standard practice, there may be a reluctance to adopt intercropping. Additional cultural practices that can reduce pest populations include use of varietal mixtures, destruction (burning) of plant debris, crop rotation, changed planting dates and high-quality, pest-free planting material (Lozano and Bellotti, 1985).

Biological control

Classical biological control has been highly successful in Africa against introduced pests. Management of many cassava pests in the Neotropics will require greater farmer involvement for the effective implementation of solutions (Bellotti *et al.*, 1999). Numerous surveys of cassava fields in various regions of the Neotropics have revealed a species-rich complex of natural enemies of important cassava pests (see sections on individual pests). CIAT maintains a taxonomic and working collection, with an accompanying computerized database, for cassava pests and their natural enemies. This information is available to producers and staff of national agricultural research and extension programmes, taxonomists and museums (Hernández *et al.*, 1995).

Results from explorations and surveys indicate that considerable natural biological control may be occurring in the Neotropics. This is to be expected because in many cropping systems, cassava is grown as a functional perennial and certain pests and associated natural enemies may be in equilibrium. A disruption of this system by, for example, the use of pesticides could cause pest outbreaks. As previously described, the CGM populations in northern South America appear to be regulated by a complex of phytoseiid predators, which, if disturbed, results in reduced yields (Braun *et al.*, 1989). The potential for enhancing the effectiveness or virulence of natural enemies through genetic engineering offers further exploitation of this rich complex.

Host plant resistance

CIAT's germplasm bank of more than 6000 landrace cassava cultivars offers entomologists and breeders a potential pool of pest-resistance genes. As previously described, differing levels of resistance have been identified for mites, whiteflies, thrips, burrower bugs, lacebugs and stemborers. The novel biotechnology tools now available facilitate access to resistance genes and more efficient, quicker manipulation at the molecular level. A considerable portion of this germplasm bank is grown continually in the field and is available for systematic evaluation for pest resistance. Techniques and methodologies are now available for mass rearing of most major cassava pests and damage and population scales have been described to identify resistant and susceptible germplasm. Accurate germplasm evaluation needs to be done in the field using natural or artificial infestations. Damage symptoms of most cassava pests are not expressed accurately on plants grown in the greenhouse or screen house, resulting in the misidentification of resistance.

Cultivars that possess multiple resistance (i.e. to more than one pest) have also been identified. For example, M Ecu 72 has high levels of resistance to whiteflies and thrips, and moderate levels to mites. One of the challenges confronting geneticists and plant breeders will be to include both disease and arthropod pest resistance in the same cultivar.

Perhaps the greatest source of pest resistance is contained in wild *Manihot* species. More than 100 of these species have been identified (Allem, 1994) and small collections exist at several locations including CIAT, EMBRAPA/Brazil and IITA. The cassava molecular genetic map has been developed (Fregene *et al.*, 1997), and this should provide a useful tool for developing transgenic cassava plants (using other *Manihot* species) having pest resistance.

Cassava IPM projects are few and the decision-making guides and strategies required for appropriate implementation of control options are seldom available to small farmers in traditional production systems (Bellotti *et al.*, 1999). It is felt strongly that for large cassava plantation systems to succeed – especially in the Neotropics where there is a large complex of arthropod pests and diseases – the implementation of an effective IPM system based on biological control and varietal resistance is critical for sustaining high yields. A promising approach to overcome the slow technology diffusion inherent with cassava producers is through the use of farmer participatory methods and the inclusion of the private sector in setting the research agenda and objectives. The successful implementation of an integrated crop and pest management pilot project with traditional farmers in northeast Brazil (Bellotti *et al.*, 1999; Ospina *et al.*, 1999) provides an example of how this can be accomplished.

References

Allem, A.C. (1994) The origin of *Manihot esculenta* Crantz (Euphorbiaceae). *Genetic Resources and Crop Evaluation* 41, 133–150.

Angel, J.C., Pineda, B.L., Nolt, B. and Velasco, A.C. (1990) Mosca blanca (Homoptera: Aleyrodidae) asociadas a transmisión de virus en yuca. *Fitopatología Colombiana* 13, 65–71.

Arias, B. (1995) Estudio sobre el comportamiento de la 'mosca blanca' *Aleurotrachellus socialis* Bondar (Homoptera: Aleyrodidae) en diferentes genotipos de yuca, *Manihot esculenta* Crantz. MS thesis. Universidad Nacional de Colombia, Palmira.

Arias, B. and Bellotti, A.C. (1984) Pérdidas en rendimiento (daño simulado) causadas por *Erinnyis ello* (L.) y niveles críticos de población en diferentes etapas de desarrollo en tres clones de yuca. *Revista Colombiana de Entomología* 10, 28–35.

Arias, B. and Bellotti, A.C. (1985) Aspectos ecológicos y de manejo de *Cyrtomenus bergi* Froeschner, chinche de la viruela en el cultivo de la yuca (*Manihot esculenta* Crantz). *Revista Colombiana de Entomología* 11, 42–46.

Barberena, M.F. and Bellotti, A.C. (1998) Parasitismo de dos razas del nemátodo *Heterorhabditis bacteriofora* sobre la chinche *Cyrtomenus bergi* (Hemiptera: Cydnidae) en el laboratorio. *Revista Colombiana de Entomología* 24, 7–11.

Bellotti, A.C. and Arias, B. (1988) *Manejo Integrado de Erinnyis ello* (L.).Centro Internacional de Agricultura Tropical (CIAT), Cali, Colombia.

Bellotti, A.C. and Kawano, K. (1980) Breeding approaches in cassava. In: Maxwell, F.G. and Jennings, P.R. (eds) *Breeding Plants Resistant to Insects*. Wiley, New York, pp. 314–335.

Bellotti, A.C. and Peña, J.E. (1978) Studies on the cassava fruit fly *Anastrepha* spp. In: Brekelbaum, T., Bellotti, A.C. and Lozano, J.C. (eds) Proceedings, Cassava Protection Workshop. Centro Internacional de Agricultura Tropical (CIAT), Cali, Colombia, pp. 203–208.

Bellotti, A.C. and Riis, L. (1994) Cassava cyanogenic potential and resistance to pests and diseases. *Acta Horticulturae* 375, 141–151.

Bellotti, A.C. and van Schoonhoven, A. (1978a) *Cassava Pests and Their Control*. CIAT, Cali, Colombia.

Bellotti, A.C. and van Schoonhoven, A. (1978b) Mite and insect pests of cassava. *Annual Review of Entomology* 23, 39–67.

Bellotti, A.C., Reyes, J.A. and Varela, A.M. (1983a) Observaciones de los piojos harinosos de la yuca en las Américas; su biología, ecología y enemigos naturales. In: Reyes, J.A. (ed.) *Yuca: Control*

Integrado de Plagas. CIAT, Cali, Colombia, pp. 313–339.

Bellotti, A.C., Vargas, O., Peña, J.E. and Arias, B. (1983b) Pérdidas en rendimiento en yuca causadas por insectos y ácaros. In: Domínguez, D. (ed.) *Yuca: Investigation, Production y Utilization.* CIAT, Cali, Colombia, pp. 393–407.

Bellotti, A.C., Mesa, N., Serrano, M., Guerrero, J.M. and Herrera, C.J. (1987) Taxonomic inventory and survey activities for natural enemies of cassava green mites in the Americas. *Insect Science Application* 8, 845–849.

Bellotti, A.C., Cardona, C. and Lapointe, S.L. (1990) Trends in pesticide use in Colombia and Brazil. *Journal of Agricultural Entomology* 7, 191–201.

Bellotti, A.C., Arias, B. and Guzmán, O.L. (1992) Biological control of the cassava hornworm *Erinnyis ello* (Lepidoptera: Sphingidae). *Florida Entomology* 75, 506–515.

Bellotti, A.C., Braun, A.R., Arias, B., Castillo, J.A. and Guerrero, J.M. (1994) Origin and management of Neotropical cassava arthropod pests. *African Crop Science Journal* 2, 407–417.

Bellotti, A.C., Smith, L. and Lapointe, S.L. (1999) Recent advances in cassava pest management. *Annual Review of Entomology* 44, 343–370.

Bento, J.M.S., Bellotti, A.C., Moraes, G.J., de, Castillo, J.A., Warumby, J.F. and Lapointe, S.L. (1999) Introduction of parasitoids for control of the cassava mealybug *Phenacoccus herreni* (Hemiptera: Pseudococcidae) in northeastern Brazil. *Bulletin of Entomological Research* 89, 403–410.

Bento, J.M.S., de Moreas, J.G., de Matos, A.P. and Bellotti, A.C. (2000) Classical biological control of the mealybug *Phenacoccus herreni* (Hemiptera: Pseudococcidae) in northeastern Brazil. *Environmental Entomology* 29, 355–359.

Bernays, E.A., Chapman, R.F., Leather, E.M., McCaffery, A.R. and Modder, W.W.D. (1977) The relationship of *Zonocerus variegatus* (L.) (Acridoidea: Pyrgomorphidae) with cassava (*Manihot esculenta*). *Bulletin of Entomological Research* 67, 391–404.

Bertschy, C., Turlings, T.C.L., Bellotti, A. and Dorn, S. (1997) Chemically-mediated attraction of three parasitoid species to mealybug-infested cassava leaves. *Florida Entomology* 80, 383–395.

Borrero, H.M. and Bellotti, A.C. (1983) Estudio biológico del chinche de encaje (*Vatiga manihotae*) y uno de sus enemigos naturales. In: Reyes, J.A. (ed.) *Yuca: Control Integrado de Plagas.* CIAT, Cali, Colombia, pp. 163–168.

Braun, A.R., Bellotti, A.C., Guerrero, J.M. and Wilson, L.T. (1989) Effect of predator exclusion on cassava infested with tetranychid mites (Acari: Tetranychidae). *Environmental Entomology* 18, 711–714.

Braun, A.R., Bellotti, A.C. and Lozano, J.C. (1993) Implementation of IPM for small-scale cassava farmers. In: Altieri, M.A. (ed.) *Crop Protection Strategies for Subsistence Farmers.* Westview, Boulder, Colorado, pp. 103–115.

Byrne, D.H., Guerrero, J.M., Bellotti, A.C. and Gracen, V.E. (1982) Yield and plant growth responses of *Mononychellus* mite resistant and susceptible cassava cultivars under protected vs. infested conditions. *Crop Science* 22, 486–550.

Byrne, D.H., Bellotti, A.C. and Guerrero, J.M. (1983) The cassava mites. *Tropical Pest Management* 29, 378–394.

Caicedo, A.M. and Bellotti, A.C. (1994) Evaluación del potencial del nemátodo entomógeno *Steinernema carpocapsae* Weiser (Rhabditida: Steinernematidae) para el control de *Cyrtomenus bergi* Froeschner (Hemiptera: Cydnidae) en condiciones de laboratorio. *Revista Colombiana de Entomología* 20, 241–246.

Calatayud, P.A., Seligmann, C.D. and Bellotti, A.C. (2002) Influence of water deficient cassava plants on parasitism success and biological characteristics of three parasitoid species to *Phenacoccus herreni* (in press).

Calvacante, M.L.S. and Ciociola, A.I. (1993) Variabilidade quanto au grau de resistência de cultivares de mandioca ao percevejo de renda em Pacajus – CE. In: *Relato Annual de Pesquisas,* 1980/92. Empresa de Pesquisa Agropecuária Ceará, Fortaleza, Brazil.

Castaño, O., Bellotti, A.C. and Vargas, O. (1985) Efecto del HCN y de cultivos intercalados sobre daño causado por la chinche de la viruela *Cyrtomenus bergi* Froeschner al cultivo de la yuca. *Revista Colombiana de Entomología* 11, 24–26.

Castillo, J. (1996) Moscas blancas (Homoptera: Aleyrodidae) y sus enemigos naturales sobre cultivos de yuca (*Manihot esculenta* Crantz) en Colombia. MS thesis, Universidad del Valle, Cali, Colombia.

CIAT (1990) *Annual Report Cassava Program, 1989.* CIAT, Cali, Colombia.

CIAT (1992) *Annual Report Cassava Program, 1987–1991.* CIAT, Cali, Colombia.

CIAT (1994) *Annual Report Cassava Program, 1993.* CIAT, Cali, Colombia.

CIAT (1999) *Annual Report: Integrated Pest and Disease Management in Major Agroecosystems.* CIAT, Cali, Colombia.

Cock, J.H. (1978) A physiological basis of yield loss in cassava due to pests. In: *Proceedings Cassava Protection Workshop,* Cali, 1977. CIAT, Cali, Colombia, pp. 9–16.

Cock, J.H., Porto, M.C.M. and El-Sharkawy, M.A. (1985) Water use efficiency of cassava. III. Influence of air humidity and water stress on gas exchange of field grown cassava. *Crop Science* 25, 265–272.

Cuéllar, M.E., Calatayud, P.A., Milo, E.L. and Bellotti, A.C. (2001) Functional response and oviposition rate of seven phytoseiid populations feeding on eggs of *Mononychellus tanajoa* (Acani: Tetranychidae). *Florida Entomologist*.

D'Almeida, Y.A., Lys, J.A., Neuenschwander, P. and Ajuonu, O. (1998) Impact of two accidently introduced *Encarsia* species (Hymenoptera: Aphelinidae) and other biotic and abiotic factors on the whitefly *Aleurodicus dispersus* (Russell) (Homoptera: Aleyrodidae), in Benin. *Biocontrol Science and Technology* 8, 163–173.

Delalibera, I., Jr, Sosa-Gomez, D.R., de Moraes, G.J., Alencar, J.A. and Farias-Araujo, W. (1992) Infection of the spider mite *Mononychellus tanajoa* (Acari: Tetranychidae) by the fungus *Neozygites* sp. (*Entomophthorales*) in Northeast Brazil. *Florida Entomology* 75, 145–147.

Diehl-Fleig, E., Silva, M.E. and Pacheco, M.R.M. (1988) Testes de patogenicidade dos fungos entomopatogênicos *Beauveria bassiana* e *Metarhizium anisopliae* em *Atta sexdens prirventris* (Santsch, 1919) em diferentes temperaturas. *Ciencia e Cultura* 40,1103–1105.

El-Sharkawy, M.A. (1993) Drought-tolerant cassava for Africa, Asia, and Latin America. *BioScience* 43, 441–451.

El-Sharkawy, M.A., Hernández, A.D.P. and Hershey, C. (1992) Yield stability of cassava during prolonged mid-season water stress. *Experimental Agriculture* 28, 165–174.

Evans, G.A. and Castillo, J.A. (1998) Parasites of *Aleurotrachelus socialis* (Homoptera: Aleyrodidae) from Colombia including descriptions of two new species (Hymenoptera: Aphelinidae: Platygasteridae). *Florida Entomologist* 81, 171–178.

Farias, A.R.N. (1985) *Hyaliodes vitreus* (Hemiptera: Miridae), un predador de *Vatiga illudens* (Drake, 1773) (Hemiptera: Tingidae) em mandioca, na Bahia. *Revista Brasileira de Mandioca* 4, 123–124.

Farias, A.R.N. (1987) Biologia de *Vatiga illudens* (Drake, 1922) (Hemiptera: Tingidae) em laboratório. *Revista Brasileira de Mandioca* 6, 17–19.

Farias, A.R.N. (1990a) *Especies de 'Mosca Blanca': Situação Atual e Perspectiva de Controle*. Empresa Brasileira de Pesquisa Agropecuária, Centro Nacional de Pesequisa em Mandioca e Fruticultura, Cruz das Almas, Bahia, Brazil.

Farias, A.R.N. (1990b) *O Mandorová da Mandioca. Problemas e Meios de Controle*. Empresa Brasileira de Pesquisa Agropecuária/Centro Nacional de Pesequisa em Mandioca e Fruticultura, Cruz das Almas, Bahia, Brazil.

Farias, A.R.N. (1994) Flutuação poblacional de *Aleurothrixus aepim* em mandioca, em São Miguel das Matas, Bahia. *Revista Brasileira de Mandioca* 13, 119–122.

Farias, A.R.N., Sousa, J.D.S. and Silveira, J.R.S. (1991) Flutuação populacional de *Bemisia tuberculata* em Maragogipe, Bahia. *Revista Brasileira de Mandioca* 10, 103–107.

França, F.H., Villas-Boos, G.L. and Branco, M.C. (1996) Occurrence of *Bemisia argentifolli* Bellow & Perring (Homoptera: Aleyrodidae) in the Federal District. *Anais da Sociedad Entomologica do Brasil* 25, 369–372.

Fregene, M., Angel, F., Gómez, R., Rodríguez, F., Chavarriaga, P., Roca, W., Tohme, J. and Bonierbale, M. (1997) A molecular genetic map of cassava (*Manihot esculenta* Crantz). *Theoretical Applied Genetics* 95, 431–441.

Frison, E.A. and Feliu, E. (eds) (1991) *FAO/IBPGR Technical Guidelines for the Safe Movement of Cassava Germplasm*. Food and Agriculture Organization of the United Nations, International Board of Plant Genetics Research, Rome.

Froeschner, R.C. (1993) The Neotropical lace bugs of the genus *Vatiga* (Heteroptera: Tingidae), pests of cassava: new synonymies and key to species. In: *Proceedings Entomological Society of Washington* 95, 457–462.

Garcia, C.A. and Bellotti, A.C. (1980) Estudio preliminar de la biología y morfología de *Cyrtomenus bergi* Froeschner. Nueva plaga de la yuca. *Revista Colombiana de Entomología* 6, 55–61.

Gold, C.S. (1993) Effects of cassava intercropping and varietal mixtures on herbivore load, plant growth, and yield: applications for small farmers in Latin America. In: Altieri, M.A. (ed.) *Crop Protection Strategies for Subsistence Farmers*. Westview, Boulder, Colorado, pp. 5, 117–142.

Gold, C.S., Altieri, M.A. and Bellotti, A.C. (1989a) Cassava intercropping and pest management: a review illustrated with a case study from Colombia. *Tropical Pest Management* 35, 339–344.

Gold, C.S., Altieri, M.A. and Bellotti, A.C. (1989b) Effects of intercrop competition and differential herbivore numbers on cassava growth and yields. *Agriculture, Ecosystems and Environment* 26, 131–146.

Gold, C.S., Altieri, M.A. and Bellotti, A.C. (1989c) The effects of intercropping and mixed varieties on predators and parasitoids of cassava whiteflies in Colombia: an examination of the 'natural enemy hypothesis'. *Bulletin of Entomological Research* 79, 115–121.

Gold, C.S., Altieri, M.A. and Bellotti, A.C. (1990) Effects of intercropping and varietal mixtures on the cassava hornworm, *Erinnyis ello* L. (Lepidoptera: Sphingidae), and the stemborer, *Chilomima clarkei* (Amsel) (Lepidoptera: Pyralidae), in Colombia. *Tropical Pest Management* 36, 362–367.

Gold, C.S., Altieri, M.A. and Bellotti, A.C. (1991) Survivorship of the cassava whiteflies, *Aleurotrachelus socialis* and *Trialeurodes variabilis* (Homoptera: Aleyrodidae) under different cropping systems in Colombia. *Crop Protection* 10, 305–309.

Henry, G. (1995) *Global Cassava Sector Constraints and Estimated Future R&D Benefits*. CIAT, Cali, Colombia.

Henry, G. and Gottret, M.V. (1995) *Global Cassava Sector Trends: Reassessing the Crop's Future*. CIAT, Cali, Colombia.

Hernández, M.P., Bellotti, A.C., Cardona, C., Lapointe, S. and Pantoja, A. (1995) Organización y utilidad de una colección de insectos para referencia y trabajo de cuatro cultivos tropicales. *Revista Colombiana de Entomología* 21, 59–62.

Herren, H.R. and Neuenschwander, P. (1991) Biological control of cassava pests in Africa. *Annual Review of Entomology* 36, 257–283.

Herrera, J.C., Van Driesche, R.G. and Bellotti, A.C. (1989) Temperature dependent growth rates for the cassava mealybug, *Phenacoccus herreni*, and two of its encyrtid parasitoids, *Epidinocarsis diversicornis* and *Acerophagus coccois* in Colombia. *Entomologia Experimentali et Applicata* 50, 21–27.

Hershey, C.H. (1987) Cassava germplasm resources. In: *Proceedings Cassava Breeding: a Multidisciplinary Review*, Manila, (Philippines), 4–7 March 1985. CIAT, Cali, Colombia, pp. 1–24.

Janzen, D.H. (1986) Biogeography of an exceptional place: what determines the saturniid and sphingiid moth fauna of Santa Rosa National Park, Costa Rica, and what does it mean to conservation biology? *Brenesia* 25/26, 51–87.

Janzen, D.H. (1987) When and when not to leave. *Oikos* 49, 241–243.

LaBerry, R. (1997) La aplicación de un programa MIP en producción industrial de yuca. In: *Memorias Congreso Fitopatología, Biodiversidad, Micorrizas*, CIAT, Cali. Asociación Colombiana de Fitopatología y Ciencias Afines, Cali, Colombia, pp. 136–137.

Lal, S.S. and Pillai, K.S. (1981) Cassava pests and their control in Southern India. *Tropical Pest Management* 27, 480–491.

Langewald, J., Thomas, M.B., Douro-Kpindon, O.K. and Lomer, C.J. (1997) Use of *Metarhizium flavoviride* for control of *Zonocerus variegatus*: a model linking dispersal and secondary infection from the spray residue with mortality in caged field samples. *Entomologia Experimentali et Applicata* 82, 1–8.

Leihner, D.E. (1983) *Management and Evaluation of Intercropping Systems with Cassava*. CIAT, Cali, Colombia.

Le Ru, B. and Calatayud, P.A. (1994) Interactions between cassava and arthropod pests. *African Crop Science Journal* 2, 419–421.

Lohr, B. (1983) Biología, ecología, daño económico y control de Chilomima clarkei barrenador de la yuca. In: Reyes, J.A. (ed.) *Yuca: Control Integrado de Plagas*. CIAT, Cali, Colombia, pp. 159–161.

Lohr, B. and Varela, A.M. (1990) Exploration for natural enemies of the cassava mealybug, *Phenacoccus manihoti* (Homopera: Pseudococcidae), in South America for the biological control of this introduced pest in Africa. *Bulletin of Entomological Research* 80, 417–425.

Lomer, C.J., Bateman, R.P., Godonou, I., Kpindou, D. and Shah, P.A. (1990) Field infection of *Zonocerus variegatus* following application of oil based formulation of *Metarhizium flavoviridae* conidia. *Biocontrol Science and Technology* 3, 337–346.

Lozano, J.C. and Bellotti, A.C. (1985) Integrated control of diseases and pests of cassava. In: Cock, J.A. and Reyes, J.A. (eds) *Cassava: Research, Production and Utilization*. CIAT, Cali, Colombia/United Nations Development Programme, pp. 575–585.

Lozano, J.C., Bellotti, A., Reyes, J.A., Howler, R. and Leihner, D. (1981) *Field Problems in Cassava*, 2nd edn. CIAT, Cali, Colombia.

Lozano, J.C., Bellotti, A.C. and Vargas, O. (1986) Sanitary problems in the production of cassava planting material. In: *Proceedings. Global Workshop on Root and Tuber Crops Propagation*. CIAT, Cali, Colombia, pp. 73–85.

Maddison, P. (1979) *Pests Associated with Cassava in the Pacific Regions: Auckland Pacific Islands Pest Survey*. Entomology Division, Department of Scientific & Industrial Research, Auckland.

Modder, W.W.D. (1994) Control of the variegated grasshopper *Zonocerus variegatus* (L.) on cassava. *African Crop Science Journal* 2, 391–406.

Montagnini, F. and Jordan, C.F. (1983) The role of insects in the productivity decline of cassava (*Manihot esculenta* Crantz) on a slash and burn site in the Amazon Territory of Venezuela. *Agriculture, Ecosystems and Environment* 9, 293–301.

Moraes de, G.J., Alencar de, J.A., Wenzel-Neto, F. and Mergulhao, S.M.R. (1990) Explorations for natural enemies of the cassava green mite in Brazil. In: *Proceedings 8th Symposium International Society of Tropical Root Crops*, Bangkok, 30 October–5 November 1988. Thai Department of Agriculture, CIAT, Centro Internacional de la Papa, Bangkok, pp. 351–353.

Múnera, D.F., de los Ríos, J. and Bellotti, A.C. (1999) Patogenicidad sobre *Erinnys ello* (Lepidoptera: Sphingidae) en condiciones de laboratorio por hongos entomopatógenos recolectados en cultivos comerciales de yuca, *Manihot esculenta* en el Valle del Cauca, Colombia. *Revista Colombiana de Entomología* 25, 161–167.

Munthali, D.C. (1992) Effect of cassava variety on the biology of *Bemisia afer* (Priesner & Hosny) (Hemiptera: Aleyrodidae). *Insect Science and its Application* 13, 459–465.

Neuenschwander, P. (1994a) Control of cassava mealybug in Africa: lessons from a biological control project. *African Crop Science Journal* 2, 369–383.

Neuenschwander, P. (1994b) Spiralling whitefly *Aleurodicus dispersus*, a recent invader and new cassava pest. *African Crop Science Journal* 2, 419–421.

Ngeve, J.M. (1995) Progress report 1992–1994 on cassava regional research project for maize and cassava. *Proceedings RRPMC/WECAMAN Joint Workshop*, Cotonou, Benin, 28 May–2 June 1995. International Institute of Tropical Agriculture (IITA), Ibadan, Nigeria, pp. 1–15.

Ospina, B., Smith, L. and Bellotti, A.C. (1999) Adapting participatory research methods for developing integrated crop management for cassava-based systems, northeast Brazil. In: Fujisaka, S. (ed.) *Systems and Farmer Participatory Research*. CIAT, Cali, Colombia, pp. 61–75.

Pegoraro, R.A. and Bellotti, A.C. (1994) Aspectos biológicos de *Pseudococcus mandio* Williams (Homoptera: Pseudococcidae) em mandioca. *Anais da Sociedade Entomologica do Brasil* 23, 203–207.

Polanía, M.A., Calatayud, P.A. and Bellotti, A.C. (1999) Comportamiento alimenticio del piojo harinoso *Phenacoccus herreni* (Sternorhyncha: Pseudococcidae) e influencia del déficit hídrico en plantas de yuca sobre su desarrollo. *Revista Colombiana de Entomología* 26, 1–9.

Peña, J.E. and Waddill, V. (1982) Pests of cassava in South Florida. *Florida Entomologist* 65, 143–149.

Porter, R.I. (1988) Evaluation of germplasm (*Manihot esculenta* Crantz) for resistance to the mealybug (*Phenacoccus herreni* Cox and Williams). PhD thesis dissertation, Cornell University, Ithaca, New York.

Renvoize, B.S. (1973) The area of origin of *Manihot esculenta*, as a crop plant – a review of the evidence. *Economic Botany* 26, 352–360.

Riis, L. (1990) The subterranean burrowing bug *Cyrtomenus bergi* Froeschner, an increasing pest in tropical Latin America: behavioural studies, population fluctuations, botanical control, with special reference to cassava. MSc thesis, Institute of Ecological and Molecular Biology, Section of Zoology, The Royal Veterinary and Agricultural University, Copenhagen.

Riis, L. (1997) Behaviour and population growth of the burrower bug, *Cyrtomenus bergi* Froeschner: Effects of host plants and abiotic factors. PhD thesis, Royal Veterinary Agricultural University, Copenhagen.

Riis, L., Bellotti, A.C. and Vargas, O. (1995) The response of a polyphagous pest (*Cyrtomenus bergi* Froeschner) to cassava cultivars with variable HCN content in root parenchyma and peel. In: *Proceedings, Second International Scientific Meeting of the Cassava Biotechnology Network*, 22–26 August 1994. Bogor, Indonesia. CIAT, Cali, Colombia, pp. 501–509.

Salick, J. (1983) Agroecology of the cassava lacebug. PhD thesis, Cornell University, Ithaca, New York.

Samways, M.J. (1980) O complexo de artropodas da mandioca (*Manihot esculenta* Crantz) em Lavras, Minas Gerais, Brasil. *Anais da Sociedade Entomologica do Brasil* 9, 3–10.

Schmitt, A.T. (1983) Ocorrência de inimigos naturais de *Erinnyis ello* (L.) no estado de Santa Catarina. *Revista Brasileira de Mandioca* 2, 59–62.

Schmitt, A.T. (1988) Uso de *Baculovirus erinnyis* para el control biológico del gusano cachón de la yuca. *Yuca Boletín Informativo* (Colombia) 12, 1–4.

Schulthess, F. (1987) The interactions between cassava mealybug (*Phenacoccus manihoti* Mat-Ferr.) populations as influenced by weather. PhD thesis, Swiss Federal Institute of Technology, Zurich.

Skovgård, H., Tomkiewicz, J., Nachman, G. and Münster-Swendsen, M. (1993) The effect of the cassava green mite *Mononychellus tanajoa* on the growth and yield of cassava *Manihot esculenta* in a seasonally dry area in Kenya. *Experimental Applied Acarology* 17, 41–58.

Smith, L. and Bellotti, A.C. (1996) Successful bio-control projects with emphasis on the Neotropics. In: *Proceedings Cornell Communications Conference on Biological Control*, 11–13 April 1996, Cornell University. Cornell University Press, Ithaca, New York. www.nysaes.cornell.edu/ent/bcconf/talks/bellotti.html

Thresh, J.M., Fargette, D. and Otim-Nape, G.W. (1994) Effects of African cassava mosaic geminivirus on the yields of cassava. *Tropical Science* 34, 26–42.

Thresh, J.M., Otim-Nape, G.W., Thankappan, M. and Muniyappa, V. (1998) The mosaic diseases of cassava in Africa and India caused by whitefly-borne geminiviruses. *Review of Plant Pathology* 77, 935–945.

Toye, S.A. (1982) Studies on the biology of the grasshopper pest *Zonocerus variegatus* (L.) (Orthoptera: Pyrgomorphidae) in Nigeria; 1911–1981. *Insect Science and its Application* 1, 1–7.

Urias López, M.A., Bellotti, A.C., Bravo Mojica, H. and Carrillo Sánchez, J.L. (1987) Impacto de insecticidas sobre tres parasitoides de *Erinnyis ello* (L.), gusano de cuerno de la yuca. *Agrociencia* 67, 137–146.

Van Driesche, R.G., Castillo, J.A. and Bellotti, A.C. (1988) Field placement of mealybug-infested potted cassava plants for the study of parasitism of *Phenacoccus herreni*. *Entomologia Experimentali et Applicata* 46, 117–123.

Van Driesche, R.G., Bellotti, A.C., Castillo, J.A. and Herrera, C.J. (1990) Estimating total losses from parasitoids for a field population of a continuously breeding insect, cassava mealybug, *Phenacoccus herreni* (Hemoptera: Pseudococcidae) in Colombia, S.A. *Florida Entomology* 73, 133–143.

van Schoonhoven, A. (1974) Resistance to thrips damage in cassava. *Journal of Economic Entomology* 67, 728–730.

Vargas, O. and Bellotti, A.C. (1981) Pérdidas en rendimiento causadas por moscas blancas en el cultivo de la yuca. *Revista Colombiana de Entomología* 7, 13–20.

Vargas, H.O. and Bellotti, A.C. (1984) Pérdidas en rendimiento causadas por *Phenacoccus herreni* Cox and Williams en dos clones de yuca. *Revista Colombiana de Entomología* 10, 41–46.

Vides, O.L., Sierra, O.D., Gómez, H.S. and Palomino, A.T. (1996) *El barrenador del tallo de la yuca Chilomima clarkei (Lepidoptera: Pyralidae) en el CRECED Provincia del Río*. Boletín CORPOICA, Santa Fe de Bogotá, Colombia.

Villegas, G.A. and Bellotti, A.C. (1985) Biología, morfología y hábitos de *Laocheirus araneiformis* Linne (Coleoptera: Cerambycidae) barrenador de la yuca en Palmira, Valle del Cauca. *Acta Agronómica* 35, 56–67.

Williams, D.J. (1985) *Pseudococcus mandio* sp. (Hemiptera: Pseudococcidae) on cassava roots in Paraguay, Bolivia and Brazil. *Bulletin of Entomological Research* 75, 545–547.

Williams, D.J. and Granara de Willink, M.C. (1992) *Mealybugs of Central and South America*. CAB International, Wallingford.

Yaninek, J.S. (1988) Continental dispersal of the cassava green mite, an exotic pest in Africa, and implications for biological control. *Experimental Applied Acarology* 4, 211–224.

Yaninek, J.S. and Animashaun, A. (1987) Why cassava green mites are dry season pests. *Proceedings Seminar Agrometeorology Crop Protection in Lowland Humid and Sub-humid Tropics*, World Meteorological Organisation/IITA, Contonou, Benin, 7–11 July 1986. IITA, Ibadan, Nigeria, pp. 59–66.

Yaninek, J.S. and Herren, H.R. (1988) Introduction and spread of the cassava green mite, *Mononychellus tanajoa* (Bondar) (Acari: Tetranychidae), an exotic pest in Africa, and the search for appropriate control methods: a review. *Bulletin of Entomological Research* 78, 1–13.

Yaninek, J.S., Mégev, B., de Moraes, G.J., Bakker, F., Braun, A. and Herren, H.R. (1992) Establishment of the Neotropical predator *Amblyseius idaeus* (Acari: Phytoseiidae) in Benin, West Africa. *Biocontrol Science and Technology* 1, 323–330.

Yaninek, J.S., Onzo, A. and Ojo, J.B. (1993) Continent-wide releases of Neotropical phytoseiids against the exotic cassava green mite in Africa. *Experimental Applied Acarology* 17, 145–160.

Yaninek, J.S., Saizonou, S., Onzo, A., Zannou, I. and Gnanvossou, D. (1996) Seasonal and habitat variability in the fungal pathogens, *Neozygites* c.f. *floridana* and *Hirsutella thompsonii*, associated with cassava mites in Benin, West Africa. *Biocontrol Science and Technology* 6, 23–33.

Chapter 12
The Viruses and Virus Diseases of Cassava

L.A. Calvert[1] and J.M. Thresh[2]

[1]*Centro Internacional de Agricultura Tropical (CIAT), A.A. 6713, Cali, Colombia;*
[2]*Natural Resources Institute, University of Greenwich, Chatham Maritime,
Kent ME4 4TB, UK*

Introduction

Crops that are propagated vegetatively are particularly prone to damage by viruses as infection tends to build up in successive cycles of propagation. Cassava is no exception to this generalization and at least 16 different viruses have been isolated from the crop. Moreover, other as yet undescribed viruses are likely to occur and may even be prevalent in some areas. This is because cassava has received far less attention from virologists than it merits as one of the world's most important and widely grown food crops.

A full list of the viruses that have been isolated from cassava is presented in Table 12.1 and key references appear in the bibliography. The viruses asterisked in the table have been detected somewhat fortuitously in studies undertaken for other reasons. There is only limited information on the properties, distribution, effects and importance of these viruses. They require further attention, but meanwhile they should be considered in operating quarantine controls on the movement of vegetative propagules between different cassava-growing areas. These viruses are not considered further here and the main emphasis is on those known to cause diseases of economic importance.

A feature of cassava viruses is that they are of diverse taxonomic groups (Table 12.1). Another is that their known distribution is largely or entirely restricted to only one of the continents in which cassava is grown, or to an even more localized geographic area. For this reason the viruses and virus diseases of Africa, South/Central America and the Indian subcontinent are considered separately.

The Viruses and Virus Diseases of Cassava in South and Central America

Cassava originated in the Neotropics and was not introduced to other regions until relatively recently. This may explain why only one of the viruses of cassava occuring in South and Central America has been found elsewhere. Moreover, several of the Neotropical viruses of cassava do not cause symptoms and have no obvious deleterious effects, which may reflect a long period of co-evolution between the host and its pathogens.

Three virus diseases justify detailed attention here. Three other viruses are listed in Table 12.1 and are considered briefly, but they do not

Table 12.1. The viruses of cassava.

Africa
  African cassava mosaic virus (*Geminiviridae: Begomovirus*)
  East African cassava mosaic virus (*Geminiviridae: Begomovirus)*
  South African cassava mosaic virus (*Geminiviridae: Begomovirus*)
  Cassava brown streak virus (*Potyviridae: Ipomovirus*)
  Cassava Ivorian bacilliform virus* (unassigned)
  Cassava Kumi viruses*
  Cassava 'Q' virus*
  Cassava common mosaic virus* (*Potexvirus*)
South/Central America
  Cassava common mosaic virus (*Potexvirus*)
  Cassava virus X (*Potexvirus*)*
  Cassava vein mosaic virus (*Caulimoviridae*)
  Cassava Colombian symptomless virus (*Potexvirus*)*
  Cassava American latent virus (*Comoviridae: Nepovirus*)*
  Cassava frogskin 'virus'
Asia/Pacific
  Cassava common mosaic virus* (*Potexvirus*)
  Indian cassava mosaic virus (*Geminiviridae: Begomovirus*)
  Cassava green mottle virus* (*Comoviridae: Nepovirus*)

Officially recognized viruses are given in italics, together with family and genus in parentheses. Viruses that are unimportant or for which little information is available are asterisked and not considered in detail in the accompanying text. (Source: Thresh *et al.*, 1994b; Thresh *et al.*, 1998c.)

cause symptoms, appear to be unimportant and are not discussed further.

Cassava common mosaic disease

History

Cassava common mosaic disease (CsCMD) was first reported in southern Brazil (Silberschmid, 1938; Costa, 1940). The disease has since been recorded in other South American countries and there is one report from Africa (Aiton *et al.*, 1988) and another from Asia (Chen *et al.*, 1981). CsCMD has no known vector and spread in the field is attributed to mechanical transmission. The disease is generally of only minor importance, although there are some areas where it is prevalent and control efforts are needed.

Symptoms

Leaves of cassava plants affected by CsCMD develop mosaic and chlorotic symptoms (Plate 1a). On some of the affected leaves there are dark and light green patches that are delimited by veins. Symptoms are most severe during relatively cool periods and cassava grown in the semitropical areas of South America is most affected by the disease. In these relatively cool conditions, the affected plants are sometimes stunted and yield losses can be up to 60% (Costa and Kitajima, 1972b).

Distribution and prevalence

CsCMD has been reported from many South American countries, but was not recorded in a survey of cassava-growing areas of Colombia (Nolt *et al.*, 1992). The disease is most prevalent in southern Brazil and Paraguay. In these regions the disease is important and phytosanitary control measures are recommended to reduce losses. More than 1000 cassava accessions in the EMBRAPA/CNPMF (Empresa Brasileira de Pesquisa Agropecuaria/Centro Nacional de Pesquisa en Mandioca y Fruticultura) collection at Cruz das Almas in Bahia, northeast Brazil, have been tested for the causal virus and the incidence was < 1%.

Aetiology

CsCMD is caused by *Cassava common mosaic virus* (CsCMV) which can infect species belonging to several families of dicotyledonous plants (Silva *et al.*, 1963; Kitajima *et al.*, 1965). The virus was ascribed originally to the potexvirus group that is now referred to as the genus *Potexvirus*.

The CsCMV virion is a semi-flexuous rod that is *c.* 15 × 495 nm (Kitajima *et al.*, 1965) and contains RNA (Silva *et al.*, 1963). Nuclear inclusions typical of the potexviruses can be found in cassava and the herbaceous host *Nicotiana benthamiana*. CsCMV is known to systemically infect cassava, *Euphorbia* spp., *Cnidoscolus aconitifolius* (chaya), *N. benthamiana* and species of several other dicotyledonous families (Costa and Kitajima, 1972a).

The viral particles of CsCMV contain a single coat protein having a relative molecular weight of 26,000 daltons (Nolt *et al.*, 1991). The CsCMV genome is single-stranded RNA and the complete sequence is known (Calvert *et al.*, 1996). The organizational structure, proteins and their predicted weights are similar to those of other potexviruses.

Epidemiology and control

There are no known vectors of CsCMV and the primary source of inoculum is infected planting material. The virus is systemic in cassava and almost all stem cuttings are infected when obtained from an infected plant. CsCMV is very stable and can be spread by mechanical transmission on machetes and other implements used to prepare cuttings. Although this mode of transmission is inefficient, it is the only known means of plant-to-plant spread.

Eliminating (roguing) plants that express CsCMV symptoms provides adequate control. The symptoms are usually obvious on the first leaves produced by infected stem cuttings. This is the best time to identify and remove diseased plants. If the plants are not rogued early they should be marked and the stems burned later after harvesting the tuberous roots. Only healthy plants should be selected as a source of vegetative propagules. To minimize the risk of mechanical transmission, cutting tools should be disinfected at regular intervals (Lozano and Nolt, 1989). With care in selecting planting material, CsCMD can be eradicated or reduced to a level of minor economic significance.

Cassava vein mosaic disease

History

The first report of cassava vein mosaic disease (CVMD) was in 1940 (Costa, 1940). The areas where this disease is most prevalent are remote and the conditions are semiarid. The region is inhabited mostly by poor rural communities and the lack of economic resources has contributed to the incomplete knowledge about this disease. Probably because the symptoms are sporadic and generally less apparent at the end of the cassava growth cycle, this disease has received inadequate attention, especially considering the large area now known to be affected.

Symptoms

The leaf symptoms of CVMD occur in flushes. After an infected stem cutting sprouts, the first four to six leaves express vein chlorosis that appears as a chevron pattern or coalesces to form ringspots (Plate 1b). Leaf deformation and epinasty are common severe symptoms. Plants then appear to 'grow out' of the infection and produce several symptomless leaves. These are followed by another series of leaves with symptoms. The expression of symptoms is influenced by the climatic conditions prevailing. Symptoms are more pronounced in the semiarid areas as compared to those expressed by the same variety grown in the wetter coastal regions of northeast Brazil. Except for the period just after sprouting, CVMD does not seem to affect plant vigour. The affected leaves senesce and fall prematurely from the plants which reduces leaf area. As infected cassava matures, it is often difficult to see any leaves with mosaic symptoms.

Distribution and prevalence

CVMD is very common in the semiarid zone of northeastern Brazil, although there are also reports from other regions of the country. The disease is common in the Brazilian States of Ceará, Pernambuco, Alagoas, Piaui and Bahia (Calvert *et al.*, 1995) and the distribution extends into some of the neighbouring states.

Aetiology

CVMD is caused by *Cassava vein mosaic virus* (CVMV). This has isometric particles, *c.* 50 nm in diameter (Kitajima and Costa, 1966) and the genome consists of ds-DNA, *c.* 8200 bases long. Initially CVMV was regarded as a tentative member of the caulimovirus group. This attribution was based on the ds-DNA genome and the particle structure (Lin and Kitajima, 1980). The complete sequence of CVMV has been determined and the genomic organization differs from that of caulimoviruses and badnaviruses (Calvert *et al.*, 1995). The virus will probably be classified as a unique genus of the plant pararetroviruses.

Epidemiology and control

Little is known about the epidemiology or control of CVMV. The only known host is cassava and the primary mode of dissemination is in infected propagules. It is not uncommon to find a farmer's variety that is totally infected. Spread occurs within fields which suggests that there is a vector, but none has yet been identified. There have been few studies on virus spread, but CVMV-infected cassava is common throughout a large area of the northeastern semiarid region of Brazil. Consequently, a vector of the virus is suspected. Until more is known about the rate of spread, the effectiveness of using 'clean' (virus-free) planting material will not be known. The virus can be latent in plants, especially during the cool, rainy seasons of the coastal regions of Brazil. 'Roguing' of planting material may be an effective control practice if diseased plants are identified and removed soon after sprouting. Most infected cassava plants appear to tolerate CVMV and produce stems of normal appearance that make good planting material. Little is known about disease loss. In the few studies that have been done, the yields of diseased plants were slightly less than from uninfected controls, but the differences were not significant statistically (Santos *et al.*, 1995). Although the full economic importance of CVMD is not well quantified, it appears that it could cause losses, especially if a drought occurs during the beginning of the growth cycle.

Cassava frogskin disease

History

Cassava frogskin disease (CFSD) is a virus-like disease that affects cassava and was first reported from southern Colombia (Pineda *et al.*, 1983). A similar disorder in northern Colombia was called Caribbean mosaic disease because of the mosaic leaf symptoms expressed by the cassava landrace 'Secundina'. In the Amazon regions of Brazil and Colombia, CFSD is called *jacare* (cayman), because of the distinctive ridges on the affected roots. Tests under uniform conditions have shown that these three disorders are manifestations of the same disease and cause the same symptoms in standard indicator varieties of cassava.

The origin of CFSD is most probably the Amazon region of Colombia, Peru or Brazil. The disease can be found in cassava grown in very isolated indigenous Amazonian Indian communities. They regard it as a physiological disorder rather than a disease and associate it with particular varieties. The native name for one variety collected from the Amazonian region of Colombia is *jacare*. Because of the geographic isolation, or the belief that the root symptoms are caused by a physiological disorder, this disease was not 'discovered' in the lowland tropics. In 1971, an apparently new disease that caused severe losses occurred in the mid-altitude Andean mountains of southern Colombia. The disorder was then recognized by scientists and named cassava frogskin disease (Hernández *et al.*, 1975).

The distribution of frogskin disease is continuing to expand. By the 1980s, it was prevalent throughout most cassava growing regions of Colombia. It has also spread to Venezuela and Costa Rica. Recently, CFSD was reported in Panama and it was established that the affected plants were grown from stem cuttings imported from Costa Rica. In Brazil, the movement of vegetative material of cassava is disseminating the disease from the Amazon region into the more semiarid areas of northeast of Brazil. CFSD also occurs in the Amazon region of Peru.

Symptoms

The expression of CFSD symptoms is influenced by temperature and host genotype. A few cassava genotypes develop mosaic symptoms on the leaves and these clones can be severely stunted. In most other genotypes, the leaves of infected plants are symptomless and appear normal. The stems of these plants may be slightly enlarged, especially near the ground. The thickening of the affected stems is associated with a lack of starch accumulation in the roots. Because of their apparent vigour, these stems are selected preferentially by farmers as they seem to provide very desirable planting material. The root symptoms range from very mild to very severe. The severity of the symptoms depends on the age of the roots and climatic factors. Hot dry conditions tend to inhibit symptom development, whereas cooler temperatures enhance symptoms. In the lowland tropics, years with above average rainfall tend to be cooler than usual and the symptoms are more severe. In hot dry years, CFSD-infected plants have few if any symptoms. The characteristic root symptoms of surface ridges develop when the root periderm and corky layers enlarge to form raised, lip-shaped fissures (Plate 1c). Severely affected roots do not accumulate starch and often show zones of constrictions. Root symptoms are most severe in plants raised from CFSD-affected stem cuttings. Newly infected plants usually have mild or no symptoms unless infected at an early stage of growth.

Aetiology

The causal agent of FSD has not been proven definitively, although isometric virus-like particles 70–80 nm in diameter can be found in thin sections of the leaves, petioles, stems and roots of affected plants, whatever the source. Viroplasm-like bodies are also found in leaves of infected plants.

At least nine species of ds-RNA are associated consistently with infected plants (Cuervo, 1989). The symptoms of hyperplasia in the root cortex are similar to the tumours caused by other plant reoviruses. The particle morphology, ds-RNA pattern and root symptoms are consistent with the causal agent being a reovirus.

CFSD is readily transmitted through grafts. To detect the disease and certify the status of a plant with regard to CFSD, a stem cutting of the plant to be tested is grafted to a plant of the indicator variety Secundina (CIAT accession M Col 2063; Calvert, 1994). The test plant is used as the rootstock and the buds of the rootstock stem should be removed to increase the likelihood of successful grafts. Plants should be grown in an area where temperatures are normally below 30°C to ensure optimum symptom expression. After 3 or 4 weeks, plants are checked, and any mosaic symptoms on the leaves of the scion indicate that the plant is affected by CFSD. The disease can be eliminated from infected plants by thermotherapy and meristem culture *in vitro* (Maffla *et al.*, 1984).

Epidemiology and control

Several studies indicate that the whitefly *Bemisia tuberculata* is the vector of CFSD (Angel *et al.*, 1990; Velasquez, 1991), although the efficiency of transmission seems to be low. In the field, the disease spreads very slowly, but progressively. In one trial to assess the rate of spread, the incidence of infected plants eventually exceeded 10%, but only after three crop cycles. The amount of spread increased as the incidence of infected plants increased.

The initial dissemination of CFSD is through the use of infected stem cuttings and spread within plantings is attributed to *B. tuberculata*. Most cassava varieties infected with CFSD express no leaf or stem symptoms and when harvesting the crop, farmers usually remove the stems before harvesting the roots. Since the stems of the diseased plants are often thicker than those of healthy ones diseased plants are often selected to provide propagules.

CFSD can be controlled by rigorous selection. Roots should be inspected for symptoms at harvest and only cuttings from apparently healthy plants that bear normal roots should be selected. This is usually adequate to maintain the disease at low levels that cause little economic loss. When the incidence of CFSD has become substantial, it is advisable to collect propagation material from a less affected source. In areas where cassava is harvested mechanically and it is not possible to inspect the roots, the use of stem cuttings from plants that are inspected and certified as being free of CFSD is very effective in controlling the disease.

Viruses that infect cassava but are not known to cause disease

Cassava virus X (CsVX) and *Cassava Colombian symptomless virus* (CCSpV) are other potexviruses that infect cassava (Lennon *et al.*, 1986b). They have only been detected in Colombia, but little effort has been made to determine if they occur elsewhere. CsVX was not detected in tests on over 1000 entries in the cassava germplasm collection of CNPMF/EMBRAPA (Cruz das Almas, Bahia, Brazil). There is only one report of *Cassava American latent virus* and little is known of its distribution (Fargette *et al.*, 1991). Since these three viruses do not cause symptoms it is difficult to determine their distribution or to evaluate their importance. The FAO/IPGRI guidelines for the safe movement of cassava germplasm (Frison and Feliu, 1991) provide additional information on these viruses.

The Viruses and Virus Diseases of Cassava in Africa

Nine viruses have been isolated from cassava in Africa (Table 12.1), but of these only *Cassava common mosaic virus* (CCMV) has been detected elsewhere. This is consistent with the view that the viruses of cassava in Africa are mainly indigenous ones that infect the crop as a consequence of spread from other hosts some time after cassava was introduced from the Neotropics in the 16th and 18th centuries.

CCMV has been detected only once in Africa (Aiton *et al.*, 1988), in material assumed to have been introduced from South America, where the virus is prevalent (see previous section). There is only very incomplete information on the occurrence and effects of four of the other viruses reported in Africa and only those causing cassava mosaic and cassava brown streak diseases are considered in detail here.

Cassava mosaic disease

History

The symptoms of what is now known as cassava mosaic disease (CMD) were first reported more than 100 years ago in what is now Tanzania

(Warburg, 1894). The disease was later identified in many other countries of sub-Saharan Africa during the early decades of the 20th century. It was particularly prevalent in Gold Coast (now Ghana), Nigeria, Cameroon, Madagascar and several of the former French Colonial territories of West and Central Africa. This led to studies on the means of spread and control. It also became apparent that some varieties of cassava were less affected by CMD than others and resistance breeding programmes began in the 1930s or 1940s in Madagascar, Tanzania and elsewhere.

In recent decades there have been major projects on the aetiology, epidemiology and control of CMD in Nigeria, Kenya, Ivory Coast and most recently in Uganda. The project in Uganda followed the onset of a particularly damaging epidemic in the late 1980s that is now affecting parts of Kenya, Tanzania and Rwanda and threatens other countries of the region (Otim-Nape *et al.*, 2000). The epidemic is the latest and most fully documented of those to have affected cassava in Africa at different times and places during the 20th century. This explains why CMD has featured so prominently and for so long in the literature on cassava in Africa. Indeed, CMD has received more attention than any other disease of an African food crop (Thresh, 1991).

Symptoms

CMD causes characteristic leaf symptoms that can usually be recognized without difficulty. The symptoms are very variable in type and severity and are of two main types that are sometimes distinguished as 'green mosaic' and 'yellow mosaic'. Leaves affected by 'green mosaic' have contrasting sectors of normal green and light green tissue. These symptoms are apparent only when the plants are examined closely and are not usually associated with an obvious decrease in leaf area, leaf number or plant size, or yield.

Leaves affected by 'yellow mosaic' are much more obvious, as they have contrasting areas of normal green and yellow tissue. Moreover, the chlorotic areas may expand less than other parts of the leaf lamina which can lead to distortion of the leaflets (Plate 1d) and rupturing of the tissues. Severe chlorosis is often associated with premature leaf abscission, a characteristic

S-shaped curvature of the petiole and an obvious decrease in growth and yield.

There are big differences between cassava varieties in the type, extent and severity of the symptoms caused by CMD and resistant varieties express much less severe symptoms than susceptible ones, especially during the late stage of crop growth when resistant varieties may become symptomless and are then said to recover. Symptom expression is also influenced by environmental factors and leaves produced during hot weather tend to be affected less than those produced at other times. Moreover, virulent strains cause more severe symptoms than avirulent ones and have greater effects on growth and yield.

There is no evidence of any consistent differences between the symptoms caused by the different cassava mosaic geminiviruses (CMGs), each of which can occur as virulent or less virulent strains. However, dual infection with two different CMGs causes more severe symptoms than either virus alone, as reported in studies in Uganda and Cameroon (Harrison *et al.*, 1997; Fondong *et al.*, 2000).

The main difficulties that arise in recording the symptoms of CMD occur when the plants being examined have been affected by pests or nutrient deficiency. The cassava green mite (*Mononychellus tanajoa*) and zinc deficiency cause particular problems. However, the damage they cause is usually similar on the different leaflets of each affected leaf, whereas CMD has less consistent effects and the two halves of a leaflet on either side of the midrib are often affected differently. This is an important distinguishing feature of CMD that should be stressed in training staff and farmers in disease recognition. However, severely damaged plants cannot be examined effectively for virus symptoms and whenever possible inspection for CMD should be made at times when the plants are growing vigorously and unaffected by drought, pests or nutrient deficiency.

In recording experiments and in screening for resistance to CMD, much use has been made of simple numerical scoring systems based on the extent and severity of the symptoms expressed. Scales of 0–4 or 1–5 have been widely used to quantify differences due to variety, season and virus strains and to assess the relationship between symptom severity and yield loss.

Distribution and prevalence

CMD occurs in all the cassava-growing areas of Africa and on the adjacent islands including Cape Verde, Zanzibar, Seychelles, Mauritius and Madagascar. There are big differences between countries in the date of the first reports (Fauquet and Fargette, 1990), which is in part related to the status of cassava in the different parts of Africa and to the amount of attention given to the crop by plant pathologists.

In many African countries there is general agreement that CMD is the most important disease of cassava (Geddes, 1990), although in some areas it is regarded as less important than cassava bacterial blight (see chapter 13). Until recently there were few data to support these assumptions. The situation changed in the 1990s when the incidence and severity of CMD were assessed in representative plantings in 13 important cassava-growing countries of Africa (Table 12.2). Surveys of this type are expensive and time-consuming and inevitably the number of plantings assessed has been small in relation to the total amount of cassava being grown. Nevertheless, surveys were undertaken in Uganda following the onset of the recent pandemic and in several other countries as part of more comprehensive assessments of pest and disease problems. The results summarized in Table 12.2 indicate the prevalence of CMD and the sometimes big differences that occur between and within particular countries.

From the results obtained, three contrasting situations have been distinguished and referred to as *epidemic*, *endemic* and *benign* (Thresh *et al.*, 1997). In the *epidemic* situation CMD is being spread very rapidly by the whitefly vector (*Bemisia tabaci*) and the symptoms are prevalent and severe. Farmers experience such serious losses that food security is threatened and it may be necessary to switch to sweet potato or other alternative food crops. Control measures are essential if production is to be restored and there is an urgent need for CMD-resistant varieties of cassava, as developed and supplied through official programmes or selected by farmers from those already available. The epidemic situation, as encountered in the 1990s in much of Uganda, has now spread to adjacent areas of western Kenya, Tanzania and Rwanda and it seems inevitable that it will soon spread to Burundi

Table 12.2. Surveys of the incidence of cassava mosaic disease (CMD) in India and 13 African countries.

Country	Organization (reference)	Year	Cassava area (million ha)	CMD % iIncidence
Uganda	NARO (Otim-Nape *et al.*, 1998b)	1990–1992	0.36	57
Uganda	NARO (Otim-Nape *et al.*, 2001)	1994	0.38	65
Uganda	NARO/ESARC (Legg *et al.*, 1999)	1997	0.34	68
Chad	US AID (Johnson, 1992)	1992	0.07	40
Malawi	NARS (Nyirenda *et al.*, 1993)	1992	0.07	21
Tanzania	NARS/NRI (Legg and Raya, 1998)	1993	0.69	26
Ghana	ESCaPP (Yaninek *et al.*, 1994; Wydra and Msikita, 1998)	1993/94	0.61	72
Benin	ESCaPP (Yaninek *et al.*, 1994; Wydra and Msikita, 1998)	1994	0.14	53
Cameroon	ESCaPP (Yaninek *et al.*, 1994; Wydra and Msikita, 1998)	1994	0.08	67
Nigeria	IITA (L.C. Dempster, unpublished)	1994	2.00	55
Nigeria	ESCaPP (Yaninek *et al.*, 1994; Wydra and Msikita, 1998)	1994	2.00	82
Zambia	NARS/SARRNET (Muimba-Kankolongo *et al.* 1997)	1995/96	0.11	41
Zanzibar	NARS/NRI (Thresh and Mbwana, 1998)	1998	NA	71
South Africa	NARS (Jericho *et al.*, 1999)	1998	< 0.01	31
Mozambique	NARS/NRI (R.J. Hillocks and J.M. Thresh, unpublished)	1999/00	0.99	20
Kenya (Western)	KARI/NRI (Legg *et al.*, 1999)	1993	< 0.01	20
Kenya (Western)	KARI/NRI (Legg *et al.*, 1999)	1996	< 0.01	56
Kenya (Western)	KARI/ESARC (Legg *et al.*, 1999)	1998	< 0.01	84
Kenya (Coastal)	NARS/NRI (T. Munga and J.M. Thresh, unpublished)	2000	< 0.01	58
India	UAS Bangalore (Mathew, 1989)	1988	0.24	19
A. Pradesh	UAS Bangalore (Mathew, 1989)	1988	NA	< 1
Karnataka	UAS Bangalore (Mathew, 1989)	1988	NA	5
Kerala	UAS Bangalore (Mathew, 1989)	1988	NA	23
Tamil N.	UAS Bangalore (Mathew, 1989)	1988	NA	30

and other parts of the region (Otim-Nape *et al.*, 2000) and beyond. Similarly unstable epidemic situations were encountered previously in the 1930s in Madagascar (Cours *et al.*, 1997) and more recently in the Cape Verde Islands and Akwa Ibom State of Nigeria (Anon., 1993).

In *endemic* areas there is a high incidence of CMD, but the symptoms are not usually very severe. The overall situation is stable and changes little from one year to the next. There is much use of infected cuttings as planting material and yields are undoubtedly impaired. Nevertheless, the losses have seldom been quantified and they are largely ignored by farmers or considered acceptable. Control measures are not regarded as essential, although they would

undoubtedly bring substantial benefits. This is the situation in much of Ivory Coast, Ghana, Nigeria and the lowland areas of Cameroon and may extend into the Democratic Republic of Congo and other areas of Central Africa.

In *benign* areas the incidence of CMD is generally low and seldom exceeds 20%. Infection is due mainly to the use of infected planting material and there is little or no evidence of spread by whiteflies. Symptoms are usually inconspicuous and not associated with obvious deleterious effects on growth or root yield. Losses are not substantial and control measures are not considered necessary and would bring little benefit. This was formerly the situation in much of Uganda and western Kenya and is encountered currently in large areas of Tanzania and Mozambique and in the mid-altitude agroecologies of Burundi, Malawi, South Africa and parts of Zambia.

There is an urgent need for information on the incidence and severity of CMD in other important cassava-growing areas of sub-Saharan Africa, including Sierra Leone, Liberia, Angola and Democratic Republic of Congo. It will then be possible to identify the areas that should receive priority in any attempts at intervention. Meanwhile, it should be appreciated that the situation can change dramatically and on a time-scale of only a few years. This is apparent from early experience in Madagascar and elsewhere (Cours *et al.*, 1997) and more recently in Uganda. There the situation changed rapidly from benign to epidemic and it is now changing to endemic as the original equilibrium between host and pathogen is being restored (Otim-Nape *et al.*, 2000).

Aetiology

For many years CMD was assumed to be caused by a virus because the disease was transmissible by grafts and by the whitefly now known as *B. tabaci*, and yet no visible pathogen was detected. The situation changed in the 1970s when a virus was transmitted mechanically from CMD-affected cassava to the herbaceous test plant *Nicotiana clevelandii*. The status of the virus isolated was at first unclear because it could not be isolated from all the CMD-affected plants tested. Hence, the virus was initially referred to as cassava latent virus and this name continues to appear occasionally in the literature. However, the name became inappropriate when an additional test plant (*N. benthamiana*) was introduced and used to isolate and differentiate between virus isolates that all caused typical symptoms of CMD when transmitted back to cassava (Bock and Woods, 1983). The different isolates were initially referred to as strains of *African cassava mosaic virus* (ACMV) and three groups or 'clusters' of strains were distinguished. These were later regarded as separate viruses (Hong *et al.*, 1993) and they are now ascribed to the genus *Begomovirus*; family *Geminiviridae*. ACMV and *East African cassava mosaic virus* (EACMV) have not been found outside Africa, whereas *Indian cassava mosaic virus* (ICMV) seems restricted to the Indian subcontinent. A fourth virus of this type (*South*

African cassava mosaic virus) has been distinguished recently in South Africa (Berrie *et al.*, 1998) and hybrid recombinant viruses have been distinguished in Uganda and Cameroon that have some of the genome properties of both ACMV and EACMV (Deng *et al.*, 1997; Zhou *et al.*, 1997).

The biological significance of the great diversity in biochemical properties of the different cassava mosaic geminiviruses has not been determined and requires investigation. Nevertheless, there is already evidence that dual infection with the hybrid recombinant virus and ACMV or with EACMV and ACMV is more damaging than any of these viruses occurring alone (Harrison *et al.*, 1997; Fondong *et al.*, 2000). The occurrence of different viruses or virus combinations in different regions could also complicate and may even undermine the effectiveness of resistance breeding programmes and quarantine controls on the movement of material between different parts of Africa. Until these issues are resolved it is important to avoid moving infected cassava between different countries or regions and especially from areas seriously affected by CMD. It is particularly important to avoid the transfer of cassava mosaic geminiviruses from Africa to the Indian subcontinent or vice versa, or from these regions to the Neotropics.

Effects on growth and yield

There is an extensive literature on the effects of CMD on the growth and yield of cassava. Data have been collected at different times and places on a wide range of cultivars using two main approaches (Thresh *et al.*, 1994a). Firstly, comparisons have been made in formal experiments established with cuttings collected from healthy and CMD-affected plants. Secondly, naturally infected and healthy plants have been identified and assessed within larger plantings at experimental stations or in farmers' fields.

Some of the main findings are:

- Varieties differ greatly in their response to infection. Some are severely stunted and produce little or no yield of foliage, stem cuttings or tuberous roots, whereas others are relatively unaffected and sustain little damage.
- There is a general relationship between symptom severity and the decrease in

growth and tuberous root yield caused by CMD.

• Plants grown from infected cuttings are more severely affected than those of the same variety infected at an early stage of growth by whiteflies; plants infected late sustain little or no damage.

• Competition and compensation effects can occur within crop stands and both healthy and diseased plants grow better along-side diseased neighbours than alongside healthy ones. Consequently, differences between the growth and yield of healthy and diseased plants are less when comparisons are made between healthy and diseased plants each having neighbours of similar health status than between plants each having neighbours of dissimilar health status.

• Some virus strains or strain combinations cause more severe symptoms and decrease growth and yield much more than others.

• CMD influences the performance and sustainability of varieties by influencing the number, viability and growth of the stem cuttings available for propagation.

Overall crop loss

The results of yield comparisons have been used to estimate the overall losses caused by CMD in whole localities, regions or countries. However, definitive estimates are only possible if detailed information is also available on the incidence and severity of the disease in different areas and on the prevalence, type, productivity and sensitivity to infection of the main varieties being grown. Such details are seldom available and the published estimates of yield loss provide only an indication of the magnitude of the damage sustained.

Watts Padwick (1956) used information from regional plant pathologists to estimate the losses caused by CMD in the former British Colonial territories of Africa. Fargette *et al.* (1988) later estimated the annual losses in Ivory Coast to be 500,000 t of roots compared to actual production at the time of 800,000 t. They assumed that all the plants being grown were affected and sustained losses in tuberous root yield of 38%, as recorded in their experiments on one of the main Ivorian varieties being grown. On similar

assumptions losses in Africa were estimated to be 30 million t compared with actual production at the time of 51 million t (FAO, 1985).

These assumptions were inappropriate because the incidence of CMD is now known to be moderate or low in some important cassava-growing areas of Africa (Table 12.2). Moreover, some widely grown varieties are much less severely affected than the variety assessed in Ivory Coast. These considerations led Thresh *et al.* (1997) to estimate total losses in Africa as 12–23 million t. This estimate was based on the assumption of an overall CMD incidence of 50–60% and a loss of 30–40% in the yield of diseased plants.

Others have estimated the losses in particular areas, as in Uganda at the height of the recent pandemic (Otim-Nape *et al.*, 2000). It was assumed that each year an area equivalent to four whole districts was rendered totally unproductive. This was equivalent to a loss of 60,000 ha, which could have been expected to produce 600,000 t of roots worth US$60 million at a conservative valuation of US$100 t[-1]. Similarly, the losses due to the epidemic in western Kenya were estimated to exceed US$10 million in 1998 alone (Legg, 1999). The losses in Kenya have since become much greater as additional areas have been severely affected.

The transmission of cassava mosaic viruses by the whitefly B. tabaci

The putative virus assumed to cause CMD in Africa was one of the first pathogens to be transmitted experimentally by whiteflies, and studies began in the 1920s when it became evident that the virus was spreading naturally and that whiteflies were the only sap-feeding insects on cassava likely to be vectors. The first transmissions were reported from Congo using adults of a species referred to as *Bemisia mosaicivecta* (Ghesquière, 1932), which was later stated to be a misprint for *B. mosaicivectura* (Storey and Nichols, 1938). The species was also referred to as *Bemisia gossypiperda* Misra & Lamba var. *mosaicivectura* (Mayné and Ghesquière, 1934). The same or a closely related species referred to as *Bemisia nigeriensis* Corbett was used in successful transmission experiments in Nigeria (Golding, 1936) and Tanzania (Storey and Nichols, 1938), where infection was achieved by transferring infective whiteflies to the youngest leaves and shoots, but not to older ones.

Later experiments on the mode of transmission were carried out in Nigeria (Chant, 1958), Ivory Coast (Dubern, 1979, 1994) and Kenya (Seif, 1981) using what seems to have been the whitefly species used earlier, but referred to as *B. tabaci* Gennadius, as in all subsequent studies. Based on current knowledge it is likely that the transmission studies in coastal East Africa (Storey and Nichols, 1938; Seif, 1981) were with EACMV and those in Congo and West Africa with ACMV (Ghesquière, 1932; Golding, 1936; Chant, 1958; Dubern, 1979, 1994). There have been no published reports of vector transmission studies with the recently distinguished Ugandan variant (UgV). The East and West African isolates are transmitted in a persistent manner and the minimum (and optimum) acquisition access, inoculation access and latent periods for successful transmission are 3 h (5 h), 10 min and 3–4 h (6 h), respectively. The virus is retained by adults for at least 9 days. It persists during moulting, but it is not transmitted transovarially (Dubern, 1979, 1994). Nymphs can transmit, but they are not of epidemiological importance because of their immobility. Up to 1.7% of the adult whiteflies were shown to be infective when collected in heavily infected cassava fields in Ivory Coast and transferred to young test seedlings of cassava (Fargette *et al.*, 1990).

Epidemiology

The whitefly-borne viruses that cause CMD have not been reported in the Neotropics and they are assumed to have spread to cassava from indigenous African plant species. Several indigenous hosts have been identified, including *Jatropha* spp., but it is uncertain whether they are the original host(s) from which spread occurred. They certainly seem to be of little or no current importance as initial sources from which virus is spread to cassava. All the spread that occurs can be attributed to viruliferous whiteflies moving between or within cassava plantings, having acquired virus from cassava plants grown from infected cuttings or infected by whiteflies at a later stage of growth. This is consistent with the findings of epidemiological studies in Ivory Coast, Kenya and Uganda that spread into and within experimental plantings is related to the number of adult whiteflies recorded and also to the incidence of CMD in the area, as indicated by surveys of farmers' fields in the district or locality (Legg *et al.*, 1997; Otim-Nape *et al.*, 1998a), or from assessments of the health status of the propagules being used (Legg and Ogwal, 1998). New plantings are soon colonized by immigrant whiteflies moving from older stands of cassava in the area. The immigrants then reproduce to reach peak populations within a few months of planting before dispersing to other, younger, cassava (Fishpool and Burban, 1994).

The distribution of immigrant whiteflies and of plants newly affected by CMD is influenced by the direction of the prevailing wind and by the effects of wind turbulence around and within stands. The incidence of whiteflies and CMD tend to be greatest at the crop margins, especially along the windward and leeward edges and environmental gradients have been observed where whitefly populations and virus incidence decrease with increasing distance from the field boundaries (Fargette *et al.*, 1985; Colvin *et al.*, 1998). Incidence is also increased by breaks or discontinuities in the crop canopy which facilitate the alighting and establishment of viruliferous vectors (Fargette *et al.*, 1985).

Control measures

There are obvious benefits to be gained by decreasing the losses caused by CMD and this can be achieved by a reduction in the incidence and/or severity of the disease. Various approaches to control are possible, as discussed in detail elsewhere (Thresh and Otim-Nape, 1994). However, the main attention has been given to the use of resistant varieties (Fargette *et al.*, 1996; Thresh *et al.*, 1998a) and phytosanitation, involving the use of CMD-free planting material and the removal (roguing) of any additional diseased plants that occur (Thresh *et al.*, 1998b).

Farmers occasionally use insecticides in attempts to restrict the spread of CMD by controlling the whitefly vector. However, the use of insecticides on cassava or other tropical root crops has received little attention from researchers in Africa and this approach is unlikely to be effective. It is also inappropriate because of the costs involved and the risks to farmers, consumers and the environment.

CROPPING PRACTICES. There are opportunities of adjusting cropping practices to decrease the losses caused by CMD. This can be done by adopting planting dates that avoid exposing young vulnerable plants to infection at times when there are likely to be the largest populations of viruliferous whiteflies (Adipala *et al.*, 1998). There are also advantages in planting away from and upwind of existing sources of infection and also in large compact blocks to minimize edge effects (Thresh and Otim-Nape, 1994). Other possibilities are to adopt close spacings or intercrops, or to interplant susceptible with resistant varieties. The benefits to be gained by adopting such practices have been established in experiments, but little or no attempt has been made to demonstrate the feasibility of these approaches. Moreover, they may be difficult for farmers to adopt within their existing cropping systems. This emphasizes the need for additional studies before attempts are made to change current farming practices.

RESISTANT VARIETIES. A feature of cassava in Africa is that many varieties are grown and there is great diversity for many different traits including susceptibility and response to CMD. Consequently, farmers who experience disease problems can usually respond by abandoning the most vulnerable varieties and adopting those that are somewhat resistant or tolerant and grow satisfactorily, even when infected. The ability of farmers to adjust to CMD in this way has long been recognized, but in the 1930s and 1940s attempts were made to breed varieties with greater levels of resistance by intercrossing cassava varieties with *Manihot glaziovii* and other species of *Manihot* (Jennings, 1994). Interspecies hybrids were backcrossed to cassava and led to the highly resistant varieties that have been developed and used in Madagascar and East Africa. Seeds of this type were also sent from East Africa to Nigeria, where selections that had been made there in the 1960s were used in the early 1970s as parents in the initial cassava improvement programme at the International Institute of Tropical Agriculture (IITA), Ibadan. This programme has been very influential and IITA clones and seeds have been widely distributed or used in National Breeding Programmes in many African countries and also by the IITA Regional Centre in Uganda (Mahungo *et al.*, 1994).

Some of the varieties produced in this way are so highly resistant to CMD that they sustain little or no damage, even under epidemic conditions. They are not readily infected and when infected usually develop inconspicuous symptoms that become even less conspicuous as growth proceeds and infected plants may eventually become symptomless. Moreover, virus is not fully systemic in highly resistant varieties and a substantial proportion of the cuttings collected from infected plants are free of virus and grow into healthy plants. This 'self-cleansing', 'reversion' phenomenon is important in restricting the progressive build-up of disease that would otherwise occur during successive cycles of vegetative propagation (Fargette *et al.*, 1994; Thresh *et al.*, 1998a).

Although highly resistant varieties of this type are available they are seldom widely grown and in many countries farmers continue to grow local varieties that have little or no resistance to CMD. This explains why the disease is so prevalent in many areas and why such serious losses have occurred during the current pandemic in East Africa. The reasons for this unsatisfactory situation and the factors influencing farmers' choices of variety are complex and not fully understood (Nweke *et al.*, 1994). In some areas little or no attempt has been made to introduce resistant varieties or to promote their use. This can be because of a lack of resources or incentive, or because CMD is not regarded as such a damaging disease that the use of resistant varieties is essential. Moreover, the resistant varieties may not be entirely satisfactory in other respects and do not always meet the exacting requirements of growers and consumers. Recent experience in Uganda is that any such defects may be overlooked or regarded as unimportant in epidemic conditions when CMD is causing serious losses and undermines food security, but not when production has been restored. Such factors as the taste, palatability and other quality characteristics of cassava varieties then become paramount (Otim-Nape *et al.*, 2000).

Undoubtedly, a greater use of CMD-resistant varieties would bring substantial benefits by decreasing the losses caused by the disease and facilitate control by other means. However, such benefits will be difficult to achieve until a full range of resistant varieties is available that meet all the requirements of producers and consumers. Until then CMD will continue to cause problems. It seems inevitable that susceptible varieties will be retained in at least some areas and that CMD will continue to cause substantial, albeit generally acceptable, losses. This emphasizes the need for management procedures that will improve the health status of susceptible varieties and enable them to be grown successfully and more productively.

PHYTOSANITATION. The use of virus-free propagules is a basic approach to the control of many virus diseases and can bring obvious advantages (Thresh and Otim-Nape, 1994; Thresh et al., 1998b). Crop establishment and initial growth are improved and there is a reduction in the number of primary sources of infection from which subsequent virus spread can occur. The yield benefits are particularly great with cassava because plants grown from infected cuttings sustain the greatest damage and much of the spread of CMD occurs during the early stages of crop growth. Moreover, whiteflies reproduce more rapidly on CMD-infected than on healthy plants and so infected plants contribute a disproportionately large proportion of the total vector population within a crop stand (Colvin et al., 1999).

Clearly, there are powerful arguments for using CMD-free planting material and this approach has been advocated repeatedly. However, it has not been widely adopted, even in official cassava improvement programmes. The reasons for this are many and complex. In some areas CMD is so prevalent that it is regarded as a normal feature of cassava, CMD-free stocks are not available and farmers simply propagate from whatever plants are available and deemed suitable to provide cuttings. Even where CMD is less prevalent and there is an opportunity to select cuttings from uninfected plants, farmers seldom do so. They may be unaware of the benefits to be gained and of the basic features of CMD and its dissemination in infected cuttings and subsequent spread by whiteflies. Moreover, even if farmers are made aware it may be difficult or even impossible for them to distinguish uninfected plants at the time cuttings are required because the plants are leafless following drought or pest attack.

These difficulties are not easily overcome and there are obvious problems in contacting and changing the practices of the millions of cassava growers in Africa, many of whom are not readily accessible and poorly educated. Nevertheless, this was done widely in Uganda during the recent pandemic (Otim-Nape et al., 2000). The effects of CMD were then so severe that farmers were very receptive to any measures that would alleviate the problem and emergency funds became available from donors for mass training programmes for farmers, extensionists and opinion leaders. Selection was shown to be feasible and was adopted widely by farmers in some of the worst affected areas who were anxious to improve the health status of their plantings as a means of restoring production. The problem in Uganda now is to ensure that farmers will continue to select 'clean' planting material as the CMD situation returns to normal. There is also a need to achieve similar results elsewhere in areas where there is no serious CMD problem and so less incentive to adopt basic control measures, or to provide special funding for training farmers. Until this is done it seems inevitable that CMD will remain prevalent in many areas and yields will be impaired because of the widespread use of infected propagules.

Cassava brown streak disease

Cassava brown streak disease (CBSD) has been recognized since early studies in the 1930s, in what is now Tanzania. It was then established that the symptoms of the disease were distinct from those of CMD and that CBSD was more important than mosaic in some coastal areas of Tanzania (Storey, 1936). There has since been research on the aetiology, transmission and other features of CBSD in Tanzania, Kenya and elsewhere in eastern and southern Africa and at laboratories in the UK. However, research has been sporadic and the aetiology of the disease has been established only recently. Many uncertainties remain, especially relating to the effects of CBSD on crop yield and the natural means of spread.

Symptoms

The symptoms of CBSD are unusual in that they can affect a wide range of organs including leaves, stems, tuberous roots and fruits. Moreover, the symptoms are very variable in type and severity and some varieties are affected much less than others and frequently express symptoms only during the early stages of growth.

The name 'brown streak' was given to CBSD because of the brown elongate necrotic lesions that develop on the young green stem tissue of affected plants. This name is not altogether appropriate because only some varieties of cassava are so affected and the symptoms may be confused with the superficial circular necrotic spots of unknown cause that develop on the stems of some varieties (Nichols, 1950). Unlike the symptoms of CBSD the affected tissue does not extend into the cortex and the condition is not graft-transmissible.

The stem symptoms of CBSD are very variable in extent and severity and may be restricted to only one or a few shoots of each affected plant. In contrast, highly sensitive varieties develop very conspicuous stem symptoms on many branches, the leaves become necrotic and absciss and the shoots die back. The most severely affected plants eventually die but others recover, especially during periods of high temperature.

The leaf symptoms of CBSD are also variable and they are quite distinct from those of CMD in type and in affecting only the mature leaves. The most easily recognizable leaf symptoms occur as a characteristic 'feathery' chlorosis closely orientated along the secondary and tertiary veins and affecting many of the leaves or leaflets (Plate 2a). The symptoms are recognized less readily if they are relatively inconspicuous and restricted to only parts of some leaflets on affected plants. Other leaf symptoms occur as yellow blotches that are not closely associated with the leaf veins (Plate 2b). These symptoms affect different proportions of the leaf and they may or may not be conspicuous. They are particularly difficult to recognize when they develop only in the oldest leaves as they begin to discolour and senesce naturally. Such leaves soon absciss and the plants may then appear to be unaffected, especially at hot times of year when younger leaves develop inconspicuous symptoms or grow normally.

CBSD causes necrosis of the tuberous roots (Plate 2c) which also develop characteristic constrictions (Plate 2d). However, some varieties do not express root necrosis or do so only at a late stage of crop growth. These varieties are damaged much less severely than those that develop extensive symptoms at an early stage.

Distribution and prevalence

In early studies on CBSD it was established that the disease occurred in coastal areas of Kenya and Tanzania and it was assumed to be present in adjacent areas of coastal Mozambique (Nichols, 1950). The disease was also reported at the time in Uganda and Malawi, especially at lower altitudes in southern Malawi towards the Mozambique border. However, there appear to have been no detailed surveys of the incidence or severity of CBSD and the overall prevalence and importance of the disease was unclear.

Information on the current incidence of CBSD has been obtained in recent surveys in Uganda, Tanzania, Mozambique and coastal Kenya. The disease was found in only one planting in Uganda (G.W. Otim-Nape and J.M. Thresh, unpublished observation) and in 62 (19%) of the 325 plantings examined in Tanzania, although the overall incidence in the country as a whole was only 6% (Legg and Raya, 1998). The incidence was much higher in the lowland coastal areas of Kenya and Tanzania and on Oguja Island of Zanzibar, as confirmed in additional detailed surveys (Thresh and Mbwana, 1998; Hillocks *et al.*, 1999; J.M. Thresh and T. Munga, unpublished).

Surveys conducted in 1999 confirmed the occurrence of CBSD in Nampula and Zambezia provinces of Mozambique, which are the two most important cassava-growing areas of the country. The overall incidence based on assessments of leaf and stem symptoms was 49% in Zambezia and 28% in Nampula, but the incidence was much higher in some districts, varieties and plantings, especially in lowland coastal areas (R. Hillocks and J.M. Thresh, unpublished). Moreover, the leaf symptoms were sometimes inconspicuous and not readily distinguished, which suggested that the results underestimate the true incidence of infection.

Symptoms also tended to be inconspicuous in reconnaissance surveys carried out in Malawi during the early 1990s (J.M. Thresh and A. Sweetmore, unpublished). CBSD was then present in many areas and was most prevalent at mid-altitudes along the northwestern shore of Lake Malawi. These areas had been used to supply planting material to many other parts of Malawi, following the severe effects of the 1990–1991 drought and thus contributed to the widespread occurrence of CBSD. There may also have been movement of planting material across the border into Zimbabwe and Zambia, where CBSD is known to occur. The disease has not been reported in South Africa or Angola, or in any of the countries of West and Central Africa.

Aetiology

From the outset CBSD was assumed to be caused by a virus because it was graft-transmissible and no visible pathogen was detected. The first evidence of a virus was obtained by sap inoculation from cassava to herbaceous hosts and back to cassava (Lister, 1959) and also by electron microscopy (Kitajima and Costa, 1964). Virus isolates in herbaceous hosts were later shown to have elongate particles 650–690 nm long (Lennon et al., 1986a). They resembled those of viruses now ascribed to the genus Carlavirus, but no serological relationship was demonstrated at the time with any definitive virus of this type.

There was later evidence that two different elongate viruses occur in CBSD-affected plants (Lennon et al., 1986a; Brunt, 1990) and isolates in herbaceous hosts were shown to induce 'pin-wheel' inclusions of the type produced by viruses now attributed to the family Potyviridae. This is consistent with the recent conclusion that CBSD is caused by a virus of the genus Ipomovirus, which is one of the four genera comprising the Potyviridae (Monger et al., 2001).

Effects on growth and yield

There is only limited information on the effects of CBSD on growth and yield. In studies on a local Kenyan variety the main effect was on the quality of the roots produced and not on root weight or number (Bock, 1994). However, yields of marketable roots were decreased in a more recent study with other varieties in Tanzania (R. Hillocks and M.D. Raya, unpublished). Apart from any such loss of yield, necrosis decreases the value of the roots produced which become unusable and unsaleable if the damage is extensive. This may necessitate farmers having to harvest prematurely before much deterioration of the roots has occurred, but this incurs a yield penalty. Additional studies are required with a wide range of varieties harvested after

different periods to establish the full significance of these effects.

Epidemiology and control

There is little information on the epidemiology and control of CBSD and there are many uncertainties which impede the development of effective management strategies. One of the problems has been the lack of assured virus-free stocks of planting material for epidemiology experiments and for use by farmers. Another has been the failure to identify the natural means of spread between plants. These issues are now being addressed in projects in Tanzania and Mozambique. Moreover, in these countries and also in Kenya and Malawi breeding lines are being assessed for resistance to CBSD, as in earlier studies in Tanzania between 1937 and 1957 (Jennings, 1957).

From experience in several countries it is apparent that much use is being made of CBSD-infected planting material which is an effective means of perpetuating and disseminating the disease. However, there is evidence of natural spread between plants as clones introduced from West Africa or other areas that are free from CBSD have become infected when grown at sites in Mozambique, Malawi, Kenya and Tanzania where infection is rife. Plants raised from seed introduced from West Africa have also become infected at these sites.

There is little evidence on temporal or spatial patterns of spread, but this is known to have been slow in an experiment at a site in coastal Kenya (Bock, 1994) and rapid in recent trials at sites in coastal Tanzania (M. Raya, K. Mtunda and R.J. Hillocks, unpublished information) and Mozambique (R. Macia and J.M. Thresh, unpublished information). This emphasizes the need for additional studies to determine the circumstances under which spread occurs and the scope for utilizing the benefits of virus-free planting material to replace the contaminated stocks now being used widely. Virus-free stocks can be produced by rigorous selection (Mtunda et al., 1999) and in future this may be facilitated by using the sensitive methods of virus-detection now being developed. It is also possible to use meristem-tip and/or heat therapy to eliminate CBSV from clones that seem to be totally infected (Kaiser and Teemba, 1979).

Natural spread of CBSD between plants is attributed to an arthropod vector or vectors as yet unidentified. However, only few transmission experiments have been done, mainly involving the aphid *Myzus persicae* and the whitefly species *B. tabaci* and *Bemisia afer* (= *Bemisia hancockii*). The two whitefly species have been considered because they are two of the few sap-feeding insects to have had a long association with cassava in Africa. Moreover, CBSV is now attributed to the same genus of the *Potyviridae* as *Sweet potato mild mottle virus* which is transmitted by *B. tabaci*. It is also notable that *B. afer* seems to be particularly common in coastal areas of eastern and southern Africa where CBSD is most prevalent. This emphasizes the need for additional studies with *B. afer* and also of insect species that visit but do not colonize and breed on cassava. At least some of the spread may be from hosts other than cassava, as CBSV has been detected only in eastern and southern Africa and it is assumed to have indigenous hosts from which it spread to cassava after the crop was introduced. The identification of a vector will help to explain the current limited geographic distribution of CBSD, which occurs mainly in the lowland coastal areas of eastern and southern Africa. Such knowledge would facilitate the development of specific control measures. Meanwhile, the emphasis has been on the use of varieties that do not develop severe root necrosis, or do so only at a late stage of crop growth. This attitude of 'living with' the disease is similar to that adopted in many areas to cassava mosaic disease and provides a means of avoiding serious losses. However, any yield penalty incurred through the widespread use of tolerant varieties has not been quantified and could be substantial. This suggests that there could be benefits in developing and exploiting virus-resistant varieties and effective methods of phytosanitation.

The Viruses and Virus Diseases of Cassava in Asia and the Pacific Regions

Cassava is grown in many countries of South-East Asia and the Pacific and these areas account for an estimated 27% of total world production. Cassava mosaic disease (CMD) is the only virus disease known to be important in the region and it seems to be restricted to India and Sri Lanka. An early report of CMD in Indonesia (Muller, 1931) has not been confirmed and the symptoms were later attributed to a mineral deficiency (Bolhuis, 1949). *Cassava green mottle virus* has been detected in cassava originating from the Pacific region (Table 12.1; Lennon *et al.*, 1987), but its prevalence and importance is not known and it is not considered further here.

Cassava mosaic disease (CMD)

CMD was not reported in India until 1966 (Alagianagalingam and Ramakrishnan, 1966), although it is known to have been present earlier (Abraham, 1956) and it has since been recorded in Sri Lanka (Austin, 1986). The disease has received much less attention in Asia than in Africa. Nevertheless, it is clear that many of the research findings from Africa as summarized in an earlier section (pp. 242–249) also apply to India and Sri Lanka.

Distribution and prevalence

There is little current information on the incidence of CMD in India and the only available data were obtained during a reconnaissance survey in 1988 (Mathew, 1989). Twenty fields were assessed in each of 18 districts, including 11 districts of Kerala State. The overall incidence of CMD was higher in the two main cassava-growing states of Kerala (23%) and Tamil Nadu (30%) than in Andhra Pradesh (< 1%) and Karnataka (5%), which are outside the main cassava-growing areas. However, the number of fields examined was limited, especially when considered in relation to the large area of cassava being grown (Table 12.2). There is a need for additional more comprehensive surveys, especially as CMD seems to have become more prevalent in recent years. This was evident on a 1996 tour of the main cassava-growing areas of Kerala and around Salem in Tamil Nadu. Many of the fields visited in the lowland areas were almost totally affected and in some localities the symptoms were unusually severe and associated with poor yields. The incidence was much less in the upland areas and in a lowland planting established with cuttings obtained from the hills (M. Thankappen and J.M. Thresh, unpublished observations).

Aetiology

The symptoms of mosaic disease on cassava in India are similar to those reported in Africa and the name cassava mosaic disease (CMD) has been adopted in some publications and Indian cassava mosaic disease (ICMD) in others. Malathi and Sreenivasan (1983) first isolated a geminivirus from CMD-affected plants in India, as in the earlier studies in Africa. Four Indian isolates were included in serological tests with isolates from coastal and western Kenya using polyclonal antisera prepared against African and Indian isolates (Malathi *et al.*, 1985, 1987). Three of the Indian isolates reacted positively with African antisera but they were distinguishable serologically from African isolates and so were regarded as being of a separate strain of ACMV. In subsequent tests using a panel of monoclonal antibodies, Indian and Sri Lankan isolates were distinguished from those from East and West Africa and later referred to as *Indian cassava mosaic virus* as described previously (p. 245).

Effects on growth and yield

There have been fewer yield loss studies on CMD in India than in Africa and no estimates have been made of overall losses in the subcontinent. Reductions in weight of tuberous roots of 84% were reported in the first experiments with a susceptible local variety (Narasimhan and Arjunan, 1974, 1976), but losses were only 19–26% in the hybrids tested and in the widely grown M4 from Malaysia (Thankappan and Chacko, 1976). In other experiments losses were 42% in the popular variety Kalikalan, ranged from 17 to 36% in nine selected hybrids and were 17% in M4 which was at the time considered to be tolerant of infection (Malathi *et al.*, 1985). Losses were even less in a later trial with M4 (7–10%) and four hybrid varieties (9–21%) and there was a positive relationship between yield loss and symptom severity scores (Nair and Malathi, 1987). These results and the low incidence of CMD in many areas suggest that the disease causes less severe losses in India than in Africa. Nevertheless, it is likely to have substantial effects in areas of India where CMD-sensitive varieties are grown and severe symptoms are prevalent.

Transmission by the whitefly B. tabaci

CMD spreads naturally in India and following earlier experience in Africa (pp. 246–247), the main attention has been on *B. tabaci* in the search for an insect vector. Successful transmissions have been reported using whiteflies transferred from infected to healthy cassava, from infected cassava to herbaceous hosts and between herbaceous hosts. High rates of transmission were achieved in some experiments, as between cassava (19%) and from cassava to *Nicotiana tabacum* cv. Jayasri (100%), *N. rosulata* (67%) and 11 other *Nicotiana* spp. (20–25%) using 50 whiteflies per test plant (Mathew and Muniyappa, 1993). However, such high rates of transmission seem to be exceptional and not readily reproducible. Much lower rates of transmission were reported in other studies (e.g. Nair, 1975), some of which were completely unsuccessful (Malathi *et al.*, 1985; Palaniswami *et al.*, 1996). Another inconsistency is that transmissions from cassava to cucumber were achieved in some trials (Menon and Raychaudhuri, 1970), but not in others (Mathew and Muniyappa, 1993). The reasons for this and the apparent difficulty experienced in transmitting Indian isolates by whiteflies compared with those in Africa, have not been determined. One possibility is that the whiteflies on cassava in India are less well adapted to their host than those in at least some parts of Africa where a cassava biotype of *B. tabaci* has been distinguished (Burban *et al.*, 1992). It certainly seems particularly difficult to transmit Indian isolates to cassava and similar difficulties have been recorded with other isolates in studies in glasshouses in temperate conditions (B.D. Harrison and P.J. Markham, personal communication). Despite these difficulties there is no reason to doubt that *B. tabaci* is the vector of ICMV and studies on epidemiology, control and whitefly population dynamics have proceeded on this assumption (e.g. Mahto and Sinha, 1978).

Epidemiology

The area of cassava grown in India is considerably less than in Africa. Nevertheless, the crop is grown in diverse environments including the lowland humid forest areas of coastal Kerala, the upland foothills of the Western Ghats and the irrigated areas of Tamil Nadu where there is a prolonged dry season.

Epidemiological studies have used virus-free stocks of selected planting material, or clones derived from meristem-tip cultures. Several cultivars were included in experiments done in three successive seasons at a site near Trivandrum, Kerala State (Nair, 1985). The final incidence of CMD did not exceed 1.3% in plots containing initial disease foci and was even less in plots without sources. There was also little or no spread in a later study where monthly plantings were made at a site near Bangalore in Karnataka State which is outside the main cassava-growing area (Mathew, 1989).

In a further trial at a site near Trivandrum, six cultivars were established in plots which contained initial sources of inoculum and CMD was also prevalent in the surrounding plantings. There was substantial spread to the susceptible cv. Kalikalan (50%), but not to the five more resistant cultivars (1–10%) (Nair, 1988). In a later more comprehensive study, there was more spread to plots which contained initial sources of inoculum (overall incidence 5.7%) than to those without (2.8%). However, the source effect was not consistent at each of the four sites or in the five cultivars and was largely due to the big difference in incidence in cv. Kalikalan at the site where most spread occurred (Nair and Thankappan, 1990).

It is not appropriate to make broad generalizations on the basis of these few experiments, but they suggest that there is considerable scope for exploiting the benefits of virus-free planting material, especially of resistant varieties and in areas of low infection pressure. Moreover, the results indicate that the high incidence of CMD in Tamil Nadu is due to the use of infected cuttings and *not* to rapid spread by whiteflies. Further studies are required to substantiate these conclusions and to establish whether they are of wide general validity. Additional evidence is also required on the importance of spread from sources within plantings and on the suggestion that this occurs more frequently in India than in Africa, where experience in Ivory Coast, Kenya and Uganda has shown that much of the spread is by infective whiteflies moving between rather than within plantings

(Bock, 1983; Fargette *et al.*, 1990; Otim-Nape, 1993).

Control

Cassava in India is grown under very different conditions from those in Africa. The relatively high productivity of cassava achieved in India is associated with the limited use of intercropping and with generally good husbandry practices. These include effective weed control, the establishment of uniform stands, the routine application of fertilizers and in some areas the use of irrigation. Moreover, the Indian crop is unaffected by either the cassava green mite or the cassava mealybug which have had such damaging effects in many parts of Africa (see Chapter 11).

In these favourable circumstances Indian farmers might be expected to give considerable attention to the health status of the planting material used and to other means of controlling CMD so as to further enhance yields and optimize production. However, their attitude towards the disease seems to be similar to that in many parts of Africa in that it is largely ignored. Little attempt is made to select cuttings from healthy plants, or to remove diseased plants from within partially infected stands. Moreover, considerable use is made of susceptible varieties even though resistant ones are available. This attitude can be explained in part by the high yields obtained, even from stands in which CMD is prevalent. Nevertheless, the disease is so widespread and has such detrimental effects on yield in some areas that productivity is affected and would be increased substantially by adopting effective control measures.

As in Africa, the main possible approaches to control are through phytosanitation and resistant varieties. Some attention has also been given in India to the use of insecticides to control the whitefly vector in attempts to reduce the spread of CMD. However, the results have been unsatisfactory and the routine use of insecticides is inappropriate on health and environmental grounds and not recommended (Malathi *et al.*, 1985).

Virus-free stocks have been obtained by rigorous selection and through the use of meristem-tip therapy. They have been used in experiments and shown to remain largely free of

CMD in areas where there is limited spread by whiteflies. Substantial increases in yield have been achieved in this way (Nair, 1990; Nair and Thankappen, 1990), but only limited attempts have been made to encourage the widespread adoption of such material.

General Discussion

From the foregoing account it is clear that the viruses and virus diseases of cassava have received considerable attention, especially those occurring in Africa. Nevertheless, the available information is very incomplete and many uncertainties remain. For example, the status, distribution and effects of several of the viruses listed in Table 12.1 have not been determined and further research may show them to be more widespread and damaging than present evidence suggests. There is also uncertainty concerning the epidemiology and mode of spread of cassava brown streak, cassava frogskin and other diseases and an urgent need to confirm the role of the whitefly or other vectors involved.

These deficiencies can be remedied by an allocation of expertise and resources commensurate with the importance of cassava as the basic staple food crop of large and populous areas of the tropics. However, a problem is likely to be encountered in achieving this because increasingly donors and grant agencies are allocating funds in response to the perceived needs of farmers, who may be totally unaware that virus problems exist. This is evident from recent experience with cassava brown streak disease in Mozambique and Tanzania and with frogskin disease in South America. In these areas farmers have created or exacerbated the problem by making extensive use of virus-infected cuttings as planting material and losses due to disease are regarded as inevitable and a normal feature of cassava in the localities affected.

These and other experiences elsewhere indicate the difficulty of achieving sustained improvements in the overall health status of cassava by adopting virus-free cuttings and by deploying resistant varieties and other research findings. It is necessary to change the attitudes and practices of millions of farmers, many of whom are remote, poorly educated and lack resources and access to extension personnel and

technical advice. There is a tendency to ignore or underestimate the importance of virus diseases unless the losses sustained are so great that rural livelihoods and food security are undermined. Relief or emergency measures are then necessary and farmers also respond by exploiting the genetic diversity available and switching to less vulnerable varieties. Once production has been restored virus diseases again receive relatively little attention even though they impair productivity and the yield penalty may be substantial.

Clearly, these difficulties will not be overcome quickly or easily and losses due to viruses are likely to continue in the foreseeable future. Indeed, they may even increase if damaging viruses, strains or strain combinations reach new areas by natural spread or through the movement of infected propagules. This emphasizes the importance of stringent quarantine controls on the movement of cassava material to maintain the present limited distribution of cassava viruses. Several of these are restricted to particular continents or regions and are likely to cause considerable damage if they are spread elsewhere. The need to prevent New World viruses reaching Africa or Asia, and Old World viruses being introduced to the Americas has long been apparent and appropriate quarantine measures have been devised and enforced (Frison and Feliu, 1991). These measures should be revised now that additional viruses of cassava have been characterized and new methods of virus detection have been developed. There is also a need to consider the implications of recent findings on the diversity and variability of cassava mosaic geminiviruses and the occurrence of particularly damaging strains or strain combinations.

Experience with cassava mosaic disease in Africa over many years and more recently with frogskin disease in South America is that the situation is labile and can change rapidly. This is also apparent from recent experience with the whitefly *B. tabaci* which seems to be adapting to cassava in different countries of South America where previously cassava was not infested. Moreover, the damaging 'B' biotype of *B. tabaci* has spread recently to parts of northern and southern Africa and could lead to increased problems. These developments emphasize the importance of continued research on the viruses and virus vectors of cassava to monitor and combat

new problems as they arise and to deal more
effectively with those already known.

References

Abraham, A. (1956) *Tapioca cultivation in India*. Farm
  Bulletin No. 17. Indian Council of Agricultural
  Research, New Delhi.
Adipala, A., Byabakama, B.A., Ogenga-Latigo, M.W.
  and Otim-Nape, G.W. (1998) Effect of planting
  date and varietal resistance on the development of
  cassava mosaic virus disease in Uganda. *African
  Plant Protection* 4, 71–79.
Aiton, M.M., Roberts, I.M. and Harrison, B.D. (1988)
  Cassava common mosaic potexvirus from mosaic-
  affected cassava in the Ivory Coast. *Report of the
  Scottish Crop Research Institute for 1987*, p. 191.
Alagianagalingam, M.N. and Ramakrishnan, K.
  (1966) Cassava mosaic in India. *South Indian
  Horticulture* 14, 71–72.
Angel, J.C., Pineda, B.L., Nolt, B. and Velasco, A.C.
  (1990) Mosca blancas (Homoptera: Aleyrodidae)
  asociadas a transmision de virus en yuca.
  Fitopatologia Colombiana 13, 65–71.
Anon. (1993) How Akwa Ibom overcame a crisis
  in cassava production. *Cassava Newsletter* 17(2),
  9–10.
Austin, M.D.N. (1986) Scientists identify cassava virus.
  Asian Agribusiness 3, 10.
Berrie, L.C., Palmer, K.E., Rybicki, E.P. and Rey, M.E.C.
  (1998) Molecular characterisation of a distinct
  South African cassava infecting geminivirus.
  Archives of Virology 143, 2253–2260.
Bock, K.R. (1983) Epidemiology of cassava mosaic
  disease in Kenya. In: Plumb, R.T. and Thresh,
  J.M. (eds) *Plant Virus Epidemiology*. Blackwell
  Scientific, Oxford, pp. 337–347.
Bock, K.R. (1994) Studies on cassava brown streak
  virus disease in Kenya. *Tropical Science* 34,
  134–145.
Bock, K.R. and Woods, R.D. (1983) Etiology of African
  cassava mosaic disease. *Plant Disease* 67,
  994–995.
Bolhuis, G.C. (1949) Waarnemingen ouer de zg
  mosaiekziekte big cassava op Java. *Buitenzoig*,
  Java General Agricultural Research Communication No. 92.
Brunt, A.A. (1990) Cassava brown streak 'carlavirus'. Cassava brown streak potyvirus. In: Brunt,
  A., Crabtree, K. and Gibbs, A. (eds) *Viruses of
  Tropical Plants*. CAB International, Wallingford,
  pp. 157–158.
Burban, C., Fishpool, L.D.C., Fauquet, C., Fargette, D.
  and Thouvenel, J.-C. (1992) Host-associated
  biotypes within West African populations of

the whitefly *Bemisia tabaci* (Genn.) (Hom.,
  Aleyrodidae). *Journal of Applied Entomology* 113,
  416–423.
Calvert, L.A. (1994) The safe movement of cassava
  germplasm. In: *International Network for Cassava
  Genetic Resources: Report of the First Meeting of
  the International Network for Cassava Genetic
  Resources*, Cali, Colombia, 18–23 August 1992.
  International Plant Genetic Resources Institute,
  Rome. International crop network series 10,
  pp. 163–165.
Calvert, L.A., Ospina, M.D. and Shepherd, R.J. (1995)
  Characterization of cassava vein mosaic virus: a
  distinct plant pararetrovirus. *Journal of General
  Virology* 76, 1271–1278.
Calvert, L.A., Cuervo, M.I., Ospina, M.D., Fauquet, C.
  and Ramirez, B.C. (1996) Characterization of cas-
  sava common mosaic virus and a defective RNA
  species. *Journal of General Virology* 77, 525–530.
Chant, S.R. (1958) Studies on the transmission of cassava mosaic virus by *Bemisia* spp. (Aleyrodidae).
  Annals of Applied Biology 46, 210–215.
Chen, C.T., Ko, N.J. and Chen, M.J. (1981) Electron
  microscopy of cassava common mosaic in
  Taiwan. *Reports of the Taiwan Sugar Research
  Inspectorate* 93, 20–27.
Colvin, J., Fishpool, L.D.C., Fargette, D., Sherington, J.
  and Fauquet, C. (1998) *Bemisia tabaci* (Hemiptera: Aleyrodidae) trap catches in a cassava field
  in Côte d'Ivoire in relation to environmental
  factors and the distribution of African cassava
  mosaic disease. *Bulletin of Entomological Research*
  88, 369–378.
Colvin, J., Otim-Nape, G.W., Holt, J., Omongo, C., Seal,
  S., Stevenson, P., Gibson, G.I., Cooter, R.J. and
  Thresh, J.M. (1999) Symbiotic interactions driv-
  ing epidemic of whitefly-borne cassava mosaic
  virus disease. In: Cooter, R.J., Otim-Nape, G.W.,
  Bua, A. and Thresh, J.M. (eds) *Cassava Mosaic
  Disease Management in Smallholder Cropping
  Systems*. Proceedings of the Workshop on CMD
  Management in Smallholder Cropping Systems.
  NARO/NRI, Chatham, pp. 77–86.
Costa, A.S. (1940) Observacóes sóbre o mosaico
  comum e o mosaico das nervuras da mandioca
  (*Manihot utilissima* Pohl.). *Journal Agronomia
  (Piracicaba)* 3, 239–248.
Costa, A.S. and Kitajima, E.W. (1972a) Cassava com-
  mon mosaic virus. C.M.I./A.A.B. In: *Descriptions
  of Plant Viruses No. 90*.
Costa, A.S. and Kitajima, E.W. (1972b) Studies on virus
  and mycoplasma diseases of the cassava plant in
  Brasil. *Proceedings of Cassava Mosaic Workshop*,
  Ibadan. International Institute for Tropical Agri-
  culture, Ibadan, Nigeria, p. 18.
Cours, G., Fargette, D., Otim-Nape, G.W. and Thresh,
  J.M. (1997) The epidemic of cassava mosaic virus

disease in Madagascar in the 1930s–1960s: lessons for the current situation in Uganda. *Tropical Science* 37, 238–248.

Cuervo, I.M. (1989) Caracterizacion de los acidos ribonucleicos de doble cadena (ARN-cd) asociados a enfermedades virales en yuca (*Manihot esculenta*, Crantz). Tesis Inginero Agriculturo Palmira. Universidad Nacional de Colombia.

Deng, D., Otim-Nape, G.W., Sangare, A., Ogwal, S., Beachy, R.N. and Fauquet, C.M. (1997) Presence of a new virus closely related to East African cassava mosaic geminivirus associated with cassava mosaic outbreak in Uganda. *African Journal of Root and Tuber Crops* 2, 23–28.

Dubern, J. (1979) Quelques propriétés de la mosaïque Africaine du manioc.1. La transmission. *Phytopathologische Zeitschrift* 96, 25–39.

Dubern, J. (1994) Transmission of African cassava mosaic geminivirus by the whitefly (*Bemisia tabaci*). *Tropical Science* 34, 82–91.

FAO (1985) *FAO Production Year Book*. Series No. 31 V 38. Rome, Italy.

Fargette, D., Fauquet, C. and Thouvenel, J.-C. (1985) Field studies on the spread of African cassava mosaic. *Annals of Applied Biology* 106, 285–294.

Fargette, D., Fauquet, C. and Thouvenel, J.-C. (1988) Yield losses induced by African cassava mosaic virus in relation to the mode and date of infection. *Tropical Pest Management* 34, 89–91.

Fargette, D., Fauquet, C., Grenier, E. and Thresh, J.M. (1990) The spread of African cassava mosaic virus into and within cassava fields. *Journal of Phytopathology* 130, 289–302.

Fargette, D., Roberts, I.M. and Harrison, B.D. (1991) Particle purification and properties of cassava Ivorian bacilliform virus. *Annals of Applied Biology* 119, 303–312.

Fargette, D., Thresh, J.M. and Otim-Nape, G.W. (1994) The epidemiology of African cassava mosaic geminivirus: reversion and the concept of equilibrium. *Tropical Science* 34, 123–133.

Fargette, D., Colon, L.T., Bouveau, R. and Fauquet, C. (1996) Components of resistance of cassava to African cassava mosaic virus. *European Journal of Plant Pathology* 102, 645–654.

Fauquet, C. and Fargette, D. (1990) African cassava mosaic virus; etiology, epidemiology and control. *Plant Disease* 74, 404–411.

Fishpool, L.D.C. and Burban, C. (1994) *Bemisia tabaci*: the whitefly vector of African cassava mosaic geminivirus. *Tropical Science* 34, 55–72.

Fondong, V.N., Pita, J.S., Rey, M.E.C., de Kochko, A., Beachy, R.N. and Fauquet, C.M. (2000) Evidence of synergism between African cassava mosaic virus and a new double-recombinant geminivirus infecting cassava in Cameroon. *Journal of General Virology* 81, 287–297.

Frison, E.A. and Feliu, E. (eds) (1991) *FAO/IBPGR Technical Guidelines for the Safe Movement of Cassava Germplasm*. Food and Agriculture Organization of the United Nations, Rome/International Board for Plant Genetic Resources, Rome.

Geddes, A.M.W. (1990) *The Relative Importance of Crop Pests in Sub-Saharan Africa*. Bulletin No. 36, Natural Resources Institute, Chatham.

Ghesquière, J. (1932) Sur la 'mycosphaerellose' des feuilles du manioc. *Bulletin Institu Col. Belge* 3, 160.

Golding, F.D. (1936) *Bemisia nigeriensis* Corb., a vector of cassava mosaic in southern Nigeria. *Tropical Agriculture, Trinidad* 13, 182–186.

Harrison, B.D., Zhou, X., Otim-Nape, G.W., Liu, Y. and Robinson, D.J. (1997) Role of a novel type of double infection in the geminivirus-induced epidemic of severe cassava mosaic in Uganda. *Annals of Applied Biology* 131, 437–448.

Hernández, A., Calderón, H., Zárate, R.D. and Lozano J.C. (1975) El cuero de sapo de la yuca (*Manihot esculenta* Crantz). *Noticias Fitopatológicas* 4, 117–118.

Hillocks, R.J., Raya, M.D. and Thresh, J.M. (1999) Distribution and symptom expression of cassava brown streak disease in southern Tanzania. *African Journal of Root and Tuber Crops* 3, 57–62.

Hong, Y.G., Robinson, D.J. and Harrison, B.D. (1993) Nucleotide sequence evidence for the occurrence of three distinct whitefly-transmitted geminiviruses in cassava. *Journal of General Virology* 74, 2437–2443.

Jennings, D.L. (1957) Further studies in breeding cassava for virus resistance. *East African Agricultural Journal* 22, 213–219.

Jennings, D.L. (1994) Breeding for resistance to African cassava mosaic geminivirus in East Africa. *Tropical Science* 34, 110–122.

Jericho, C., Thompson, G.J., Gerntholtz, U. and Viljoen, J.C. (1999) Occurrence and distribution of cassava pests and diseases in South Africa. In: Akoroda, M.O. and Teri, J.M. (eds) *Food Security and Crop Diversification in SADC Countries: the Role of Cassava and Sweet Potato*. Proceedings of the Scientific Workshop of the Southern African Root Crops Research Network (SARRNET), Lusaka, Zambia, 17–19 August 1998. SADC/IITA/CIP, pp. 252–262.

Johnson, A. (1992) *Report Lake Chad Farmer Training and Agricultural Development Project*. American Organizations for Rehabilitation Through Training, U.S. Agency for International Development.

Kaiser, W.J. and Teemba, L.R. (1979) Use of tissue culture and thermotherapy to free East African cassava cultivars of African cassava mosaic and cassava brown streak diseases. *Plant Disease Reporter* 63, 780–784.

Kitajima, E.W. and Costa, A.S. (1964) Elongated particles found associated with cassava brown streak. *East African Agricultural and Forestry Journal* 29, 28–30.

Kitajima, E.W. and Costa, A.S. (1966) Particulas esferoidais associadas ao virus do mosaico das nervuras da mandioca. *Bragantia* 25, 211–221.

Kitajima, E.W., Wetter, C., Oliveira, A.R., Silva, D.M. and Costa, A.S. (1965) Morfologia do virus do mosaico comum da mandioca. *Bragantia* 24, 247–260.

Kufferath, H. and Ghesquière, J. (1932) La mosaïque du manioc. *Compte-rendu de la Société Belge* 109, 1146.

Legg, J.P. (1999) Emergence, spread and strategies for controlling the pandemic of cassava mosaic virus disease in east and central Africa. *Crop Protection* 18, 627–637.

Legg, J.P. and Ogwal, S. (1998) Changes in the incidence of African cassava mosaic virus disease and the abundance of its whitefly vector along south-north transects in Uganda. *Journal of Applied Entomology* 122, 169–178.

Legg, J.P. and Raya, M.D. (1998) A survey of cassava virus diseases in Tanzania. *International Journal of Pest Management* 44, 17–23.

Legg, J.P., James, B., Cudjoe, A., Saizonou, S., Gbaguidi, B., Ogbe, F., Ntonifor, N., Ogwal, S., Thresh, J.M. and Hughes, J. (1997) A regional collaborative approach to the study of ACMD epidemiology in sub-Saharan Africa. *African Crop Science Conference Proceedings* 3, 1021–1033.

Legg, J.P., Sseruwagi, P., Kamau, J., Ajanga, S., Jeremiah, S.C., Aritua, V., Otim-Nape, G.W., Muimba-Kankolongo, A., Gibson, R.W. and Thresh, J.M. (1999) The pandemic of severe cassava mosaic disease in East Africa: current status and future threats. In: Akoroda, M.O. and Teri, J.M. (eds) *Food Security and Crop Diversification in SADC Countries: the Role of Cassava and Sweet Potato.* Proceedings of the Scientific Workshop of the Southern African Root Crops Research Network (SARRNET), Lusaka, Zambia, 17–19 August 1998. SADC/IITA/CIP, pp. 236–251.

Lennon, A.M., Aiton, M.M. and Harrison, B.D. (1986a) Cassava viruses from Africa: third country quarantine of cassava. *Annual Report 1985.* Scottish Crop Research Institute, Dundee, pp. 168–169.

Lennon, A.M., Aiton, M.M. and Harrison, B.D. (1986b) Cassava viruses from South America. In: *Annual Report 1985.* Scottish Crop Research Institute, Dundee, p. 167.

Lin, M.T. and Kitajima, E.W. (1980) Purificão e serologia do virus de mosaico das nervaduras da mandioca. *Fitopatologia Brasileira.* 5, 419. (Abstr.)

Lister, R.M. (1959) Mechanical transmission of cassava brown streak virus. *Nature* 183, 1588–1589.

Lozano, J.C. and Nolt, B.L. (1989) Pest and pathogens of cassava. In: Kahn, R.P. (ed.) *Plant Protection and Quarantine: Selected Pests and Pathogens of Quarantine Significance,* Vol. 2. CRC Press, Boca Raton Florida, pp. 174–175.

Maffla, G., Roa, J.C. and Roca, W.M. (1984) Erradicacion de la enfermedad cuero de sapo de la yuca Manihot esculenta, pro medio del dultive de meristemos. Efecto de la termoterapia, y del tamano del explante sobre la tasa de saneamiento. In: Perea, D. and Angarita, Z.A. (eds) *Congreso Nacional de Cultivo de Tejidos Vegetales,* Bogota, Colombia, Memorias. Bogota, Universidad Nacional de Colombia, pp. 171–175.

Mahto, D.N. and Sinha, D.C. (1978) Evaluation of insecticides for the control of whitefly, *Bemisia tabaci* Genn. in relation to the incidence of mosaic of cassava. *Indian Journal of Entomology* 40, 316–319.

Mahungo, N.M., Dixon, A.G.O. and Kumbira, J.M. (1994) Breeding cassava for multiple pest resistance in Africa. *African Crop Science Journal* 2, 539–552.

Malathi, V.G. and Sreenivasan, M.A. (1983) Association of gemini particles with cassava mosaic disease in India. *Journal of Root Crops* 9, 69–73.

Malathi, V.G., Nair, N.G. and Shantha, P. (1985) *Cassava Mosaic Disease.* Technical Bulletin Series 5, Central Tuber Crops Research Institute, Trivandrum.

Malathi, V.G., Thankappan, M., Nair, N.G., Nambisan, B. and Ghosh, S.P. (1987) Cassava mosaic disease in India. In: *Proceedings of the International Seminar on African Cassava Mosaic Disease and its Control,* Côte d'Ivoire, 4–8 May 1987. CTA/FAO/ORSTOM/IITA/IAPC, pp. 189–198.

Mathew, A.V. (1989) Studies on Indian cassava mosaic virus diseases. PhD Thesis, University of Agricultural Sciences, Bangalore.

Mathew, A.V. and Muniyappa, V. (1993) Host range of Indian cassava mosaic virus. *Indian Phytopathology* 46, 16–23.

Mayné, R. and Ghesquière, J. (1934) Hémiptères nuisibles aux végétaux du Congo Belge. *Annales Gembloux* 1934, 41.

Menon, M.R. and Raychaudhuri, S.P. (1970) Cucumber, a herbaceous host of cassava mosaic virus. *Plant Disease Reporter* 54, 34–35.

Monger, W.A., Seal, S., Isaac, A.M. and Foster, G.D. (2001) Molecular characterization of *Cassava brown streak virus* coat protein. *Plant Pathology* 50, 527–534.

Mtunda, K., Mahungu, N.M., Thresh, J.M., Kilima, M.S. and Kiozya, H.C. (1999) Cassava planting material sanitation for the control of cassava brown streak disease. In: Akoroda, M.O. and Teri, J.M. (eds) *Food Security and Crop Diversification in SADC Countries: the Role of Cassava and Sweet Potato*. Proceedings of the Scientific Workshop of the Southern African Root Crops Research Network (SARRNET), Lusaka, Zambia, 17–19 August 1998. SADC/IITA/CIP, pp. 300–304.

Muimba-Kankolongo, Chalwe, A.A., Sisupo, P. and Kanga, N.C. (1997) Distribution, prevalence and outlook for control of cassava mosaic disease in Zambia. *Roots* 4(1), 2–7.

Muller, H.R.A. (1931) Mozaiekziekte bij cassave. *Institut voor Plantenziekten Bulletin* 24, 17.

Nair, N.G. (1975) Transmission trials on cassava using white fly (*Bemisia tabaci*). Annual Report Central Tuber Crop Research Institute, Trivandrum, India.

Nair, N.G. (1985) Nature and extent of spread of cassava mosaic disease. In: Ramarujam *et al. Tropical Tuber Crops: Proceedings of the National Symposium in 1985*, CTCRI, Trivandrum, pp. 175–178.

Nair, N.G. (1988) Influence of inoculum source and varietal susceptibility on the field spread of Indian cassava mosaic disease. *Journal of Root Crops* 14, 5–9.

Nair, N.G. (1990) Performance of virus-free cassava (*Manihot esculenta* Crantz) developed through meristem tip culture. *Journal of Root Crops* 16, 123–131.

Nair, N.G. and Malathi, V.G. (1987) Disease severity and yield loss in cassava (*Manihot esculenta* Crantz) due to Indian cassava mosaic disease. *Journal of Root Crops* 13, 91–94.

Nair, N.G. and Thankappan, M. (1990) Spread of Indian cassava mosaic disease under different agroclimatic locations. *Journal of Root Crops*. ISRC National Symposium Special Volume, CTCRI, Trivandrum, pp. 216–219.

Narasimhan, V. and Arjunan, D. (1974) Effect of mosaic disease on cassava. Tapioca Research Station Report, Salem, Tamil Nadu, India, p.10.

Narasimhan, V. and Arjunan, D. (1976) Mosaic disease of cassava – loss in yield and tuber splitting. *Indian Phytopathology* 29, 428–429.

Nichols R.F.W. (1950) The brown streak disease of cassava: distribution, climatic effects and diagnostic symptoms. *East African Agricultural Journal* 15, 154–160.

Nolt, B.L., Velasco, A.C. and Pineda, B. (1991) Improved purification procedure and some serological and physical properties of cassava common mosaic virus from South America. *Annals of Applied Biology* 118, 105–113.

Nolt, B.L., Pineda, L.B. and Velasco, A.C. (1992) Surveys of cassava plantations in Colombia for virus and virus-like diseases. *Plant Pathology* 41, 348–354.

Nweke, F.I., Dixon, A.G.O., Asiedu, R. and Folayan, S.A. (1994) Cassava varietal needs of farmers and the potential for production growth in Africa. *Collaborative Study of Cassava in Africa, Working Paper* No. 10.

Nyirenda, G.K.C., Munthali, D.C., Phiri, G.S.N., Sauti, R.F.N. and Gerling, D. (1993) Integrated pest management of *Bemisia* spp. whiteflies in Malawi. Report Makoka Research Station, Thondwe, Malawi.

Otim-Nape, G.W. (1993) The epidemiology of the African cassava mosaic geminivirus disease in Uganda. PhD thesis, University of Reading.

Otim-Nape, G.W., Thresh, J.M., Bua, A., Baguma, Y. and Shaw, M.W. (1998a) Temporal spread of cassava mosaic virus disease in a range of cassava cultivars in different agro-ecological regions of Uganda. *Annals of Applied Biology* 133, 415–430.

Otim-Nape, G.W., Thresh, J.M. and Shaw, M.W. (1998b) The incidence and severity of cassava mosaic virus disease in Uganda: 1990–1992. *Tropical Science* 38, 25–37.

Otim-Nape, G.W., Bua, A., Thresh, J.M., Baguma, Y., Ogwal, S., Ssemakula, G.N., Acola, G., Byabakama, B., Colvin, J., Cooter, R.J. and Martin, A. (2000) *The Current Pandemic of Cassava Mosaic Virus Disease in East Africa and its Control*. Natural Resources Institute, Chatham.

Otim-Nape, G.W., Alicai, T. and Thresh, J.M. (2001) Changes in the incidence and severity of cassava mosaic virus disease, varietal diversity and cassava production in Uganda. *Annals of Applied Biology* 138, 313–327.

Palaniswami, M.S., Radhakrishnan, R., Nair, R.G., Pillai, K.S. and Thankappan, M. (1996) Whiteflies on cassava and its role as vector of cassava mosaic disease in India. *Journal of Root Crops* 22, 1–8.

Pineda, B., Jayasinghe, U. and Lozano, J.C. (1983) La enfermedad 'Cuero de Sapo' en yuca (*Manihot esculenta* Crantz). ASIAVA 4, 10–12.

Santos, A.A., Gonçalves, J.A., Queiroz, G.M. and Lima, R.N. (1995) Efeito do vírus do mosaico das nervuras sobre os componentes produtivos da mandioca, cv Pretinha, em duas regioes fisiográficas do Ceará. *Fitopatologia Brasileira* 20, 506–508.

Seif, A.A. (1981) Transmission of cassava mosaic virus by *Bemisia tabaci*. *Plant Disease* 65, 606–607.

Silberschmid, K. (1938) O mosaico da mandioca. *O Biologico* 4, 177–182.

Silva, D.M., Kitajima, E.W. and Oliveira, A.R. (1963) Obtencao do virus do mosaico comun da mandioca purificado. *Ciencia e cultura* 15, 304.

Storey, H.H. (1936) Virus diseases of East African plants: VI. A progress report on studies of the disease of cassava. *East African Agricultural Journal* 2, 34–39.

Storey, H.H. and Nichols, R.F.W. (1938) Studies of the mosaic diseases of cassava. *Annals of Applied Biology* 25, 790–806.

Thankappan, M. and Chacko, C.I. (1976) Effect of cassava mosaic on the different plant parts and tuber yield in cassava. *Journal of Root Crops 2*, 45–47.

Thresh, J.M. (1991) The ecology of tropical plant viruses. *Plant Pathology* 40, 324–339.

Thresh, J.M. and Mbwana, M.W. (1998) Cassava mosaic and cassava brown streak virus diseases in Zanzibar. *Roots* 5(1), 6–9.

Thresh, J.M. and Otim-Nape, G.W. (1994) Strategies for controlling African cassava mosaic geminivirus. *Advances in Disease Vector Research* 10, 215–236.

Thresh, J.M., Fargette, D. and Otim-Nape, G.W. (1994a) Effects of African cassava mosaic geminivirus on the yield of cassava. *Tropical Science* 34, 26–42.

Thresh, J.M., Fargette, D. and Otim-Nape, G.W. (1994b) The viruses and virus diseases of cassava in Africa. *African Crop Science Journal* 2, 459–478.

Thresh, J.M., Otim-Nape, G.W., Legg, J.P. and Fargette, D. (1997) African cassava mosaic virus disease: the magnitude of the problem. *African Journal of Root and Tuber Crops* 2, 13–19.

Thresh, J.M., Otim-Nape, G.W. and Fargette, D. (1998a) The components and deployment of resistance to cassava mosaic virus disease. *Integrated Pest Management Reviews* 3, 209–224.

Thresh, J.M., Otim-Nape, G.W. and Fargette, D. (1998b) The control of African cassava mosaic

virus disease: phytosanitation and/or resistance? In: Hadidi, A., Khetarpal, R.K. and Koganezawa, H. (eds) *Plant Virus Disease Control*. American Phytopathological Society, St Paul, Minnesota, pp. 670–677.

Thresh, J.M., Otim-Nape, G.W., Thankappan, M. and Muniyappa, V. (1998c) The mosaic diseases of cassava in Africa and India caused by whitefly-borne geminiviruses. *Review of Plant Pathology* 77, 935–945.

Velasquez, M.R. (1991) Estudio del complejo de moscas blancas (Hom:Aleyrodidae) y trips (Thysan, Thripidae) como posibles vectores de la enfermedad del cuero de sapo en yuca, (*Manihot esculenta* Crantz). Tesis Ing. Agr. Palmira. U. Nacional de Colombia.

Warburg, O. (1894) Die kulturpflanzen usambaras. *Mitt. Dtsch. Schutzgeb* 7, 131.

Watts Padwick, G. (1956) Losses caused by plant diseases in the Colonies. Phytopathological Papers No. 1. The Commonwealth Mycological Institute, Kew, Surrey.

Wydra, K. and Msikita, W. (1998) An overview of the present situation of cassava diseases in West Africa. In: Akoroda, M.O. and Ekanayake, I.J. (eds) *Proceedings: Sixth Triennial Symposium of the International Society for Tropical Root Crops: Africa Branch.* pp. 198–206.

Yaninek, J.S., James, B.D. and Bieler, P. (1994) Ecologically sustainable cassava plant protection (ESCaPP): a model for environmentally sound pest management in Africa. *African Crop Science Journal* 2, 553–562.

Zhou, X., Liu, Y., Calvert, L., Munoz, C., Otim-Nape, G.W., Robinson, D.J. and Harrison, B.D. (1997) Evidence that DNA-A of a geminivirus associated with severe cassava mosaic disease in Uganda has arisen by interspecific recombination. *Journal of General Virology* 78, 2101–2111.

Chapter 13
Bacterial, Fungal and Nematode Diseases

Rory J. Hillocks[1] and Kerstin Wydra[2]

[1]Natural Resources Institute, University of Greenwich, Chatham Maritime, Kent ME4 4TB, UK; [2]Institut für Pflanzenpathologie und Pflanzenschutz, Georg August Universität, Grisebachstr. 6, D-37077 Göttingen, Germany

Introduction

Cassava is affected by a wide range of diseases caused by viruses, bacteria, fungi and nematodes (Table 13.1). With the exception of some of the virus diseases (see Chapter 12) and bacterial blight, most of the other diseases have been regarded as of minor or local importance. This may not always be the case and this perception may be at least partly because cassava in many countries is a subsistence crop, where yields are limited more by other major factors, such as low soil fertility and moisture stress, than by plant diseases. At present, there is often too little information on losses caused by these diseases to draw any conclusions about their importance. There is considerable variation among cassava cultivars in susceptibility to the 'minor' diseases, and where susceptible cultivars are grown and conditions favour disease development, losses can be substantial. This is particularly true of the root rot diseases, where cassava is grown in soil with a high water table, or, on land newly cleared from forest or bush. Brown leaf spot is common almost everywhere that cassava is grown, but usually regarded as unimportant, yet the defoliation it causes may have a significant effect on yield, where cassava is grown intensively for commercial production.

The diseases considered of economic importance vary to some extent between countries and between continents. Cassava mosaic virus disease occurs wherever the crop is grown in Africa but is absent from South America. Bacterial blight occurs in Africa, South America and Asia. In both Africa and South America the next most important group of diseases are the root rots. In Nigeria, Cameroon and Benin the pathogens causing root diseases of economic importance are *Sclerotium rolfsii*, *Botryodiplodia theobromae*, *Fomes lignosus*, *Rosellinia necatrix*, *Rhizoctonia solani*, *Phytophthora* spp. and *Fusarium* spp. (Arene *et al.*, 1990; Afouda and Wydra, 1996). In Brazil, *Phytophthora* is probably the most important root pathogen but in some areas *Fusarium* spp. are also a problem (Lozano, 1991).

Bacterial Diseases

Cassava bacterial blight

Causal organism, distribution and importance

Cassava bacterial blight (CBB) is the most important bacterial disease of cassava and in Africa, it is second to cassava mosaic virus disease as a

Table 13.1.   Pathogens associated with cassava diseases worldwide.

Pathogen	Disease	Selected references
Fungi and bacteria		
Agrobacterium tumefaciens	Stem gall	CIAT (1978)
Armillariella mellea (*Armillaria mellea*)	Dry root rot	Makambila and Koumouno (1994); Mwenje *et al.* (1998)
Botryodiplodia theobromae	Stem rot, root rot	Lozano and Booth (1976); Afouda and Wydra (1996, 1997)
Cercospora caribaea	White leaf spot	Chevaugeon (1956); Lozano and Booth (1976)
Cercospora henningsii	Brown leaf spot	Lozano and Booth (1976); Ayesu-Offei and Antwi-Boasiako (1996)
Cercospora vicosae	Diffuse leaf spot	Lozano and Booth (1976)
Colletotrichum gloeosporioides f.sp. *manihotis*	Anthracnose	Lozano *et al.* (1981); Ikotun and Hahn (1994); Fokunang *et al.* (1997)
Cochliobolus lunatus	Stem rot	Msikita *et al.* (1997a)
Elsinoe brasiliensis	Superelongation disease	Lorenzo *et al.* (1981)
Erisyphe manihotis	Ash disease (powdery mildew)	Ferdinando *et al.* (1968); Lozano and Booth (1976)
Erwinia carotovora subsp. *carotovora*	Soft rot of stems and roots	Lozano and Bellotti (1979); Daniel *et al.* (1981)
Fomes lignosus	White thread	Lozano and Booth (1976)
Fusarium moniliforme	Wet root rot	Osai and Ikotun (1993); Msikita *et al.* (1996)
Fusarium oxysporum	Wet root rot	Afouda and Wydra (1996, 1997); Msikita *et al.* (1996)
Fusarium semitectum	Wet root rot	Afouda and Wydra (1996)
Fusarium solani	Wet root rot	Msikita *et al.* (1997b)
Leptoporus lignosus	Root rot	Makambila and Koumouno (1994)
Phaeolus manihotis	Root rot	Makambila and Koumouno (1994)
Phoma sp.	Ring leaf spot	Ferdinando *et al.* (1968); Lozano *et al.* (1981)
Phytophthora drechsleri	Root rot	Lima *et al.* (1995); Poltronieri *et al.* (1997)
Ralstonia solanacearum	Sudden wilt	Nishiyama *et al.* (1980); Machmud (1986)
Rhizoctonia solani	Root rot	Afouda *et al.* (1995)
Rosellinia necatrix	Dry root rot	Lozano and Booth (1976)
Sclerotium rolfsii (*Corticium rolfsii*)	Root/stem rot	Martin (1970); Osai and Ikotun (1993); Afouda and Wydra (1997)
Scytalidium lignicola	Black rot	Laranjeira *et al.* (1994)
Sphaerostilbe repens	Stem/root rot	Osai and Ikotun (1993); Makambila and Koumouno (1994); Afouda *et al.* (1995)
Uromyces sp.	Rust	Normanha (1970); Santos *et al.* (1988)
Xanthomonas campestris pv. *manihotis*	Bacterial blight	Boher and Verdier (1994); Laberry and Lozano (1992)
Xanthomonas campestris pv. *cassavae*	Angular leaf spot	Onyango and Mukunya (1982); Janse and Defranq (1988)
Nematodes		
Helicotylenchus erythrinae	Spiral nematode	McSorley *et al.* (1983)
Helicotylenchus dihystera	Spiral nematode	Caveness (1980)
Meloidogyne arenaria	Root-knot	Tanaka *et al.* (1979)
Meloidogyne hapla	Root-knot	Tanaka *et al.* (1979)
Meloidogyne incognita	Root-knot	Caveness (1982); Bridge *et al.* (1991); Crozzoli and Hidalgo (1992)
Meloidogyne javanica	Root-knot	Hogger (1971); Caveness (1980); Jatala and Bridge (1990)
Pratylenchus brachyurus	Lesion nematode	Charchar and Huang (1981); McSorley *et al.* (1983); Coyne (1994)
Rotylenchulus reniformis	Reniform nematode	Caveness (1980); Jatala and Bridge (1990)
Scutellonema bradys	Yam nematode	Caveness (1980)

cause of yield loss (Centro Internacional de Agricultura Tropical; CIAT, 1996). The causal agent of CBB is *Xanthomonas campestris* pv. *manihotis* (Berthet and Bondar, 1915) Dye 1978 (*Xcm*), proposed by Vauterin *et al.* (1995) to be reclassified as *Xanthomonas axonopodis* pv. *manihotis*. The bacterium can induce symptoms on cassava (*Manihot esculenta*) and related, wild *Manihot* species (*Manihot apii, Manihot glaziovii* and *Manihot palmata*), and after artificial inoculation, on (poinsettia) *Euphorbia pulcherrima* and *Pedilanthus tithymaloides* (Dedal *et al.*, 1980). Epiphytic survival for up to 1 month was demonstrated on several weed species occurring in cassava fields in Latin America and Africa (Marcano and Trujillo, 1984; Fanou *et al.*, 1998). *Xcm* is a Gram-negative rod and has a whitish to cream colony colour which is atypical for xanthomonads. Information on cassava bacterial blight was reviewed by Laberry and Lozano (1992) and by Boher and Verdier (1994).

Xcm originated in Latin America, where bacterial blight has been known since the early 1900s and the disease spread to the cassava growing regions of Africa and Asia during the 1970s (Bradbury, 1986; Boher and Verdier, 1994; Wydra and Msikita, 1998). CBB caused severe epidemics in Africa when it was first introduced, but the disease remains of minor importance in Asia. Severe infection, causing losses of up to 75% (Ohunyon and Ogio-Okirika, 1979), were reported in the 1970s from some areas in Ghana (Aklé and Gnouhoué, 1979) and Benin (Korang-Amoakoh and Oduro, 1979) and from Togo, Cameroon, Congo, Central African Republic, Nigeria, as well as from East Africa – Tanzania, Rwanda, Burundi and Kenya (Hahn and Williams, 1973; Terry and MacIntyre, 1976; Persley and Terry, 1977; Terry, 1977; Maraite and Perreaux, 1979). In Democratic Republic of Congo, CBB led in some areas to total loss of yield and availability of planting material, causing widespread famine. After the introduction of quarantine measures in some countries (e.g. Ghana) and the identification and planting of resistant varieties, the disease was contained. Further surveys were rarely conducted until the late 1990s (see below), so the current situation in respect of the distribution and severity of CBB remains unclear in several countries.

Surveys in 1994/1995 revealed a widespread occurrence of CBB in West Africa.

Epidemics occurred in the transition forest and in the moist and dry savannah zones, where the disease occurred respectively at 50, 60 and 42% of sites at a mean disease incidence (% plants infected) of 16, 34 and 16%. CBB occurred only sporadically in the rainforest zone (Wydra and Msikita, 1998). Average disease severity of infected plants was high, with scores between 2.4 (savannah zones) and 3.1 (transition forest) on a scale of 1 to 5. Observations from Benin revealed a disease incidence of 19% in the transition forest zone, 77% in the Southern Guinea (moist) savannah, 35% in the Northern Guinea (dry) savannah and 84% in the Sudan savannah, with a high percentage of plants scored in classes 4 and 5 (Wydra and Verdier, unpublished). This level of severity indicates severe systemic infection which causes losses in root and leaf yield, and in planting material for the subsequent season. Most strains of the pathogen collected from all five ecozones were highly virulent (Wydra *et al.*, 2001c).

Strains of *Xcm* were collected from cassava growing in different edaphoclimatic zones and the population structure was evaluated by analysis of DNA polymorphisms and virulence variation. The population from South America was grouped into five clusters (ribotypes), each largely composed of strains from the same edaphoclimatic zone. Strains varied in virulence but this was unrelated to origin and was not correlated with DNA polymorphisms (Restrepo and Verdier, 1997). By contrast, the African population was very uniform, all strains belonging to one of the five ribotypes identified in South America, reflecting its more recent introduction in Africa (Verdier *et al.*, 1993, 1997). Care must be taken therefore, when introducing planting material from South America to Africa (see below).

Importance

Yield losses to CBB in the major cassava production zone, the humid lowlands of Africa, are estimated at 3.2 million t with 60% of the area affected. For Africa as a whole, it is estimated that up to 7.5 million t are lost annually due to CBB (CIAT, 1996). Comparing losses in five ecozones of West Africa, 13–50% loss of root yield was recorded with the highest losses in the dry savannah zone (Wydra *et al.*, 2001a). Crop losses of 75% and of 90–100% were reported from Nigeria (Ohunyon and Ogio-Okirika, 1979) and Uganda (Otim-Nape, 1980). Strong genotype–environment interactions occurred and losses varied with variety, ecozone and year (Wydra et al., 2001a,b). Disease epidemics and yield losses vary from year to year but may cause the same percentage loss, averaged over a number of years as has been reported for diseases with a steady, more predictable effect on yield. The unpredictability of disease outbreaks and of associated losses makes the disease a major risk for cassava production by subsistence farmers. Due to vascular infection, 30% of cuttings taken from an infected plantation may be lost in the first season of planting into a previously clean field. The losses can reach 80% by the third year of production (Restrepo and Verdier, 1997).

Symptoms and management

CBB symptoms appear with the first rains after the dry season and reach their maximum during the peak of the rainy season. Infected plants show typical water-soaked, angular leaf spots (Fig. 13.1), leaf blight and leaf fall, and systemic symptoms, resulting in the formation of cankers on the stems. In severe cases, CBB causes shoot die-back (Fig. 13.2), showing the typical 'candle stick' symptom. The pathogen invades the plant systemically, entering the stem and the seeds, but often producing no symptoms initially. The

bacterium may then survive for a considerable period within the seed (Lozano et al., 1989). Symptoms of CBB were described in detail by Maraite (1993).

Integrated control strategies comprise of improved cultural methods, crop sanitation, resistant cultivars and quarantine measures to prevent introduction of highly aggressive strains to areas with low or no infection (Wydra et al., 1998; Fanou, 1999b; Wydra and Rudolph, 1999). A PCR-based diagnostic test has been developed which is sensitive down to the level of 10×2^2 colony forming units ml^{-1}, in cassava stem and leaf tissue (Verdier et al., 1998).

Crop rotation to completely break the life cycle of the pathogen, and burying or burning infected debris, can provide some control, especially under dry conditions, when the pathogen may survive in crop residues for up to 5 months (Fanou et al., 1998). Other control measures which may contribute to control of CBB, are weeding and avoiding bush fallow around cassava fields, control of grasshoppers – vectors of the disease (Fanou et al., 1998; Fanou, 1999a) – mixed cropping associating cassava with maize to suppress the disease (Fanou, 1999b), avoiding the peak time of the epidemic by shifting planting date towards the end of the rainy season (International Institute of Tropical Agriculture; IITA, 1998), and application of potash fertilizer to reduce disease severity (Odurukwe and Arene, 1980).

Fig. 13.1. Cassava bacterial blight (*Xanthomonas campestris* pv. *manihotis*) – leaf symptoms showing typical water-soaked, angular spots.

Most cassava cultivars planted in Africa are sensitive to *Xcm* (Boher and Verdier, 1994). Several lines derived from inter-specific crosses between *M. glaziovii* and *M. esculenta* display good resistance to bacterial blight (Hahn, 1978; Hahn *et al.*, 1980). Varieties with considerable resistance to CBB now exist in the germplasm collection of IITA (Hahn, 1979; Wydra *et al.*, 1998; Fanou, 1999b). Nevertheless, varieties selected for stable resistance in various environments tend to lose some of their resistance over time, probably due to natural selection for more aggressive strains of the pathogen (Verdier *et al.*, 1994; Wydra *et al.*, 1998). Periodic reselection for resistance is advisable, following inoculation with freshly collected, virulent strains of the pathogen.

Screening for resistance can be performed by observing symptom development in the field under strong disease pressure over several crop cycles. This takes account of inoculum remaining in the stem (Boher and Verdier, 1994). Less time-consuming screening methods have been developed. Inoculation of young shoots on cuttings enables selection to be done in 2 months (Pacambaba, 1987; Boher and Agbobli, 1992; Wydra *et al.*, 2001c). The resistant response to *Xcm* has been associated with physical and chemical barriers in the host which restrict bacterial growth within the phloem (Boher *et al.*, 1996). An *in vitro* method was reported by Flood *et al.* (1995) which could identify resistance to *Xcm* in embryonic cell suspensions of the host,

based on more rapid electrolyte leakage from cells of the susceptible lines.

Sanitary measures and quarantine precautions

During the cropping season, removal of infected leaves can significantly reduce disease severity (Fanou, 1999b). However, this is labour-intensive and is not a practical option for small-scale farmers. Planting material should be selected from fields free of CBB symptoms when inspected at the end of the preceding rainy season. New plantations should not be established near old cassava fields. Movement of man and animals in cassava fields, especially after rain, and use of contaminated tools contribute to the dissemination of the pathogen (Lozano, 1986).

For international germplasm exchange, the risk of introducing new strains of *Xcm* to Africa are greatest with seeds and this should generally be avoided. However, seeds can be disinfected by heat-treatment (Persley, 1979; Lozano *et al.*, 1989; Frison and Feliu, 1991; Fanou, 1999b) and tested for contamination using semi-selective media (Fessehaie *et al.*, 1999) or serological methods (Fessehaie, 1997; Fessehaie *et al.*, 1997). Vegetative propagating material should be exchanged by meristem cultures originating from heat-treated shoot tips of healthy plants. The FAO/IBPGR technical guidelines for the safe movement of cassava germplasm provide details on exchange of germplasm (Frison and Feliu, 1991).

Fig. 13.2. Cassava bacterial blight (*Xanthomonas campestris* pv. *manihotis*) – shoot dieback.

Other bacterial diseases

Angular leaf spot

Angular leaf spot (or bacterial necrosis) of cassava induced by *X. campestris* pv. *cassavae* Wiehe & Dowson (*Xcc*) is reported only from East Africa (Onyango and Mukunya, 1982) and Southern Africa, with one unconfirmed exception from Niger (Janse and Defrancq, 1988). Typical symptoms are angular leaf spots which develop less rapidly than leaf spots induced by *Xcm*, and leaf blight does not develop. Bright yellow exudates from infected leaf tissue during periods of high humidity are characteristic of this disease (Maraite and Perreaux, 1979). In severe cases bacterial necrosis can lead to defoliation of the plant, although the disease does not become systemic and the pathogen invades only the cortex of stems, not the vascular tissues (Maraite and Weyns, 1979). In contrast to *Xcm* with whitish colonies, *Xcc* colonies have a yellow colour which is typical for xanthomonads. Pathological, physiological and biochemical characteristics were described in detail by Maraite and Weyns (1979) and Van den Mooter *et al.* (1987). The disease occurs mainly on poor soils and after rainstorms injure the plants (Mostade and Butare, 1979). Fertilization is reported to retard disease development (Butare and Banyangabose, 1982).

Soft rot of stems and roots

Erwinia carotovora ssp. *carotovora* (*Ecc*) has been reported from Latin America, causing internal rotting of stems and branches, dark lesions and external cankers, necrosis of roots, wilting of young shoots and tip dieback, associated with injuries by insects, among them the cassava fruitfly (*Anastrepha* spp.; Lozano and Bellotti, 1978; Hernandez *et al.*, 1986). In the Democratic Republic of Congo and the Central African Republic, soft rot of harvested roots due to *Ecc* was observed (Daniel *et al.*, 1981). Infection causes root yield loss and loss of planting material (Cock, 1978). Planting uninfested, healthy cuttings of varieties resistant to the fruitfly and use of insecticides is recommended to control the disease (Lozano and Bellotti, 1978; Guevara *et al.*, 1992).

Sudden wilt, leaf drop and linear discoloration of stems and roots, and soft rot of roots

were attributed to *Ralstonia solanacearum* [*Pseudomonas solanacearum*] in India and Indonesia (Nishiyama *et al.*, 1980; Machmud, 1986), whereas Lozano (1979) does not consider cassava as a host of *R. solanacearum*. *Agrobacterium tumefaciens* has been reported to cause stem gall on cassava (CIAT, 1978).

Fungal Diseases of the Leaf

Brown leaf spot

Causal organism, distribution and importance

Several *Cercospora* spp. and related fungi induce leaf spots on cassava but the most important of these is brown leaf spot (BLS) caused by *Cercosporidium henningsii* Allesch. [syn. *Cercospora henningsii, Cercospora manihotis*] (Powell, 1972). The fungus has a relatively wide host range, affecting in addition to *M. esculenta, M. glaziovi* (ceara rubber), *Manihot piauhynsis* and by inoculation, sweet potato (*Ipomea batatas*; Golato, 1963; Powell, 1968, 1972; Golato and Meossi, 1971). Lozano and Booth (1976) provide a detailed description of pathogen morphology. Another disease known in Brazil as brown large spot (diffuse spot) is caused by *Cercospora vicosae* (see below).

BLS is of worldwide distribution and occurs in most cassava fields in the lower canopy of crops more than 5 months old. It is favoured by high temperature and humidity. The optimum conditions for spore production were reported to be free water on the leaf surface and a temperature of 25–32°C (Ayesu-Offei and Antwi-Boasiako, 1996). Extensive surveys conducted in several West African countries showed that BLS was widely distributed with site incidence varying between 78% (wet savannah) and 98% (transition forest) and incidence within sites of 41% and 85%, respectively. However, average disease severity score was low (Wydra and Msikita, 1998).

The importance of BLS may be underestimated because it is often confined to the lower-canopy leaves, but it can cause leaf chlorosis and extensive defoliation. The effect of this defoliation, particularly when infection is followed by a period of drought-stress is difficult to quantify. It has been reported that in Africa, in areas of high rainfall, the disease can cause yield losses on individual plants of up to 20% (Théberge, 1985). The effect on yield of a single leaf spot pathogen is often difficult to assess as they commonly occur as a leaf spot complex. In Brazil, the combination of *C. henningsii* and *C. vicosae* can cause up to 30% yield loss (Takatsu *et al.*, 1990).

Symptoms and management

The leaf symptoms on cassava are visible on both sides of the lamina but are more pronounced on the upper surface, where the spots are a uniform brown with a distinct darker border (Fig. 13.3). During humid conditions, the presence of conidia gives the underside of the spots a greyish appearance. Spots are generally 3–12 cm in diameter and roughly circular but becoming more irregular as they expand. On some cultivars there may be a yellow halo around the spots and as the disease progresses leaves turn yellow, dry and eventually fall (Lozano and Booth, 1976). The older leaves are more susceptible to infection than younger leaves, but in highly susceptible cultivars, leaves in the upper canopy can also be infected (Chevaugeon, 1956).

BLS can be managed by wider spacing to reduce humidity and by the use of copper fungicides (Golato and Meossi, 1966, 1971). Disease management measures are rarely required in practice but where the crop is grown in high rainfall areas, the use of less susceptible cultivars is advised. Differences in susceptibility between cultivars has been reported from Africa and in the extensive germplasm collection at CIAT, Colombia (Lozano and Booth, 1976).

White leaf spot

Causal organism, distribution and importance

Cercospora caribaea Chupp & Ciferri has a worldwide distribution but white leaf spot (WLS) which it causes, is less common than BLS and is favoured by cooler conditions. A survey which included several West African countries showed WLS to be widely distributed in all the ecozones, although severity was generally low. The percentage of plants with symptoms ranged from only 1% in the dry savannah to 62% in the rainforest zone (Wydra and Msikita, 1998). Sporulation occurs readily on the surface of the lesions during humid weather and the conidia are distributed by wind and rain (Lozano and Booth, 1976). Little is known about the effect of the disease on the cassava plant but it generally appears to be negligible. There does, however, seem to be some differences between cultivars in susceptibility and considerable defoliation can occur in the most susceptible cultivars (Chevaugeon, 1956).

Fig. 13.3. Brown leaf spot (*Cercosporidium henningsii*).

Symptoms and management

First symptoms of the disease are circular chlorotic areas on the upper lamina which later develop small (1.5–2 mm) circular white lesions in the centre (Fig. 13.4) which are also visible on the underside of the leaf. The spots enlarge to 3–5 mm and may have a purple border, retaining the chlorotic halo (Lozano *et al.*, 1981; Théberge, 1985). Sporulation occurs mainly on the underside and conidia are elongated and tapering, typical of *Cercospora* spp.

Diffuse leaf spot

Causal organism, distribution and importance

Cercospora vicosae Muller and Chupp, causes diffuse leaf spot (DLS). Again of worldwide distribution, it is more prevalent in the warmer and wetter areas where BLS is also common, particularly in Brazil and Colombia (Lozano and Booth, 1976) and in West Africa (Théberge, 1985). Wydra and Msikita (1998) refer to *Cercospora* leaf spot and their surveys in West Africa showed it to be one of the most widely occurring leaf spot diseases on cassava. Site incidences varied from 63% in the dry savannah zone to 100% in the mountain zone. Plant incidence was 97% in the mountain zone compared to 79% in the lowland rainforest zone

which is surprising in view of other reports that the disease is more prevalent in warm, wet areas. The disease can cause severe defoliation in the more susceptible cultivars but this tends to occur in plants more than 6–9 months old and yield losses are believed to be slight.

Symptoms

The symptoms are quite distinct from the other two leaf spots described above, being large and diffuse without definite borders. Each spot may cover more than one-fifth of the leaf lobe (Fig. 13.5). They are uniformly brown on the upper side but on the underside, the centre of the spot takes on a greyish appearance, when sporulation is induced by humid weather (Lozano and Booth, 1976).

Ring leaf spot

Causal organism, distribution and importance

The exact aetiology of ring leaf spot (RLS) is unknown. A number of fungi originally identified as *Phyllosticta* spp. have been associated with the leaf lesions. The more generally accepted taxonomy for these fungi is now *Phoma* spp. (Lozano and Booth, 1976). RLS is more common in Latin America (Viegas, 1943; CIAT, 1972) than elsewhere, but has been reported

Fig. 13.4. White leaf spot (*Cercospora caribaea*).

also from India (Ferdinando *et al.*, 1968) and Africa (Vincens, 1915). There do not seem to be any more recent reports from Africa and the disease is not mentioned in the IITA field guide to pests and diseases of root and tuber crops in Africa (Théberge, 1985). The disease is favoured by temperatures below 22°C and is more common at high altitudes or in lowland areas during prolonged periods of cool, wet weather. In Latin America the disease is known to cause defoliation of the more susceptible cultivars and can, under optimum conditions for disease development, attack the young shoot to cause dieback and even in severe cases, death of the plant. Considerable losses are attributed to the disease where cassava is grown in cooler areas (Lozano *et al.*, 1981).

Symptoms and management

RLS is characterized by large brown spots visible on both sides of the lamina that are roughly circular, 1–3 cm in diameter and commonly found at the edges of the leaf lobes or associated with the midribs. On the upper surface, lesions bear a pattern of concentric rings due to the production of pycnidia. These may be washed off older lesions by rain, so that the spots resemble those of diffuse leaf spot. No control measures are known but there may be differences in susceptibility between cultivars as the disease usually affects the younger leaves but has also been observed in the lower canopy in some cultivars.

Ash disease

Causal organism, distribution and spread

The causal organism of ash disease is a powdery mildew fungus, *Oidium manihotis* Henn. and the sexual stage has been described as *Erisyphe manihotis* (Ferdinando *et al.*, 1968). Although this disease is widespread and of common occurrence (Lozano and Booth, 1976), the lesions are superficial and it is not considered to be damaging. The disease is usually observed during the dry season in warmer cassava growing areas.

Symptoms

The first symptoms of the disease are seen as small white patches of mycelium on the upper leaf surface. Beneath the mycelium the cells are killed to form diffuse yellow patches within which pale brown angular water-soaked spots develop and become necrotic. Mature, fully expanded leaves appear to be the most susceptible.

Fungal Diseases of the Stem

Anthracnose

Causal organism, distribution and importance

The causal organism of anthracnose on cassava is described as *Glomerella manihotis* Chev.

Fig. 13.5.   Diffuse leaf spot (*Cercospora vicosae*).

However, the *Colletotrichum* state [*Colleto-trichum gloeosporioides* f. sp. *manihotis* Henn. (Penz.) Sacc] is more commonly referred to than the perfect state. The disease occurs worldwide but is more common in the wetter areas of Latin America (CIAT, 1972; Lozano *et al.*, 1981) and West Africa (Chevaugeon, 1950; Affran, 1968; IITA, 1972; Akonumbo and Ngeve, 1998). Disease surveys in West Africa showed disease incidence and severity to be greatest in cassava growing in the lowland rain forest, where 64% of all plants examined showed symptoms and the wet savannah zones, decreasing towards the drier ecozones (Wydra and Msikita, 1998). Although the disease is less conspicuous in the dry season, anthracnose causes dieback and affected plants have fewer leaves than unaffected plants. During surveys, extensive areas were observed where cassava crops had been defoliated by anthracnose (Wydra and Msikita, 1998). However, one of the main effects of the disease in Africa is in reducing the quality of stem cuttings as planting material (Théberge, 1985).

Symptoms and management

In Latin America anthracnose symptoms are usually found on the leaves but in Africa, the disease mainly affects the stem. Leaf spots are about 10 mm in diameter and, as with brown spot, they are produced at the base of the leaves. In Latin America, stems may also be attacked, causing a wilt of young plants and cankers on older ones. New leaves produced at the beginning of the rains are reported to be most susceptible (Irvine, 1969). In Africa, the stem canker is the most important symptom but the fungus also causes twig and fruit cankers as well as leaf spots and shoot-tip dieback. Tip dieback was reported to be a common symptom in Ghana (Moses *et al.*, 1996). The disease begins on the stems as slightly depressed oval lesions, 1–1.5 cm in diameter and a patch of greenish tissue in the centre of the lesion soon turns dark brown (Fig. 13.6). On older stems, raised fibrous lesions develop into deep cankers, causing brittleness of the stems.

Fokunang *et al.* (1997) have reported that the pathogen is transmitted in cassava seed in Nigeria with up to 40% incidence in some genotypes. Spore dispersal is mainly by wind

and water but in Africa, a sap-sucking insect (*Pseudotheraptus devastans* Dist.) is believed to facilitate infection with the fungus by stem puncturing (Boher *et al.*, 1983). In the Republic of Congo, anthracnose on cassava stems is associated with the feeding punctures (Makambila, 1994). The fungus then invades the necrotic tissue and infection develops under conditions of high relative humidity. Some cassava cultivars are more sensitive to attack by the insect than others.

There appears to be some variability in the reaction of some cultivars to anthracnose. Screening experiments in Nigeria identified several cultivars in which the onset of first symptoms was delayed until after tuberization had started and as a result, yield was little affected (Ikotun and Hahn, 1994). The parameters considered to be most useful in differentiating the less susceptible cultivars were number of cankers per plant, size of cankers

Fig. 13.6. Anthracnose (*Colletotrichum gloeo-sporioides* f. sp. *manihotis*) – raised lesions and cankers on cassava stem.

and distance from soil level to the first stem canker.

Glomerella stem rot

This is mainly a problem in stored cassava cuttings and is caused by *Colletotrichum* sp. which may be the same as that causing anthracnose and has been identified as being within the broad grouping of *Glomerella cingulata* (Stonem.) Spauld. Schrenk.

This problem seems to be more common in Latin America than elsewhere (Lozano and Booth, 1976) and cuttings may be predisposed to this infection if they are taken during wet periods. The rot appears at the cut end of the stem and spreads throughout the cutting during storage. The vascular strands become blackened and dark-coloured blisters which contain the perithecia appear on the bark.

Botryodiplodia stem rot

A similar stem rot of stored cuttings to that described above is caused by *Botryodiplodia theobromae* Pat. Symptoms are similar except that the blisters on the bark contain pycnidia instead of perithecia. This disease is frequently found in Nigeria, Benin and Cameroon (Afouda and Wydra, 1996, 1997).

Rust

Rust is caused on cassava by *Uromyces* sp. and has been reported from Brazil and Colombia (Normanha, 1970). Although six rust species have been reported on cassava in different parts of the world (Lozano *et al.*, 1981), rust does not seem to be common in Africa or elsewhere. The orange to reddish brown pustules are most often seen on the green stems but also on the petioles and veins on the underside of the leaf. Occasionally the pustules are surrounded by a yellow halo. Affected leaves may be distorted but the disease is not generally considered of economic importance. Of 485 genotypes screened in Brazil, 139 were highly resistant and 122 showed some resistance (Santos *et al.*, 1988).

Superelongation disease

Causal organism, distribution and importance

Superelongation disease was first reported during the 1970s and appears to be confined to parts of Colombia (Lozano, 1972; Lozano and Booth, 1976). The causal organism was thought to be a species of *Taphrina* or *Sphaceloma*, but is now reported to be *Elsinoe brasiliensis*. The disease is reported to cause considerable losses in plantations where susceptible cultivars are grown (Lozano *et al.*, 1981). Symptoms are more common in the wet season than at other times, when the infection spreads rapidly as spores are distributed by wind and rain. High humidity is required for spore germination.

Symptoms and management

The disease is observed in the field as abnormal elongation of the internodes on young stems which appear thin and weak. Infected plants are taller than surrounding healthy ones. Young shoots, leaves and petioles become distorted and bear eye-shaped cankers appear along the midribs and veins. White, irregular spots may occur on the leaf lamina. There may also be dieback and defoliation. The disease is believed to be spread through planting of infected cuttings so it is recommended in endemic areas to select planting material from disease-free plants and to treat cuttings with captafol solution (3000 p.p.m.; Lozano *et al.*, 1981). Field observations suggest that some cultivars may be resistant (Lozano and Booth, 1976). Cultivars with some resistance to superelongation disease have been identified by CIAT in Colombia. Resistance was associated with thickened cuticles and Bonilla *et al.* (1992) suggested that this could be used to identify resistant genotypes.

Other stem diseases

Cochliobolus lunatus was reported to be the causal organism of a stem disease which was severe on some cultivars in south Ghana and southeast Nigeria. Disease incidence per field ranged from 0 to 80% (Msikita *et al.*, 1997a). A fungal complex was associated with a rot of mini-stems grown in polythene bags in Nigeria, consisting of *B. theobromae, Corticium rolfsii* and *Sphaerostilbe repens* (Osai and Ikotun, 1993).

Fungal Diseases of the Root

Phytophthora root rot

Causal organism, distribution and importance

A number of *Phytophthora* species have been associated with soft rots of cassava roots and these often occur with a number of other soil-borne fungi, particularly *Pythium* spp. and *Fusarium* spp. *Phytophthora drechsleri* Tucker has been reported from Latin America and *Phytophthora erythropseptica* Pethybr. from Africa. Soft rot in general is a worldwide problem and cool wet conditions and root damage predispose the tuberous roots to infection. In areas close to drainage ditches or in poorly drained soils, losses can be up to 80% (Théberge, 1985). In Brazil the pathogens associated with soft rot were identified as *Phytophthora drechsleri, Phytophthora richardii, Phytophthora parasitica* and *Pythium scleroteichum* (Poltronieri *et al.*, 1997).

Symptoms and management

Young roots initially show water-soaked patches which later turn brown and the feeder roots die. As the rot progresses, the starch-bearing tissue disintegrates and the affected roots have a pungent odour. Root dysfunction causes dieback of the terminal shoots and leads to sudden wilting in advanced stages of root decay.

Cultivars differing in resistance to *Phytophthora* have been identified in Brazil (Lima *et al.*, 1995). Differences in susceptibility were identified using stem inoculation but not using root inoculation. A greenhouse inoculation method was developed based on inoculation of the stem of 40 day-old plants with a plug of mycelial culture of the fungus and their evaluation 7 days later. Only three of 96 cultivars screened were considered to be resistant but a further 23 were moderately resistant.

White thread disease

Causal organism, distribution and importance

White thread is caused by *Fomes lignosus* (Klot.) Bres (*Rigidoporus lignosus* Johansen and Ryv.). Lozano and Booth (1976) state that this is the most widespread and serious root disease of cassava in Africa but that it is less common in Latin America. However, the disease is not described in the IITA field guide to pests and diseases of root crops in Africa (Théberge, 1985). The disease tends to be a problem only in cassava planted immediately after forest clearance or near virgin forest. The fungus survives on dead tree roots and can grow through the soil to attack cassava plants. *F. lignosus* has a wide host range among woody species growing in the humid tropics.

Symptoms

The disease can be recognized by the white mycelial mat under the bark of the roots and by the characteristic white cotton-like mycelial threads on the exterior of the root up to the stem base. Internal tissues exhibit a dry rot which is associated with a smell of rotting wood. Occasionally young plants are affected causing sudden wilting (Lozano and Booth, 1976).

Sclerotium root rot

This root rot is caused by *Sclerotium rolfsii* Sacc., which is common in tropical soils, causing root, crown and stem rot diseases of diverse crops. In the humid tropics, particularly West Africa, *Sclerotium* rot is among the most common diseases of cassava roots (Afouda *et al.*, 1995), generally affecting older plants. In Latin America, it is reported to affect young cuttings and as a surface coating on mature tuberous roots (Martin, 1970; CIAT, 1972). The disease can be identified by the white mycelium on affected roots which sometimes penetrates the epidermis to cause necrosis and occasionally root rot, but this usually occurs in combination with other pathogens. Spherical brown sclerotia may be visible associated with the mycelium on roots close to the soil surface during moist weather.

Dry root rot

Causal organism, distribution and importance

Dry root rot in cassava may be caused by *Rosellinia necatrix* (Hartig) Berl. or *Armillariella mellea* (Vahl.) Pat. (*Armillaria mellea* (Vahl) Fr.), or by both fungi together. Both of these pathogens have been recorded on cassava in different parts of the world, mainly where the crop is growing in moist soils high in organic matter. Both pathogens have a wide host range among woody perennials. Like white thread disease, the dry root rot fungi normally attack cassava planted after forest clearance and can completely destroy the roots of affected plants. However, incidence is normally low and the disease is not regarded as a serious problem. It can be managed by planting annual crops after forest clearance before planting cassava. In the Republic of Congo, *A. mellea* and *Armillaria heimii* were identified, as members of the dry root rot complex in the Republic of Congo. In farms cleared from the bush within the previous 3 years, about half the farms were affected by root rot caused by *A. heimii* but the incidence increased with increasing age of the farms and was 69% in the 2–3-year-old group (Makambila and Koumouno, 1986). Both fungi produce rhizomorphs which appear as thickened mycelial strands on the outside of the roots. They are at first white later turning black. Infected roots are discoloured and exude a watery liquid when squeezed. Rhizomorphs penetrate into the infected tissues. Above-ground plants wilt but do not shed their leaves, eventually desiccating to assume a scorched appearance (Théberge, 1985). Although *Armillaria* has a wide host range, there is evidence for some degree of host specialization. Mwenge *et al.* (1998) showed that in Zimbabwe, where *Armillaria* isolates have been separated into three distinct groups, only isolate from groups I and III were pathogenic to cassava.

Other root rot fungi

Fusarium moniliforme Sheldon, *Fusarium oxysporum* Schlecht and *Fusarium semitectum* were reported from Benin, Nigeria and Cameroon as cause of root, stem and storage rot (Osai and Ikotun 1993; Afouda and Wydra, 1996; Msikita *et al.*, 1996). *F. moniliforme* and *F. oxysporum* were isolated from 44–55% of rotted roots and crowns and from discoloured chips and were re-isolated from cassava plants showing symptoms of wilting and necrosis, 6–10 days after inoculation (Msikita *et al.*, 1996). Lozano (1992) lists *Fusarium solani* (Mart.) Sacc. and *F. oxysporum* Schlecht as contributing to root rots in Colombia. *F. solani* is associated with wet root rots also in Africa where an *in vitro* system has been developed to screen for resistance (Msikita *et al.*, 1997b). Root rot incidence in the humid forest area of the Republic of Congo was 30% in 20-month-old cassava and the pathogens identified were *Armillaria* spp., *S. repens* B. and Br., *F. lignosus* and *Phaeolous manihotis* Heim (Makambila, 1994; Makambila and Koumouno, 1994). In Brazil, a disease known as black rot, affecting roots and shoots has been attributed to *Scytalidium lignicola* Pesante (Laranjeira *et al.*, 1994).

Nematodes

Numerous nematode species have been associated with cassava roots and several of these multiply on cassava roots to reach high populations. Extensive lists of these nematodes have been produced by Hogger (1971), Caveness (1980) and McSorley *et al.* (1983) but there is little evidence that they have a significant effect on yield. Jatala and Bridge (1990) consider that the nematodes with the greatest effect on cassava production are two species of root-knot nematode (*Meloidogyne* spp.), the lesion nematode (*Pratylenchus brachyurus* Filipjev and Strekhoven), the spiral nematodes (*Helicotylenchus* spp.) and the reniform nematode (*Rotylenchulus reniformis* Lindford & Oleveira). In addition, *Scutellonema* spp., particularly *Scutellonema bradys*, are commonly associated with cassava roots in large numbers, although this can also be said for a wide range of crops growing in tropical sandy loam soils. Nevertheless, *Scutellonema* spp. may decrease yields and contribute to postharvest deterioration as they do in other root and tuber crops.

Root-knot nematodes

The most widely reported parasitic nematodes on cassava are the root-knot nematodes (RKN), which occur in Latin America (da Ponte et al., 1980; Crozzoli and Hidalgo, 1992), the USA (McSorley et al., 1983), West Africa (Caveness 1979, 1982; Sikora et al., 1988), East Africa (Saka, 1982; Bridge et al., 1991; Van den Oever and Mangane, 1992) and the Pacific (Bridge, 1988).

The most important species are *Meloidogyne incognita* (Kofoid & White) Chitwood and *Meloidogyne javanica* (Treub.) Chitwood. Galls are produced on cassava roots but there appears to be a wide range of susceptibility to galling between cultivars, ranging from immunity to high susceptibility (da Ponte et al., 1980; Nwauzor and Nwanko, 1989). However, there is no information on the relationship between susceptibility to galling and yield loss. The most widespread severe galling due to RKN was reported from Uganda (Bridge et al., 1991), where 94% of 88 fields examined were affected and 17% of the damaged roots were in the severe category (Coyne and Namaganda, 1994).

Direct effects of *Meloidogyne* spp. on root yield have been difficult to demonstrate. In Uganda yields were found to be consistently lower in fields with greater root-knot damage and for two of the more susceptible cultivars Bukalasa 11 and TMS 30337, yields were decreased by 24–38% (Coyne, 1994). In Nigeria, RKN is regarded as an important pest, causing galls exceeding 1 cm diameter on susceptible cultivars. In severe attacks, the feeder root system is greatly reduced, causing stunting and decrease in stem diameter to which root yield losses of 17–50% have been attributed (Théberge, 1985). RKN has other indirect effects on cassava production. Severe galling causes reductions in plant height and weight which decreases the quality and quantity of planting material (Gapsin, 1980, 1981; Caveness, 1981, 1982). The greatest effect of the nematode may be on the storability of the harvested roots. Caveness (1982) reported postharvest losses of up to 87% due to rapid root deterioration under severe nematode attack.

The evidence presently available indicates that damage caused by root-knot nematodes does not warrant control measures. Studies in Uganda, however, show that losses can be substantial if nematode populations build-up and highly susceptible cultivars are grown. Although yield increases have been obtained in Latin America when nematodes have been controlled by soil fumigation (da Ponte and Franco, 1981), such measures are inappropriate and would rarely provide an economic return and root-knot is best managed by the use of less susceptible cultivars and crop rotation to avoid excessive nematode population increases.

Lesion nematodes

The lesion nematode *Pratylenchus brachyurus* is the second most important parasitic nematode species affecting cassava and occurs on the crop in many parts of the world including the USA (McSorley et al., 1983), East Africa (Coyne, 1994) and Latin America (Charchar and Huang, 1981). Greenhouse tests conducted in Brazil, showed an eightfold increase in populations of *P. brachyurus* after 3 months and a gradual decline in production over several years was attributed to the nematode (Charchar and Huang, 1981). McSorley et al. (1983) reported yield increases after nematicidal treatment of fields where cassava was affected by a nematode complex consisting of *P. brachyurus* and *Helicotylenchus erythrinae* (Zimmerman) Golden. If necessary, control can be achieved with resistant cultivars as it has been shown that there is considerable variability between cassava cultivars in susceptibility to *P. brachyurus* (Corbett, 1976).

References

Affran, D.K. (1968) Cassava and its economic importance. *Ghana Farmer* 12, 172–178.

Afouda, L. and Wydra, K. (1996) Virulence analysis of root and stem rot pathogens of cassava (*Manihot esculenta* Crantz) in West Africa and development of methods for their biological control with antagonists. *Mitteilungen aus der Biologischen Bundesanstalt* 50, Deutsche Pflanzenschutztagung, Münster, Germany, September 1996, p. 634.

Afouda, L. and Wydra, K. (1997) Pathological characterization of root and stem rot pathogens of cassava and evaluation of antagonists for their biological control. DPG Working Group 'Plant

Protection in the Tropics and Subtropics', Berlin, July 1997. *Phytomedizin* 27, 43–44.

Afouda, L., Wydra, K. and Rudolph, K. (1995) Root and stem rot pathogens from cassava and their antagonists, collected in Cameroon, Nigeria and Benin. XIII International Plant Protection Congress, The Hague, The Netherlands, 2–7 July 1995. *European Journal of Plant Pathology* Abstract 573.

Aklé, J. and Gnouhoué, H. (1979) CBB development in Benin. In: Cassava bacterial blight in Africa: past, present and future. *Report of an Interdisciplinary Workshop* IITA, Ibadan, Nigeria, 26–30 June 1978. Centre for Overseas Pest Research, London, p. 43.

Akonumbo, D.N. and Ngeve, J.M. (1998) Characterisation of isolates of *Colletotrichum gloeosporioides* Penz. f. sp. *manihotis* from susceptible cassava cultivars in Cameroon. In: Akoroda, M.O. and Ekanayake, I.J. (eds) *Root Crops for Poverty Alleviation. Proceedings of the Sixth Triennial Symposium of the International Society for Tropical Root Crops (ISTRC)*, Lilongwe, Malawi, October 1995. ISTRC, IITA and Government of Malawi, pp. 163–166.

Arene, O.B., Hahn, S.K. and Caveness, F.E. (1990) Advances in integrated control systems of economic diseases of cassava in Nigeria. In: Hahn, S.K. and Caveness, F.E. (eds) *Proceedings of the Workshop on the Global Status and of Prospects for IPM of Root and Tuber Crops*, Ibadan, Nigeria, 25–30 October 1987. IITA, Ibadan, Nigeria, pp. 169–175.

Ayesu-Offei, E.N. and Antwi-Boasiako, C. (1996) Production of microconidia by *Cercospora henningsii* Allesch., cause of brown leaf spot of cassava. *Annals of Botany* 78, 653–657.

Berthet, A. and Bondar, G. (1915) Molestia bacteriana da mandioca. *Boletim de Agricultura, Sao Paulo* 16, 513–524.

Boher, B. and Agbobli, C.A. (1992). La bactériose vasiculaire du manioc au Togo. Characterisation, répartition géographique et sensibilité variétale. *Agronomie Tropicale* 2, 131–136.

Boher, B., Daniel, J.-F., Fabres, G. and Bani, G. (1983) Effect of *Pseudotheraptus devastans* (Dist) (Het. Coreidae) and *Colletotrichum gloeosporioides* Penz. on the development of cankers and loss of leaves in cassava. *Agronomie* 3, 989–994.

Boher, B., Nicole, M., Calatayud, P. and Geiger, J. (1996) Cytochemistry of defense responses in cassava infected by *Xanthomonas campestris* pv. *manihotis*. *Canadian Journal of Microbiology* 42, 1131–1143.

Boher, K. and Verdier, V. (1994) Cassava bacterial blight in Africa: the state of knowledge and implications for designing control strategies. *African Crop Science Journal* 4, 505–509.

Bonilla, X., Laberry, R. and Lozano, J.C. (1992) Morphological differences between clones resistant and susceptible to *Elsinoe brasiliensis*, a causal agent of superelongation. *Fitopatologi Colombiana* 16, 158–164.

Bradbury, J.F. (1986) *Guide to Plant Pathogenic Bacteria*. CAB International, Wallingford, UK.

Bridge, J. (1988) Plant parasitic nematode problems in the Pacific Islands. *Journal of Nematology* 20, 173–183.

Bridge, J., Otim-Nape, G.W. and Namaganda, J. (1991) The root-knot nematode, *Meloidogyne incognita* causing damage to cassava in Uganda. *Afro-Asian Journal of Nematology* 1, 116–117.

Butare, I. and Banyangabose, F. (1982) Effects of soil fertility on cassava bacterial blight in Rwanda. In: *Root Crops in Eastern Africa. Proceedings of a Workshop*, Kigali, Rwanda, 23–27 November 1980. International Development Research Centre, Ottawa, Canada, pp. 53–55.

Caveness, F.E. (1979) Cowpea, lima bean, cassava, yams and *Meloidogyne* spp. in Nigeria. In: Lamberti, F. and Taylor, C.E. (eds) *Root-knot Nematodes (Meloidogyne spp.); Systematics, Biology and Control*. Academic Press, London, pp. 295–300.

Caveness, F.E. (1980) Plant parasitic nematodes on cassava. In: Ezumah, H.C. (ed.) *Proceedings of the Workshop on Cassava Production and Extension in Central Africa*, Mbanza-Ngungu, Zaire. IITA, Ibadan, 4, 83–106.

Caveness, F.E. (1981) Root-knot nematodes on cassava. *Annual Report IITA*. IITA, Ibadan, Nigeria, pp. 64–65.

Caveness, F.E. (1982) Root-knot nematodes as parasites of cassava. *IITA Research Briefs* 3, 2–3.

Charchar, J.M. and Huang, C.S. (1981) Circulo de hospedeiras de *Pratylenchus brachyurus*. 3. Plantas diversas. *Fitopatologia Brasileira* 6, 469–473.

Chevaugeon, J. (1950) Maladies cryptogramique du manioc en Côte d'Ivoire. Observations preliminaires sur la necroses des somites. *Revue de Pathologie Végétale et d'Entomologie Agricole de France* 29, 3–9.

Chevaugeon, J. (1956) Le maladie cryptogamique du manioc en Afrique Occidentale. *Encycopedie Mycologique* I28, 1–205.

CIAT (1972) *Annual Report for 1972*. CIAT, Cali, Colombia.

CIAT (1978) Cassava Production Systems Program. In: *Annual Report 1977*. CIAT, Cali, Colombia.

CIAT (1996) Global cassava trends. Reassessing the crop's future. In: Henry, G. and Gottret, V. (eds) Working document no. 157. CIAT, Cali, Colombia.

Cock, J.H. (1978) A physiological basis of yield loss in cassava due to pests. In: Brekelbaum, T., Bellotti, A. and Lozano, J.C. (eds) *Proceedings, Cassava*

Protection Workshop, CIAT, Cali, Colombia, 7–12 November 1977, pp. 9–16.

Corbett, D.C.M. (1976) *Pratylenchus brachyurus. C.I.H. Descriptions of Plant Parasitic Nematodes Set 6 No 89.* Commonwealth Agricultural Bureaux, St Albans.

Coyne, D.L. (1994) Nematode pests of cassava. *African Crop Science Journal* 2, 355–359.

Coyne, D.L. and Namaganda, J.M. (1994) Root-knot nematodes, *Meloidogyne* spp., incidence on cassava in two areas of Uganda. *Roots Newsletter* 1, 2–3.

Crozzoli, P.R. and Hidalgo, S.O. (1992) Response of ten cassava cultivars to the nematode *Meloidogyne incognita. Fitpatalogia Venezolana* 5, 20–22 (Abstract).

Daniel, J.F., Boher, B. and Kohler, F. (1981) Les maladies bactériennes du manioc (*Manihot esculenta* Crantz) en République Populaire du Congo et en République Centrafricaine. *Agronomie* 1, 751–757.

Dedal, O.I., Palomar, M.K. and Napiere, C.M. (1980) Host range of *Xanthomonas manihotis* Starr. *Annals of Tropical Research* 2, 149–155.

Dye, D.W. (1978) Genus *Xanthomonas* Dowson (1939) In: Young, J.M., Dye, D.W., Bradbury, J.F., Panagopoulos, C.G. and Robbs, C.F. (eds) *A Proposed Nomenclature and Classification for Plant Pathogenic Bacteria. New Zealand Journal of Agricultural Research* 21, 153–177.

Fanou, A. (1999a) Epidemiological studies on the role of weeds, plant debris and vector transmission in survival and spread of *Xanthomonas campestris* pv. *manihotis*, causal agent of cassava bacterial blight. MSc thesis, University of Göttingen, Germany.

Fanou, A. (1999b) Epidemiological and ecological investigations on cassava bacterial blight and development of integrated methods for its control in Africa. PhD thesis, University of Göttingen, Germany.

Fanou, A., Wydra, K., Zandjanakou, M. and Rudolph, K. (1998) Epidemiological studies on the role of weeds, plant debris and vector transmission in survival and spread of *Xanthomonas campestris* pv. *manihotis*, causal agent of cassava bacterial blight. International Congress of Plant Pathology, International Society for Plant Pathology/British Society for Plant Pathology, Edinburgh 1998 (Abstract 2.5.22).

Fanou, A., Wydra, K., Zandjanakou, M., LeGall, P. and Rudolph, K. (2001) Studies on the survival mode of *Xanthomonas campestris* pv. *manihotis* and the dissemination of cassava bacterial blight through weeds, plant debris and an insect vector. In: Akoroda, M.O. and Ngeve, J.M. (eds) *Proceedings of the 7th Triennial Symposium of the International Society of Tropical Root Crops Africa Branch*

(ISTRC-AB), Cotonou, Benin. ISTRC-AB and Government of Benin, pp. 569–580.

Ferdinando, G., Tokeshi, H., Carvhalo, P., Balmer, E., Kimati, H., Cardosa, C. and Salgado, C.L. (1968) *Manual de Fitopatologia Doencas das Plantas e seu Controle.* Biblioteca Agonomica, Ceres, São Paolo, Brazil.

Fessehaie, A. (1997) Biochemical/physiological characterization and detection methods of *Xanthomonas campestris* pv. *manihotis* (Berthet-Bondar) Dye 1978, causal agent of cassava bacterial blight. PhD thesis, University of Göttingen, Germany.

Fessehaie, A., Wydra, K. and Rudolph, K. (1997) Development of a semi-selective medium and use of serology for fast and sensitive detection of *Xanthomonas campestris* pv. *manihotis*, incitant of bacterial blight of cassava. In: Dehne, H.-W., Adam, G., Diekmann, M., Frahm, J., Mauler-Machnik, A. and van Halteren, P. (eds) *Diagnosis and Identification of Plant Pathogens. Developments in Plant Pathology*, Vol. 11. Kluwer, Dordrecht, pp. 137–140.

Fessehaie, A., Wydra, K. and Rudolph, K. (1999) Development of a new semi-selective medium for isolating *Xanthomonas campestris* pv. *manihotis* from plant material and soil. *Phytopathology* 89, 591–597.

Flood, J., Cooper, R.M., Deshappriya, N. and Day, R.C. (1995) Resistance of cassava to *Xanthomonas* blight *in vitro* and *in planta. Aspects of Applied Biology* No. 42, 277–284.

Fokunang, C.N., Ikotun, T., Dixon, A.G.O. and Akem, C.N. (1997) First report of *Colletotrichum gloeosporioides* f. sp. *manihotis* cause of cassava anthracnose disease, being seed-borne and seed-transmitted in cassava. *Plant Disease* 81, 695.

Frison, E.A. and Feliu, E. (1991) FAO/IBPGR Technical guidelines for the safe movement of cassava germplasm. Food and Agriculture Organization of the United Nations/International Board for Plant Genetic Resources, Rome, Italy.

Gaspin, R M. (1980) Reaction of golden yellow cassava to *Meloidogyne* spp. inoculation. *Annals of Tropical Research* 2, 49–53.

Gaspin, R.M. (1981) Control of *Meloidogyne incognita* and *Rotylenchulus reniformis* and its effect on the yield of sweet potato and cassava. *Annals of Tropical Research* 3, 92–100.

Golato, C. (1963) *Cercospora henningsii* sulla manioca in Nigeria. *Rivista di Agricultura Subtropicale e Tropicale* 57, 60–66.

Golato, C. and Meossi, E. (1966) Una nueva malattis fogliare della manioca in Ghana. *Rivista di Agricultura Subtropicale e Tropicale* 60, 21–26.

Golato, C. and Meossi, E. (1971) Una grave infezione fogliare de manioca in Ghana. *Rivista di Agricultura Subtropicale e Tropicale* 65, 21–26.

Guevara, Y., Rondon, A., Arnal, E., Suarez, Z., Solorzano, R. and Navas, R. (1992) Bacterial stem rot of cassava in Venezuela. *Fitopatologia Venezolana* 5, 33–36.

Hahn, S.K. (1978) Breeding cassava for resistance to bacterial blight. *PANS* 24, 480–485.

Hahn, S.K. (1979) Breeding of cassava for resistance to cassava mosaic disease (CMD) and bacterial blight (CBB) in Africa. In: Maraite, H. and Meyer, J.A. (eds) *Diseases of Tropical Food Crops, Proceedings of an International Symposium*, UCL, 1978. Université Catholique de Louvain, Louvain-la-Neuve, Belgium, pp. 211–219.

Hahn, S.K. and Williams, R.J. (1973) Investigations on cassava in the Republic of Zaire. *Rapport au Commissaire d'Etat a l'Agriculture*. République du Zaire. IITA, Ibadan, Mimeo.

Hahn, S.K., Howland, A.K. and Terry, E.R. (1980) Correlated resistance of cassava to mosaic and bacterial blight diseases. *Euphytica* 29, 305–311.

Hernandez, J.M., Laberry, R. and Lozano, J.C. (1986) Observations on the effect of inoculating cassava (*Manihot esculenta*) plantlets with fluorescent pseudomonads. *Journal of Phytopathology* 117, 17–25.

Hogger, C.H. (1971) Plant parasitic nematodes associated with cassava. *Tropical Root and Tuber Crops Newsletter* 4, 4–9.

IITA (1972) *Report of Root, Tuber and Vegetable Improvement Program for 1972*. IITA, Ibadan, Nigeria.

IITA (1998) *Annual Report*. Plant Health Management Division. IITA, Ibadan, Nigeria.

Ikotun, T. and Hahn, S.K. (1994) Screening cassava for resistance to the cassava anthracnose disease. *Acta Horticulturae* 380, 178–183.

Irvine, F.R. (1969) Cassava (*Manihot esculenta*). In: *West African Agriculture 2. West African Crops*. Oxford University Press, London, 153–195.

Janse, J. and Defrancq, M. (1988) Characterization of bacterial strains isolated from *Manihot esculenta* and of strains of *Xanthomonas campestris* pv. *oryzae* and *X. campestris* pv. *ricini* from Niger. *Phytopathologia Mediterranea* 27, 182–185.

Jatala, P. and Bridge, J. (1990) Nematode parasites of root and tuber crops. In: Luc, M., Sikora, R.A. and Bridge, J. (eds) *Plant Parasitic Nematodes in Subtropical and Tropical Agriculture*. CAB International, Wallingford, pp. 137–180.

Korang-Amoakoh, S. and Oduro, K.O. (1979) Present situation of cassava bacterial blight disease in Ghana. In: *Cassava Bacterial Blight in Africa: Past, Present and Future. Report of an Interdisciplinary Workshop*, IITA, Ibadan, Nigeria, 26–30 June 1978. Centre for Overseas Pest Research, London, p. 58.

Laberry, R. and Lozano, J.C. (1992) Bacterial blight of cassava caused by *Xanthomonas campestris* pv.

manihotis: present situation. *ASCOLFI Informa* 18, 6–10.

Laranjeira, D., Santos, E.O. dos, Mariano, R. de L.R. and Barros, S.T. (1994) Occurrence of cassava black rot caused by *Scytalidium lignicola* in the state of Pernambuco, Brasil. *Fitopatologia Brasileira* 19, 466–469.

Lima, M.F., Takatsu, A. and Reifschneider, F.J.B. (1995) Reaction of cassava genotypes to *Phytophthora dreschleri*. *Fitopatologia Brasileira* 20, 406–415.

Lozano, J.C. (1972) Status of virus and mycoplasma-like diseases of cassava. In: *Proceedings of the Third International Symposium of Tropical Root and Tuber Crops*. IITA, Ibadan, Nigeria, pp. 290–292.

Lozano, J.C. (1979) Overview of cassava pathology. In: Maraite, H. and Meyer, J.A. (eds) *Diseases of Tropical Food Crops, Proceedings of an International Symposium*, UCL, 1978. Université Catholique de Louvain, Louvain-la-Neuve, Belgium, pp. 13–26.

Lozano, J.C. (1986) Cassava bacterial blight: a manageable disease. *Plant Disease* 70, 1089–1093.

Lozano, J.C. (1991) Integrated control of cassava diseases. *Fitopatologia Venezolana* 4, 30–36.

Lozano, J.C. (1992) Overview of integrated control of cassava diseases. *Fitopatologia Brasileira* 17, 18–22.

Lozano, J.C. and Bellotti, A. (1978) *Erwinia carotovora* var. *carotovora*, causal agent of bacterial stem rot of cassava: etiology, epidemiology and control. *PANS* 24, 467–479.

Lozano, J.C. and Booth, R.H. (1976) *Diseases of Cassava*, 2nd edn. CIAT, Cali, Colombia.

Lozano, J.C., Bellotti, A., Reyes, J.A., Howeler, R., Leihner, D. and Doll, J. (1981) *Field Problems in Cassava*. CIAT, Cali, Colombia.

Lozano, J.C., Nolt, B.L. and Khan, R.P. (1989) Pests and pathogens of cassava. Plant protection and quarantine. In: Kahn, R.P. (ed.) *Selected Pests and Pathogens of Quarantine Significance*, Vol. II. CRC Press, Boca Raton, Florida, pp. 169–182.

Machmud, M. (1986) Bacterial wilt in Indonesia. In: Persley, G.J. (ed.) *Proceedings ACIAR, Bacterial Wilt Disease in Asia and the South Pacific*. Canberra, Australia. *ACIAR* 13, 32–34.

Makambila, C. (1994) The fungal diseases of cassava in the Republic of Congo, Central Africa. *African Crop Science Journal* 2, 511–517.

Makambila, C. and Koumouno, B.L. (1986) Les pourridiès à *Armillaria* sp., *Spaerostilbe repens* et *Phaeolus manihotis* sur le manioc. *L'Agronomie Tropicale* 4, 258–264.

Makambila, C. and Koumouno, B.L. (1994) Cassava root rot in Congo: first evaluation of damage and identification of pathogenic agents. *Acta Horticulturae* 380, 184–186.

Maraite, H. (1993) *Xanthomonas campestris* pathovars on cassava: cause of bacterial blight and bacterial necrosis. In: Swings, J.G. and Civerolo, E.L. (eds) *Xanthomonas*. Chapman and Hall, London, pp. 18–25.

Maraite, H. and Perreaux, D. (1979) Comparative symptom development in cassava after infection by *Xanthomonas manihotis* or *X. cassavae* under controlled conditions. In: Terry, E.R., Persley, G.J. and Cook, S.C.A. (eds) *Cassava Bacterial Blight in Africa; Past, Present and Future. Report of an Interdisciplinary Workshop*, IITA, Ibadan, Nigeria, 1978. Centre for Overseas Pest Research, London, pp. 17–24.

Maraite, H. and Weyns, J. (1979) Distinctive physiological, biochemical and pathogenic characteristics of *Xanthomonas manihotis* and *X. cassavae*. In: Maraite, H. and Meyer, J.A. (eds) *Diseases of Tropical Food Crops. Proceedings of an International Symposium*, UCL, 1978. Université Catholique de Louvain, Louvain-la-Neuve, Belgium, pp. 103–117.

Marcano, M. and Trujillo, G. (1984) Role of weeds in relation to survival of bacterial blight of cassava. *Revista de la Facultad de Agronomia, Universidad Central de Venezuela* 13, 167–181.

Martin, F.W. (1970) Cassava in the world of tomorrow. In: *Proceedings Second International Symposium of Tropical Root and Tuber Crops*. International Society for Tropical Root Crops/International Institute of Tropical Agriculture, Hawaii, pp. 53–82.

McSorley, R., Ohair, S.K. and Parrado, J.L. (1983) Nematodes of cassava, *Manihot esculenta* Crantz. *Nematropica* 13, 261–287.

Moses, E., Nash, C., Strange, R.N. and Bailey, J.A. (1996) *Colletotrichum gloeosporioides* as the cause of stem tip dieback of cassava. *Plant Pathology* 45, 864–871.

Mostade, J.M. and Butare, I. (1979) Symptomatologie et epidemiologie de la necrose bacterienne du manioc causée par *Xanthomonas cassavae* au Ruanda [Symptomatology and epidemiology of cassava necrosis caused by *X. cassavae* in Rwanda]. In: Maraite, H. and Meyer, J.A. (eds) *Diseases of Tropical Food Crops. Proceedings of an International Symposium*, UCL 1978. Université Catholique de Louvain, Louvain-la-Neuve, Belgium, pp. 95–101.

Msikita, W., Yaninek, J.S., Ahounou, M. and Fagbemissi, R. (1996) First report of *Fusarium moniliforme* causing cassava root, stem and storage rot. *Plant Disease* 80, 823.

Msikita, W., Yaninek, J.S., Ahounou, M., Baimey, H. and Fagbemissi, R. (1997a) First report of *Curvularia lunata* associated with stem disease of cassava. *Plant Disease* 81, 112.

Msikita, W., Yaninek, J.S., Fioklu, M., Baimey, H., Ahounou, M. and Fagbemissi, R. (1997b) A

system to screen and select for resistance to *Fusarium solani*. *African Journal of Root and Tuber Crops* 2, 59–61.

Mwenge, E., Ride, J.P. and Pearce, R.B. (1998) Distribution of Zimbabwean *Armillaria* groups and their pathogenicity on cassava. *Plant Pathology* 47, 623–634.

Nishiyama, K., Achmad, N.H., Wirtono, S. and Yamaguchi, T. (1980) Causal agents of cassava bacterial wilt in Indonesia. *Contributions to Central Research Institute for Agriculture*, Bogor, Indonesia (59).

Normanha, S.E. (1970) General aspects of cassava root production in Brazil. In: *Proceedings Second International Symposium on Tropical Root and Tuber Crops*. International Society for Tropical Root Crops/International Institute of Tropical Agriculture, Hawaii, pp. 61–63.

Nwauzor, E.C. and Nwankwo, O.C. (1989) Reaction of some cassava cultivars/lines to root-knot nematode, *Meloidogyne incognita* race 2. *International Nematology Network Newsletter* 6, 40–42.

Odurukwe, S.O., Arene, O.B. (1980) Effect of N, P, K fertilizers on cassava bacterial blight and root yield of cassava. *Tropical Pest Management*, 26, 391–395.

Ohunyon, P.U. and Ogio-Okirika, J.A. (1979) Eradication of cassava bacterial blight/cassava improvement in the Niger delta of Nigeria. In: *Cassava Bacterial Blight in Africa: Past, Present and Future. Report Interdisciplinary Workshop*. IITA, Ibadan, Nigeria, pp 55–57.

Onyango, D.M. and Mukunya, D.M. (1982) Distribution and importance of *Xanthomonas manihotis* and *X. cassavae* in East Africa. In: *Root Crops in Eastern Africa. Proceedings of a Workshop*, IDRC, Kigali, Rwanda, 23–27 November 1980.

Osai, E.O and Ikotun, T. (1993) Micro-organisms associated with cassava ministem rot. *International Journal of Tropical Plant Diseases* 11, 161–166.

Otim-Nape, G.W. (1980) Cassava bacterial blight in Uganda. *Tropical Pest Management* 26, 274–277.

Pacambaba, R.P. (1987) A screening method for detecting the resistance against cassava bacterial blight disease. *Journal of Phytopathology* 119, 1–6.

Persley, G.J. (1979) Studies on the survival and transmission of *Xanthomonas manihotis* on cassava seed. *Annals of Applied Biology* 93, 159–166.

Persley, G.J. and Terry, E.R. (eds) (1977) *Proceedings IDRC/IITA Workshop on Cassava Bacterial Blight*. 1976. IITA, Ibadan, Nigeria.

Poltronieri, L.S., Trindade, E.R., Silva, H.M. and Albuquerque, F.C. de (1997) *Fitopatologia Brasileira* 22, 111.

Ponte, J.J. da and Franco, A. (1981) Manipueira un nematicidanao convencional de comprovada potencialidade. In: Lordello, L.G.E. (ed.) *Trabalhos*

Apresentados a V Reuniao Brasiliera de Nematologia, Sao Paolo. Sociedade Brasiliera de Nematologia, Piracicaba, SP, Brazil, pp. 25–33.

Ponte, J.J. da, Torres, J. and Simplicio, M.E. (1980) Behaviour of cassava cultivars in relation to root-knot nematodes. In: Lordello, L.G.E. (ed.) *Trabalhos Apresentados a IV Reuniao Brasiliera de Nematologia, Sao Paolo.* Sociedade Brasiliera de Nematologia, Piracicaba, SP, Brazil, pp. 107–133.

Powell, P.W. (1968) *The* Cercospora *Leaf Spots of Cassava.* University of Cornell, Ithaca, USA.

Powell, P.W. (1972) The *Cercospora* leaf spots of cassava. *Tropical Root and Tuber Crops Newsletter*, 6, 10–14.

Restrepo, S. and Verdier, V. (1997) Geographical differentiation of the population of *Xanthomonas axonompodis* pv. *manihotis* in Colombia. *Applied and Environmental Microbiology* 63, 4427–4434.

Saka, V.W. (1982) International *Meloidogyne* project report in Malawi. In: *Proceedings of the 3rd Research Planning Conference on Root-knot Nematodes. Regions IV and V.* North Carolina State University Graphics, Raleigh, USD, pp. 31–36.

Santos, A.A., Melo, Q.M.S., Teixeira, L.M.S. and Fukuda, W.M.G. (1988) Occurrence of cassava rust *Uromyces manihotis* in the state of Ceara. *Fitopatologia Brasileira* 13, 73.

Sikora, R.A., Reckhaus, P. and Adamou, E. (1988) Presence, distribution and importance of plant parasitic nematodes in irrigated agricultural crops in Niger. *Mededelingen van de Faculteit Landbouwwetenschappen Rijksuniversiteit Gent* 53, 821–834.

Takatsu, A., Fukuda, S., Hahn, S.K. and Caveness, F.E. (1990) Integrated pest management for tropical root and tuber crops. In: Hahn, S.K. and Caveness, F.E. (eds) *Proceedings of the Workshop on the Global Status and of Prospects for IPM of Root and Tuber Crops,* Ibadan, Nigeria, 25–30 October 1987. IITA, Ibadan, Nigeria, pp. 127–131.

Tanaka, M.A. de S., Chalfoun, S.M. and Abreu, M.S. de (1979) Doencas de manioca en su controle. *Informe Agropecuario* 5, 70–78.

Terry, E.R. (1977) Factors affecting the incidence of cassava bacterial blight in Africa. In: *Proceedings of the Symposium of the International Society for Tropical Root Crops,* 4th, Cali, Colombia, 1976. IDRC, Ottawa, Canada, pp.179–184.

Terry, E.R. and MacIntyre, R. (eds) (1976) In: *The International Exchange and Testing of Cassava Germplasm in Africa. Proceedings of an Interdisciplinary Workshop,* IITA, Ibadan, Nigeria, 1975. IDRC, Ottawa, Canada.

Théberge, R.I. (1985) (ed.) *Common African Pests and Diseases of Cassava, Yam, Sweet Potato and Cocoyam.* IITA, Ibadan, Nigeria.

Van den Mooter, M., Maraite, H., Meiresonne, L., Swings, J., Gillis, M., Kersters, K. and De Ley, J. (1987) Comparison between *Xanthomonas campestris* pv. *manihotis* (ISPP List 1980) and *X. campestris* pv. *cassavae* (ISPP List 1980) by means of phenotypic, protein electrophoretic, DNA hybridization and phytopathological techniques. *Journal of General Microbiology* 133, 57–71.

Van den Oever, H.A.M. and Mangane, S.E. (1992) A survey of nematodes on various crops in Mozambique. *Afro-Asian Journal of Nematology* 2, 74–79.

Vauterin, L., Hoste, B., Kersters, K. and Swings, J. (1995) Reclassification of *Xanthomonas*. *International Journal of Systematic Bacteriology* 45, 472–489.

Verdier, V., Dongo, P. and Boher, B. (1993) Assessment of genetic diversity among strains of *Xanthomonas campestris* pv. *manihotis*. *Journal of General Microbiology* 169, 2591–2601.

Verdier, V., Boher, B., Maraite, H. and Geiger, J.P. (1994) Pathological and molecular characterization of *Xanthomonas campestris* strains causing diseases of cassava (*Manihot esculenta*). *Applied and Environmental Microbiology* 60, 4478–4486.

Verdier, V., Restrepo, S., Boher, B., Nicole, M., Geiger, J.P., Alvarez, E. and Bonierbale, M. (1997) *African Journal of Root and Tuber Crops* 2, 64–68.

Verdier, V., Mosquera, G. and Assigbetse, K. (1998) Detection of the cassava bacterial blight pathogen, *Xanthomonas axonopodis* pv. *manihotis* by polymerase chain reaction. *Plant Disease* 82, 79–83.

Viegas, A.P. (1943) Alguns fungos de manidoca. I. *Bragantia* 3, 1–19.

Vincens, F. (1915) Une maladie cryptogamique de *Manihot glaziovii*, abre a caoutchouc du Ceara. *Boletin de la Société de la Pathologie Végétale de France* 2, 22–25.

Wydra, K. and Msikita, W. (1998) An overview of the present situation of cassava diseases in West Africa. In: Akoroda, M.O. and Ekanayake, I.J. (eds) *Root Crops for Poverty Alleviation. Proceedings of the Sixth Triennial Symposium of the International Society for Tropical Root Crops (ISTRC),* Lilongwe, Malawi, 22–28 October 1995. ISTRC (International Society for Tropical Root Crops), IITA (International Institute of Tropical Agriculture) and Government of Malawi, pp 163–166.

Wydra, K. and Rudolph, K. (1999) Development and implementation of integrated control methods for major diseases of cassava and cowpea in West Africa. *Göttingen Beiträge zur Lanct und Forstuirtschaft in den Tropen und Subtropen* 133, 174–180.

Wydra, K., Zinsou, V. and Fanou, A. (1998) The expression of resistance against *Xanthomonas campestris* pv. *manihotis*, incitant of cassava bacterial blight, in a resistant cassava variety compared to a susceptible variety. In: Mahadevan, A. (ed.) *Plant Pathogenic Bacteria, Proceedings IX International Conference*, Madras, India, pp. 583–592.

Wydra, K., Fanou, A. and Rudolph, K. (2001a) Effect of cassava bacterial blight on cassava growth parameters and root yield in different ecozones and influence of the environment on symptom development. In: Akoroda, M.O. and Ngeve, J.M. (eds) *Proceedings 7th Triennial Symposium of International Society of Tropical Root Crops, Africa Branch (ISTRC-AB), Cotonou, Benin*. ISTRC-AB and Government of Benin, pp. 562–569.

Wydra, K., Fanou, A., Yaninek, J.S. and Rudolph, K. (2001b) Distribution of bacterial blight in different ecozones in West Africa and the interaction of environment, symptom development and growth parameters of cassava. In: *International Congress of Plant Pathology*, Edinburgh 1998, International Society for Plant Pathology/ British Society for Plant Pathology (Abstract 4.7.7).

Wydra, K., Fessehaie, A., Fanou, A., Sikirou, R., Janse, J. and Rudolph, K. (2001c) Variability of strains of *Xanthomonas campestris* pv. *manihotis*, incitant of cassava bacterial blight, from different geographic origins in pathological, physiological, biochemical and serological characteristics. In: Mahadevan, A. (ed.) *Plant Pathogenic Bacteria, Proceedings IX International Conference*, Madras, India, pp. 317–323.

Chapter 14
Cassava Utilization, Storage and Small-scale Processing

Andrew Westby

Natural Resources Institute, University of Greenwich, Chatham Maritime, Kent ME4 4TB, UK

Introduction

The importance of cassava in the world is mainly a reflection of the agronomic advantages of the crop. However, if the contribution that cassava can make to the livelihoods of poor people is to be increased, there is a need to consider also its postharvest handling, processing and marketing.

There are three major limitations to the increased utilization of cassava roots: poor shelf-life, low protein content and their naturally occurring cyanogens. Rapid postharvest deterioration means that processing is more important than for any of the other root crops. In addition to producing storable products, processing can also add value to the crop and provide employment opportunities.

Both cassava roots and leaves can be used as food, but economically the roots are usually more important, although in parts of some African countries, the leaves may be as important or more important than the roots. The tuberous roots can be 15–100 cm in length and reach a weight of 0.5–2.0 kg (Knoth, 1993). They have no function in vegetative propagation, which is done using stem cuttings. The roots contain large carbohydrate reserves, mainly as starch. Although the function of the starch-bearing tissue is not completely clear, it is assumed that it helps the plant to survive unfavourable conditions such as drought.

This chapter considers world cassava utilization and the nutritional value of the root and leaves for human consumption. The current state of knowledge on cassava storage and small-scale processing is presented with the emphasis on sub-Saharan Africa.

World Cassava Utilization

Total world cassava use is expected to increase from 172.7 million t to 275 million t in the period 1993–2020 using the International Food Policy Research Institute's (IFPRI's) baseline data. A higher prediction of demand and production growth puts the 2020 production at 291 million t (Scott *et al.*, 2000). In both projections cassava use in Africa is equivalent to 62% of total world production.

Cassava consumption

Cassava consumption has remained relatively constant in the period 1983–1996 (Table 14.1). Consumption per capita is highest in sub-Saharan Africa and increased between 1983 and 1996 which is remarkable given the region's high population growth rate of nearly 3% year^{-1} (Scott *et al.*, 2000).

Cassava's main contribution to the human diet is as a source of carbohydrate. Table 14.2 shows the percentage contribution that cassava makes to the total energy consumption in the major consuming countries. Many of these nations are among the poorest in the world.

Cassava utilization

Cassava utilization patterns vary considerably in different parts of the world as indicated in Table 14.3. In Africa the majority of cassava produced (88%) is used for human food, with over 50% used in the form of processed products. Animal feed and use for starch are only minor uses of the crop. In the Americas animal feed is far more important, accounting for approximately one-third of consumption, and human food represents only 42% of production. Starch also represents an important use of cassava in South America. The situation in Asia is greatly influenced by the export of cassava chips by

Thailand to the European Community for use as animal feed. If Thailand is disregarded, then it can be seen that consumption of fresh roots is the most important use of the crop (46% of production), starch is relatively important at over 10% of production and animal feed and export are minor uses.

Although the majority of data available for cassava relates to the roots, cassava leaves are important in some countries. In the Democratic Republic of Congo cassava leaves have greater market value than roots (Lutaladio and Ezumah, 1981). It has been estimated that cassava leaves account for approximately 68% of all vegetable output in the country (Tshibaka and Lumpungu, 1989).

Cassava's nutritional contribution to the diet

The composition of cassava roots is shown in Table 14.4. Cassava roots are a rich source of carbohydrate. Most of the carbohydrate is present as starch (31% of fresh weight; Table 14.4) with smaller amounts of free sugars (less than 1% of fresh weight). Cassava roots are low in protein (0.53%), although higher concentrations of 1.5% have been reported by Ekpenyong (1984), and fat (0.17%). Protein from other sources is therefore needed if cassava is to be part of a balanced diet.

Cassava is generally considered to have a high content of dietary fibre, magnesium, sodium, riboflavin, thiamin, nicotinic acid and citrate (Bradbury and Holloway, 1988). Iron and vitamin A are considered to be low. There are,

Table 14.1. Per capita consumption (kilograms per capita) of cassava as food and feed, 1983 and 1996.

Region/country	1983	1996
Latin America	29	25
South-East Asia	2	1
India	7	6
Other South Asia	2	1
Sub-Saharan Africa	102	106
Developing countries	20	21
World	15	16

Data adapted from Scott *et al.* (2000).

Table 14.2. Percentage contribution of cassava roots to the total energy intake in populations of the principal consuming nations in the period 1990–1992.

> 25%		15–25%		10–15%		5–10%	
DR Congo	54.0	Benin	21.8	Paraguay	14.3	Burundi	9.5
Mozambique	38.5	Tanzania	21.7	Comores	13.1	Zambia	9.3
Congo	35.2	Liberia	19.8	Cameroon	13.0	Chad	8.7
Angola	27.3	Togo	19.0	Ivory Coast	12.4	Rwanda	8.2
Ghana	26.0	Uganda	17.1	Gabon	11.1	Tonga	6.9
RCA	25.9	Madagascar	16.3	Guinea	10.8		
		Nigeria	15.4				

Source: Treche (1995).

Table 14.3. World utilization patterns of cassava. Figures are percentage of total production.

Area	Human food – fresh	Human food – processed	Animal feed	Starch	Export	Waste	Stock
World	30.8	33.8	11.5	5.5	7.0	10.0	1.4
Africa	37.9	50.8	1.4	< 1	< 1	9.5	< 1
Americas	18.5	23.9	33.4	9.6	< 1	14.0	< 1
Asia	33.6	21.7	2.9	8.6	23.0	6.3	3.9
Asia (without Thailand)	45.7	27.9	3.9	11.7	2.3	8.6	< 1

Source: Cock (1985).

Table 14.4. Composition of cassava roots in the Pacific Islands as reported by Bradbury and Holloway (1988).

Component	Roots	Leaves
Moisture (%)	62.8	74.8
Energy (kJ 100 g^{-1})	580	
Protein (%)	0.53	5.1
Fat (%)	0.17	2.0
Starch (%)	31.0	–
Sugar (%)	0.83	–
Dietary fibre (%)	1.48	5.1
Ash (%)	0.84	2.7
Minerals (mg 100 g^{-1})		
Calcium	20	350
Potassium	302	56
Phosphate	46	–
Magnesium	30	–
Iron	0.23	–

Values for leaves are from Gomez and Valdivieso (1985), but have been recalculated by Bradbury and Holloway (1988) to a fresh-weight basis.

however, some varieties that are yellow in colour and these contain a significant concentration of β-carotene, up to 1 mg 100 g^{-1} on a dry-weight basis (McDowell and Oduro, 1983).

Cassava leaves, in contrast to the roots, are high in protein (5.1% on a fresh-weight basis, which exceeds 20% on a dry-matter basis). There is therefore much to be gained from the expanded consumption of cassava leaves.

Cassava Cyanogens

Cassava contains cyanogenic glucosides, which, together with their breakdown products (cyanohydrins and free HCN) formed during processing, can cause health problems. Acute intoxication, manifested as vomiting, dizziness or even death, can occur under very rare conditions. Such poisoning occurs when food shortage and social instability induce shortcuts in established processing methods, or when high cyanogen varieties are introduced into an area lacking appropriate processing techniques (Bokanga *et al.*, 1994). It is well established that thiocyanate resulting from dietary cyanide exposure can aggravate iodine exposure deficiency expressed as goitre and cretinism (Bokanga *et al.*, 1994). There is also strong evidence for a causal link between cyanide and the paralytic disease *konzo* (Tylleskar, 1994) and tropical ataxic neuropathy (Osuntakun, 1994). The removal of cyanogens during processing is discussed in the section on small-scale processing.

Cassava varieties are often described as being bitter or sweet. Although the description of bitter and sweet mainly refers to the taste of the raw roots, there is some correlation between high cyanogen/bitter roots and low cyanogen/sweet roots. Concentrations of cyanogens in roots, is however, affected by environmental growth conditions. This means that some varieties generally considered sweet can have a high cyanogenic potential under certain conditions.

Storage and Handling of Fresh Cassava

Postharvest deterioration

Roots as living organs of the plant continue to metabolize and respire after harvest. Cassava's roots are used only to store energy, unlike the roots of sweet potato and yam that are reproductive organs. Despite their agronomic advantages, root crops are far more perishable than the other main staple food crops, the cereals. Once out of the ground, some root crops have a shelf-life of only a few days (Wenham, 1995). There are three main approaches to overcoming the problem of perishability: (i) conventional breeding of varieties with roots having longer shelf-lives; (ii) use of genetic modification to bring about targeted changes in metabolism; and (iii) the use of improved storage techniques. Breeding and genetic modification are long-term strategies, whereas improved storage is likely to have a more immediate impact, but the extent of the improvement will be limited by the roots inherent perishability.

Cassava has a shelf-life that is generally accepted to be of the order of 24–48 h after harvest. Two types of postharvest deterioration are recognized: primary physiological deterioration that involves internal discoloration and is the initial cause of loss of market acceptability, and secondary deterioration due to microbial spoilage (Booth and Coursey, 1974). Physiological deterioration is thought to be a consequence of tissue damage during harvesting. In most cases it is seen as a blue–black discoloration of the vascular tissue referred to as vascular streaking. These initial symptoms are followed by a more general discoloration of the starch-bearing tissue.

Physiological deterioration is a complex process, which is still not fully understood. The process is considered to resemble a typical wounding response in which the healing process is inadequate (Beeching et al., 1998). Physiological deterioration shares features of wound responses in other plants: increased activity of enzymes such as phenylalanine ammonia lyase and polyphenyl oxidase, the synthesis of lignin and suberin or secondary metabolites from the phenylpropenoid or terpenoid pathway and the synthesis of free radicals. There is also an accumulation of phenolic compounds, including coumarins, catechins and flavonoids (Buschman et al., 2000). The wound healing response is, however, slower than in other crops, but when it does occur, it suppresses physiological deterioration (Wenham, 1995).

Traditional marketing and storage systems have adapted to the perishability of root crops (Wenham, 1995). In the case of cassava, these adaptations range from use of in-ground storage,

processing into storable forms at farm level and the general practice of traders is to deal in only small quantities of roots. These ensure that physical losses are minimized. Marketing systems for cassava are also limited by the perishability of the fresh roots and this has financial implications. For example, in Tanzania, delay in getting the crop to market has been reported to be an important factor in determining the level of price discounting at the various stages of marketing. At some stages in the marketing chain, economic losses were greater than 90% of initial value (Fig. 14.1). These economic losses are related to the age of the crop rather than the physical condition of the roots. The perishability of the crop also limits the potential for farmers distant from markets to sell their produce.

Means of overcoming perishability

In-ground storage

The simplest means of preserving cassava is to delay harvesting until the crop is needed. This flexibility in harvesting is one of the most important features of the crop when used for food security.

Cassava roots have an optimum harvest age after which there is a loss in yield. At the same time the roots become woody and there can be impairments to flavour (Lancaster and Coursey, 1984). During storage, there is the danger that roots will be infested by pathogens. There is also the problem that this form of storage ties up large amounts of land that could be used to grow other crops. This is a significant problem in densely populated areas (Knoth, 1993).

Traditional storage structures

Knoth (1993) reviewed traditional methods of cassava storage. Roots can be buried in the soil. It is said that this method can be used for storing roots from one season to the next. In West Africa and India, roots that cannot be consumed or processed immediately are piled into heaps and watered daily. Roots can also be coated with a loam paste to attain storage of 4–6 days. Knoth (1993) also mentions the work of Baybay (1981) who tested various traditional methods of storage in the Philippines. He concluded that all the traditional processes tested could only

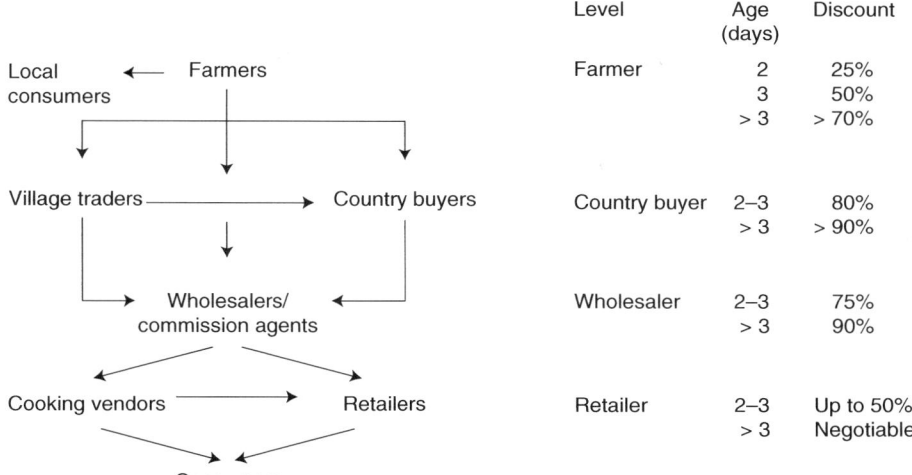

Fig. 14.1. Fresh cassava marketing chain and levels of losses associated with time delays at different stages (modified from Ndunguru *et al.*, 1998).

prolong shelf-life by a few days. Only storage in trench silos was more successful in extending the storage period.

Improved cassava storage

The Natural Resources Institute (NRI) collaborated with the International Center for Tropical Agriculture (CIAT) to gain an understanding of the process of physiological deterioration (Booth, 1976; Rickard and Coursey, 1981; Rickard, 1985; Wenham, 1995). This research led to an understanding of which storage structures were most effective (Booth and Coursey, 1974; Booth, 1977) and the development of the low-cost storage technique detailed below (Wheatley, 1989).

Initial work focused on the clamp silo. Roots were piled up on a layer of straw in conical heaps weighing between 300 and 500 kg (Rickard and Coursey, 1981). These were covered with straw and soil and openings were left for ventilation. With such a system, it was possible to store roots for up to 4 weeks without significant weight loss or microbial deterioration. The systems required relatively high labour inputs and management of the stores required a degree of experience. These factors may have limited adoption of the system.

Research work on the storage of cassava roots in pits containing sand/soil at 15%

moisture content has been conducted in India (Balagopalan, 2000). After 2 months of storage 80–85% of roots were recovered undamaged. Roots lost 15–20% of their starch content after 2 months of storage, which was equivalent to 1 week of storage under ambient conditions. There were also significant reductions in root cyanogen content.

Another method that has been evaluated is the storage of cassava in wooden crates containing damp sawdust. If the sawdust is too moist it promotes fungal growth and if too dry the roots deteriorate quickly. Lining the crates with plastic foil prevents drying out of the sawdust resulting in a storage period of 4–8 weeks (Rickard and Coursey, 1981). The availability and expense of crates, their limited capacity in comparison with the low value and bulky nature of the commodity, together with the high labour requirement of the system, limited its applicability.

The storage of cassava treated with a fungicide in polythene bags was a technique developed by CIAT and NRI. The technique was dependent upon the curing effect of storing roots in polythene bags combined with a fungicide (thiabendazole) to prevent secondary microbial deterioration (Wheatley, 1989). This technique extended the shelf-life from 1–2 days to between 2 and 3 weeks. Although trials with this method were successful in Colombia, it was less successful when it was modified for use in Ghana by

Bancroft and Crentsil (1995). A limitation to adoption of the technology in Ghana was the high costs of polypropylene sacks and the fungicide. For this reason, the polythene bags were replaced with other tightly woven bags such as rice or cocoa sacks and the technology was tested without the use of the fungicide. With these modifications, storage times of 7–10 days were achievable that were adequate for Ghanaian marketing systems (Gallat *et al.*, 1998). The modified technique was evaluated with a number of potential stakeholders and it was found to be particularly useful for local food retailers and itinerant traders. The technique has subsequently been transferred to Tanzania (Westby *et al.*, 1999).

Advanced methods of overcoming physiological deterioration

Refrigeration is not a viable method for preserving cassava in many developing countries, but may be practical for high-value markets. The most favourable temperature for storing fresh cassava is 3°C. At this temperature, the total weight loss after 14 days was 14% and was 23% after 4 weeks (Rickard and Coursey, 1981). Alternatively, roots, or more usually pieces of root, can be stored frozen. Freezing changes the texture making it somewhat spongier, but the flavour is preserved (Rickard and Coursey, 1981). This technique is already used commercially by the world's major exporter of cassava, Costa Rica.

Coating of cassava roots with paraffin wax is another means of extending the shelf-life of fresh roots. This has been done with or without a fungicide. Shelf-lives of up to 2 months have been reported (Knoth, 1993). Waxing is currently the most common way of treating fresh cassava for export.

Varietal selection as a means of overcoming physiological deterioration

Breeding or genetic manipulation are alternative means of overcoming the problem of cassava physiological deterioration. The various approaches were reviewed at an expert consultation at FAO (Wenham, 1995). It was considered that it was possible to use conventional breeding by recurrent selection methods to produce cultivars with resistance to physiological

deterioration. However, tremendous efforts would be required to incorporate the trait into different cultivars without altering the characteristics of the parent genotypes.

Genetic manipulation was considered most appropriate to resolve the problem since it should be possible to add new traits to elite genotypes without altering other desired characteristics. Nevertheless, there was no information available on genes involved in the biochemical pathways that are associated with physiological deterioration of cassava. However, because of the their implication in the process of physiological deterioration, the genes and gene products associated with the synthesis and degradation of phenylpropanoids were considered to be the principal targets.

Subsequent work has been carried out on the biochemistry, molecular biology and genetics of physiological deterioration in order to identify potential avenues for its control. Although this has led to more new knowledge on physiological deterioration, there is a need for a concerted global research effort to address the problem.

Small-scale Processing

This section focuses on small scale processing with a specific emphasis on sub-Saharan Africa where processed products are the most important (Table 14.3). This complements the information in chapter 15.

Traditional processing systems

The processing of cassava into more storable forms offers an opportunity to overcome the perishability of the fresh produce. A wide variety of products are produced, especially in Africa and South America. The best overview of the complex nature of cassava processing in Africa has come from the Collaborative Study of Cassava in Africa (COSCA) as described by Nweke (1988). Details of the three most important products in 233 villages in six countries (Côte d'Ivoire, Ghana, Nigeria, Democratic Republic of Congo, Uganda and Tanzania) were collected (NRI, 1992). Across the countries, 147 different product names were used to describe the

623 products for which details were collected. Through a process of examining the key processing steps, this complex array of products was rationalized in eight main groups (Table 14.5; NRI, 1992; Poulter *et al.*, 1992).

Further analysis of the COSCA data (Westby, 1993) has enabled more detailed characterisation of products according to the processing steps involved. This analysis is shown schematically in Fig. 14.2 and is quantified in Table 14.6. Slight discrepancies between Tables 14.5 and 14.6 are due to the more accurate manual form of classification used for the latter table.

Stages in traditional processing

A detailed description of all of the possible cassava products is not possible in this chapter (see Chapter 15). However, some of the key common stages in cassava processing are described below.

Harvesting and transportation of roots

Transportation of cassava roots from the field to the roadside or household is one of the major limitations in postharvest processing. Roots are typically transported in a bowl carried on the head. In some places bicycles are used. The contribution of harvesting and transportation to the labour requirement of processing a mould-fermented cassava product, *udaga*, in Tanzania

is shown in Fig. 14.3. Improvements to intermediate forms of transport would have a large impact in many African communities.

Root preparation (peeling, slicing)

Cassava roots are usually peeled prior to processing. Mechanical peelers are not generally available, although technology exists in Brazil for the debarking of cassava roots for the processing of *farinha de mandioca* and extraction of cassava starch (Westby and Cereda, 1994). For many products, peeling is considered one of the most labour intensive processes (see for example Fig. 14.3).

Size reduction (grating)

Size reduction of fresh roots is usually by grating. In many locations in West Africa this is a mechanized process often carried out at a communal facility. There are many designs of grater ranging from punched metal discs to ones that use nails punched through wood. Grating is also a necessary part of starch extraction and similar machines are often used. Where machines are not available, grating is done by hand but this is a very labour-intensive process.

Drying

Sun-dried products are the most common types of processed product in Africa (see Table 14.6). Drying over a fire is practised in some places.

Table 14.5. Product types by country for the first three ranked products in each of 233 villages, COSCA Phase 1.

Product type	Côte d'Ivoire	Ghana	West Nigeria	East Nigeria	Tanzania	Uganda	Zaire	Total	%
Cooked roots	35	20	–	11	9	33	–	108	17
Roasted granules	7	19	18	24	–	–	–	68	11
Steamed granules	30	1	–		–	–	1	32	5
Flours/dry pieces	21	27	17	35	61	52	66	279	45
Fermented pastes	4	10	19	21	1	–	20	75	12
Leaves	–	–	–	–	1	3	2	6	1
Drinks	–	–	–	–	–	6	–	6	1
Sedimented starch	22	–	3	3	–	–	–	28	4
Unclassified	–	5	4	4	2	2	4	21	3
Total								623	100

Note: The figures in the columns indicate the number of times a particular product type was ranked as one of the first three most important in the 233 surveyed villages.
Source: NRI (1992).

One problem of sun drying is that drying times are long. In a study conducted by Wareing *et al.* (2001) on a dried product in Ghana, *kokonte*, it took 7–12 days to dry during the dry season and 8–14 days during the rainy season. Mould growth is common on such products and may be a concern in respect of mycotoxin formation (see pp. 292–293).

Various methods are available for improving drying to produce a better quality product. These include modifications to the size and shape of cassava pieces, use of inclined trays or concrete drying floors (e.g. Best, 1978; Balagopalan, 2000). In Ghana, a system of dried cassava chip production has been developed whereby cassava is chipped into small pieces using a machine developed by the International Institute of Tropical Agriculture (IITA) in Nigeria (Jeon and Halos, 1991). The combination of the chipper, drying on raised trays for 1 day and on polythene sheeting for 1 day produces high-quality product at minimum cost (Westby and Gallat, 1999).

Fermentation of cassava

Fermentation is an important processing technique for cassava, especially in Africa. Three major types of fermentation are recognized: the grated root fermentation, fermentation of roots under water and mould fermentation of roots in heaps.

The grated root fermentation method is important in the processing of many West African products including the roasted granules (*gari*), steamed granules (*attieke* from Côte d'Ivoire) and some of the fermented pastes (*agbelima* and *placali* from Ghana and Côte d'Ivoire respectively). Typically grated roots are allowed to ferment in sacks for 3–5 days, which encourages a lactic acid fermentation and a

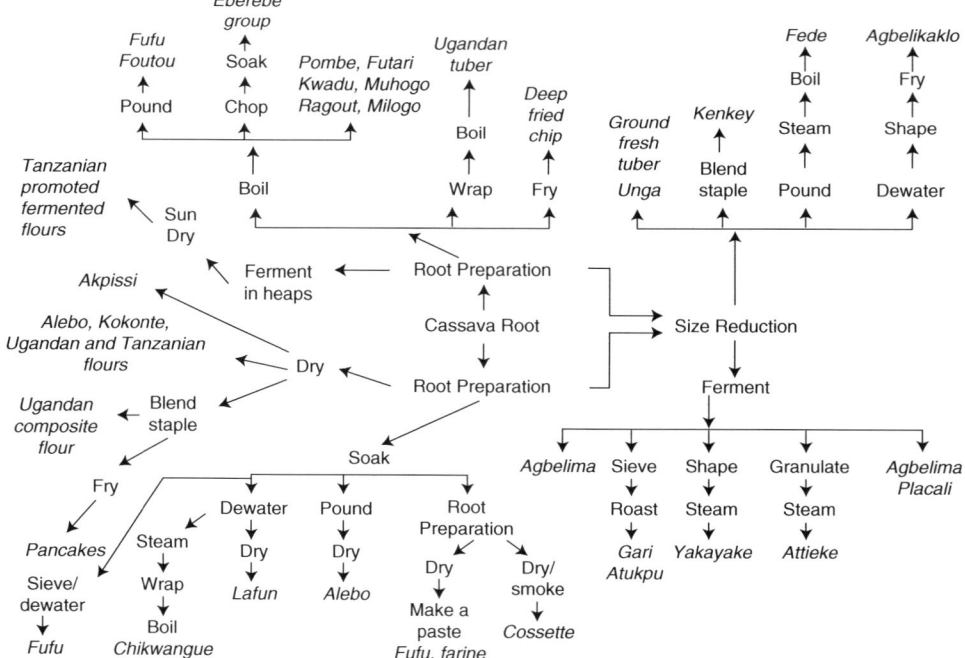

Fig. 14.2. Interrelationship of cassava products from 233 villages in six African countries (Henry *et al.*, 1998).

Footnote for Table 14.6 opposite
[a]The number after the number of villages ranking the product third is the number of villages where the ranking was not recorded.

Table 14.6.   Distribution of cassava products using the categories provided in Table 14.5 and Fig. 14.2.

Product group/ product type	No. of alternative names	Country	No. of villages where ranked			Total no. villages (% of surveyed in country)
			1st	2nd	3rd[a]	
1. Fresh roots						108
Ererebe group	6	Nigeria	0	1	10	11 (18%)
Foutou/fufu	2	Côte d'Ivoire	16	9	6 + 1	32 (80%)
	1	Ghana	10	3	2	15 (50%)
Tuber	12	Uganda	29	2	0	31 (97%)
Other		Various				19
2. Roasted granules						78
Gari	2	Côte d'Ivoire	1	2	4 + 1	8 (20%)
		Ghana	7	13	2	22 (73%)
		Nigeria	25	22	1	48 (79%)
3. Steamed granules						35
Attieke	1	Côte d'Ivoire	15	12	7	34 (85%)
Others	1	Ghana				1
4. Dried flours/pieces						267
Acid soaked						
Alebo	6	Nigeria	21	1	3	25 (40%)
Cossette	1	Zaire	15	16	0	33 (92%)
Fufu	2	Zaire	7	12	7 + 4	30 (83%)
Lafun	1	Nigeria	2	6	4	12 (20%)
Others	3	Nigeria				6
Air dried						
Alebo	5	Nigeria	10	1	2	13 (20%)
Kabalagala	2	Uganda	0	7	4	11 (34%)
Kokonte	2	Ghana	9	8	11	28 (93%)
		Côte d'Ivoire	3	8	5 + 2	18 (45%)
Cassava flour (Tz)	12	Tanzania	6	10	5 + 7	28 (93%)
Cassava Flour (Ug)	5	Uganda	0	14	7	21 (66%)
Composite flour	5	Uganda	1	5	2	8 (25%)
Others	2	Various				5
Mould fermented						
Tanzanian		Tanzania	12	5	3 + 8	28 (93%)
Others	1	Uganda				1
5. Fermented pastes						47
Grated roots						
Agbelima	2	Ghana	3	3	3 + 1	10 (33%)
Placali	2	Côte d'Ivoire	4	8	11	23 (58%)
Soaked roots						
Akpu (fufu)	6	Nigeria	8	13	19	40 (63%)
Chikwangue	3	Zaire	12	2	5 + 5	24 (64%)
6. Products from leaves						
Total	5	Zaire, Ug, Tz				7
7. Drinks						
Total	14	Zaire, Uganda				22
8. Sedimented starches						
Starch	1	Nigeria	0	2	2 + 1	5 (8%)
9. Unclassified						
Total	5					5

Footnote opposite

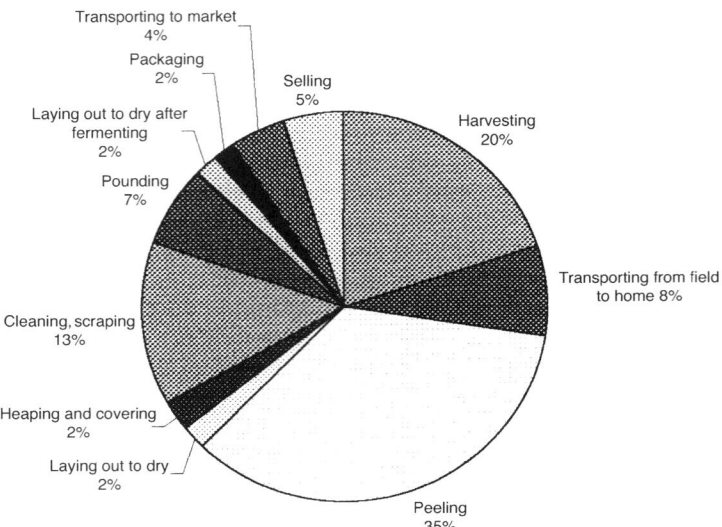

Fig. 14.3. Estimated distribution of labour during the household level processing of *udaga*, a heap-fermented cassava product in Tanzania (from van Oirschot *et al.*, 2001).

consequent reduction in pH value to less than 4.0. Starter cultures are only common in Côte d'Ivoire where some pre-mould fermented roots are added during grating. This is said to improve smoothness. Although there is probably a succession of organisms, the grated root fermentation is dominated by lactic acid bacteria (Okafor, 1977; Abe and Linsay, 1978; Ngaba and Lee, 1979).

Fermentation of cassava roots under water is conducted across Africa from Sierra Leone to Tanzania. A variety of products are produced including wet paste (such as *akpu*, *fufu* or *chikwangue*) and dried flours (such as *lafun*). Roots are soaked in water with or without peeling for typically 3–5 days. The fermentation causes the roots to soften (Westby and Choo, 1994) which means that they can be easily broken up by hand into small pieces and sun dried or passed through a sieve to remove fibre, leaving a smooth paste. At the start of the fermentation there is a mixed microbial flora consisting of *Bacillus* spp., *Leuconostoc* spp., *Klebsiella* spp., *Corynebacterium* spp., *Lactobacillus* spp., *Aspergillus* spp., *Candida* spp. and *Geotrichum* spp. The final fermentation is, however, dominated by lactic acid bacteria and yeasts (Oyewole and Odunfa, 1988). *Clostridium* is thought to be the origin of butyrate, which imparts a typical odour to the product (Brauman *et al.*, 1995). They may

also play other roles such as the production of pectic enzymes and this deserves further investigation.

Heap fermented cassava products are produced in Tanzania (Ndunguru *et al.*, 1999), Uganda and Mozambique (Essers, 1995). This type of fermentation is achieved by heaping peeled roots and leaving them to ferment naturally. Essers and Nout (1989) reported the isolation of *Rhizopus* spp., *Mucor* spp., *Penicillium* spp. and *Fusarium* spp. Studies in Tanzania (M.N. Kendall, personal communication) have indicated that *Rhizopus* spp., *Neurospora sitophilia* and *Penicillium* spp. can be isolated from the *udaga* fermentation. Market studies in the Lake Zone of Tanzania have indicated that different moulds can have an impact on the value of the commodity. Ndunguru *et al.* (1999) reported that visible mould is disliked, but some types of mould were disliked more than others. Average valuation discounts are 10–15% for orange-coloured mould, 20–25% for green-coloured mould and 35–40% for black-coloured mould.

Cyanogen removal during processing

The cyanogenic glucosides present in fresh cassava roots are linamarin (93%) and lotaustralin (7%) (Nartey, 1978). Linamarin is stored in the

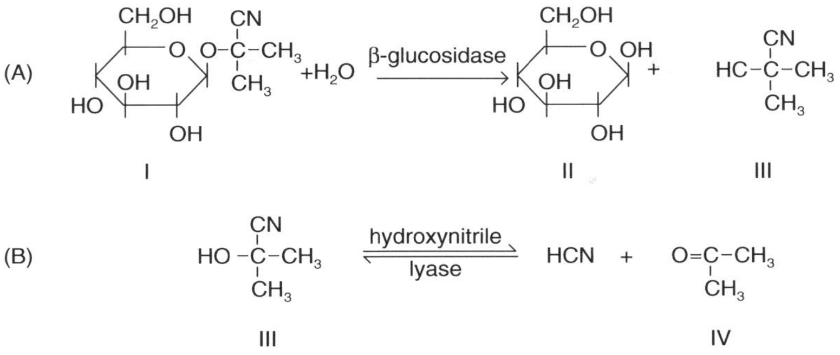

Fig. 14.4. Enzymatic breakdown of linamarin, Compound 1 (from Conn, 1994). Compound III, (acetone cyanohydrin) also breaks down at a rate dependent upon pH and temperature.

vacuoles of the cassava cells (McMahon *et al.*, 1995). It is hydrolysed to the corresponding ketone (acetone cyanohydrin) and glucose by the endogenous enzyme, linamarase (Fig. 14.4), when cellular damage occurs (de Bruijn, 1973; Nartey, 1978). Linamarase is situated in the cell wall (Mkpong *et al.*, 1990) physically separated from linamarin. Cyanohydrins breakdown non-enzymically at a rate dependent upon pH and temperature (Cooke, 1978), with their stability increasing at acidic pH values (Fig. 14.4). Acetone cyanohydrin can also be broken down by an enzyme hydroxynitrile lyase (HNL), but it has been demonstrated that expression of the HNL gene is mainly in the leaves with little activity in the roots (White *et al.*, 1998).

The low levels of HNL expression in the roots is very significant to the medical condition *konzo*. Tylleskar *et al.* (1992) have shown that certain processed flours contain high levels of acetone cyanohydrin, but little linamarin or HCN. The low activity of HNL in the roots will contribute to this accumulation.

Chemically linamarin is stable, soluble in water and resists boiling in acid. Acetone cyano-hydrin is also soluble in water and has a boiling point of the order of 82°C. Free HCN is volatile at 25.7°C and so is rapidly volatilized at tropical ambient temperatures.

Efficient processing to ensure that cyano-gens are reduced to a safe level is based on the above knowledge. The essential features of good processing are sufficient tissue disruption to allow endogenous linamarase to react with linamarin and then favourable conditions for the breakdown of acetone cyanohydrin, or,

conditions under which the compound will volatize spontaneously. It was the development of an assay method for the different cyanogenic compounds (Cooke, 1978, modified by O'Brien *et al.*,1991 and later by Essers *et al.*,1993) that allowed the mechanisms of cyanogen reduction during processing to be elucidated.

The production of the Nigerian product *gari* (similar to *farinha de mandioca* in Brazil) is an example of efficient cassava processing. Proces-sing involves the grating of peeled roots, holding of the roots at ambient temperature to facilitate fermentation. Water is squeezed out of the fer-mented material to bring the moisture content down to 50%. The final product is roasted to form a granular dried product. The changes in cyano-gens during the *gari* fermentation were investi-gated by Vasconcelos *et al.* (1990). Plant and microbial enzymic activities were distinguished by comparing a natural fermentation with irra-diated grated root material incubated under the same conditions. In both cases, more than 95% of the initial linamarin content was hydrolysed within 3 h of grating (Fig. 14.5). Grating is therefore important for bringing linamarin into contact with linamarase. Significant concen-trations of cyanohydrin and free HCN were left in the paste after fermentation. The cyanohydrin is stable under acidic conditions and it is now known that roots lack hydroxynitrilase activity.

Assuming efficient grating and an accept-able level of endogenous enzymatic activity, linamarin reduction is not the constraint and so processing has to be geared to reducing the concentrations of cyanohydrin and free HCN. In

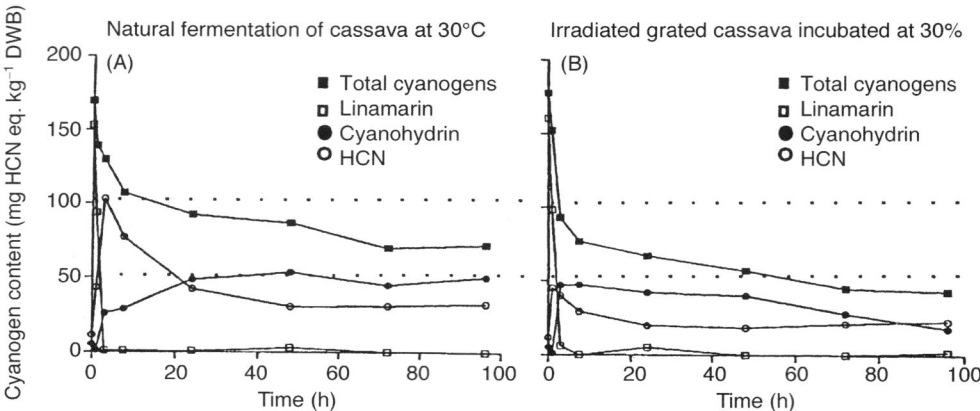

Fig. 14.5. Changes in cyanogens during the fermentation of grated cassava and incubation of grated irradiated cassava at 30°C. Roots were grated at 0 h. The effects of roasting the product are not shown (Westby and Choo, 1994).

the case of *gari* processing, roasting is efficient at volatilizing HCN and cyanohydrin leaving low residual concentrations (HCN 3.4 mg CN equivalents kg^{-1} and cyanohydrin 2.2 mg CN equivalents kg^{-1} on a dry-weight basis in the natural fermentation).

Where roots are soaked in water, the fermentation enables softening of the roots which has the combined effect of enabling linamarin and linamarase to mix and also to enable leaching of the cyanogens (Westby and Choo, 1994). The efficiency of these processes and of any post-fermentation processes in reducing cyanogens dictates the safety of the product. For dried products the efficiency of drying is important to ensure that most of the residual cyanogens are volatilized.

In the case of the heap-fermented products, microbial growth contributes to cyanogen reduction by softening the cassava roots which enhances the contact between endogenous linamarin and linamarase (Essers, 1995). The efficiency of subsequent drying is important to ensure volatilization of residual cyanogens.

Storage of Dried Cassava Products

Dried cassava products are intrinsically more stable than fresh roots. Their deterioration is caused primarily by exogenous factors such as fungi, bacteria, insects and rodents (McFarlane,

1982). Fungal deterioration is discussed later in respect of mycotoxin formation.

Insect infestation of dried cassava is the most important form of damage and well dried products are unable to support microbial growth. A wide range of insect species have been associated with dried cassava products and these were reviewed by McFarlane (1982). A significant problem of stored cassava is *Prostephanus truncatus* (Horn), the larger grain borer. It was introduced into Africa in the 1970s and has become a major pest of both cassava and maize (Hodges, 1986). Losses in Togo from the pest have been estimated at up to 30% after 6 months of storage (Wright, 1993).

Parker *et al.* (1981) suggested that standards and practices similar to those applied to cereal grains should be used for dried cassava since the spectrum of infestation is similar.

Mycotoxin Contamination of Cassava Products

Fungal growth is common on dried cassava products (Clerk and Caurie, 1968; Essers and Nout, 1989; Jonsyn, 1989), especially in Africa where this class of processed product is the most common. Fungal growth occurs at three stages in processing: during slow drying, during storage under humid conditions, or in some specific products that are fungally fermented. When growth of potentially mycotoxigenic fungi occurs, there is the possibility of mycotoxin formation. Mycotoxins are extracellular zootoxic metabolites (exotoxins) produced by filamentous fungi (moulds) in foods consumed by man or animals.

A number of potentially mycotoxigenic storage fungi have been isolated from cassava, for example: *Aspergillus flavus* (Clerk and Caurie, 1968; Shank *et al.*, 1972; Mota and Lourenco, 1974; Masimango *et al.*, 1977), *Aspergillus ochraceus* (Masimango *et al.*, 1977), *Aspergillus versicolor* (Clerk and Caurie, 1968) and *Penicillium* spp. (Clerk and Caurie, 1968; Mota and Lourenco, 1974). The potential mycotoxin risk associated with this contamination has not been fully assessed. Mycotoxin contamination of cassava has been documented (Mota and Lourenco, 1974; Brudzynski *et al.*, 1977; Constant *et al.*, 1984; Sajise and Ilag, 1987), although the findings have not always been subjected to confirmatory tests.

Scopoletin, a coumarin compound, can accumulate in cassava roots during postharvest physiological deterioration (Wenham, 1995). It fluoresces in a similar way and has a similar Rf to aflatoxin B_1 in some common thin layer chromatography (TLC) systems. It is therefore possible that some reports of aflatoxin contamination of cassava chips may instead be due to the presence of scopoletin (Wheatley, 1984). Reports in which the presence of aflatoxin has not been confirmed, or scopoletin not removed, should be treated with caution. Other mycotoxins have been reported. For example, zearalenone in cassava meal from Indonesia and Thailand, at 90 µg kg^{-1} and 3 mg kg^{-1} respectively (Bottalico *et al.*, 1980) and ochratoxin A in two out of 33 samples of cassava flour from Brazil, 32 and 65 µg kg^{-1} respectively (Soares and Rodriguez-Amaya, 1989).

In a recent study of dried cassava in Ghana (Wareing *et al.*, 2001), 125 households in 19 villages producing dried cassava products were interviewed. Mould growth during processing or storage was a problem during June and July, the rainy season. Most producers and market traders preferred non-mouldy *kokonte* (the dried cassava product), although many (59%) would consume a mouldy product. There was a price premium for non-mouldy *kokonte*. The most commonly isolated microorganisms were yeasts and *Cladosporium* spp. (44 out of 49 samples). Other fungi isolated included *Aspergillus* spp. (20 samples), *Penicillium* spp. (15 samples) and *Fusarium* spp. (30 samples). Sterigmatocystin was detected in ten samples at 0.17–1.67 mg kg^{-1}; patulin in four samples at 0.55–0.85 mg kg^{-1}; cyclopiazonic acid in four samples at 0.08–0.72 mg kg^{-1}; penicillic acid in five samples at 0.06–0.23 mg kg^{-1} and tenuazonic acid in three samples at 0.02–0.34 mg kg^{-1}. The authors concluded that mycotoxin contamination of mouldy cassava was a potential problem and improvements in processing were necessary to improve the speed of drying in order to avoid mould growth.

Mould growth on fermented cassava products has so far not been associated with mycotoxin formation. Essers (1995) reported that in his investigation of the solid substrate fermentation of cassava in Uganda, the Ames test for mutagenicity and cytotoxicity was negative in all of the tested flours and aflatoxins were absent in ten screened samples.

Development of Cassava Processing Beyond Household Scale

There is a progression of cassava processing at the small-scale level beyond traditional products that starts at the production of high-quality flour, chips and eventually starch and starch products.

Commercialization of traditional products

Traditional products can themselves be scaled up. This can be done in several ways, but the most common is to focus an enterprise around a machine (a grater or mill). This is common for *gari* processing. Individual unit operations, such as peeling, sieving or roasting of *gari* are carried out in a similar way to that at household level. The only difference is that hired labour may be brought in to do the tasks. Also common are situations were groups of people, often women, come together to market a processed commodity.

Attempts have been made to produce *gari* on a larger scale, but have not always been successful because of problems of raw material supply and in producing a product that is competitive with that produced at the household/small enterprise level.

The situation is different in Latin America where traditional *farinha de mandioca* (cassava flour prepared by toasting grated cassava) processing and *polvilho azedo* (fermented cassava starch) have been scaled up and in many cases are mechanized.

Some forms of commercialization require a modified product. Sanni *et al.* (1998) in their analysis of the current state of *fufu* (a wet fermented paste) processing concluded that if the product is to compete with *gari*, a dried form of the product is required that will appeal to urban consumers.

Accessing new market opportunities – using cassava flour to replace wheat flour

In broad terms, there are three major new market opportunities that farmers and processors can access. These are high quality cassava flour as a replacement for wheat flour, cassava starch as a raw material for food and non-food industries and cassava chips for either the domestic livestock feed sector or for export (Bokanga, 1995). In addition, cassava flour or starch can be further processed to serve as a raw material for sweeteners, ethanol, etc. (see Chapter 15). These market opportunities, when developed, provide new outlets for farmers' produce. In some cases, through on-farm processing, value is added to the commodity before it is sold into the marketing system.

Cassava flour is common in Africa and, provided the quality is good, there is potential to substitute wheat flour in a number of products including bread, biscuits and cakes. Ouraga-Djoussou and Bokanga (1998) have shown that, with a 15% substitution rate of wheat flour with cassava, Nigeria could save up to US$14.8 million in foreign exchange annually. US$12.7 million would go to cassava processors and US$4.2 million to cassava farmers. Bokanga (1998) summarizes the use of cassava flour in bread. He cites a recent survey in Nigeria and Côte d'Ivoire where it was shown that the majority of the bread consumed in the survey area was from composite flour (wheat mixed with cassava, sorghum or maize flour).

In order to address a need to diversify the range of cassava products in Tanzania, bakery products made from cassava instead of wheat (doughnuts, cakes, biscuits, croquettes and chinchin) that had been developed at IITA (Onabolu *et al.*, 1998) were evaluated. Kapinga *et al.* (1998) adopted a cautious approach to the dissemination of these products in the Lake Zone. This involved the following stages: (i) identification of the initial need to diversify cassava utilization; (ii) a feasibility study; (iii) an interactive pilot phase where information was obtained on the factors that would facilitate sustainable uptake of the technology; and (iv) a wider dissemination phase.

There was potential for some new products, but not for others. This was reflected in the high take-up rates in both the pilot and wider dissemination phases of only certain products. The most effective dissemination route for these products was through church and women's groups and it was necessary to provide technical support during adoption of the technology (Kapinga *et al.*, 1998). Returns to labour investment when using cassava were significantly improved (Kapinga *et al.*, 1998).

Cassava as a Spur for Rural Development

As indicated in this chapter and elsewhere in this book, cassava is important to the livelihoods of many poor people in the world. This importance is largely a consequence of cassava's agronomic advantages, particularly its high yield of carbohydrate even on poor soils, good tolerance to drought, is relatively resistant to pest infestation and disease, and because it can be stored in the ground until required (DGIS, 1991).

If cassava's contribution to poverty alleviation is to be enhanced, improvements have to made in the postharvest sector. The best way of making any improvements is a subject of much debate, but clearly with limited resources, means have to be found to ensure interventions have the maximum impact. Reasons for introducing improvements to existing systems include: enhanced financial returns through more efficient processing or a higher quality product, introduction of a new product to meet a market opportunity, reduced drudgery or reduced environmental impact. To conclude this chapter, we consider two specific but complementary approaches: use of 'needs assessment' and the use of a 'demand driven' approach as proposed in the Global Cassava Development Strategy.

The use of needs assessment

Informal needs assessment (NA) is a term used to describe a range of qualitative diagnostic methods such as rapid rural appraisal and participatory rural appraisal (Cropley and Gilling, 1993). Overviews of participatory research approaches are given by Chambers (1992) and Chambers and Gildyal (1985). Their essence is that they facilitate scientists to allow farmers to participate in the formulation of the research agenda. A common criticism of previous post-harvest research, and indeed agricultural research in general, is that technical innovation has been high, but adoption has been poor. The use of NA can improve this situation by actively involving beneficiaries in the key phases of the project or research and development cycle in which priorities for research are set or in which technology choices are made. By ensuring the relevance of research and subsequent technical interventions, the prospects for adoption, and therefore impact, are greatly improved.

The tools used in NA include: review of secondary information, direct observation, semi-structured interviewing, scoring, ranking, diagramming and the use of case-studies. The techniques are described in detail in Kleih *et al.* (1997). It is not necessary, or even in many cases desirable, to use all of the tools in all circumstances. The tools should only be used as means to facilitate a dialogue between scientists and those involved in the system. As in all scientific research, hypothesis formulation and testing are central and take place through an iterative process of discussion and explanation.

NA has been used for a number of years to target technical interventions. Examples of how the techniques have been used are described in Westby *et al.* (1998).

Demand-driven approach as proposed in the Global Cassava Development Strategy

A Global Cassava Development Strategy (Plucknett *et al.*, 2000) was validated at an international forum in 2000. The strategy presents a vision that cassava will spur rural industrial development and raise incomes for producers, processors and traders and it will contribute to the food security status of its producing and consuming households. The essence of the strategy is to use a demand-driven (market-orientated) approach to promote and develop cassava-based industries with the assistance of a coalition of groups and individuals interested in developing the cassava industry.

The strategy consists of identifying, in a systematic manner, the opportunities and constraints of cassava at each level of the supply chain. This can be done by groups and individuals interested in developing the cassava industry – producers, processors and consumers of cassava – as well as associated national, international and non-governmental organizations (NGOs). Concepts of business development and management as well as international economic cooperation are important tools in implementing the strategy. Scientific support is also essential to help overcome important problems within the production–processing–marketing continuum. Adaptive research is also important to ensure that existing and evolving knowledge is harnessed in an appropriate and useful fashion.

The strategy suggests the utilization of 'industry analysis'. Industry analysis consists of identifying, in a systematic manner, the opportunities and constraints at each stage of the supply chain. Industry analysis involves stakeholders in a participatory effort to identify strengths, weaknesses and opportunities. Industry analysis is a demand-driven approach to technical change through:

1. Explicitly considering stakeholders as equal partners in determining the needs and future plans for a dynamic cassava industry.
2. Building a practical, shared vision for cassava development.

3. Helping make action plans for the industry, including the who, what, why, and how, plus the question, with whose money?
4. Building better linkages with private sector organizations.
5. Better links with and among public-sector institutions.
6. Co-stewardship of research and service outputs with users.
7. Rapid introduction of high-impact technologies through public and private sector partnerships.

The initiation of this strategy will require 'catalysts' capable of identifying marketing opportunities, and bringing these to the attention of stakeholders and 'champions', capable of providing support and resources for the growth and development of cassava markets. Even if the stakeholders agree that there is a growth market for cassava, there may still be need for research and development, provision of infrastructure and investments, and changes in policies to grasp the new opportunity.

A necessary first step in the development of a market-driven global cassava strategy is the identification of markets that are growing or could potentially grow. A second step is the provision of a consistent supply of a relatively uniform product. A third step, related to step two, is to provide the market with a competitively priced product that meets the consumers' requirements. A fourth step is to secure the cooperation of those associated with the market opportunity.

The development path for cassava will be product, location and time specific. Nevertheless, it would appear that if the market growth potential exists because of a structural change in the economy (e.g. decreasing number of farmers and increasing number of urban consumers of cassava products, resulting in market growth) it would be expected that NGOs and national governments would be in the best position to act as champions and catalysts. If, on the other hand, the market growth exists because cassava is price competitive then both national and international agencies may act as champions and catalysts.

The Global Strategy should be seen as comprising both 'bottom–up' and 'top–down' approaches. It is an amalgamation of national, regional and continental strategies and plans, augmented by global efforts to identify and stimulate markets. The national efforts will be the action-sites for implementing the global strategy by undertaking specific investment projects. The global effort assists through the promotion and diffusion of vital product, market and technological information (as a global public good). Moreover, at the global level, the validated strategy will be promoted to key players in both private and public sectors (industries, governments, finance and development agencies, etc.).

Conclusion

In this chapter the use of cassava as food, and aspects of storage and small-scale processing have been considered. The perishability of fresh cassava means that postharvest interventions are probably more important than for any other crop. Although there has been an emphasis on the technical side of storage and processing, it can be seen that if practical interventions in cassava postharvest systems are to have an impact, there is a need for a multi-disciplinary approach. Whatever the approach, it has to be demand/needs driven and be economically viable. In this way, cassava can be a spur to rural development.

References

Abe, M.O. and Lindsay, R. (1978) Evidence for the lactic streptococci role in acidic cassava (*Manihot esculenta* Crantz). *Journal of Food Science* 42, 781–784.

Balagopalan, C. (ed.) (2000) *Integrated Technologies for Value Addition and Post Harvest Management in Tropical Tuber Crops*. Central Tuber Crops Research Institute, Kerela, India.

Bancroft, R.D. and Crentsil, R.D. (1995) Application of a low-cost storage technique for fresh cassava (*Manihot esculenta*) roots in Ghana. In: Agbor Egbe, T., Brauman, A., Griffon, D. and Treche, S. (eds) *Transformation Alimentaire du Manioc*. ORSTOM Editions, Paris, France, pp. 547–555.

Baybay, D.S. (1981) Storage of some root crops and other perishable farm products. *Philippine Agriculture* 10, 423–440.

Beeching, J.R., Han, Y.H., Gomez-Vasquez, R., Day, R.C. and Cooper, R.M. (1998) Wound and defense responses in cassava as related to postharvest physiological deterioration. *Phytochemical Signals and Plant-Microbe Interactions, Recent Advances in Phytochemistry* 32, 231–249.

Best, R. (1978) Cassava processing for animal feed. In: Weber, E.J., Cock, J.H. and Chouinard, A. (eds) *Cassava Harvesting and Processing: Proceedings of a Workshop held at CIAT, Colombia, 24–28 April 1978.* IDRC, International Development Research Centre, Ottawa, pp. 12–20.

Bokanga, M. (1995) Cassava: opportunities for the food, feed and other industries in Africa. In: Agbor Egbe, T., Brauman, A., Griffon, D. and Treche, S. (eds) *Transformation Alimentaire du Manioc.* ORSTOM Editions, Paris, pp. 557–569.

Bokanga, M. (1998) Cassava in Africa: the root of development in the twenty-first century. *Tropical Agriculture (Trinidad)* 75, 89–92.

Bokanga, M., Essers, A.J.A., Poulter, N.H., Rosling, H. and Tewe, O. (1994) International Workshop on Cassava Safety, Ibadan, Nigeria, 1–4 March 1994. *Acta Horticulturae* 375, 11–119.

Booth, R.H. (1976) Storage of fresh cassava (*Manihot esculenta*). 1. Postharvest deterioration and its control. *Experimental Agriculture* 12, 103–111.

Booth, R.H. (1977) Storage of fresh cassava (*Manihot esculenta*). 2. Simple storage techniques. *Experimental Agriculture* 13, 119–128.

Booth, R.H. and Coursey, D.G. (1974) Storage of cassava roots and related postharvest problems. In: Araullo, E.V., Nestel, B. and Campbell, M. (eds) *Cassava Processing and Storage. Proceedings of an Interdisciplinary Workshop.* IDRC, Ottawa, pp. 43–49.

Bottalico, A., Lerario, P. and Frisullo, S. (1980) Presenza di aflatossine, di zearalenone e di ceppi di *Aspergilli* produttori di aflatossine in campioni di farina di mandioca. *Zootecnia e Nutrizione Animale* 6, 209–214.

Bradbury, J.H. and Holloway, W.D. (1988) Chemistry of Tropical Root Crops: Significance for Nutrition and Agriculture in the Pacific. ACIAR Monograph No 6, Australian Centre for International Agricultural Research, Canberra.

Brauman, A., Keleke, S., Mavoungou, O., Ampe, F. and Miambi, E. (1995) Etude cinetique du rouissage traditionnel des racines de manioc en Afrique centrale (Congo). In: Agbor Egbe, T., Brauman, A., Griffon, D. and Treche, S. (eds) *Transformation Alimentaire du Manioc.* ORSTOM Editions, Paris, France, pp. 287–305.

Brudzynski, A., Pee, W. Van and Kornaszewski, W. (1977) The occurrence of aflatoxin B1 in peanuts, corn and dried cassava sold at the local market in Kinshasa, Zaire; its coincidence with high hepatoma morbidity among the population. *Zeszyty Problemowe Nauk Rolniczych* 189, 113–115.

de Bruijn, G.H. (1973) The cyanogenic character of cassava (*Manihot esculenta*). In: Nestel, B.L. and MacIntyre, R. (eds) *Chronic Cassava Toxicity.* IDRC, Ottawa, pp. 43–48.

Buschmann, H., Rodriguez, M.X., Thome, J. and Beeching, J.R. (2000) Qualitative and quantitative changes of phenolic compounds of cassava (*Manihot esculenta* Crantz) roots during postharvest physiological deterioration. In: Carvalho, L.J.C.B., Thro, A.M. and Vilarinhos, A.D. (eds) *Cassava Biotechnology: IV International Scientific Meeting.* EMBRAPA/CBN, Brasilia, pp. 517–525.

Chambers, R. (1992) Rural appraisal: rapid, relaxed and participatory. *Institute of Development Studies, Discussion Paper No. 311.* The IDS, University of Sussex, Brighton.

Chambers, R. and Ghildyal, B.P. (1985) Agricultural research for resource-poor farmers: the farmer first and last model. *Agricultural Administration* 20, 1–30.

Clerk, G.C. and Caurie, M. (1968) Biochemical changes caused by some *Aspergillus* species in root tuber of cassava (*Manihot esculenta* Crantz) *Tropical Science* 10, 149–154.

Cock, J.H. (1985) *Cassava, New Potential for a Neglected Crop.* Westview Press, Boulder, Colorado.

Conn, E.E. (1994) Cyanogenesis – a personal perspective. *Acta Horticulturae* 375, 31–43.

Constant, J.-L., Kocheleff, P., Carteron, B., Perrin, J., Bedere, C. and Kabondo, P. (1984) Distribution geographique des aflatoxines dans l'alimentation humaine au Burundi. *Sciences des Aliments* 4, 305–315.

Cooke, R.D. (1978) An enzymatic assay for the total cyanide content of cassava (*Manihot esculenta* Crantz). *Journal of the Science of Food and Agriculture* 29, 345–352.

Cropley, J. and Gilling, J. (1993) Needs assessment for agricultural development: Practical issues in informal data collection. NRI Socio-economic series 1. NRI, Chatham.

DGIS (1991) *Cassava and Biotechnology. Proceedings of a Workshop,* Amsterdam, 21–23 March 1990. Directorate for International Co-operation, The Hague.

Ekpenyong, T.E. (1984) Composition of some tropical tuberous foods. *Food Chemistry* 15, 31–36.

Essers, A.J.A. (1995) Removal of cyanogens from cassava roots: studies on domestic sun-drying and solid substrate fermentation in rural Africa. PhD thesis, Wageningen Agricultural University, The Netherlands.

Essers, A.J.A. and Nout, M.J.R. (1989) The safety of dark moulded cassava flour compared with white – a comparison of traditionally dried cassava pieces in North East Mozambique. *Tropical Science* 29, 261–268.

Essers, A.J.A., Bosveld, M., van der Grift, R.M. and Voragen, A.G.J. (1993) Studies on the quantification of specific cyanogens in cassava products and introduction of a new chromogen. *Journal of the Science of Food and Agriculture* 63, 287–296.

Gallat, S., Crentsil, D. and Bancroft, R.D. (1998) Development of a low cost cassava fresh root storage technology for the Ghanaian market. In: Ferris, R.S.B. (ed.) *Postharvest Technology and Commodity Marketing. Proceedings of a Postharvest Conference*, 2 November–1 December 1995, Accra, Ghana. IITA, Ibadan, Nigeria, pp. 77–84.

Gomez, G. and Valdivieso, M. (1985) Cassava foilage: chemical composition, cyanide content and effect of drying on cyanide elimination. *Journal of the Science of Food and Agriculture* 36, 433–441.

Henry, G., Westby, A. and Collinson, C. (1998) *Global Cassava End-Uses and Markets: Current Situation and Recommendations for Further Study*. FAO, Rome. www.globalcassavastrategy.net/phase _i_market

Hodges, R.J. (1986) The biology and control of *Prostephanus truncatus* – a destructive pest with an increasing range. *Journal of Stored Products Research* 22, 1–14.

Jeon, Y.W. and Halos, L.S. (1991) Technical performance of a root crop chipping machine In: Ofori, F. and Hahn, S.K. (eds) *Tropical Root Crops in a Developing Country. Proceedings of the Ninth Symposium of the International Society for Tropical Root Crops* 20–26 October 1991, International Society for Tropical Root Crops, Accra, Ghana, pp. 94–100.

Jonsyn, F.E. (1989) Fungi associated with selected fermented foods in Sierra Leone. *MIRCEN Journal* 5, 457–462.

Kapinga, R., Westby, A., Rwiza, E., Bainbridge, Z. and Nsanzugwanko, A. (1998) Diversification of cassava utilisation in the Lake Zone of Tanzania: a case study. *Tropical Agriculture (Trinidad)* 75, 125–128.

Kleih, U., Digges, P. and Westby, A. (1997) *Assessment of the Needs and Opportunities in Postharvest Systems of Non-grain Starch Staple Food Crops*. NRI, Chatham.

Knoth, J. (1993) Traditional storage of yams and cassava and its improvement. *Deutsche Gesellschaft fur Technische Zusammenarbeit (GTZ) GmbH.*

Lancaster, P.A. and Coursey, D.G. (1984) Traditional postharvest technology of perishable tropical staples. *FAO Agricultural Services Bulletin* 59, FAO, Rome.

Lutaladio, N.B. and Ezumah, H.C. (1981) Cassava leaf harvesting in Zaire. In: Terry, E.R., Oduri, K.A. and Caveness, F. (eds) *Tropical Root Crops:*

Research Strategies for the 1980s. IDRC, Ottawa, Canada, pp. 134–136.

Masimango, N., Ramaut, J.J.-L. and Remacle, J. (1977) Aflatoxins and toxigenic moulds in foods consumed in Zaire. *Revue des Fermentations et des Industries Alimentaires* 32, 164–170.

McDowell, I. and Oduro, K.A. (1983) Investigation of the β-carotene in yellow varieties of cassava (*Manihot esculenta*). *Journal of Plant Foods* 5, 169–171.

McFarlane, J.A. (1982) Cassava storage part 2: storage of dried cassava products. *Tropical Science* 24, 205–236.

McMahon, J.M., White, W.L.B. and Sayre, R.T. (1995) Cyanogenesis in cassava (*Manihot esculenta* Crantz). *Journal of Experimental Botany* 46, 731–741.

Mkpong, O.E., Yan, H., Chism, G. and Sayre, R.T. (1990) Purification, characterization and localization of linamarase in cassava. *Plant Physiology* 93, 176–181.

Mota, T.P. and Lourenco, M.C. (1974) Farinha de mandioca de Mocambique. *Agronomia Mocambicana* 8, 47–59.

Nartey, F. (1978) *Cassava – Cyanogenesis, Ultrastructure and Seed Germination*. Munksgaard, Copenhagen.

(NRI) Natural Resources Institute (1992) *COSCA phase I processing component. COSCA Working Paper no. 7.* IITA, Ibadan, Nigeria.

Ndunguru, G.T., Modaha, F., Bancroft, R.D., Digges, P.D., Kleih, U., Westby, A. and Mashamba, F. (1998) The use of needs assessment methodologies to focus technical interventions in root and tuber crop systems: a case study to improve the marketing and postharvest handling of cassava entering Dar es Salaam, Tanzania. In: Akoroda, A.O. and Ekanayake, I.J. (eds) *Proceedings of the 6th Symposium of the International Society for Tropical Root Crops – Africa Branch.* International Society of Tropical Root and Tuber Crops–Africa Branch, Ibadan, Nigeria, pp. 76–82.

Ndunguru, G.T., Thomson, M., Waida, T.D.R., Rwiza, E., Jeremiah, S. and Westby, A. (1999) Relationship between quality and economic value of dried cassava products in Mwanza, Tanzania. In: Akoroda, M.O. and Terri, J. (eds) *Food Security and Crop Diversification in SADC Countries. The Role of Cassava and Sweetpotato.* SARRNET, Malawi, pp. 408–414.

Ngaba, P.R. and Lee, J.S. (1979) Fermentation of cassava (*Manihot esculenta* Crantz). *Journal of Food Science* 44, 1570–1571.

Nweke, F.I. (1988) COSCA project description. *COSCA Working Paper no. 1.* Collaborative Study of Cassava in Africa, IITA, Ibadan, Nigeria.

O'Brien, G.M., Taylor, A.J. and Poulter, N.H. (1991) Improved enzymic assay for cyanogens in fresh and processed cassava. *Journal of the Science of Food and Agriculture* 56, 277–289.

Onabolu, A., Abass, A. and Bokanga, M. (1998) *New Food Products from Cassava*. IITA, Ibadan, Nigeria.

Okafor, N. (1977) Microorganisms associated with cassava fermentation for 'gari' production. *Journal of Applied Microbiology* 42, 279–284.

Osuntakan, B.O. (1994) Chronic cyanide intoxication of dietary origin and a degenerative neuropathy in Nigerians. *Acta Horticulturae* 375, 311–321.

Ouraga-Djoussou, L.H. and Bokanga, M. (1998) Cassava and wheat consumption in Africa: New opportunities for cassava in the 21st century. In: Akoroda, A.O. and Ekanayake, I.J. (eds) *Proceedings of the 6th Symposium of the International Society for Tropical Root Crops – Africa Branch*. International Society of Tropical Root and Tuber Crops – Africa Branch, Ibadan, Nigeria, pp. 328–332.

Oyewole, O.B. and Odunfa, S.A. (1988) Microbiological studies on cassava fermentation for 'lafun' production. *Food Microbiology* 5, 125–133.

Parker, B.L., Booth, R.H. and Haines, C.P. (1981) Arthropods infesting stored cassava (*Manihot esculenta* Crantz) in Peninsular Malaysia. *Protection Ecology* 3, 141–156.

Plucknett, D.L., Philips, T.P. and Kagbo, R.B. (2000) A global development strategy for cassava: transforming a traditional tropical root crop. Paper presented at the FAO/IFAD Validation Forum for the Global Cassava Development Strategy, April 2000, Rome.

Poulter, N.H., Westby, A. and Gilling, J. (1992) Survey of village processing of cassava in Africa. *Tropical Agriculture Association Newsletter* September, 13–14.

Rickard, J.E. (1985) Physiological deterioration in cassava roots. *Journal of the Science of Food and Agriculture* 36, 167–176.

Rickard, J.E. and Coursey, D.G. (1981) Cassava storage, part 1: storage of fresh cassava roots. *Tropical Science* 23, 1–32.

Sajise, C.E. and Ilag, L.L. (1987) Incidence of aflatoxin contamination in cassava (*Manihot esculenta* Crantz). *Annals of Tropical Research* 9, 137–156.

Sanni, L.O., Akingbala, J.O., Oguntunde, A.O., Bainbridge, Z.A., Graffham, A.J. and Westby, A. (1998) Processing of *fufu* from cassava in Nigeria: problems and prospects for development. *Science, Technology and Development* 16, 58–71.

Scott, G.J., Rosegrant, M.W. and Ringler, M.W. (2000) Roots and tubers for the 21st century: trends, projections and policy options. *Food, Agriculture and the Environment Discussion Paper 31*. International Food Policy Research Institute.

Shank, R.C., Wogan, G.N. and Gibson, J.B. (1972) Dietary aflatoxin and human liver cancer. I.

Toxigenic moulds in foods and foodstuffs of tropical South-East Asia. *Food and Cosmetics Toxicology* 10, 51–60.

Soares, L.M.V. and Rodriguez-Amaya, D.B. (1989) Survey of aflatoxins, ochratoxin A, zearalenone, and sterigmatocystin in some Brazilian foods by using multi-toxin thin-layer chromatographic method. *Journal of the Association of Official Analytical Chemists, USA* 72, 22–26.

Treche, S. (1995) Importance du manioc en alimentation humaine dans differentes regions du monde. In: Agbor Egbe, T., Brauman, A., Griffon, D. and Treche, S. (eds) *Transformation Alimentaire du Manioc*. ORSTOM Editions, Paris, pp. 25–35.

Tshibaka, T. and Lumpungu, K. (1989) *Trends and Prospects for Cassava in Zaire*. Working Paper no. 4 on Cassava, International Food Policy Research Institute, Washington.

Tylleskar, T. (1994) The association between cassava and the paralytic disease konzo. *Acta Horticulturae* 375, 321–331.

Tylleskar, T., Banea, M., Bikangi, N., Cooke, R.D., Poulter, N.H. and Rosling, H. (1992) Cassava cyanogens and konzo, an upper motor neuron disease in Africa. *The Lancet* 339, 208–211.

van Oirschot, Q., White, J. and Westby, A. (2001) Case study on the technical and economic aspects of small-scale cassava processing in a selected village in the Lake Zone of Tanzania. Technical report. FAO, Rome (in press).

Vasconcelos, A.T., Twiddy, D.R., Westby, A. and Reilly, P.J.A. (1990) Detoxification of cassava during gari preparation. *International Journal of Food Science and Technology* 25, 198–203.

Wareing, P.W., Westby, A., Gibbs, J.A., Allotey, L.T. and Halm, M. (2001) Consumer preferences and fungal mycotoxin contamination of dried cassava products. *International Journal of Food Science and Technology* 36, 1–10.

Wenham, J.E. (1995) Postharvest deterioration of cassava, a biotechnology perspective. *FAO Plant Production and Protection Paper 130*. FAO, Rome.

Westby, A. (1993) Collaborative Study of Cassava in Africa. Phase I – Village-level survey. Presentation of data on processed products. Unpublished report. NRI, Chatham.

Westby, A. and Cereda, M.P. (1994) Production of sour cassava starch (polvilho azedo) in Brazil. *Tropical Science* 34, 203–210.

Westby, A. and Choo, B.K. (1994) Cyanogen reduction during the lactic fermentation of cassava. *Acta Horticulturae* 375, 209–215.

Westby, A. and Gallat, S. (1999) Cassava chip processing in Ghana: participatory postharvest research and technology transfer in response to new market opportunities. In: Kwarteng, J. (ed.) *Enhancing*

*Postharvest Technology Generation and Dissemina-
tion in Africa.* Sasakawa Africa Association, Mexico City, pp. 61–65.

Westby, A., Kleih, U., Hall, A., Bockett, G., Crentsil, D.,
Ndunguru, G., Graffham, A., Gogoe, S., Hector,
D., Nahdy, S. and Gallat, S. (1998) Improving the
impact of postharvest research and development
on root and tuber crops: the needs assessment
approach. *Tropical Agriculture (Trinidad)* 75,
143–146.

Westby, A., Kleih, U., Hall, A., Ndunguru, G., Crentsil,
D., Bockett, G. and Graffham, A. (1999) Needs
assessment in postharvest research and development. In: Grant, I.F. and Sear, C.B. (eds) *Decision*

Tools in Sustainable Development, 6. NRI, Chatham,
pp. 143–160.

Wheatley, C.C. (1984) Aflatoxins in cassava . . . is it
a real problem? *Cassava Newsletter* 8, 2–14.

Wheatley, C.C. (1989) *Conservation of Cassava in
Polyethylene Bags. CIAT Study Guide 04SC-07.06.*
CIAT, Cali, Colombia.

White, W.L.B., Arias-Garzon, D.I., McMohon, J.M. and
Sayre, R.T. (1998) Cyanogenesis in cassava: the
role of hydroxynitrile lyase in root cyanogen
production. *Plant Physiology* 116, 1219 -1225.

Wright, M. (1993) Contributions on the ecology and
control of *Prostephanus truncatus* (Horn) in dried
cassava. PhD thesis, University of Reading.

Chapter 15

Cassava Utilization in Food, Feed and Industry

C. Balagopalan

Crop Utilization and Biotechnology, Central Tuber Crops Research Institute, Sreekariyam, Trivandrum 695 017, Kerala, India

Introduction

Cassava contributes significantly to the nutrition and livelihood of up to 500 million people and thousands of processors and traders around the world. Besides serving as the primary staple food of millions of people in the tropics and subtropics, it can also be used as a carbohydrate source in animal feed. Cassava is used as a raw material in the manufacture of processed food, animal feed and industrial products. Wider utilization of cassava products can be a catalyst for rural industrial development and raise the incomes for producers, processors and traders. It can also contribute to the food security status of its producing and consuming households (Plucknett *et al.*, 1998).

Cassava in Foods

Fresh cassava

In Africa and some other countries where the crop is grown, it is a common practice to eat cassava raw, after removing the skin and rind. 'Sweet' cassava cultivars are grown for this purpose, close to the homestead to discourage theft. Slices of fresh tubers in oil form a common snack in many countries. Cassava cultivars with a high content of cyanogens must be cooked before consumption. The skin and rind are removed from fresh cassava tubers and cut into slices before boiling in water for 10–40 min, depending on the cultivar. The water is decanted after cooking and the boiled tubers are consumed with a suitable dish. Draining the water helps to remove cyanogenic compounds present in the fresh roots. Although boiling destroys the enzyme linamarase and eliminates the hydrocyanic acid, prolonged eating of high-cyanide cassava, because of the presence of linamarin B, can cause chronic cyanide toxicity when cassava is consumed without sufficient protein. Fish is therefore considered as an ideal accompaniment to cassava and this combination is common in coastal and other areas with access to fish.

Fresh cassava may also be cooked by baking. After baking, the charred skin is removed and the cooked flesh eaten. In Brazil, a sweet food is prepared by cooking peeled roots in sugar syrups. It is also a practice in Brazil to make a soup called *sacncocho* or *cocido* by boiling cassava tubers with other vegetables (Balagopalan *et al.*, 1988).

Culinary uses of cassava around the world

In South America there are many methods adopted for cassava utilization as food (Lancaster *et al.*, 1982). Cassava tubers are grated on palm roots or spiny palm trunks and pounded into a pulp. The pulp is squeezed by hand and cooked in a variety of ways. The pulp is shaped into pies or cakes and then baked in fire after wrapping in a protective covering of leaves. Balls of the pulp are sun dried, wrapped in leaves and placed in baskets or buried in the ground. The cassava pulp is boiled either by dropping pies or dumplings made from the pulp into boiling water or by stirring the pulp into water to form a porridge.

Fufu

Peeled roots after cooking by steaming or boiling are pounded and the sticky dough eaten with soups made out of fish or vegetables. This is common in West Africa, particularly in Ghana.

Mingao

This drink is prepared in the Amazon region by dissolving fermented starch in boiling water and simmering for a period. In order to mask the unpleasant taste it is flavoured with palm fruits, pineapple or bananas.

Manicuera

This is a boiled slightly sweet cassava drink available in the northwest Amazon region.

Dumby

This Liberian food is prepared by placing boiled cassava tubers in a wooden mortar and beating with a heavy pestle. In order to prevent sticking, the pestle is dipped into water. In about 45 min pounding is complete. The *dumby* is cut into pieces and put in meat or vegetable soup and then swallowed whole.

Farina

In South America and West Indies *farina* is a common food. Fresh tubers are pressed in a wooden screw press, forcing the pulp through a sieve and finally, roasting it on a slow fire. It is preserved for several months and consumed as a cereal in combination with several other foods. In the Amazon region of Brazil, yellow bitter varieties of cassava are soaked in water for 2 or 3 days, then peeled and grated. The resulting

mash is mixed with fresh roots and allowed to ferment for several days before toasting.

Cassareep/tucupay

The juice pressed out from tubers during the preparation of *farina* is concentrated and spices added to make the sauce known as *cassareep* that is prepared in the West Indies and *tucupay* in Brazil.

Ampesi

The boiled roots may be eaten alone, mashed and eaten with sauce. In Brazil, *ampesi* is also cooked with vegetables and meat.

Landang or cassava rice

Fresh cassava roots are squeezed and after expelling the juice the pulp is made into pellets which are called *landang* in the Philippines, or cassava rice.

Macaroni

This is prepared by blending cassava flour and groundnut flour with wheat semolina in the ratio 60 : 12 : 15. It contains about 12% protein. Enriched macaroni containing 12–18% proteins and fortified with vitamins and minerals has also been developed for feeding children and vulnerable groups.

Cassava pudding

Cassava roots are grated and mixed with coconut or banana as a pudding.

Tiwul

This is an Indonesian preparation made from *gaplek* (dried cassava) after pulverizing and sieving. The meal is kneaded along with a little water into paste, mixed with sugar and steamed. This gritty material is served as a substitute for rice (Setyona *et al.*, 1991).

Oyek

Cassava roots are peeled and soaked in water for about 1 week and then drained and ground. The ground cassava is kneaded with a little water, steamed and sun dried to prepare this Indonesian dish (Setyona *et al.*, 1991).

Gatot

In Indonesia, *gaplek* (dried cassava) is cut into pieces, steamed and spread out on a bamboo mat. The pieces are kept wet for 2–3 days by continuously sprinkling water over them. The pieces turn black in colour and they acquire a characteristic taste. They can be served after steaming (Setyona *et al.*, 1991).

Cassava rava *and pre-gelatinized cassava starch (yuca rava and yuca porridge)*

Rava is a wheat-based convenience food used for the preparation of various breakfast recipes like *uppuma* and *kesari*. Conventionally, wheat semolina is used for this purpose. The method to be followed for cassava *rava* requires controlled gelatinization of starch. The conditions for controlled gelatinization and swelling of starch for the preparation of *rava* were developed for cottage and small-scale industrial programmes. The process for producing cassava *rava* consists of partial gelatinization of chips of cassava tubers, drying and powdering. The fine-grade pre-gelatinized cassava starch can be used to make an instant energy drink (yuca porridge) using hot milk or hot water. Two teaspoons of pre-gelatinized starch may be added to hot milk or water, after adding sugar to taste and served to infants and invalids as an energy drink. Addition of cardamom powder to yuca porridge adds flavour to the product (Padmaja *et al.*, 1996).

Pappad

Cassava *pappad* is an important snack food item prepared from the flour. The preparation involves gelatinization of the flour with a minimum quantity of water, spreading out the paste on a mat or some similar surface and drying in the sun. After drying, it is packed in polythene covers. The *pappad* is cooked by deep frying in oil, usually coconut oil. The final product undergoes two to three times expansion on frying. It is crisp and can be consumed as a side dish. The *pappad* loses its crispness if stored in the presence of moisture and the fried product should be stored in closed containers.

Sago wafers

Sago wafer is an important product produced as a cottage industry. The wafers are deep fried in oil and consumed as an accompaniment. Preparation involves packing the sago in round aluminium trays and the trays are then steamed for 20 min. The gelatinization induced by steaming makes the sago pearls adhere to each other to form a round-shaped dough. The trays are then sun dried and the resulting wafers are peeled out. Natural colours and salt are added to taste.

Wafers

Wafers are made from cassava starch similar to sago wafer. Here the starch cake containing approximately 40% moisture is used instead of sago. They can be made in different shapes and sizes. The product on frying undergoes three- to fourfold increase in size.

Fried chips

Fried chips are made by deep fat frying thin finger chips of cassava. The tubers are washed thoroughly and the peel and rind are removed. The tubers are then sliced as thinly as possible. The quality of the chips depends greatly on the thickness of the slices and the age of the crop. Tubers of the correct maturity with relatively low dry matter may be used. In addition, the tubers may be subjected to some blanching. The slices may be dipped in salt, or sodium bisulphite solution for 5–10 min, and then taken out, washed with water, and surface dried on a filter paper or cloth. The chips are fried in oil (preferably coconut oil which has been heated to nearly boiling temperature and to which salt solution had been sprinkled). The frying may take 5–10 min. The fried chips are removed from the oil, and the oil allowed to drip off before packing in polythene covers. The covers are sealed to prevent the entry of moisture and air.

Extrusion cooking of cassava flour

Extrusion processing has become an increasingly popular procedure in food industries, for the development of many successful products, including snacks and baby foods. Food extrusion is the process in which a low-water powder is pressed and heated simultaneously in a shear field, converted into a plastic mass, forced through a shaping die and rapidly hardened by cooling.

The cassava flour is cleaned and sieved through ISS mesh no. 70 (0.710 mm) to remove clots and other foreign materials. The moisture content of the flour is determined and the desired moisture content is obtained by adding a calculated amount of distilled water. The most acceptable cassava flour-extruded snack food product can be obtained at a moisture content of 16% (dry-weight basis), extrusion temperature of 110°C and a screw speed of 100 r.p.m. Cassava flour after mixing with salt, spices, fish and meat powder is extruded and a variety of foods are made in this way.

Fermented foods and drinks

Cassava forms a substrate for a wide variety of fermented foods and drinks in Africa, Asia and Latin America. As with Ugandan cassava beer, fermented drinks such as *beiju*, *banu* or *ula* and *kasili* are made after fermenting grated cassava and are common in the tribal belts of South America (Lancaster *et al.*, 1982).

Sour flour and cassava bread

In northwestern America, a traditional preparation of bread called *cazabe* is prepared by the Indians. The freshly harvested roots are washed and peeled. The tubers are then pulped normally by grating, but sometimes by crushing in a mortar or between stones. Pressing separates the roots into liquids, starch and fibre. This is followed by grating and straining. The resulting pulp is placed in a large basket, rinsed with water, squeezed, kneaded and pressed against the strainer to squeeze out the liquid. The starch from the extracted liquid is allowed to stand in a clay pot and is allowed to settle. The starch and fibre is allowed to undergo slight fermentation and the fermented mash is used for making bread. Cassava sour starch is a product of traditional rural industry in Latin America. Breads such as *pandebono* and *pan de yuca* in Colombia and *pao de queijo* in Brazil are made from sour starch. The traditional method consists of wet-process extraction of starch from cassava roots. The starch is then stored in 0.5–5 t capacity tanks and fermented for 20–60 days, according to climatic conditions (temperatures may range from 15 to 25°C). Lactic fermentation takes place and the starch pH drops to 3.5–4.0. It is then sun dried on drying tables or on black plastic sheets placed on the ground. Both fermentation and sun drying give cassava starch its bread-making potential. Fermentation also causes substantial modifications to the organoleptic and physico-chemical characteristics of the starch (Dufour *et al.*, 1996).

Sour starch is the main ingredient in traditional high-swelling breads. Before making the bread the starch is mixed with fats or cheese, eggs and salt. Such breads do not contain wheat flour. They do not undergo yeast fermentation before baking. The dough is baked immediately after kneading with no 'rising' or ' proofing' time. Rising does not involve a protein–gluten network nor the production of carbon dioxide by yeasts.

Gari

In Ghana, Nigeria, Guinea, Benin and Togo, *gari* is one of the most important foods. Pulp made from cassava is placed in cloth bags or sacks made from jute and allowed to ferment for 3–10 days. *Corynebacterium manihot* and *Geotrichum candidum* are the two organisms which help in the fermentation of cassava in two stages. After the fermentation period, the partially dried cassava pulp is taken out and sieved to remove the fibrous material. It is then heated in shallow iron pans and stirred continuously until it becomes light and crisp. Palm oil can be added to prevent burning. Good quality *gari* is usually creamy yellow in colour with uniformly sized grains and should swell to three times its volume when placed in water (Balagopalan *et al.*, 1988).

Polvilhoazedo

Sour cassava starch, known as *polvilhoazedo*, is a typical Brazilian product obtained by fermentation of raw cassava starch for a period of about 30 days. The fermented starch, which has a strong and characteristic flavour, has many applications in local cookery and in the manufacture of biscuits and cheese breads.

Meduame-M-bong

This is a preparation of Cameroon. Cassava tubers are peeled, washed and cut into large pieces and boiled for 30 min to 1 h. After discarding the water, the roots are cut into small pieces and soaked in running water for 12–36 h. The final product Meduame-M-bong is then eaten with meat, fish, groundnuts, green leaves, etc.

Attieke

Roots are peeled, placed in water and ground into a paste and the paste is left to ferment for 2 days in jute sacks and then pressed. Finally the paste is removed from the sacks, crumbled by hand and steamed. The final product is eaten with milk, meat or vegetables. This is a common food in Cameroon.

Chick-wangue

After removing the rind and skin, cassava is soaked in water, pounded and made into a paste. The paste is sun dried or smoke dried. The wet paste is made into balls and packed in leaves on a screen over the hearth and left for 15 days. The leaves are then removed and the black coating formed is scraped off. The dried paste is ground and sieved to produce flour. This food is found in several West African countries.

Kapok pogari

This mid-western Nigerian food is similar to *gari* preparation. The only difference is that the grated and fermented mass is not sieved before roasting. The resultant product has bigger particles. *Kapok pogari* is consumed with fish, coconut or meat.

Peujeum

This is a traditional food of Java. The peeled roots are steamed until tender and allowed to cool and then dusted with finely powdered *ragi* (a mixture of flour and spices in which fungi and yeasts are active). The cassava mash mixed with *ragi* is wrapped in banana leaves in an earthenware pot and left for 1–2 days to ferment aerobically. This product has a refreshing acidic and alcoholic flavour. It is eaten as it is or it may be baked.

Lafun

This fermented food is available in Nigeria. Unpeeled roots are soaked in pots of water for 4 days, after which the peels are removed and the roots are sun dried for 4 days.

Wayana cassava cakes

The Wayana Indians of the Amazon, process high cyanide cassava by peeling and grating the root into mush, then squeezing the mush in a tubular wicker press hung from overhead beams. Once the poisonous juice has been extracted, the mush is turned into flour known as cassava and served as pancakes.

Tape/*flour* tape

Cassava roots are cut into pieces 5–10 cm in length, washed until quite clean and then half cooked. The cooked cassava is then fermented by inoculating with *ragi tape* or yeast (*Clamidomucer oryzae* and *Rhizopus oryzae*) and covered with banana leaves for 2–3 days. During fermentation the starch is converted into simple sugars by an enzyme produced by *Rhizopus oryzae*. *Tape* contains 0.5% protein, 0.1% fat, 42.5% carbohydrates and 56% moisture. This is an Indonesian preparation (Setyona *et al.*, 1991).

Cassava in Animal Feed

The potential of cassava in animal feed has been investigated extensively by researchers worldwide. Various parts of the cassava plant including tubers, stems and leaves are used for animal feeding. The importance of cassava in tropical livestock nutrition arises because of the deficiency of dietary energy in the form of soluble carbohydrates. This deficiency is more acute in the tropics where forage crops are more fibrous, coarse, bulky and less palatable than in the temperate zones. One of the specific features of cassava root products is their lower amylose content compared with other starchy carriers. The high-energy value of cassava makes it a very attractive carbohydrate ingredient in animal diet (Omole, 1977). The low protein content of cassava tubers (0.7–1.3% fresh weight) is a disadvantage, restricting the use of cassava as animal feed, but this can be overcome by upgrading the feed with protein additives, such as soya, or, by using microbial techniques, although the latter is probably uneconomic.

The aerial part of the plant comprising of stems, branches and leaves has been shown to have a protein content as high as 17%. Foliage can be cut from the plant at 4 months after sprouting and then every 60–75 days, to give up to 4 t ha^{-1} year^{-1} of crude protein. By using cassava products for 70% of livestock rations, it is estimated that 100 kg of poultry meat, 5 kg of pork and 200 eggs per person could be supplied to 2.3 billion people in the tropics, using only a fraction of the area that would be required for an equivalent production of animal products based on cereals and oil seed crops (Montaldo, 1977).

Fresh cassava tubers are very often fed to cattle, either raw or in the boiled form. Feeding of fresh tubers may cause cyanide toxicity, depending on the cyanogen concentration in the tubers. It has been observed that up to 10 kg day^{-1} of fresh cassava tubers can be fed to dairy animals and replacement of cereals with cassava at a 50–100% level did not affect the milk quantity or quality (Mathur et al., 1969). Higher milk yield by 19.5% has been also recorded as a result of the increased energy from cassava (11.9–14.6 MJ kg^{-1}). The performance of growing calves, goats, sheep, steer and poultry have been improved after incorporation of cassava in the diet (Castillo et al., 1964; Yoshida et al., 1966; Chou and Muller, 1972; Khajarren and Khajarren, 1977; Gomez, 1991).

Cassava leaves

Cassava leaves are used for animal feed in some parts of the world (Ravindran and Blair, 1991). In Brazil, leaves are considered valuable as forage, especially in the dry season when other feeds are scarce. There is some resistance to the use of cassava leaves to feed ruminants, due to the relatively low leaf yields obtained during harvesting at maturity, the possibility of hydrocyanic acid poisoning and inadequate appreciation of the relatively high crude protein content of the leaves.

Silage preparation from cassava

Poor shelf-life of fresh cassava roots and the bulky nature of the dried product demand improved processing, enabling their storage round the year for animal feeding. Silage preparation emerges as one of the best techniques for preserving the nutritive value of cassava, enhancing shelf-life and increasing the palatability through lactic acid enrichment. One of the main problems encountered during ensiling of cassava is the release of large quantities of silage effluent which leads to loss of essential nutrients and also results in a poor quality watery silage with very low shelf-life. Another problem which demands modification of the silage-making process is the residual cyanogens in cassava silage. Padmaja et al. (1994) observed that rice straw served as good absorbent of silage effluent when mixed with cassava at 10%. Ensiling leads to considerable decrease in pH due to lactic acid enrichment within 2 days, although this rapid decrease in pH helped to stabilize the process and produced good quality silage. In order to reduce the level of cyanogens, exposure of chopped cassava roots to sunlight is recommended.

In South America, silage preparation is done in silage pits near the animal sheds. Silage made of grass and cassava is stored for off-season use. Whole-plant cassava silage is made in closely tied polythene bags after chopping the plants into smaller pieces mechanically.

Cassava chips/pellet industry for animal feeding

Cassava is processed in different ways for the export market. Thailand is the largest exporter of cassava to Europe in the form of dried chips. Cassava chips are produced simply by slicing fresh cassava roots into small pieces using a chipping machine. The fresh chips are dried on a large concrete floor for about 2–3 days, depending on the intensity of solar radiation, until the moisture content is reduced to 14%. Most of the cassava chips are marketed directly to factories for the manufacture of feed pellets. The standard specifications for export of cassava chips are as follows: starch 65% minimum, raw fibre 5% maximum, sand 3% minimum and moisture 14% maximum. In Indonesia, dried cassava or *gaplek* is mainly exported as animal feed.

Cassava pellets

Cassava pellets are produced from dried cassava chips by machine. The small dried chips are preheated with steam, then passed through a die having several hundred 7–8-mm diameter holes. At this stage, the pellets are soft and warm and are air-cooled to harden them. The traditional pellets, called 'native pellets' are no longer produced, because of the dust pollution on handling these pellets at the port of destination. Dust-free hard pellets are now produced with improved machinery. The standard specifications for hard pellets are starch 65% minimum, raw fibre 5% maximum, sand 3% maximum, moisture 14% maximum, hardness 1.92 kg cm^{-2} force minimum (by Kahl hardness tester), meal 8% maximum (1 mm sieve) and foreign matter nil. The demand for cassava pellets is rapidly increasing in the European market as a result of the high animal:land ratio in European farming. The performance of animals fed on cassava pellets is good, there is less pollution, and the pellets are less bulky than some other feeds, so that transportation costs are lower (Phillips, 1974).

Improving the nutritional value of cassava products using microbial techniques

The successful experiments of Brook *et al.* (1969), Stanton and Wallbridge (1969) and Gray and Abou-El-Seound (1966) led to the development of new fermentation techniques for the protein enrichment of cassava products.

Gray and Abou-el-Seound (1966), Strasser *et al.* (1970), Gregory *et al.* (1977) and Mikami *et al.* (1982) showed that protein-enriched cassava could be produced for animal feeding by submerged fermentation, using the organisms *Cladosporium eladosporoids*, *Candida utilis* and *Cephalosporium eichhorniae*. Muindi and Hanssen (1981) described a fermentation procedure to increase the protein content of cassava root meal by growing *Trichoderma harzianum* in 4% cassava root meal (CRM) medium. The estimated efficiency of conversion of CRM into CRM/biomass was shown to be 30%.

Several organisms and fermentation methods have been investigated to increase the protein content of cassava and cassava residues using solid-state fermentation (Varghese *et al.*, 1976; Raimbault *et al.*, 1985; Daubresse *et al.*, 1987). A solid-state fermentation process for the protein enrichment of cassava flour and cassava starch factory wastes using the fungus *Trichoderma pseudokoningii* Rifai was developed by Balagopalan and Padmaja (1988) and

Balagopalan (1996). The highest increase in protein content was observed, i.e. 14.32 g 100 g^{-1} dry matter from an initial 1.28 g 100 g^{-1} dry matter, where cassava flour was the sole ingredient. Feeding experiments on poultry showed the potential for protein-enriched cassava feed using microbial techniques (Balagopalan *et al.*, 1991).

Industrial Processing of Cassava

Cottage-scale starch extraction

Cassava roots are washed by hand and peeled with hand knives. The roots are then manually rasped to a pulp on a stationary grater which is simply a tin or mild steel plate perforated by nails, so as to leave projecting burrs on one side. The pulp is collected on a piece of fabric fastened on four poles and washed vigorously with water by hand. Finally, the fibre is squeezed out while the starch milk collects in a bucket. When the starch granules settle out, the supernatant water is decanted and the moist starch is crumbed and dried in a tray or on a bamboo mat. In some places, the starch milk is squeezed through a closely woven thick fabric to trap the starch granules, or, hung overnight to remove water by gravity, followed by sun drying. This simple process is used in many rural areas of the tropics.

Microbial techniques to extract starchy flour with modified functional attributes from cassava

In order to facilitate the enzymatic cleavage of the cell walls of cassava for starch separation, a simple low-cost technology using mixed inoculum of microorganisms was developed. Through selection and repeated enrichment, a mixed culture inoculum has been developed to soften cassava tubers, facilitating easy extraction of starchy flour (Mathew *et al.*, 1991).

Small-scale industrial cassava starch production

Most of the industrial starch production from cassava is in the small-scale sector. Before processing to produce starch, the tubers are washed and peeled manually to remove the outer skin and rind. The peeled roots are washed and then rasped. Effective disintegration of tubers is obtained by the rasper. The rasp is a wooden drum with a steel shaft and cast-iron ends. A sheet of metal, perforated with nails, is clamped around the drum with the protrusions facing outwards. The drum rotates in a housing with a hopper at the top for feeding the peeled and washed tubers and with a perforated metallic plate underneath, through which the pulp has to pass into a sump below. Water is applied in small quantities continuously to the rasper. The entire rasping and activating of shaking screens is carried out with the help of an electric motor. Against the sharp protrusions of the rasper surface, the cell walls are torn up and the whole of the root flesh is turned into a fine pulp in which most but not all of the starch granules are released. After rasping, pulp from the sump is pumped on to a series of flat, slightly inclined, vibrating screens of diminishing mesh size. The screens used are usually three in number, of 80, 150 and 260 mesh, respectively, with the first retaining the coarse fibre and the other fine particles. Usually, a small spray of water is applied to assist the separation of starch granules from their fibrous matrix and to keep the screen meshes clean. Starch granules carried with the water fall to the bottom of the tank in which the sieves are placed. The starch milk is then channelled for gravitational sedimentation. Sometimes a

final washing is carried out manually over a 300-mesh screen. Nylon, phosphor bronze or stainless steel wire meshes are used for screens.

In separating the pulp from the free starch, large quantities of water must be added to the pulp and then stirred vigorously. Residual pulp from the screening is considered a by-product of the cassava starch industry. The oldest practice for settling starch from its suspension in water is to let the starch milk stand for a period of 8 h in tanks with plugged effluent outlets at varying heights. The starch settles at the bottom and the supernatant liquor is run off. During the process of dehydration, a number of tanks are filled in succession (Fig. 15.1).

After the removal of free water from the starch by sedimentation, a cake is obtained containing 35–40% moisture. Usually, the starch cake is crumbled into small lumps (1–3 cm) and spread out in thin layers on large open areas for sun drying; drying generally takes 24–120 h. The crude starch is considered sufficiently dry when the lumps are too hard to be crumbled by hand and the moisture content is between 15 and 20% (Radley, 1976; Grace, 1977).

Extraction of starch from dried cassava

To extract starch, the dried roots are washed and grated and the starch is separated by cylindrical sieves. However, this practice is expensive and the starch produced is of inferior quality.

Fig. 15.1.   Settling of starch in a small factory.

The brown skin, which contains chlorophyll and coagulated proteinaceous substances, adheres strongly to the ligneous tissues, imparting a dark colour to the starch of the dried roots. This problem can be solved by using cassava chips prepared from tubers that were peeled prior to drying.

Large-scale cassava starch production

In large factories, cassava tubers are immediately peeled and washed by mechanical scrubbing. The washing machine is a perforated drum partially immersed in a water bath. The roots are propelled forward by a series of paddle arms, or a spiral brush attached to a central rotating shaft. A counter-current flow of water through the bath ensures continuous removal of dirt. In some designs, high-pressure water sprays from nozzles may also act on the roots. The combined action of the high-pressure water jets and abrasion of the tubers, against the drum walls and against each other, remove most of the skin.

The Jahn-type rasper, used in the modern process, consists of a rotating drum of 40–50 cm length, with longitudinally arranged saw-tooth blades in grooves milled around the circumference. Blades have between eight and ten saw teeth per cm and are spaced 6–10 mm apart projecting about 1 mm above the surface. The optimum speed is 100 r.p.m., corresponding to a linear velocity of about 25 m s^{-1}. In many mills, the coarse pulp retained on the first shaking screen is re-ground in a secondary rasper with finer blades, having a greater number of teeth per unit length of blade (10–12 cm^{-1}), and then returned for re-screening (Radley, 1976).

Shaking screens can be used on a larger scale in a series of increasing fineness, such as 80-, 150- and 260-mesh phosphor bronze, aided by gentle washing in water sprays. However, the modern practice is to use sieve bends or dutch static mill screens, working in three to six stages in a series. The slurried pulp is sprayed at a right angle to the wedges. Flowing down across the screen, the smaller starch granules pass through the slots and larger fibrous material is separated continuously. A counter-current system of overflow performs the extraction and sieving, requiring no fresh water for washing. Rotating screens, usually horizontal sieve cones, by the action of centrifugal force make the retained fibre slide over the screen and fall out. Rotating screens are also used sometimes in large-scale operations and may operate in batches or continuously.

After being separated from fibres, the starch needs to be dehydrated. Mechanical dehydration is commonly done either on vacuum filters or centrifuges. A vacuum within the cylinder sucks the water while starch adheres to the cloth screen, which is scraped out continuously. Sedimentation is achieved using various types of disk centrifuge and peeler centrifuge.

In the case of over-dilution during screening, the starch suspension may be pre-concentrated by gravitational sedimentation or by hydrocyclones, thus reducing the volume and size of final equipment required. Tray dryers, rotary dryers and belt-and-tunnel dryers have been used followed by grinding of the dried product to speed up the drying operation. However, for large-scale operations, flash or pneumatic dryers are used. Finally, the damp starch is transported in a vertical stream of hot air at a temperature of about 150°C, to a cyclone filter, where dried starch granules are separated from the air. Residence time is a few seconds and the starch from a flash dryer is a fine powder, with a final moisture content of 10–13%.

Cassava Products Based on Starch

Starch-based adhesives

Adhesives are made from cassava starch using simple, low-cost technologies. These include gums made by gelatinizing starch by heat treatment without any additives and those made by adding different materials.

Gums without additives

The simplest liquid starch pastes are made by cooking starch with water and preservatives are added later. These are useful in bill pasting, bag making and in tobacco products. The starch is cooked in stainless steel or wooden vats with excess water until all the starch has gelatinized. The consistency of the paste is measured by its appearance and flowability; it should flow freely and come out as a long, continuous stream. On cooling, it becomes more viscous. Copper sulphate is added to impart resistance to micro-bial damage. Cassava starch is preferred for paste manufacture in view of its excellent cohesiveness and clarity, and its bland flavour allows it to be used in food packaging.

Gums prepared using different chemicals

Various chemicals are added during the prepa-ration of gums. These include inorganic salts like calcium and magnesium chlorides, borax, urea, glycerol and carboxymethyl cellulose. The chemicals act in a number of ways by increasing the viscosity, increasing flowability and humid-ity control and they are added with stirring, while the starch is being gelatinized, to prevent lump formation. The gums are useful in various applications such as lamination of papers, wall-paper printing, water-resistant formulations, pasting labels and other stationery applications.

Dextrin

The steps involved in dextrin production depend on the type of product desired. However gener-ally they can be catagorized as: (i) pre-drying; (ii) acidification; and (iii) conversion.

PROCESS FOR THE PRODUCTION OF DEXTRIN.

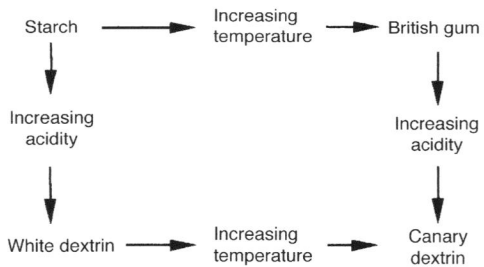

An aqueous solution of dextrin can be used to form films capable of bonding similar or dissimi-lar surfaces. Although these films are not as strong as starch films, they have wide-ranging applications. They can be used at a higher con-centration than starch and hence dry faster and provide a better bond. The adhesive industry is the major consumer of dextrins; they are used as gums for envelopes, as bottle-labelling adhesives, as adhesives in re-moistening gummed tapes, postage stamps, lined cardboard boxes and pho-tographic mounting materials.

Sago

Originally sago was derived from the palm, *Metroxylon* sp. found in Malaysia and Thailand. However, sago is now manufactured using cassava starch. The initial steps are similar to starch production, i.e. peeling and washing, disintegration, and settling. The starch from the settling tank is spread over cemented yards and partially dried in the sun (Fig. 15.2). The partially dried material (40–45% moisture) is made into small granules by shaking in power-driven shakers or granulators which consist of wooden trays with cloth flooring. The oscillatory movement enables the starch gran-ules to adhere together and form into globules. The next step is partial gelatinization. This is carried out by roasting the granules to gelatinize the surface. Wet globules are placed on shallow metal pans made of aluminium or iron which are smeared with small quantities of oil, and heated by fire. The granules are stirred continu-ously throughout the operation for around 15 min. The granules are then dried in the sun or in a hot air oven (40–50°C). The dried mass is passed through a polisher to separate the large clumps into small granules. They are then graded according to size, colour and degree of roasting, and packed in gunny bags. The yield of sago is around 25% of the weight of fresh roots.

Sago contains about 12% moisture, 0.2% protein, 0.2% fat, 87% carbohydrates, and has a calorific value of 351 calories 100 g^{-1}. Sago is used mainly as infant and invalid food, and in preparation of puddings. Small-scale factories make *pappad*, *vadam*, etc., using sago.

Liquid glucose and dextrose

Starch is a polymer of glucose and hence is the raw material for glucose. The hydrolysis of starch to glucose can be carried out by acid hydrolysis or enzyme hydrolysis. Starch is suspended in water, of approximately 25–30% solids, to which sufficient HCl is added to bring it to a normality of 0.01–0.02 HCl. It is heated in a converter under a pressure of 0.35 kg cm^{-2} for 15 min (140–160°C). The reaction mixture is tested periodically for residual starch by iodine staining. When no colour develops with iodine, the heating is stopped, pressure is released, and

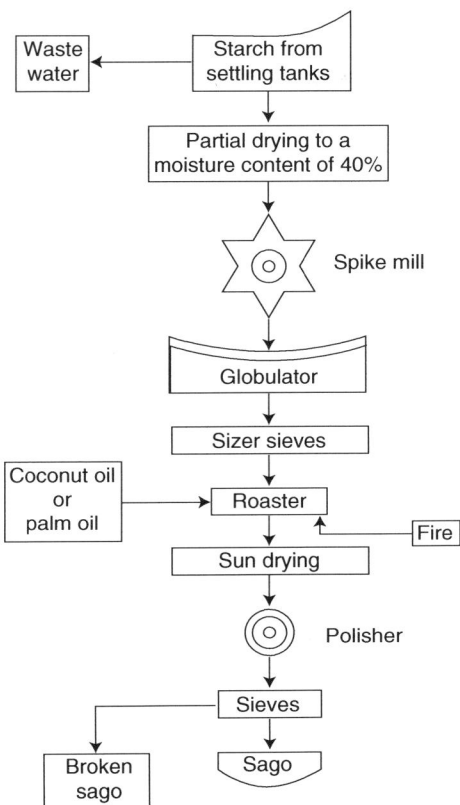

Fig. 15.2. Flow diagram for the production of sago.

liquid is transferred to a neutralization tank where it is neutralized to pH 7.0 with soda ash. The mixture is passed through a filter press, the filtrate is decolorized by activated carbon and the clear filtrate concentrated in a triple-effect evaporator. The liquid is again treated with carbon and concentrated in a triple-effect evaporator and after another treatment with carbon it is concentrated in a vacuum. The concentrated syrup (40–45°C) is quickly cooled and transferred to drums. The product contains 43% dextrose on a dry-weight basis. The syrup can be used for various confectionery purposes, and after further purification is used for pharmaceutical purposes. In the production of crystalline D-glucose monohydrate, the solution is evaporated under a vacuum to 70–88% solids, cooled to around 45°C and fed into 10,000 gallon crystallizers. The mass is now slowly cooled to 20–30°C over a period of 3–5 days and at the end, about 60% of the solids are crystallized as D-glucose monohydrate. Glucose syrup is used widely in the confectionery, the pharmaceutical industry and for energy foods (Balagopalan *et al.*, 1988).

Fructose syrup

Fructose syrup has gained importance in view of the fluctuating prices of sugars and the harmful effects of synthetic sweeteners. Fructose is 1.7 times sweeter than sucrose and four times sweeter than glucose. The conversion of glucose to fructose can be achieved by alkali or by the enzyme glucoisomerase. Whereas, an alkali gives only low levels of isomerization, isomerase can bring 40–45% glucose to fructose conversion. The conversion of starch is the same as for the production of liquid glucose. The next step involved is the decolorization of the syrup which is achieved by either activated carbon treatment or passing through ion exchange columns. The decolorized sample is then subjected to isomerization with glucose isomerase in glass-lined tanks. The optimum temperature for the isomerization is 62°C and the reaction is carried out for 6 h; the pH is maintained at 8.0. The syrup is stirred continuously and samples are taken at regular intervals to check for the conversion by estimation of the fructose content. When there is no further increase in the fructose content, the solution may be concentrated in a vacuum to the desired levels of solid content. Pure fructose crystals can be obtained by separation on ion-exchange columns, concentration and seeding out with crystals of pure fructose (Fig. 15.3).

Maltose

Maltose is a disaccharide formed from two glucose units and is a reducing sugar. It can be obtained commercially from starch by enzyme treatment. There are three types of commercial maltose syrups: high-maltose syrups, extremely high-maltose syrups and high-conversion syrups.

The production of various syrups involves two steps. The first being liquefaction in which a suspension of starch is gelatinized by heat and

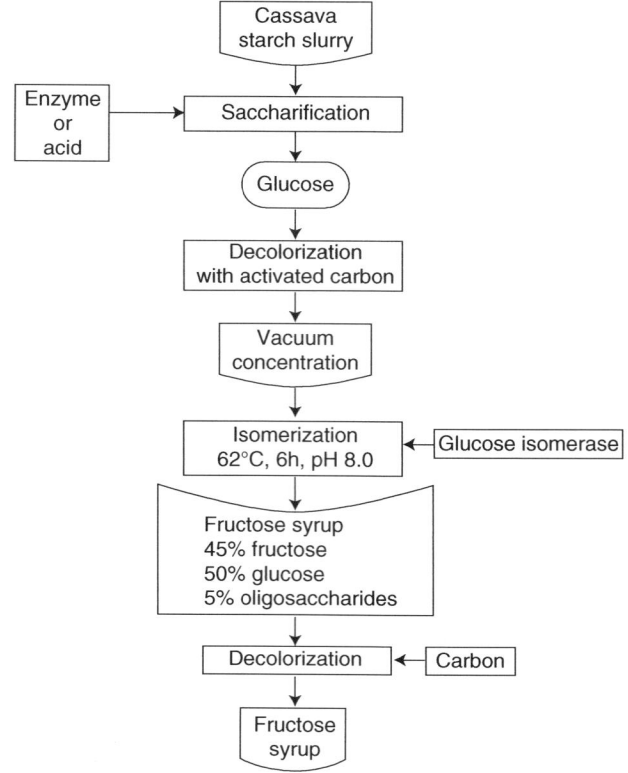

Fig. 15.3. Flow diagram for the production of high fructose syrup.

partly hydrolysed by a thermostable α-amylase. The second step is saccharification using microbial β-amylase or fungal α-amylase; the use of a different enzyme system gives rise to different products. Purification and refining of the maltose syrup is accomplished by first filtering to remove insoluble matter such as fat and denatured protein. The syrup is then refined by means of activated carbon and ion exchange, which removes colour, ash and other minor impurities. The clarified solution is then concentrated.

In the Vietnamese artisan process, 10-day-old rice seedlings rich in α-amylases are crushed into boiled cassava starch and then hydrolysed to produce maltose (Queynh and Cecil, 1996).

Maltodextrin

Maltodextrins are partially hydrolysed starch with dextrose equivalent less than 20. They are approved as food ingredients and manufactured

by the action of α-amylase on starch. Maltodextrin is becoming increasingly important as it is not as sweet as glucose and also acts as a thickening agent.

Modified starches

The main aim of modified starches is to improve the functionality of the starch for industrial applications such that it can be cooked at higher concentration. The modifications are commonly called conversions and involve treatment of the starch granules by chemical or physical means to cause rupturing of some or all of the starch molecules. This increases the number of granules, decreasing their capacity to swell on pasting or cooking in water and decreasing the size of the molecules. As a result, the viscosity is lowered, permitting the converted starches to be dispersed at higher concentrations than unmodified starch.

Acid-modified starches

The acid-modified method uses a temperature range of 40–60°C, more dilute acid (0.5–3%) and short reaction times (0.5–14 h). Because of their clarity and stability, acid-modified starches may be used in the preparation of adhesives for the production of remoistening gum tapes and other applications requiring starch which may be dispersed at higher concentrations and provide stable suspensions. They are also suitable for industrial applications requiring film formation and adhesion, such as warp sizing, bag adhesives, etc., and in the preparation of candies and other confectionery items.

Oxidized starches

Although oxidation can be carried out using various chemicals such as permanganate or persulphate, only hypochlorite or chlorine oxidation is commercially useful. Although often called chlorinated starches, they are actually oxidized starches. Oxidized starches find extensive application in the paper industry in view of their lower viscosity, better film strength and clarity. They also find use in textiles for sizing warps of cotton, spun rayon and other synthetics. Other applications include their use in textile and laundry finishes, fabrication of construction materials, and in the manufacture of starch derivatives.

Cross-linked starches

The starch molecule contains a number of hydroxyl groups. Each anhydroglucose unit is a mode of two secondary hydroxyl groups, and a large majority contain primary hydroxyl groups. These groups are able to react with any chemical capable of reacting with alcoholic hydroxyls. They include acid anhydrides, organic chloro-compounds, aldehydes, epoxy, ethylenic compounds, etc. When the reagent contains two or more moieties capable of reacting with hydroxyl groups, there is a possibility of reacting at two different hydroxyl sites, resulting in cross-linking between hydroxyls on the same molecule or on different molecules. The cross-linking by interaction with bi- or poly-functional reagents is carried out to thicken or reduce solubility or insolubilize their solutions or films. Cross-linked starches are required for a number of industrial applications such as preparation of wet-rub-resistant starch paper coatings, permanent textile sizes, wet strength and water-resistant adhesives, etc.

Acetylated starches

Highly acetylated starches are of interest in view of their solubility in solvents such as acetone and chloroform and their thermo-plasticity. Their advantage is the preservation of granule structure throughout the course of the reaction, purification by washing with water and recovery by centrifugation or filtration. This enables production of a high-purity, low-ash content product, as required in pharmaceuticals and food. The application of this process lies in ease of formation and stability of colloidal suspensions of starch. Another use of acetylated starches is to adjust the colloidal properties of a compound to the requirement of the application. The various acetylating agents used for acetylation include acetic anhydride, acetic anhydride–pyridine, acetic anhydride–acetic acid mixture, ketene, vinyl acetate and acetic acid.

Cationic starches

Cationic starches are derivatives of starch after treatment with reagents having amino, imino, ammonium, sulphonium or phosphonium groups which have a positive charge. The most important ones are tertiary amino and quarternary ammonium starch ethers. The cationic starches are useful additives in the paper industry to provide strength and a glaze to the paper, surface sizes and coating binders. Cationic starches improve sheet strength by promoting fibre bonding through a combination of ionic bonding and additional hydrogen bonding. In addition, inorganic esters of starches such as starch phosphates, methyl and ethyl starches with a wide range of applications in industry are also manufactured.

Biodegradable plastic from starch

The process for production of starch-based plastics involves mixing and blending starch with suitable synthetic polymers (namely low density polyethylene and linear low density polyethylene) as stabilizing agents and suitable amounts of appropriate coupling, gelatinizing and plasticizing agents. Compounding of the blend prior to extrusion film blowing is adopted to attain proper melt mixing. Successful extrusion film blowing is possible with formulations containing up to 40% starch and appropriate amounts of suitable gelatinizing, plasticizing and coupling agents. Synthetic polymers grafted or blended with starch, either in its native form or modified, have been reported to impart biodegradability to the fabricated plastic goods. Incorporation of low-cost starch into synthetic polymers also provides a potential method for expanding its applications as well as improving the economics of the plastics. Their superior utility has been deployed in specific applications such as short-life agricultural mulch, single-use disposable packaging and controlled release in soil or other growth medium of pesticides, pheromones, growth regulators, fertilizers, etc.

Fermented commodity chemicals from cassava starch

Cassava alcohol

Cassava starch/flour is gelatinized first by cooking and further converted to simpler sugars by a process called saccharification, accomplished with the help of mild acids or amylase enzymes. Cassava starch, having lower swelling and gelatinization temperature, can be easily saccharified to simple sugars. The main advantage of cassava over any other crop for this purpose is the presence of highly fermentable sugars after saccharification. Large volumes of the saccharified starch are fed into fermentation vessels and inoculated with actively growing yeast (*Saccharomyces cerevisiae*). Usually, 5–10% of the total volume is fermented aerobically in stages from a pure culture. The optimum concentration of sugars for fermentation is 12–18%. The pH of the mash for fermentation is optimally 4–4.5 and the temperature range is 28–32°C. Alcohol is recovered from the fermented mash after 48–72 h by distillation (Fig. 15.4).

SORBITOL.   Sorbitol is made from glucose prepared from cassava starch by hydrogenation in a high pressure reactor. Because it readily absorbs moisture, it can replace glycerin in the manufacture of tooth paste, cosmetics and oil-based paints. It also serves as a raw material for fermentation to produce vitamin C, first into hygric acid and then into ascorbic acid. Approximately 2.7 t of sorbitol is required to produce 1 t of ascorbic acid (Ren, 1996).

MANNITOL.   This is another hexanol with little moisture absorption capacity. It is produced commercially by hydrogenizing fructose (prepared out of starch), of which 50% converts into mannitol which is then purified by crystallization. Mannitol has wider applications as a dehydrating agent in blood vessel diastolic preparations and in the treatment of cerebral thrombosis and other circulating disorders. It is also used for the production of polyester, polyethylene and solid foam plastics.

MALTOL.   The sugar alcohol, maltol is produced by incomplete hydrolysis of starch using the enzyme maltase and subsequent hydrogenation. It is used in confectionery (Ren, 1996).

Citric acid

In China, citric acid is produced from cassava starch using certain strains of *Aspergillus niger*. Cassava starch after gelatinization and liquefaction is subjected to fermentation for 4 days, by which time the citric acid content exceeds 15%. An extraction rate of more than 92% is possible. The short fermentation period and ease of liquefying the starch and extracting the acid keep the production costs low (Ren, 1996).

Lactic acid

Lactic acid fermentation is important in many traditional foods, silage and animal feed. Cassava starch can be utilized for the production of

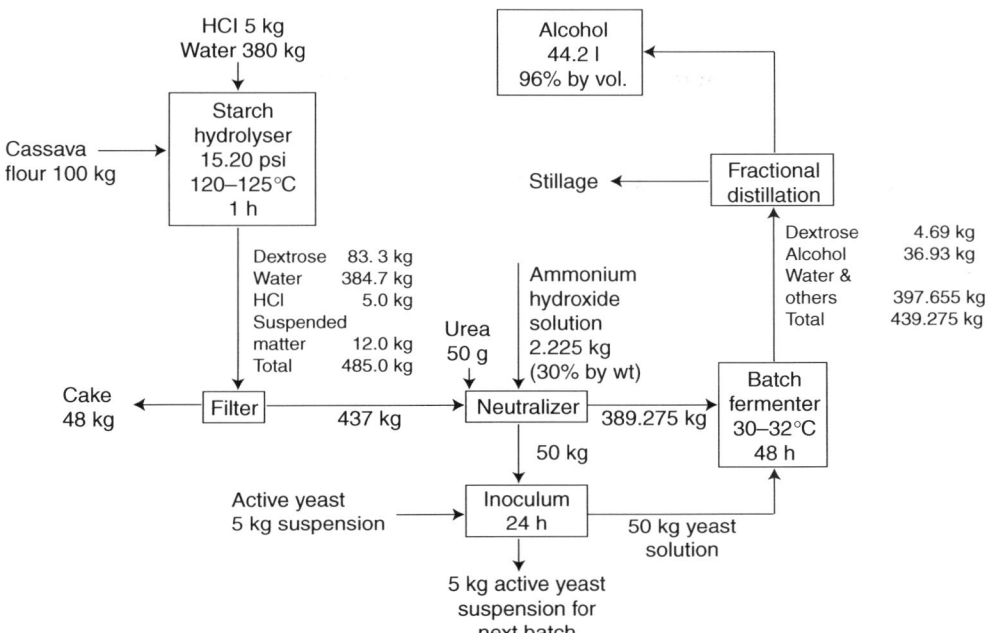

Fig. 15.4. Flow diagram for the production of ethylalcohol from cassava.

lactic acid. The starch has to be saccharified into sugars before fermentation. The techniques of saccharification are the same as those described for ethanol. The bacteria *Lactobacillus plantarum*, *Lactobacillus bichmaina*, *Lactobacillus mesenterioides* and *Lactobacillus delbruiki* can all be used for fermenting sugars to produce lactic acid.

Waste Management in Cassava Starch Factories

The wastes generated from cassava processing may be solid or liquid. The brown peel of cassava roots, known as the periderm, varies between 2% and 5% of the root total. The solid waste is made up of fibrous root material and contains starch that physically could not be extracted. The solid residues can be used as animal feed. Cassava waste can replace a part or all of the feed components.

The process of starch extraction from cassava tubers requires large quantities of water resulting in the release of a significant quantity of waste waters (effluents). It is common for factories to discharge the effluents into the nearby rivers, drainage channels, crop fields or to the land adjacent to the factories. These effluents pose a serious threat to the environment and quality of life in rural areas. Wide variations are observed in the physical and chemical constituents of primary and secondary effluents obtained from cassava starch factories. Manilal *et al.* (1991) observed that the chemical oxidation demand (COD) ranged between 33,600 and 38,223 mg l^{-1} in the primary effluents, whereas in the secondary effluents the range was only 3800–9050 mg l^{-1}. The biological oxidation demand (BOD) was in the range of 13,200–14,300 mg l^{-1} in the primary effluents. The corresponding figures for the secondary effluents were 3600–7050 mg l^{-1}. The acidity of the effluents ranged between pH 4.5 and 4.7. Nitrogen and phosphorus are the main nutrients contributing to the stability of organic wastes and the analysis revealed low nitrogen content, indicating the necessity for enrichment of effluents to reduce the BOD and COD. Balagopalan and Rajalekshmy (1998) observed that the concentration of total cyanoglucosides in the effluents ranged between 12.9 and 66.6 mg l^{-1} in the case of initial samples, whereas in the case of final waste water samples, the concentration ranged between 10.4 and 274 mg l^{-1}. A high concentration of cyanide was observed in the groundwater sources near the processing factories, ranging between 1.2 and 1.6 mg l^{-1}. Initial settling, anaerobiosis, filtration through sand and charcoal and aeration can reduce the pollution load to the desired level (Balagopalan et al., 1994).

Potential Markets for Cassava Products

If cassava is processed and sold only at the primary level, the prospects for cassava as a source of income are limited. The conventional products from cassava that are traded on the world markets are starch and pellets for animal feed. Statistics of international trade in cassava starch show only a marginal increase in recent years due to competition from starch derived from maize. In the animal feed sector, however, the European Union and the Russian Federation have created opportunities for the import of cassava products.

Diversification of starch into value-added products seems to be a way to increase the demand and this market has been exploited, particularly by Thailand. In the late 1970s, USA and Thailand collaborated to produce modified starches for export and this was followed by further joint ventures with European and Japanese companies. The animal feed companies also integrated modified starch production units into their factories. The impressive economic growth of Thailand between 1980 and 1990 was partly due to investment in the technology of cassava processing. Moreover, Thailand was able to circumvent restrictions on import of cassava pellets by developing the industry for modified starch.

Production statistics for monosodium glutamate (MSG) in South-East Asian countries and China, using cassava as the raw material, showed a gradual increase in the 1980s and 1990s. The export potential for MSG for the growing fast-food industries in the developing and developed countries is considerable and the market for cassava starch is expected to increase to meet this demand.

In China, citric acid is produced from cassava starch and this could be an area of expansion. The market could expand for a range of ready-made foods based on cassava flour, produced mainly in The Netherlands and Thailand. Domestic markets for sago in India, gari in West Africa and cassabe and other sour starch products in South America have also grown in recent years. The Latin American domestic animal feed industry depends largely on cassava and cassava wastes. These countries have been able to find markets for cassava starch in the USA and Japan.

Investment in research and development for diversification of cassava utilization is required in order to exploit these markets and may hold some promise for currently marginalized cassava growers.

References

Balagopalan, C. (1996) Improving the nutritional value of cassava by solid state fermentation: CTCRI Experiences. Journal of Scientific and Industrial Research 5, 479–482.

Balagopalan, C. and Padmaja, G. (1988) Protein enrichment of cassava flour by solid state fermentation with Trichoderma pseudokoningii Rifai for cattle feed. In: Howeler, R.H. (ed.) Proceedings of Eighth Symposium of the International Society for Tropical Root Crops, Bangkok, Thailand. CIAT, Cali, Colombia, pp. 426–432.

Balagopalan, C. and Rajalakshmy, L. (1998) Cyanogen accumulation in environment during processing of cassava (Manihot esculenta Crantz) for starch and sago. Water, Air and Solid Pollution 102, 407–413.

Balagopalan, C., Padmaja, G., Nanda, S.K. and Moorthy, S.N. (1988) Cassava in Food, Feed and Industry. CRC Press, Boca Raton, Florida.

Balagopalan, C., Padmaja, G. and Mathew G. (1991) Improving the nutritional value of cassava products using microbial techniques. In: Machin, D. and Nyrold, S. (eds) Proceedings of the Expert Consultation on the Use of Roots, Tubers, Plantations and Bananas in Animal Feeding. Cali, Colombia, 21–25 January 1991, pp. 127–140.

Balagopalan, C., Ray, R.C., Sheriff, J.T. and Rajalakshmy, L. (1994) Biotechnology for the value addition of waste waters and residues from cassava processing industries. In: Proceedings of the Second International Scientific Meeting of the Cassava Biotechnology Network, Bogor, Indonesia, 22–26 August, CBN, Cali, Colombia, pp. 690–701.

Brook, E.J., Stanton, W.R. and Wallbridge, A. (1969) Fermentation methods for protein enrichment of cassava by solid substrate fermentation in rural conditions. *Biotechnology and Bioengineering* 11, 1271–1284.

Castillo, L.S. Aglibut, F.B., Javier, T.A., Gerpacio, A.L., Puyyoan, R.B. and Remin, B.B. (1964) Carrot and cassava tuber silage as replacement for corn in swine growing fattening rations. *Philippine Agriculture* 47, 460–462.

Chou, K.C. and Muller, Z. (1972) Complete substitution of maize by tapioca in broiler rations. *Proceedings of Australian Poultry Science Convention.* World Poultry Science Association, Auckland, New Zealand, pp. 149–160.

Daubresse, P., Ntibashirwa, S. and Gheysen, A. (1987) A process for protein enrichment of cassava by solid substrate fermentation in rural conditions. *Biotechnology and Bioengineering* 29, 962–968.

Dufour, D., Larsonneur, S., Alarcon, F., Barbet, C. and Chuzel, G. (1996) Improving the bread-making potential of cassava sour starch. In: Dufour, D., O'Brien, G.M. and Best, R. (eds) *Cassava Flour and Starch: Progress in Research and Development.* CIAT, Cali, Colombia, pp. 133–142.

Gomez, G.G. (1991) Use of cassava products in pig feeding. In: Machin, D. and Nyrold, S. (eds) *Proceedings of the Expert Consultation on the Use of Roots, Tubers, Plantains and Bananas in Animal Feeding.* CIAT, Cali, Colombia, 21–25 January 1991.

Grace, M.R. (1977) Cassava processing. *Plant Production and Protection Series No. 3.* FAO, Rome.

Gray, W.D. and Abou-el-Seound, M.D. (1966) Fungal protein for food and feeds, 3. Manioc as a potential crude raw material for tropical areas. *Economic Botany* 20, 251–255.

Gregory, K.F., Reade, A.F., Santosnumez, J.C., Smitch, R.E. and Machean, S.J. (1977) Further thermotolerant fungi for the conversion of cassava starch to protein. *Animal Feed Science Technology* 2, 7–19.

Khajarren, S. and Khajarren, J.M. (1977) Use of cassava as a food supplement for broiler chicks. In: Cock, J. and MacIntyre, R. (eds) *Proceedings of the Fourth Symposium of the International Society for Tropical Root Crops.* IDRC, Ottawa, Canada, IDRC-080e, pp. 246–250.

Lancaster, P.N., Ingram, J.S., Lin, H.Y. and Coursey, D.G. (1982) Traditional cassava based foods, survey of processing techniques. *Economic Botany*, 36, 12–25.

Manilal, V.B., Naryanan, C.S. and Balagopalan, C. (1991) Cassava starch factory effluent treatment with concomitant SCP production. *World Journal of Microbiology and Biotechnology* 7, 185–190.

Mathew, G., Padmaja, G. and Moorthy, S.N. (1991) Enhancement of starch extractability from cassava (*Manihot esculenta* Crantz) tubers through fermentation with a mixed culture inoculum. *Journal of Root Crops* 17(1) 1–9.

Mathur, M.L., Sampath, S.R. and Ghosh, S.N. (1969) Studies on tapioca: effect of 50 and 100 percent replacement of oats by tapioca in the concentrate mixture of dairy cows. *Indian Journal Dairy Science* 22, 193–199.

Mikami, Y., Gregory, K.F., Levadoux, W.L., Balagopalan, C. and Whitwell, S.T. (1982) Factors affecting yield and safety of protein production from cassava by *Cephalosporium eichhorniae. Applied and Environmental Microbiology* 43, 403–411.

Montaldo, A. (1977) Whole plant utilization of cassava for animal feed. In: Nestel, B. and Graham, M. (eds) *Cassava as Animal Feed.* IDRC, Ottawa, Canada, IDRC-095e, pp. 95–106.

Muindi, P.J. and Hanssen, J.F. (1981) Protein enrichment of cassava root meal by *Trichoderma harzianum* for animal feed. *Journal of Science of Food and Agriculture* 32, 655–661.

Omole, T.A. (1977) Cassava in the nutrition of layers. In: Nestel, B. and Graham, M. (eds) *Cassava as Animal Feed.* IDRC, Ottawa, Canada, IDRC-095e, pp. 51–55.

Padmaja, G., Geroge, M. and Balagopalan, C. (1994) Ensiling as an innovative biotechnological approach for conservation of high cyanide cassava tubers for feed use. In: *Proceedings of the Second International Scientific Meeting of Cassava Biotechnology Network*, Bogor, Indonesia, CBN, Cali, Colombia, pp. 784–794.

Padmaja, G., Balagopalan, C., Moorthy, S.N. and Potty, V.P. (1996) Quality evaluation and functional properties of two novel cassava food products 'Yuca Rava' and 'Yuca Porridge'. In: Dufour, D., O'Brien, G.M. and Best, R. (eds) *Cassava Flour and Starch: Progress in Research and Development.* CIAT, Cali, Colombia, pp. 323–333.

Philipps, T.P. (1974) *Cassava Utilization and Potential Markets.* IDRC, Ottawa, Canada, IDRC-020e, pp. 1–182.

Plucknett, D.L., Phillips, T.P. and Kagho, R.B. (1998) A global development strategy for cassava: transforming a traditional tropical root crop. Paper presented at Asian Cassava Stakeholders' consultation on a global cassava development strategy at Bangkok, Thailand, 23–25 November 1998.

Queynh, N.C. and Cecil, J. (1996) *Sweetness from Starch, a Manual for Making Maltose from Starch.* FAO, Rome, p. 37.

Radley, J.A. (1976) *Starch Production Technology.* Applied Science Publishers, London, p. 587.

Raimbault, M., Revah, S., Pina, F. and Villalobos, P. (1985) Protein enrichment of cassava by solid substrate fermentation using moulds isolated from traditional foods. *Journal of Fermentation Technology* 63, 395–399.

Ravindran, V. and Blair, R. (1991) Feed resources for poultry production in Asia and the Pacific.1. Energy sources. *World's Poultry Science Journal* 47, 213–231.

Ren, J.S. (1996) Cassava products for food and chemical industries: China. In: Dufour, D., O'Brien, G.M. and Best, R. (eds) *Cassava Flour and Starch: Progress in Research and Development*. CIAT, Cali, Colombia, pp. 48–54.

Setyona, A., Damardjati, D. and Malian, H. (1991) Sweet potato and cassava development: present status and future prospects in Indonesia. In: Scott, G.J., Wiersema, S. and Ferguson, P.I. (eds) *Product Development for Root and Tuber Crops*. CIP, Lima, Peru, pp. 29–40.

Stanton, W.R. and Wallbridge, A. (1969) Fermented food process. *Process Biochemistry* 4, 45–51.

Strasser, J.A., Abbott, J.A. and Battey, R.F. (1970) Process enriches cassava with protein. *Food Engineering* 4, 112–116.

Varghese, G., Thambirajah, J.J. and Wong, F.M. (1976) Protein enrichment of cassava by fermentation with micro fungi and the role of natural nitrogenous supplements. In: Cock, J., MacIntyre, R. and Graham, M. (eds) *Proceedings of the Fourth International Symposium of the International Society for Tropical Root Crops*. IDRC, Ottawa, Canada, IDRC-095e, pp. 250–255.

Yoshida, M., Hoshu, H., Kosaka, K. and Morimoto, H. (1966) Nutritive value of various energy sources for poultry feed. IV. Estimation of available energy of cassava meal. *Japanese Poultry Science* 3, 29–34.

Index

Page numbers in **bold** refer to figures and tables